IMPLICIT LARGE EDDY SIMULATION

The numerical simulation of turbulent flows is a subject of great practical importance to scientists and engineers. The difficulty in achieving predictive simulations is perhaps best illustrated by the wide range of approaches that have been developed and are still being used by the turbulence modeling community. In this book the authors describe one of these approaches: implicit large eddy simulation (ILES). ILES is a relatively new approach that combines generality and computational efficiency with documented success in many areas of complex fluid flow. This book synthesizes the current understanding of the theoretical basis of the ILES methodology and reviews its accomplishments. Here ILES pioneers and lead researchers combine their experience to present the first comprehensive description of the methodology. This book should be of fundamental interest to graduate students, basic research scientists, and professionals involved in the design and analysis of complex turbulent flows.

Fernando F. Grinstein and Len G. Margolin are theoretical and computational physicists in the Applied Physics Division of Los Alamos National Laboratory.

William J. Rider is theoretical and computational physicist in the Computational Physics Research and Development Department of Sandia National Laboratories.

To our parents,
our wives, Julia, Holly, and Felicia,
our children, Frederic, Rachel, and Jackson,
and the many contributors to this volume.

Implicit Large Eddy Simulation

COMPUTING TURBULENT FLUID DYNAMICS

Edited by

FERNANDO F. GRINSTEIN
Los Alamos National Laboratory

LEN G. MARGOLIN
Los Alamos National Laboratory

WILLIAM J. RIDER
Sandia National Laboratories

CAMBRIDGE
UNIVERSITY PRESS

CAMBRIDGE UNIVERSITY PRESS
Cambridge, New York, Melbourne, Madrid, Cape Town, Singapore,
São Paulo, Delhi, Dubai, Tokyo, Mexico City

Cambridge University Press
32 Avenue of the Americas, New York, NY 10013-2473, USA

www.cambridge.org
Information on this title: www.cambridge.org/9780521172721

First published 2007
First paperback edition 2010

A catalog record for this publication is available from the British Library

Library of Congress Cataloging in Publication data

Implicit Large Eddy Simulation : computing turbulent fluid dynamics /
edited by Fenando F. Grinstein, Len G. Margolin, William J. Rider.
    p.   cm.
Includes bibliographical references and index.
ISBN 0-521-86982-x (hardcover)
1. Turbulence – Mathematical models.   2. Eddies – Mathematical models.
1. Grinstein, Fenando F.   II. Margolin, Len G.   III. Rider, William.   IV. Title.
TA357.5.T87147   2006
532'.0527 – dc22   2006031149

ISBN 978-0-521-86982-9 Hardback
ISBN 978-0-521-17272-1 Paperback

Additional resources for this publication at www.cambridge.org/9780521172721

Contents

Plate section follows p. 242*
*These plates are available for download in color from
www.cambridge.org/9780521172721

Preface

This book represents the combined efforts of many sponsors. Most of the basic planning and organization was carried out while one of us (F. F. Grinstein) was the 2003–2004 Orson Anderson Distinguished Visiting Scholar at the Institute for Geophysics and Planetary Physics (IGPP) at Los Alamos National Laboratory (LANL). It is very important that we acknowledge the critical role played by the implicit large eddy simulation (ILES) workshops at LANL in January and November of 2004. These workshops took place under the auspices of IGPP and with partial support from the Center for Nonlinear Studies at LANL. They provided us with an ideal forum to meet and exchange ILES views and experiences, and to extensively discuss their integration within the book project. At the personal level, special thanks go to IGPP's Gary Geernaert and to the U.S. Naval Research Laboratory's (NRL's) Jay Boris and Elaine Oran for their continued encouragement and support. Last but not least, continued support of F. F. Grinstein's research on ILES during his tenure at NRL from the U.S. Office of Naval Research through NRL and from the U.S. Department of Defense High-Performance Computing Modernization Program is also greatly appreciated.

This book has evolved far beyond the early plan of merely putting together a collection of review papers on ILES authored by the lead researchers in the area. Several very useful collaborations have quite spontaneously occurred in the process of integrating the material, and we now have an active ILES working group that is focusing on a variety of timely research projects.

Much of the work on ILES reported here was accomplished despite the lack of acceptance and support of the turbulence modeling community. We hope that our readers will recognize the quality of these results and will be encouraged to do their own experiments and evaluations.

Fernando F. Grinstein, Len G. Margolin, and William J. Rider

List of Acronyms

ALE	Arbitrary Lagrangian Eulerian
AMR	Adaptive Mesh Refinement
BBC	code name
CFD	Computational Fluid Dynamics
CV	Control Volume
DES	Detached Eddy Simulation
DNS	Direct Numerical Simulation
ENO	Essentially Non-Oscillatory
ENZO	code name
EULAG	code name
EV	Eddy Viscosity
FCT	Flux-Corrected Transport
FLASH	code name
FLIC	Fluid in Cell
FV	Finite Volume
ILES	Implicit LES
KH	Kelvin-Helmholtz
KRAKEN	code name
LES	Large-Eddy Simulation
MEA	Modified Equation Analysis
MILES	Monotone (or Monotonically) Integrated LES
MM	Mixed Model
MPDATA	Multidimensional Positive Definite Advection Transport Algorithm
MUSCL	Monotonic Upstream-Centered Scheme for Conservation Laws
NFV	Non-Oscillatory Finite Volume
NSE	Navier-Stokes Equations
ODE	Ordinary Differential Equation
OEEVM	One-Equation EV Model
PDE	Partial Differential Equation
PIC	Particle in Cell
PPM	Piecewise Parabolic Method
RANS	Reynolds-Averaged Navier-Stokes
Re	Reynolds Number

RM	Ritchmyer-Meshkov
RT	Rayleigh–Taylor
SGS	Sub-Grid Scale
SHASTA	code name
SMAG	Smagorinsky
TGV	Taylor Green Vortex
TURMOIL3D	code name
TVD	Total Variation Diminishing
VC	Vorticity Confinement
VH-1	code name
WENO	Weighted ENO

List of Contributors

Prof. Dr. Nikolaus A. Adams
Lehrstuhl für Aerodynamik
Technische Universität München
Boltzmannstrasse 15
85748 Garching, Germany

Dr. Niklas Alin
The Swedish Defence Research Agency
  (FOI)
SE-172 90 Stockholm, Sweden

Dr. Magnus Berglund
The Swedish Defence Research Agency
  (FOI)
SE-172 90 Stockholm, Sweden

Dr. Jay P. Boris
Laboratory for Computational
  Physics & Fluid Dynamics
Naval Research Laboratory
Washington, DC 20375

Dr. Carl R. DeVore
Laboratory for Computational
  Physics & Fluid Dynamics
Naval Research Laboratory
Washington, DC 20375

Bill Dietz
University of Tennessee
  Space Institute
Tullahoma, TN 37388

Professor J. A. Domaradzki
Department of Aerospace and
  Mechanical Engineering
University of Southern California
Los Angeles, CA 90089

Professor Dimitris Drikakis
Cranfield University
School of Engineering
Befordshire MK43 0AL
United Kingdom

Meng Fan
University of Tennessee Space Institute
Tullahoma, TN 37388

Dr. Christer Fureby
Department of Weapons and Protection
The Swedish Defence Research
  Agency (FOI)
SE-172 90 Stockholm, Sweden

Dr. Fernando F. Grinstein
Applied Physics Division, MS B259
Los Alamos National Laboratory
Los Alamos, NM 87545

Marco Hahn
Cranfield University
School of Engineering
Befordshire MK43 0AL
United Kingdom

Dr. S. Hickel
Institute of Aerodynamics
Technische Universitaet Muenchen
85747 Garching, Germany

Prof. Doyle D. Knight
Department of Mechanical and
   Aerospace Engineering
Rutgers University – The State
   University of New Jersey
Piscataway, NJ 08854

Dr. Marco Kupiainen
Department of Numerical Analysis and
   Computer Science
Royal Institute of Technology
S-144 00 Stockholm, Sweden

Dr. Mattias Liefvendahl
The Swedish Defence Research Agency
   (FOI)
SE-172 90 Stockholm, Sweden

Dr. Eric Lillberg
The Swedish Defence Research Agency
   (FOI)
SE-172 90 Stockholm, Sweden

Dr. Nicholas Lynn
University of Tennessee Space Institute
Tullahoma, TN 37388

Dr. Len G. Margolin
Applied Physics Division, MS F644
Los Alamos National Laboratory
Los Alamos, NM 87545

Dr. Oskar Parmhed
The Swedish Defence Research Agency
   (FOI)
SE-172 90 Stockholm, Sweden

Dr. Gopal Patnaik
Laboratory for Computational
   Physics & Fluid Dynamics
Naval Research Laboratory
Washington, DC, 20375

Dr. Leif Persson
The Swedish Defence Research Agency
   (FOI)
Department of NBC Protection
SE 901 82 Umeå, Sweden

Dr. Tobias Persson
Chalmers University of Technology
Department of Shipping and Marine
   Technology
SE 412 96 Göteborg, Sweden

Prof. David H. Porter
Department of Astronomy
University of Minnesota
Minneapolis, MN 55455

Dr. William J. Rider
Computational Physics Research and
   Development Department
Sandia National Laboratories
PO Box 5800, Albuquerque,
   NM 87185-0183

Prof. Pierre Sagaut
Laboratoire de Modélisation
   en Mécanique
Université Pierre et Marie Curie
   Centre National de la
   Recherche Scientifique
75252 Paris, Cedex 05, France

Dr. Piotr K. Smolarkiewicz
Mesoscale and Microscale
   Meteorology Division
National Center for Atmospheric
   Research
PO Box 3000, Boulder, CO 80307

Prof. John Steinhoff
University of Tennessee Space Institute
Tullahoma, TN 37388

Dr. Urban Svennberg
The Swedish Defence Research Agency
   (FOI)
SE-172 90 Stockholm, Sweden

Dr. Lesong Wang
University of Tennessee Space Institute
Tullahoma, TN 37388

Prof. Paul R. Woodward
Department of Astronomy
University of Minnesota
Minneapolis, MN 55455

Wenren Yonghu
University of Tennessee Space Institute
Tullahoma, TN 37388

Ted R. Young
Laboratory for Computational
    Physics & Fluid Dynamics
Naval Research Laboratory
Washington, DC, 20375

Dr. David L. Youngs
Atomic Weapons Establishment
Building E3 Aldermaston
Reading Berkshire RG7 4PR
United Kingdom

Introduction

Fernando F. Grinstein, Len G. Margolin, and William J. Rider

The numerical simulation of turbulent fluid flows is a subject of great practical importance to scientists and engineers. The difficulty in achieving predictive simulations is perhaps best illustrated by the wide range of approaches that have been developed and that are still being used by the turbulence modeling community. In this book, we describe one of these approaches, which we have termed *implicit large eddy simulation* (ILES).

ILES is remarkable for its simplicity and general applicability. Nevertheless, it has not yet received widespread acceptance in the turbulence modeling community. We speculate that this is the result of two factors: the lack of a theoretical basis to justify the approach and the lack of appreciation of its large and diverse portfolio of successful simulations. The principal purpose of this book is to address these two issues.

One of the complicating features of turbulence is the broad range of spatial scales that contribute to the flow dynamics. In most examples of practical interest, the range of scales is much too large to be represented on even the highest-performance computers of today. The general strategy, which has been employed successfully since the beginning of the age of computers, is to calculate the large scales of motion and to introduce models for the effects of the (unresolved) small scales on the flow. In the turbulence modeling community, these are called *subgrid-scale (SGS) models*.

In ILES, we dispense with explicit subgrid models. Instead, the effects of unresolved scales are incorporated implicitly through a class of nonoscillatory finite-volume (NFV) numerical fluid solvers. This class includes such well-known methods as flux-corrected transport (FCT), the piecewise parabolic method (PPM), and total variation diminishing (TVD) algorithms. In general, NFV methods have been a mainline methodology in the computational fluid dynamics community for more than 30 years. However, the recognition of the ILES property is much more recent.

The opening section of the book, Motivation, recounts the early history of ILES and proposes a rationale for the approach. One can imagine the initial resistance of the turbulence modeling community, for whom the creation of new and more complex subgrid-scale models has become its *raison d'etre*. Further, this community has

1

historically emphasized the importance of distinguishing physical model from numerical error, leading to the belief that NFV methods are inappropriate for turbulence modeling. One should also realize that the earliest pioneers of ILES worked in relative isolation, being unaware of each other's work. This made it difficult to identify the essential elements of the ILES approach and to develop a physical justification.

More recently, it has been recognized that the ILES property originates in the blurring of the boundaries between physical modeling and numerical approximation. This is the subject of the second section of the book, Capturing Physics with Numerics, where the idea of building physics into numerical methods is developed. To be precise, NFV methods implement the mathematical property of preserving monotonicity into the integral form of the governing equations. This leads to a computable model in which energy is dissipated on the resolved scales of the simulation at the physically correct rate, due to the inertially dominated processes involved. In this section, we describe several concepts that lead to NFV approximations. We also explore the relationship between ILES and more conventional subgrid-scale models.

The most compelling argument for ILES lies in the diversity of its portfolio of successful applications. In the third section, Verification and Validation, we collect many results aimed at verifying and validating our models by comparison with theory and experiment. The engineering and laboratory experiments range in Reynolds number (Re) from several hundreds to several tens of thousands. The simulations in this section depend on a variety of NFV methods; our purpose here is to demonstrate the underlying generality of ILES, rather than to distinguish among different NFV methods.

In the fourth and final section, Frontier Flows (Re $> 10^6$), we offer examples of more complex simulations for which there are, at present, little on no data available for validation. In principle, the lack of parameters of the ILES approach implies the increased generality and predictive capabilities of the ILES approach. In practice, we believe it is important to quantify the limits of applicability of this approach by continuing with testing and making efforts to understand its foundations.

Section A: Motivation

In Chapter 1, Jay Boris, the originator and first proponent of the ILES approach, recounts his early development of monotone integrated LES (MILES), which is the monotonicity-preserving version of the ILES approach. He reflects on the initial reactions of the traditional turbulence modeling community to his work. In Chapter 2, we introduce our survey of ILES methods with a synthesis of the main components of our current theoretical understanding of ILES for the simulation of turbulent flows. We emphasize the importance of numerical methods that are built upon a physical understanding and are not simply numerical approximations. In fact, the best numerical approximations encode a substantial amount of physics directly into the integration procedures. This recognition is a key to further progress in modeling complex turbulent flows through ILES.

Section B: Capturing physics with numerics

Whenever a physical system is modeled computationally, the desire is to represent the physical world with as much fidelity as possible. One of the key aspects of achieving this goal is the physical modeling that is numerically integrated. The methods described in this book are no different, but the approach to achieving this goal is.

Rather than separate the physical modeling from the numerical integration, the two enterprises are conducted simultaneously. We can refer to this as *capturing physics*, which we regard as a natural extension of the idea of shock capturing. We prefer this approach because it circumvents the problem of first composing a model and then worrying about integrating it successfully. The model and the integration are seamlessly coupled, and the basic issues of numerical stability and model consistency are handled as a matter of course.

In the following chapters these basic ideas are expanded upon in a sequence of expositions that describe the basic models and their physical basis, their basic mathematical properties, their fundamental performance, and the details of the algorithms. In Chapter 3, Sagaut provides a whirlwind overview of the world of large eddy simulation (LES) modeling of which ILES is but a part. The presentation begins with a discussion of the derivation of the equations to be solved. Various approaches to conducting modeling and calculations are discussed. The basic requirements for LES methods and models are detailed, along with a description of the basic modeling approaches.

The description of the various methods used for ILES is presented in Chapter 4. The numerical methods for ILES are nonlinear and provide nonlinear stability for numerical integration. The analysis uncovers the models implicitly included through the utilization of these techniques. With the use of standard techniques in numerical analysis, the various methods are studied and classified, and their behavior is predicted. This produces a common perspective to view the methods and their relation to standard LES modeling.

Chapter 4a, by Drikakis et al. discusses the details of flux-limiting methods as used for ILES. This represents an enormous class of hybrid methods, including FCT and many TVD algorithms. This chapter focuses on the commonality of the various methods and their basic performance on turbulent flows, which is then demonstrated in a canonical test case that has been traditionally used to study transition and turbulence decay. This performance is related to the functional forms implicitly included through the nonlinear numerical algorithms. Modified equation analysis is used to show that the truncation error terms introduced by such methods provide implicit SGS models similar in form to those of conventional mixed SGS models.

One important class of methods for ILES consists of high-resolution Godunov-type methods. Chapter 4b, by Woodward, discusses one important method in this class, PPM. The basics of the piecewise constant and linear Godunov methods are introduced, followed by PPM. This introduction covers the basic method and more modern extensions that are particularly useful for modeling turbulent flows.

In the Lagrangian remap method by Youngs (Chapter 4c), a simple finite-difference formulation is used for a Lagrangian phase in conjunction with an accurate interface reconstruction method that is based on using a van Leer monotonic advection method. This relatively simple technique has been very successfully used for a wide range of complex problems, providing an effective way of calculating many relatively complex turbulent flows.

Many ILES models are distinquished by the physical insight that is inherent in their composition. Chapter 4d, by Smolarkiewicz and Margolin, explores algorithms that use different physical insights than other methods discussed in Chapters 4a and 4b. Despite these differences, the methods described in this chapter perform at a high level. The chapter also provides a guide to the aspects of these methods that are common among the other approaches.

The underlying idea of the vorticity confinement method presented in Chapter 4e is similar to that used in shock capturing, where intrinsically discrete equations are satisfied in thin, modeled regions. This direct, grid-based modeling approach is an effective alternative to formulating a partial differential equation model for the small-scale, turbulent vortical regions and then discretizing it.

A common framework for numerical regularization, analysis, and understanding of ILES is discussed in Chapter 5 by Rider and Margolin. The analysis provided is founded on the modified equation analysis technique, which produces a differential equation that the numerical method solves more accurately than the original differential equation. The prototypical example is the advection–diffusion equation produced by upwind differencing for a purely advection equation. In the case of modern high-resolution methods, the numerical method provides an effective differential equation that has many explicit LES models embedded implicitly in the solution technique. This approach also provides a rational approach to understanding ILES through the common language of continuum mechanics.

Approximate deconvolution is used in traditional LES modeling. Researchers have found that it provides a framework to connect traditional LES models to implicit models. Chapter 6, by Adams, Hickel, and Domaradzki, describes this technique and its power in connecting these two seemingly disparate approaches. Much of the machinery associated with modern high-resolution methods can be recast as an approximate deconvolution. This approach provides an effective bridge between many recent explicit LES models and a broad family of implicit models associated with ILES. This is based on the recognition that one must add physically realizable information to close a differential equation whether one is solving the equation through explicit modeling or a high-resolution numerical method. The approach described in Chapter 6 is an innovative contribution to this general approach.

Section C: Verification and validation

In this section we present results aimed at the verification and validation of ILES by comparison with analytic models of idealized turbulence, with direct numerical simulation of simple flows, or with laboratory studies, as appropriate. Implementation issues and examples of ILES for the simulation of turbulent flows of practical interest are also addressed in this section. The performance of the approach is demonstrated and assessed in a variety of fundamental engineering applications.

In Chapter 7, Porter and Woodward begin the section by presenting a detailed and systematic evaluation of ILES applications for canonical and simple flows for which direct numerical simulation data, theoretical data, or high-quality laboratory data are available. These flows include forced and decaying homogeneous isotropic turbulence.

Next, Chapters 8 and 9, by Grinstein and Drikakis, respectively, discuss new insights generated in ILES studies of the dynamics of high-Renolds-number flows driven by Kelvin–Helmholtz instabilities. The studies in Chapter 8 are relevant to global instabilities, vortex dynamics and topology governing the shear layer development, and transition to turbulence from laminar conditions. Chapter 9 examines symmetry breaking and nonlinear bifurcation phenomena, and shows that NFV numerical schemes exhibit a variety of behavior in the simulation of these phenomena, depending on the nonlinear dissipation and dispersion properties of the particular scheme.

The application of ILES in the study of wall-bounded flows is addressed in Chapters 10 and 11. In Chapter 10 Fureby et al. first discuss the application of ILES to incompressible flows such as (i) fully developed turbulent channel flows, (ii) flow over a cylinder, (iii) flow over a sudden expansion, (iv) flow over a surface-mounted cube, and (v) the flow past a prolate spheroid at different angles of attack. Next, in Chapter 11, Fureby and Knight discuss the ILES applications in compressible bounded turbulent shear flow regimes. Issues associated with the computational grid are discussed, including requirements for resolution of the turbulence production mechanism in the viscous sublayer. The implications of scaling LES to higher Reynolds numbers are addressed. Specific applications are presented, including compressible adiabatic and isothermal zero-pressure-gradient boundary layers and shock wave boundary layers. The extension of these techniques to general compressible bounded shear flows is discussed, and examples such as the supersonic base flow are presented.

ILES based on vorticity confinement is examined by Steinhoff et al. in Chapter 12. The method is especially well suited to treat flow over blunt bodies, including attached and separating boundary layers, and resulting turbulent wakes. Results are presented for three-dimensional flows over round and square cylinders and a realistic helicopter landing ship.

Youngs discusses Rayleigh–Taylor and Richtmyer–Meshkov mixing in Chapter 13. ILES is successfully applied to the three-dimensional simulation of Rayleigh–Taylor and Richtmyer–Meshkov mixing at high Reynolds number. The author argues that some form of monotonicity-preserving method is needed for these problems where there are density discontinuities and shocks. He shows how high-resolution ILES is currently making major contributions to understanding the Rayleigh–Taylor and Richtmyer–Meshkov mixing processes, constituting an essential tool for the construction and validation of engineering models for application to complex real applications.

Section D: Frontier flows

For the extremely complex flows of geophysics, astrophysics, and engineering discussed in this section, whole-domain scalable laboratory studies are impossible or very difficult. Deterministic simulation studies are very expensive and critically constrained by difficulties in (1) modeling and validating all the relevant physical subprocesses, and (2) acquiring all the necessary and relevant boundary condition information.

The section begins with a presentation by Smolarkiewicz and Margolin of ILES studies of geophysics in Chapter 14. The difficulties of modeling the dynamics of the global ocean and atmosphere (i.e., geostrophic turbulence) are compounded by the broad range of significant length scales (Re $\sim 10^8$) and by the relative smallness of the vertical height of these boundary layers in comparison with their horizontal extent, which accentuates the importance of the backscatter of energy to the larger scales of motion. In this chapter the authors demonstrate the ability of an ILES model based on high-order upwinding to reproduce the complex features of the global climate of the atmosphere and ocean.

Next, in Chapter 15, Porter and Woodward discuss their ILES-based studies of astrophysics. Homogeneous decaying and driven compressible turbulence (Re $\sim 10^{12}$), local area models of stellar convection, global models of red giant stars, and the Richtmyer–Meshkov mixing layer are examined in this context.

The section continues with Chapter 16, by Alin et al., on complex engineering turbulent flows, where they discuss the application of ILES to a variety of complex engineering-type applications ranging from incompressible external flows around typical naval applications to external and internal supersonic flows in aerospace applications. Cases examined include flows such as (1) the flow around a model scale submarine, (2) multi-swirl cumbustion flows, (3) solid rocket motor flows, and, (4) the flow and wave pattern around a modern surface combatant with transom stern.

Large-scale urban simulations are discussed by Patnaik et al. in Chapter 17. Airborne contaminant transport in cities presents challenging new requirements for computational fluid dynamics. The unsteady flow physics is complicated by very complex geometry, multiphase particle and droplet effects, radiation, latent and sensible heating effects, and buoyancy effects. Turbulence is one of the most important of these phenomena, and yet the overall problem is sufficiently difficult that the turbulence must be included efficiently with an absolute minimum of extra memory and computing time. This chapter describes a MILES methodology used as a simulation model for urban contaminant transport, and addresses the very difficult validation issues in this context.

Finally, outlook and open research issues are presented in Chapter 18.

SECTION A

MOTIVATION

1 More for LES: A Brief Historical Perspective of MILES

Jay P. Boris

1.1 Introduction to monotone integrated large eddy simulation

Turbulence is proving to be one of nature's most interesting and perplexing problems, challenging theorists, experimentalists, and computationalists equally. On the computational side, direct numerical simulation of idealized turbulence is used to challenge the world's largest computers, even before they are deemed ready for general use. The Earth Simulator, for example, has recently completed a Navier–Stokes solution of turbulence in a periodic box on a $4096 \times 4096 \times 4096$ grid, achieving an effective Reynolds number somewhat in excess of 8000. Such a computation is impossible for nearly every person on the planet. Further, periodic geometry has little attraction for an engineer, and a Reynolds number of 8000 is far too small for most problems of practical importance.

The subject of this chapter is monotone integrated large eddy simulation (LES), or MILES – monotonicity-preserving implicit LES (ILES), a class of practical methods for simulating turbulent high-Reynolds-number flows with complicated, compressible physics and complex geometry. LES has always been the natural way to exploit the full range of computer power available for engineering fluid dynamics. When the dynamics of the energy-containing scales in a complex flow can be resolved, it is a mistake to average them out. Doing so limits the accuracy of the results, because uniform convergence to the physically correct answer, insofar as one exists, is automatically voided at the scale where the averaging has been performed. Even if the computational grid is refined repeatedly, the answer can get no better. At the same time, the overall resolution of a computation suffers when many computational degrees of freedom are expended unnecessarily on unresolved scales. Solving extra equations, for example, to define unresolved subgrid quantities is expensive and limiting.

Fortunately, we now know that a wide class of efficient fluid dynamics methods, generally based on monotone convection algorithms, has a built-in subgrid "turbulence" model that is coupled continuously to the grid-scale errors in the computed fluid dynamics. When the fluid dynamics embodies positivity-preserving or monotone (nonlinear) convection algorithms, the original name MILES applies. There are now hundreds of such algorithms and variations in use throughout the world. MILES works because the necessary physics, that is, conservation, monotonicity (positivity), causality, and locality, are built into the underlying fluid dynamics.

Professor John Lumley hosted a meeting at Cornell University in 1989, called Whither Turbulence: Turbulence at the Crossroads (Lumley 1990). The conference was focused on controversial questions in turbulence research that John felt had not been satisfactorily resolved. The program was structured as a dialog, with position papers on six major questions that were critiqued by several reviewers, reviewed, and subsequently published as a book. Professor W. C. Reynolds led the discussion of one of these questions with a keynote paper entitled "The Potential and Limitations of Direct and Large Eddy Simulations" (Reynolds 1990). This address summarized the situation in turbulence modeling at that time and was the jumping off point for my MILES paper. His exposition is highlighted in Section 1.2.

As a responding presenter, in my paper I focused on giving the evidence and reasons for the surpising success of monotone *no-model* LES methods. I named this approach to turbulence simulation *monotone integrated large eddy simulation*, or MILES, as an up-front reminder that monotone methods come with an integrated LES turbulence capability built in: "These monotone integrated LES algorithms are derived from the fundamental physical laws" (Boris 1990). The MILES hypothesis was formulated to explain "strong evidence suggesting that monotone convection algorithms (e.g., FCT), designed to satisfy the physical requirements of positivity and causality, in effect have a minimal LES filter and matching subgrid model already built in. The positivity and causality properties ... seem to be sufficient to ensure efficient transfer of the residual subgrid motions, as they are generated by resolved field mechanisms, off the resolved grid with minimal contamination of the well-resolved scales by the numerical filter."

Looking back before the Cornell conference, my plunge into turbulence simulation actually dates to the early 1970s when flux-corrected transport (FCT) was invented. It soon became clear that FCT was good for much more than dynamic shocked flows, but recognizing just how much was going on under the surface in the computed solutions took almost another two decades, until 1989. Section 1.3 discusses the origins of FCT and the early "turbulence" computations performed with it. Section 1.4 extends this discussion to the properties of monotone methods that lend themselves to effective LES. Widespread acceptance that FCT and other monotone methods have a perfectly functional implicit subgrid turbulence model is taking as long again. This long delay is due in part to the fact that the result seems almost too good to be true. Even engineers and scientists with no previous LES persuasion have long been trained to decry the possibility of a free ride. Section 1.5 explains why it should not be surprising that MILES works well, and Section 1.6 considers some of the early tests performed to show how it was working.

Today, MILES' growing acceptance is based in no small part on the work reported in the following chapters. With theoretical understanding and insight provided by the various authors, it is now easier to see why MILES works and to understand that there is a well-defined subgrid turbulence model in place. This chapter traces this evolving understanding, in a brief personal way. Here is a situation in which the complexity of the physical problem contributes to the ease of solution. Imitating the physics with the numerics, rather than cranking on a more formal mathematical approach, actually brings a big win. The MILES hypothesis evolved more as a growing realization than an invention. With 20:20 hindsight, perhaps it isn't surprising that it is taking others a long time to realize that this is one of those special cases where there can be a free ride.

1.2 The numerical simulation of turbulence

Turbulent flows are common in nature. They occur on space scales ranging from millimeters to megaparsecs. Such flows are extremely important for a number of reasons. Turbulence provides an efficient way for distinct, initially separate materials to interpenetrate and mix rapidly, and it provides for the rapid transport of heat and momentum to and from surfaces. In a typical turbulent flow, the important space and time scales can span many orders of magnitude, and these scales all coexist simultaneously in the same volume of fluid.

According to Reynolds (1990), the goals of LES are "to compute the three-dimensional, time-dependent details of the largest scales of motion (those responsible for the primary transport) using a simple model for the smaller scales. LES is intended to be useful in the study of turbulence physics at high Re, in the development of turbulence models, and for predicting flows of technical interest in demanding complex situations where simpler model approaches (e.g., Reynolds stress transport) are inadequate." The computational challenge is to resolve a wide enough range of scales to study the underlying physical mechanisms and provide a predictive capability for important practical applications.

Understanding LES is not possible without considering the Kolmogorov spectrum (Kolmogorov 1941, 1962). The kinetic energy density of turbulent motions is $\mathcal{E}(k)dk$ in the wave-number range from k to $k + dk$. This energy density, a function of the wave number, is expected to follow the power law

$$\mathcal{E}(k) = \mathcal{E}(1/L)(kL)^{-5/3}.$$

This Kolmogorov spectrum can be derived purely from dimensional arguments (Batchelor 1956) and describes how the energy density of turbulent structures of size $\eta = 1/k$ decreases rapidly with increasing wave number k. This relationship is generally valid only for an intermediate range of scales, called the *inertial range*, between the system size L and the small Kolmogorov scale, η_K, at which the viscous dissipation dominates the inertial flow of the fluid:

$$\eta_K = (v^3/\epsilon)^{1/4}.$$

This Kolmogorov scale is generally very small and may be micrometers, millimeters, or kilometers where the system length is of the order of meters, kilometers, or parsecs.

The turbulence energy density $\mathcal{E}(k)$ decreases with increasing k. It results from a process in which the dynamic structures in the fluid (called *eddies*) of size $1/k$ interact with each other and with the more energetic larger-scale eddies to generate smaller eddies. These smaller-scale eddies, in turn, populate eddies of yet smaller size. This generally downward transfer of energy from large to small scales is called the *inertial energy cascade* or just *turbulent cascade* and is dissipationless. This cascade terminates at small scales η_K where an eddy is so small that it diffuses appreciably as a result of viscosity in the time it takes to rotate. Another way of saying this is that the local Reynolds number, defined as $\text{Re} = UL/\mu$, for a Kolmogorov eddy is unity. Because the characteristic speed in this small eddy scales as $\eta^{1/3}$, it is possible to relate η_K to Re for the macroscopic flow through

$$\eta_K/L = \text{Re}^{-3/4},$$

where U is the the maximum or characteristic large-scale fluid velocity in the system (Frisch and Orszag 1990). This means that when the macroscopic Reynolds number is 10^4, η_K is about one thousandth of the system size.

Numerical solutions of the Navier–Stokes equations are always "filtered" through the numerical representation of the equations and through the algorithms used to solve this representation. The algorithms affect the solution directly, but there is always the difficulty of knowing what these numerical effects are. At the grid or discretization scale we need to be particularly concerned about these numerical errors, both for direct numerical simulation (DNS) and LES. Today, computational research on turbulence is pursued as a two-step, bootstrap process. A numerical simulation that aims to resolve the full range of physical scales in the chosen flow is a DNS. These DNS computations augment theory and experiment by providing information about the important nonlinear mechanisms in turbulence that can be inspected by dissecting the details of the computed flow. What is just as important, however, is that DNS provides benchmark data sets for simple, relatively low-Reynolds-number flows that can be used to calibrate the accuracy and the performance of a more efficient type of engineering model – the LES.

For DNS, it is necessary to resolve all of the scales of the problem down to η_K. The larger the computational grid and the finer the resolution used, the larger the range of physical scales in the turbulent flow that can be computed (resolved) accurately. Adequate numerical resolution for DNS must ensure that the physical viscosity dominates any residual numerical diffusion. A linear, fixed-grid algorithm can only guarantee positivity-preserving solutions for convection when the diffusion in the numerics is large enough. The relation that must be satisfied is $L/\delta x \geq \text{Re}$, where δx is the grid spacing in the calculation (e.g., Oran and Boris 2001). This is the cell Reynolds number restriction. A linear algorithm requires at least Re cells in each direction of a DNS to guarantee an accurate solution of the equations on a microscopic scale.

Using the definition of η_K above, this restriction is $\eta_K/\delta x \geq \text{Re}^{1/4}$. Thus, the computational cell size must be appreciably smaller than η_K to ensure physical positivity of the solutions at short wavelengths. For example, when the physical Reynolds number is 10^4, the cell size for a linear algorithm has to be at least 10 times smaller than η_K despite the fact that nothing of consequence to the flow happens at this scale. In other words, a factor of 1000 or more could be wasted in DNS because of this cell Reynolds number condition.

One way around this problem for nonmonotone (nonpositive) methods is to simulate flows with low Reynolds number. Another is to abandon the fluid dynamic positivity-preserving property and live with nonphysical structure in the DNS solutions at short and even moderate wavelengths. When linear convection methods are used for DNS, correct and fully accurate solutions of all scales may only be obtained for flows where the physical viscosity is large enough to suppress grid-scale fluctuations inherent in low-dissipation numerical methods. To some extent, however, this stringent condition on the cell size can be relaxed when there is only a small amount of energy in the small scales.

LES models are attractive because they do not try to compute all scales. Instead, they represent only the largest, so-called *resolved* scales. It is necessary to solve the problem with a grid size considerably larger than η_K, so some assurance is needed that not resolving the small-scales has an acceptably small detrimental effect on the large-scale behavior that is being computed. The usual LES approach results in additional terms in the numerical model, often in the form of a phenomenological eddy diffusivity, to represent the effects of the unresolved small scales on the resolved motions. The important issue for LES models has always been how to account satisfactorily for these unrepresented small scales, a process called *subgrid turbulence modeling*. In LES, unlike DNS, it is important to minimize the dissipation of the fluid dynamic convection algorithm at the short wavelengths corresponding to the computational cell size, because the physical flows themselves have no dissipation at these scales. The turbulent kinetic energy at these discretization scales is actually transforming (cascading) nonlinearly to smaller scales with no dissipation. This process is not well represented as a viscosity. In strong distinction to the more viscous flows being treated by DNS, an LES flow should show more and more structure as the resolution is improved, not less and less. Therefore, algorithms optimal for DNS are not necessarily optimal for LES.

As Reynolds (1990) said, "Clark, Ferziger, and Reynolds (1979) developed an approach for using DNS to test LES models (now called a priori testing). The idea is to filter a velocity field obtained by DNS and compare the residual stresses with the residual stress model." This two-step computational procedure could only gain acceptance as computers became more powerful and respect for the accurately simulated solutions could be appreciated. In 1989, this approach was beginning to become accepted. Reynolds further noted that although the comparisons of DNS and LES using their a priori testing procedures did not give large correlation coefficients for the residual stress, nevertheless the macroscopic statistical predictions of LES were "in rather good agreement with those of DNS." This is an important theme in this and following chapters: The physically important aspects of the fluid dynamics of turbulent flow can be notably insensitive to the small-scale details of how it is computed.

Many efforts to understand turbulence were being treated as opportunities for intellectual creativity; certainly that was the research climate. Theoreticians massaged the Navier–Stokes equations, took multiple moments, created hierarchies of nonlinear equations, intuited closure approximations, filtered and averaged the equations in various ways, and even converted them to spectral representations. The computational fluid dynamicist was then called upon to implement the complicated residues of this mathematical–intellectual prowess. Nature really doesn't seem to resort to any of this. It has also become quite clear that this isn't the best way to apply high-performance computing. Using modern computers to solve the fundamental set of equations directly – or as directly as possible – is actually easier and more productive. This alternate philosophy on the use of computers was not so widely understood or accepted in 1989, but John Lumley considered these potentials and limitations to be one of the key questions for the Cornell conference. Using computers properly ensures that every improvement in performance results in better answers. For LES, a factor of 2 increase in spatial resolution is probably better than any likely improvements to a subgrid model.

There is now a widespread effort to delve more deeply into these resolution and accuracy issues for the monotone fluid dynamics algorithms. These algorithms were originally designed to compute compressible flows with high resolution and to capture shocks and material discontinuities accurately within one or two cells. More recently, they have been applied to subsonic compressible and reactive flows where high resolution of small eddies and narrow shear layers is equally valuable, as illustrated by several of the examples in Section 1.4. Typical monotone methods include FCT (Boris 1971; Boris and Book 1976a), MUSCL (van Leer 1973, 1979), PPM (Colella and Woodward 1984; Woodward and Colella 1984) TVD (Harten 1983; Harten et al. 1987; Sweby 1984; Yee, Warming, and Harten 1985), and the second-order Godunov method (Colella 1985; Colella and Glaz 1985). Some general information on and comparison of these methods is given by Zalesak (1987) and Oran and Boris (1987, 2001). Some of the numerical results reported in support of MILES in the convenient conspiracy theory (Oran and Boris 1993) were, in fact, obtained from full solutions of the Navier–Stokes equations using monotone algorithms.

1.3 Flux-corrected transport: Our monotone method of choice

The MILES hypothesis for LES of turbulence had its beginnings in early applications of the FCT convection algorithm. In 1970, the qualitative deficiencies of strong, multidimensional shock computations using available methods were hampering important programs at the U.S. Naval Research Laboratory (NRL). Thus FCT was invented to improve the quality of numerical convection (Boris 1971) in computational fluid dynamics (CFD) applications where dynamic convection of strong gradients plays an important role. The key word here is *quality*; the issue was not one of higher-order accuracy. The issue was to get the physics of strongly shocked flows at least approximately right. FCT became our algorithm of choice ever since.

The traditional CFD approaches were not faring well. Primary fluid quantities such as mass density and chemical species number densities would become negative – a physical impossibility – near shocks and steep gradients. Talented people had been working on this problem for years, but FCT turned out to be a qualitatively better solution and thus broke new ground. It took a few years, however, to realize that the advantages of FCT for computing shock and blast problems carried over for applications to unstable and turbulent flows. Two notable, independent, and contemporary research efforts moving toward positivity-preserving algorithms for fluid dynamics were the development of MUSCL by Bram van Leer (1973) and artificial compression by Ami Harten (1974).

The key new feature in FCT was using a nonlinear method to solve an intrinsically linear but constrained problem. The nonlinear component of the flux-correction algorithm came to be known as the *flux limiter*. I now discuss the considerations that led to this statement of the underlying FCT principle: "The antidiffusion stage should generate no new maxima or minima in the solution, nor should it accentuate already existing extrema" (Boris 1971; Boris and Book 1973). Keith Roberts, the Director of the Culham Plasma Physics Laboratory in England, offered me a visiting appointment at Culham Laboratory as a member of Princeton's research staff immediately upon graduation. John Greene of Princeton and Klaus Hain of NRL were instrumental in arranging this.

Keith's general approach to computational physics, which I adopted and have used ever since, was to build in as much of the physics as you can into the model and let the mathematics take care of itself. We began looking at ways to improve multidimensional CFD – as we had promised Klaus and John. During some particularly discouraging tests, Keith introduced me to Godunov's (1959) theorem – that second- and higher-order algorithms could not preserve the physical positivity property of convection. It was certainly clear that high-order schemes were not necessarily bringing greater accuracy, so physics was going to have to step in to shore up the ailing numerics.

By the fall of 1970 at NRL, computing strong, dynamic shocks was a primary roadblock. Urged to concentrate on the problem, I developed the first FCT convection algorithm in the winter of 1971. It was called SHASTA (sharp and smooth transport algorithm), and its results were sufficiently encouraging that both Keith and Klaus urged me to publish the new capability at the Computing as a Language of Physics workshop, held in Trieste, Italy, in the summer (Boris 1971). David Book began working with me to generalize the FCT technology. Our *Journal of Computational Physics (JCP)* articles (Boris and Book 1973, 1976b; Boris, Book and Hain 1975) extended the Trieste paper to include solutions of complete sets of equations for strong shocks and other fluid flows and to use a number of different linear convection algorithms as a basis for FCT.

SHASTA was constructed as a layered set of corrections, each layer being added to mitigate the errors left from the previous layers. The starting point was general dissatisfaction with the negative densities and nonphysical wiggles near sharp gradients observed in classical, second-order convection algorithms and a deep suspicion of the excess diffusion introduced in first-order donor-cell (upwind) algorithms. The first layer of the SHASTA algorithm is a conservative Lagrangian displacement and compression of linear trapezoids of fluid, allowing the physical principles of conservation, positivity, causality, and locality to be preserved. It was easy to program but led to jagged sequences of mismatched trapezoids that quickly spawned many of the problems of a fully Lagrangian approach. The second layer, therefore, was a conservative Eulerian remap of the displaced, distorted trapezoids of fluid back onto the original fixed grid.

This introduced a zero-order numerical diffusion whose coefficient, 1/8, was generally worse than donor cell diffusion! As a result, the third layer of SHASTA was an explicit, linear antidiffusion to subtract the excess diffusion introduced by the remap. However, without the nonlinear flux-correction process, the resultant solutions were just as bad as other linear, second-order algorithms.

Rather than adding a strong numerical diffusion ("viscosity") controlled by numerical derivatives that would be meaningless near discontinuities and sharp gradients, I began looking in detail at the physics of local profiles of the density being convected – before and after the antidiffusion stage. In each specific case it was clear just how much of the antidiffusive correction flux had to be thrown away to prevent a particular cell value from drifting past the monotonicity limits imposed by the neighboring values. However, the expression of this "flux correction," now usually called a flux limiter or *slope limiter*, was different in every case. Finally, with recognition that the sign function of the density difference across a cell interface could be used to collapse the behavior near maxima and near minima into a single formula, the basic nonlinear max–min expression emerged in March 1971.

At NRL we recognized that the underlying linear convection algorithm could be just about anything. The SHASTA algorithm in the first working FCT code was just one particular vehicle and maybe not even a particularly good one. Numerous variants of Lax–Wendroff, leapfrog, and the other simple linear algorithms were tried. The result was the set of three papers in the Journal of Computational Physics (JCP). Increasing the zero-order diffusion coefficient to 1/6 reduced the phase error in the convection from second order to fourth order with improved fidelity in convected structures. We also found that the initial density profile could be made to emerge unscathed, when the interface velocities were zero, despite the diffusion and antidiffusion stages. David called this *phoenical FCT*, and we use this trick today. Perhaps the best of the early FCT publications, though not the most cited, was that by Boris and Book (1976b). The first journal article in the *JCP* series was reprinted as the most cited article in the 30th Anniversary Issue of *JCP* with an introduction by Zalesak (1997).

Monotone algorithms should be designed with minimum dissipation while maintaining monotonicity. If less dissipation is required to get a substantially correct answer, the only solution is to increase the spatial resolution. Using a CFD algorithm with lower dissipation is either unstable or leads to nonphysical results. FCT algorithms have been analyzed theoretically and have been shown by Ikeda and Nakagawa (1979) to converge to the correct solution of the underlying continuity equation being solved. This means that increasing resolution, even without any added subgrid transport model, will lead to a converged solution to the target high-Reynolds-number problem once the residual, resolution-based numerical dissipation has become smaller than the eddy diffusivity from all unresolved scales. Section 1.7 shows a test of the MILES hypothesis that demonstrates this.

Our colleagues made significant contributions to this FCT technology. Steve Zalesak (1979, 1987) invented the multidimensional limiter, and this was modified and extended by Rick DeVore (1989, 1991) to include magnetohydrodynamic flows where enforcing div $B = 0$ is an important consideration. Zalesak (1981), McDonald, Ambrosiano, and

Zalesak (1985), and Karniadakis (1998) showed us that FCT could be added to spectral algorithms, spawning a fast fourier transform FCT and a (linearly) reversible FCT algorithm. These variants, however, generally do not preserve the locality or causality properties of the physics and thus are generally not good choices for shock problems or MILES applications. Theodore Young, Jr. (Young, Landsberg, and Boris 1993), and Alexandra Landsberg (Landsberg, Young, and Boris 1994, 1997) were major contributors to vectorizing and parallelizing FCT and to extending the capability to arbitrary complex geometry, leading to our current FAST3D models. Elaine Oran, Sandy Landsberg, and John Gardner participated in a project to make FCT generally useful and available (Boris et al. 1993) through the widely distributed LCPFCT modules. Rainald Löhner (Löhner et al. 1987a, 1987b, 1988) carried FCT into the finite-element world with a practical, general geometry formulation and adaptive meshing capability, and Gopal Patnaik (Patnaik et al. 1987; Patnaik and Kailasanath 1996) extended FCT to an implicit formulation called BICFCT for slow but fully compressible reacting flows. Progress on FCT algorithms has continued to this day, including multidimensional FCT with adaptive gridding (Ogawa and Oran 2004) and the urban airflow modeling subsequently reported by Patnaik, Grinstein, and Boris (2006); see also Chapter 17. A summary of this effort and view of the current status of FCT is given by Kuzmin, Löhner, and Turek (2005).

1.4 Using monotone methods for turbulent flow problems

Almost immediately upon their introduction, scientists began to apply FCT and other monotone methods to turbulence and mixing problems beyond the shock dynamics studies that fostered the technology. Communities specializing in shock-generated turbulence (e.g., weapons effects), turbulent combustion (flames and detonation), Rayleigh–Taylor and Richtmyer–Meshkov instabilities and mixing, compressible and incompressible turbulence (both for free and wall-bounded flows), and atmospheric modeling began using the MILES models. David Youngs (1982) has developed a monotone, multi-dimensional methodology for Rayleigh–Taylor and Rictmyer–Meshkov mixing studies and Piotr Smolarkiewicz (1983, 1984) has pioneered a comparable positive definite advection scheme whose main application has been geophysical flows. This section shows examples of several of these early applications. Later work populates the following chapters.

As computers became more powerful, three-dimensional shock computations (blasts and explosions) became practical. Unsteadiness and turbulence are intrinsic to these problems. Figure 1.1 shows the interaction of three large blast waves (fireballs) computed in a $100 \times 100 \times 50$ grid on an IBM 360/91 in 1976. Such computations are invariably turbulent after the shock transients have passed, with turbulent velocities as high as Mach 0.3 (e.g., Boris and Oran 1981; Oran and Boris 1981). Even smaller computers were becoming useful. Figure 1.2 shows an early computation of a reflected shock in an inert, ideal gas performed on a VAX 11/780 in 1978. An expanding coordinate system was used so the expected similarity solution would have the opportunity to be stationary. The flow is not truly turbulent but is clearly in transition on the slip line

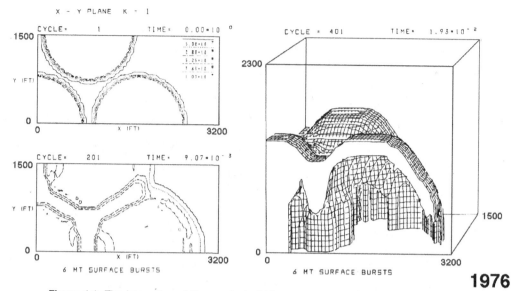

Figure 1.1. The interaction of three spherical blast waves computed using FCT. The initial condition (upper left) shows the shock fronts almost touching. The lower left panel show the results after 200 time steps at 91 ms. The panel on the right shows a cut-away 3D plot of the density in the expanding, interacting shells at 193 ms (Boris and Oran 1981).

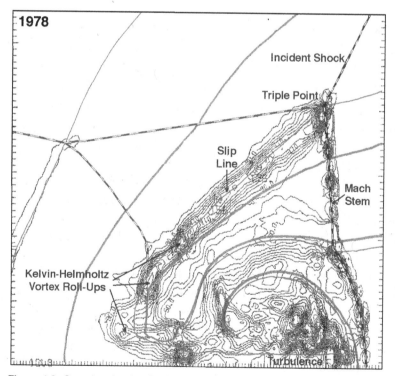

Figure 1.2. Complex shock reflection from an ideal flat surface computed with time dependence in an uniformly expanding (similarity solution) coordinate system. The Mach stem is being deformed by a region of turbulence formed by vortex roll-up. The slip line originating at the shock triple point is Kelvin–Helmholtz unstable (Boris 1978, unpublished).

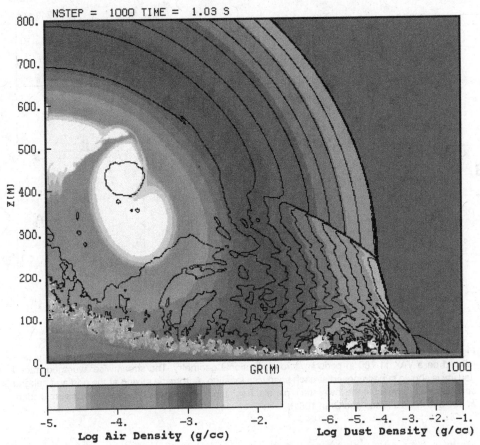

Figure 1.3. Complex, reflecting shock structures induce a turbulent, multiphase flow as they interact with a dusty ground plane. In this computational result, the heavy black lines indicate the major shock structures propagating to the right. The gray and white lines show the vorticity generated by the flow. Flows behind the shock that impinge on the ground cause the surface layer with entrained dust to lift, giving rise to strong interactions between a free shear layer and a baroclinic boundary layer (Kuhl et al. 1993).

originating at the triple point. This particular *class of problem* has been the subject of similarity solutions and other steady-state theory even though the triple-point slip line is clearly unstable and leads to turbulence. Book et al. (1980) considered a number of strongly shocked flows in which the innate unsteadiness must give rise to turbulence.

Other turbulent flows simulated by MILES include beam–channel interactions such as occur in lightening (Picone et al., 1981), charged-particle beam propagation (Picone and Boris 1983), laser ablation in nonuniform media (Emery et al. 1984), and laser and beam deposition in dusty gases (Boris et al. 1992b). More recently we have treated geometry-generated turbulence in open-air munition-destruction devices (Lind et al. 1996, 1997). These flows are all compressible, and shocks were present in most of the problems. Often the geometry or boundary conditions are complex. As computer power increased, three-dimensional simulations became the norm, as required to properly represent most turbulence. Figure 1.3 shows a high-resolution, multiphase flow problem

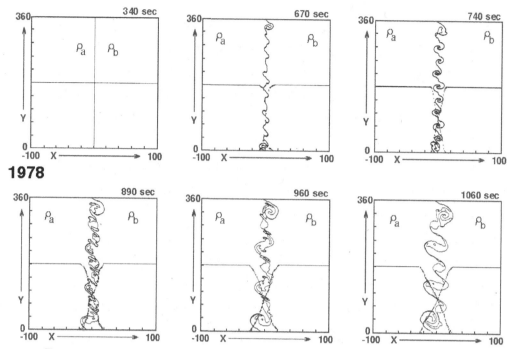

Figure 1.4. Classical Kelvin–Helmholtz instability between two different fluids of equal density, performed on a VAX 11/780 in periodic, two-dimensional geometry. The streamwise averaged traced density is shown superimposed at each spanwise location. Automatic mixing at the grid and subgrid scale is clearly evident in the lower three panels. Beginnings of longer-wavelength roll-ups are evident at 960 and 1060 s (Boris and Oran 1981).

employing adaptive gridding to treat the dust entrained by a large-scale expanding fireball. The turbulence in the dust layer just above the ground is clearly evident (Kuhl et al. 1993). A number of organizations had comparable success in computing complex shock dynamics and blast–structure interactions; some of these efforts are discussed later in this volume. Again, no explicit turbulence model was needed in these computations.

Early FCT and other monotone methods were also being used to study a number of the classical turbulence and transition-to-turbulence problems treated by earlier methods. Figure 1.4 shows an early computation of Kelvin–Helmholtz instability performed on a VAX 11/780 in 1978. Six cross sections of the evolving two-dimensional shear layer are shown (Boris and Oran 1981). Figure 1.5 shows a gray-scale contour plot of a mass density cross section through an FCT computation of a heavy fluid (dark) falling through a lighter fluid (light) in the late nonlinear stages of the Rayleigh–Taylor mixing phase. (Boris 1992). Understanding the nonlinear phases of Rayleigh–Taylor and Richtmyer–Meshkov mixing, including configurations with a number of distinct fluids, has been a major application of MILES methods. Recent applications and tests in this problem class are discussed in later chapters.

The sets of large-scale computations performed by Professor Kunio Kuwahara and his colleagues in the late 1980s tackled standard low-Mach-number-fluid–structure interaction problems at high Reynolds numbers. These computations should probably

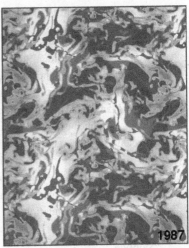

Figure 1.5. Late nonlinear Rayleigh–Taylor mixing of two fluids. This computation, performed on the GAPS, seeks an asymptotic interpenetration rate of the two fluids. Equal and opposite accelerations were applied to the two fluids in a double-period laboratory frame of reference (Boris 1992c).

be classed as ILES since the Kuwahara model features a high-order linear dissipation for stability (no monotone limiter per se has been acknowledged). See, for example, Tsuboi, Tamura, and Kuwahara (1989), Suzuki and Kuwahara (1989), and Tamura and Kuwahara (1989). Figure 1.6 illustrates a more recent NRL MILES application. It shows three frames from one of a series of simulations of enhanced mixing in subsonic jets emerging from square and rectangular nozzles (Grinstein and DeVore 1992; Grinstein 2001; Grinstein, Gutmark, and Parr, 1995). This calculation of a flow with Re > 120, 000 could not possibly have been performed by DNS. Comparison of Kuwahara computations and these FCT simulations with a number of different experimental results have shown uniformly excellent results, despite the complete absence of an explicit subgrid turbulence model.

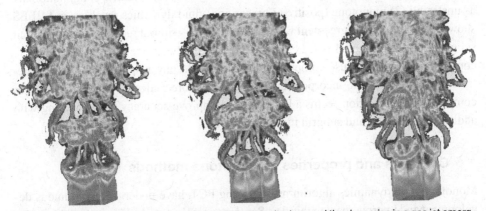

Figure 1.6. Sequential visualizations of the vorticity magnitude reveal the dynamics in a gas jet emerging from a square nozzle. The gray scale in these visualizations ranges from semi-transparent to opaque $(0.05\Omega_{peak}-0.60\Omega_{peak})$. The figure depicts the large-scale vorticity dynamics controlling the jet development and approach to turbulence, characterized by vortex rings undergoing self-deformation, hairpin vortices originated by vortex induction and stretching in the initial shear layer, and, farther downstream, a more disorganized flow regime characterized by tube-like vortices (Grinstein and DeVore 1992).

The Kuwahara computations were specifically reviewed by Reynolds:

One sees remarkable pictures of turbulent flows that look very realistic, obtained from simulations with no explicit residual turbulence model at all. Typically these are some sort of higher-order upwind difference scheme (Tsuboi, Tamura, and Kuwahara 1989), and are put forth as a high Reynolds number result. In real turbulence the dissipation is set by the non-linear cascade in the inertial range, and the only role of viscosity is to set the smallest scales of motion. Is it conceivable that in LES one needs only to calculate the inertial eddies properly, allowing whatever dissipation mechanism is present (i.e. upwind differencing) to set the smallest scale?

This approach might actually work if the dissipative process is confined to small scales. In this case the simulation could produce useful results in flows where Reynolds number effects are not important and separation points are set by sharp edges rather than by viscous fluid dynamics. But as there is no relationship between numerical viscosity and fluid viscosity, *this approach simply can not predict important Reynolds number effects* and consequently should be used with caution.

Professor Kuwahara simply commented that "Fluid dynamics is easy on a large enough computer" – and his team had the largest computers available. How many others in the late 1980s had inklings that something important was going on with "no model" LES is unknown. MILES certainly wasn't a popular notion with the established turbulence community, but practical people began using it. They really had little choice if they needed to solve realistic engineering problems with time dependence.

From these wide-ranging computations without an explicit subgrid turbulence model, one concludes that the physics of turbulent cascade seems to be controlled by the macroscopic, energy-containing scales of the flow and the dissipation of this energy that is due to molecular viscosity occurs at scales considerably larger than the Kolmogorov scale. Indeed, the simple conclusions, spanning three decades and four orders of magnitude increase in computer performance, are these: (1) Because turbulent fluids are intrinsically unsteady, they should be represented computationally as unsteady; (2) monotone (positivity-preserving) fluid dynamics resulting in MILES shows uniformly good agreement with data and theory without recourse to any explicit subgrid turbulence model or eddy viscosity; (3) MILES models lend themselves to complex problems because compressibility, multifluid physics, and complex geometry are relatively easy to incorporate; (4) including more scales by an improvement in computational resolution seems a better way to improve accuracy than tinkering with additional equations and subgrid turbulence models.

1.5 Concepts and properties of monotone methods

Monotone fluid dynamics algorithms, including FCT, have undergone continuous development and improvement to this day. See, for example, Smolarkiewicz and Margolin (1998), Belotserkovskii (1999), Kuzman, et al. (2005), or the chapter by Patnaik, Grinstein, and Boris below. In simulating convection, two types of error are introduced by the solution algorithm: numerical diffusion and numerical dispersion. A third type of error is introduced by the finite resolution itself: Gibbs phenomenon oscillations. These errors interact and complicate each other. When a higher-order spectral, pseudospectral, or spectral-element representation is used to decrease the numerical diffusion and numerical dispersion, the Gibbs oscillations become more evident, especially at steep gradients. When coupled continuity equations are solved to describe fluid dynamics, other coupling difficulties can arise. The practical question is how to minimize the combined effect of these errors to obtain an accurate solution (Oran and Boris 1987, 2001).

Nonlinear, monotone convection algorithms are designed to minimize these purely numerical problems while enforcing the important physical properties of global conservation, monotonicity (positivity), causality, and locality. Replicating these physical properties is an important determinant of an algorithm's suitability as a basis for MILES. *Conservation* in Eulerian formulations implies that the total quantity of a convected quantity such as mass density $\rho(x, y, z, t)$, when integrated over the spatial domain, is not changed by the convection algorithm. The conserved integral of ρ cannot be changed by convection or compression. Conservation is easily enforced by writing the conservation equations in flux form; every flux of a conserved variable that enters one volume must leave another. The conservation integral can only change by unbalanced fluxes into or out of the overall domain and by explicit physical sources or sinks within the domain. Conservation is important to MILES because it ensures that kinetic energy that cascades out of the resolved scales automatically appears as heat (temperature). Even the instant conversion of subgrid-scale kinetic energy into heat, it turns out, is not a bad assumption (Section 1.6), given the shape of the Kolmogorov spectrum.

Monotonicity means that the numerical algorithm does not add unphysical oscillations to ρ – despite the presence of numerical dispersion effects and finite-resolution Gibbs effects in the numerical algorithm. *Positivity* refers to that particular instance of monotonicity when ρ is a positive-definite convected quantity, such as mass density, that cannot physically become negative. Discussion and analysis of monotonicity and positivity have been extensive. Ensuring this property in multidimensions also raises interesting issues. Some MILES algorithms are positive, for example, but not completely monotone. Ensuring numerically that solutions of the continuum continuity equation cannot become negative anywhere, if they are initially positive or zero, is the important condition. This physical property, built into monotone methods, avoids a number of numerical pitfalls by construction. Even nonmonotone methods must add terms to ensure at least minimal positivity – or pay a serious toll in numerical instability and questionable accuracy of the computed solutions.

Causality requires that a fluid being convected from point A to point B must pass through all the cells on a flow path between them. Convection does not allow mass, momentum, or energy to jump across an intervening space discontinuously; so the numerical methods should have this property. Fluid dynamic turbulence in one region travels to another by convection of vorticity across the space between. The magical appearance of vorticity in free space without an approximate source term should be forbidden. Causality and conservation are usually implemented by using locally computed fluxes to move fluid dynamic quantities between adjacent cells.

Locality is a closely related requirement to causality. The region used to determine the derivatives (gradients, curls, and divergences) in the fluid dynamic equations should be quite limited; otherwise, the computational fluid will exhibit one form or another of nonphysical action at a distance. Evaluation of the fluxes used to compute these derivatives should also be through local algorithms. Though nonlocal evaluation of derivatives can be more accurate in a mathematical sense, causality and often conservation are compromised when the algorithm is nonlocal. The key to maintaining monotonicity, and thus the excellent properties of a monotone algorithm at the grid scale, is to adjust the amount of numerical diffusion locally while still ensuring these physical properties.

To understand why these monotone conservation equation algorithms work so well on turbulence problems, researchers have performed various tests on selected numerical problems to study the actual numerical dissipation left in the solution after the flux-correction procedure has been completed (Book et al. 1991; Grinstein and Guirguis 1992). These test computations support the following conclusion: The local, time-dependent dissipation in nonlinear monotone algorithms behaves as a subgrid turbulence model for scales smaller than several grid sizes. This subgrid model properly connects the large, energy-containing scales with the unresolved subgrid scales of motion. Local dissipation, though highly nonlinear, was found to scale with wave number roughly as $k^{3.3}$ to $k^{3.8}$, depending on the flow configuration.

Monotone methods provide accurate computations of high-speed compressible flows and good resolution of shocks, gradients, and mixing. In addition, they are flexible with respect to adding variables and using nonuniform and adaptive spatial grids to treat oddly shaped chambers and obstacles in the flow. It is also easy to add additional physical effects such as chemical, atomic, or nuclear reactions and radiation transport. Since the late 1970s, scientists and engineers have been applying monotone methods to Navier–Stokes and Euler problems involving unstable turbulent flows. When monotone algorithms are used to simulate complex turbulent or transitional flows, even without the explicit molecular viscosity terms of the Navier–Stokes equations, they have repeatedly done an excellent job in computing the large-scale flows, faithfully reproducing important structures and behavior on scales considerably larger than η_K.

1.6 Why should MILES work?

The wide range of space and time scales in turbulent flows and the nonlinear superposition of all these scales in the same macroscopic volume could be viewed as Nature's attempt to make accurate numerical simulation of turbulence impossible. Actually, Nature seems to be trying to give us a break. In this section I want to make it at least plausible that MILES does work and explain qualitatively why, based on the physics and numerics. The brief explanation for the success of MILES is that the correct equations are being solved in a suitable representation by appropriate numerical algorithms. For people trained in the classical treatments of turbulence and the usual approaches to LES, however, the notion of relying on the grid-scale numerics of an algorithm to provide all the features of an externally constructed subgrid turbulence model is hard to swallow. The various analyses and validation studies in the chapters to follow will back this explanation up, showing how and how well MILES and ILES work.

Five related considerations, a set of convenient physical and computational circumstances, appear to be conspiring with us, rather than against us, to make the MILES simulation of turbulent flows practical on today's computers – despite the very wide range of scales involved. The first three circumstances are based on the behavior of the Kolmogorov spectrum of the turbulent kinetic energy. As the wavelength decreases, the spectrum drops off so quickly that the shortest neglected scale lengths do not contain very much of the energy of the flow. Nevertheless, they are energetic enough to mix the larger scales rather quickly. In addition, the small scales are not very important dynamically, because the cascade of energy from large scales to short scales moves through the intervening scales.

A fourth result from theory, experiment, and computation shows that there are few features of the flow that are important at scales as small as η_K or even 10 times larger. The final observation relates to the numerical properties of the nonlinear monotone fluid dynamic algorithms. Because these methods properly connect the flow at the smallest computed scale to the flow at the smaller unresolved scales, the resulting solution is expected to be reliable down to the small resolved scales. Taken all together, these five fortunate circumstances combine to make it possible to compute turbulent flows on the larger scales as if the very small scales were present, and thus to capture the relevant physics on today's computers. Let's inspect these five considerations a little more closely.

The first two circumstances relate to the fact that the spectrum of the turbulent kinetic energy density, the Kolmogorov spectrum $\mathcal{E}(k) \approx \mathcal{E}(1/L) \cdot (kL)^{-5/3}$, drops off at a very convenient rate at small scales. The energy in eddies of progressively smaller size in turbulence decreases fast enough that most of the energy resides in eddies that can be resolved in three-dimensional simulations performed today. If the spectrum were flatter, for example if it scaled as $(kL)^{-1}$, most of the energy would be in the unresolved small-scale eddies. It would then be necessary to resolve many more of the small scales to capture most of the energy. This would be a kind of *ultraviolet catastrophe* with the large-scale features of the flow receiving less and less energy as the Reynolds number increases. The small scales would dominate the dynamics, and small changes in initial conditions might well be expected to have macroscopic effects on the computed solutions.

The Kolmogorov spectrum, on the other hand, is also not too steep. The characteristic velocity of the small eddies is large enough that their rotation rate increases rapidly with decreasing eddy size. Though relatively insignificant energetically, small eddies turn over considerably faster than the larger eddies. This means that the small scales mix large-scale inhomogeneities as fast as the large eddies can produce them. An example of this can be seen in Figure 1.7. Small, secondary instabilities grow, saturate, and mix fluid

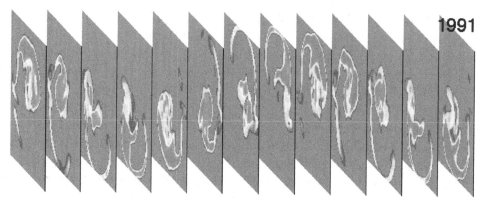

Figure 1.7. Visualization of an intermediate stage nonlinear evolution of a helically perturbed Mach 0.45 jet after 2.0 ms (4000 time steps). The simulation was performed on the NRL Connection Machine using the FAST3D–FCT model. A 256 × 128 × 128 grid was used with a tracer fluid along the high-speed jet core to mark the mixing region (Boris et al. 1992a).

during the linear growth phase of the primary Kelvin–Helmholtz instability of a helically perturbed jet. If the Kolmogorov spectrum were steeper, molecular mixing would not occur quickly inside a fully turbulent large-scale eddy and resolved kinetic energy cascading through the grid cutoff would not quickly appear as heat – as predicted by LES energy conservation. More complex, phenomenological models would be needed that depend on the unresolved details. The only way to do this would be to resolve many more of the subgrid scales to determine the rate of molecular mixing for combustion problems, or to postulate probability distribution function (PDF) phenomenologies to describe the important but unresolved details.

The third useful consideration, supported by both theory and DNS, is that the turbulent energy being transferred by the inertial-range cascade passes through all the intervening, local scales to short wavelengths where it is eventually dissipated. Spectral DNS of the nature of the dominant couplings between the large- and small-scale Fourier modes in homogeneous turbulence (Yeung and Brasseur 1991) show two types of interaction: (1) distant interactions, among one relatively long wavelength mode and two short wavelength modes, and (2) local interactions, coupling three modes with about the same wavelength. Energy transfer in the inertial range is dominated by the local interactions (Domaradzki 1992). In practice, virtually all of the energy extraction from a given scale occurs as a result of interactions with eddies less than an order of magnitude smaller.

This is in contrast to the possibility that the energy might skip directly from the large, energy-containing scales to the small scales. If the very smallest scales were dynamically significant in this way, extracting energy directly from the large scales, there would be no easy alternative to resolving all the scales at once. Instead, the actual, relatively smooth, local cascade of energy guarantees that numerical simulations resolving the energy-containing scales of the flow can give a reasonably accurate representation of the energy transfer out of those scales by using only the nonlinear effects of the resolved scales and the necessary grid-scale dissipation provided by monotone methods.

The fourth consideration is also purely physical: the apparent lack of important dynamics occurring at scales a factor of 10 or more larger than the classically defined Kolmogorov scale. Evidence of this comes from experiment, theory, and simulations. Vuillermoz and Oran (1992a, 1992b) conducted a set of very finely resolved two-dimensional FCT simulations with a corresponding physical viscosity to look at scales down in the Kolmogorov range. They found very little structure in the flow even at $10 \cdot \eta_K$. Three-dimensional simulations (Zohar et al. 1990) show essentially the same result. The most probable length scale of the small-scale eddies, the scale where there is maximum dissipation, and the point in the Kolmogorov spectrum where the curve bends sharply downward are all considerably larger than η_K. Concentration of the small-scale structures around the scale of peak dissipation, which is well above η_K, is corroborated in experiments (Dahm, Southerland, and Buch 1991). In addition to being quite fortunate from the perspective of modeling turbulent cascade, this result is perfectly reasonable: η_K is relatively unimportant dynamically because the turbulence energy has already been dissipated at longer wavelength before it can reach η_K.

This makes the cell Reynolds number limit even more odious and reduces the need for spatial resolution, allowing even nominally resolved DNS to stretch the cell Reynolds number limitation by using bigger cells, longer time steps, and smaller viscosity. For LES this means that the extent of the subgrid-modeled inertial range and the amount of energy in this region is considerably smaller than might otherwise be thought. However, as the following chapters will consider in some detail, the question of how fluid dynamics, and the numerics intended to reproduce the fluid dynamics, deals with the transition from the inertial range into the dissipation range of the Kolmogorov spectrum still has tantalizing unanswered questions. Experimentalists, theoreticians, and computationalists are currently debating the "bump" often seen in the Kolmogorv spectrum at small scales just before it plunges into dissipation. This bump is generally seen in well-resolved MILES computations at the small but still resolved end of the inertial range.

The fifth useful consideration concerns the properties of the class of fluid dynamic convection algorithms called *nonlinear monotone methods*. These algorithms are based on principles of conservation, positivity (monotonicity), causality, and locality. As discussed in Section 1.5, enforcing these simple physical requirements on convection and fluid dynamics seems to sufficiently constrain the numerical solutions that resolved kinetic energy "cascades off" the grid into heat in a physically reasonable way. The consequence of these properties for fluid dynamics is that the local, nonlinear dissipation in the algorithm properly connects the large, energy-containing scales to the unresolved subgrid scales of motion. This subject receives high-level attention in the next few chapters by Margolin, Rider, Fureby, Grinstein, and others. If we use these monotone methods for DNS, we can believe the results of computations that need to resolve eddies only down to 5 or 10 η_K. If we use these methods for LES, we can rely on their computing the large scales correctly without significant contamination from numerical errors at the grid scale and from the segment of the physical inertial range that has been truncated.

Because the fluid dynamics at each scale in Nature is quite insensitive to the details of the appreciably smaller scales of motion, it should not be the least bit surprising that the corresponding numerical models also display this insensitivity. The small scales do what they need to do, in the context of the larger-scale eddies in which they are embedded, to ensure that their details do not generally need to be resolved to compute the larger picture. At one point we characterized this behavior and the circumstances leading to it as a *convenient conspiracy* (Oran and Boris 1993).

1.7 Testing the MILES concepts

This volume contains chapters devoted to comparisons of MILES and other LES approaches among themselves and with experimental data. This is the traditional way of validating numerical methods, so many of these tests are the standard problems used by the broader turbulence modeling and LES communities. Such comparisons illuminate the relative effectiveness of different methods and highlight limitations and problems, but it is difficult to draw much useful information about the MILES hypothesis itself from these problems. This is due in part to difficulties in performing unequivocal laboratory experiments and in part to the fact that many methods appear to do an acceptable job on at least some problems. This difficulty is also due in no small part to the fact that "there is no truth," as expressed by Len Margolin (private communication, January 2004), to be compared with an LES computation in detail. There is no well-defined answer when a turbulent, high-Reynolds-number flow is being treated. This situation differs from well-posed, steady-state flow problems in which convergence to a unique, well-defined answer can be expected with increasing grid resolution. Any reduction of a turbulent flow problem to one with a steady solution is highly questionable because the physical problems are intrinsically unsteady. In this regard, Nature does seem to be working against us, making it difficult to see how well we are doing.

How to demonstrate *convergence* for an LES method is a perplexing concern. Every increase in spatial resolution uncovers more structure in the computed flow as unresolved scales become resolved. A high Reynolds-number flow is unstable, even when the full turbulent cascade is being resolved down to the Kolmogorov scale. Therefore, two solutions computed by using a perfect procedure (zero round-off, zero-resolution error, and zero-truncation error) will deviate progressively from each other as a result of arbitrarily small initial or boundary conditions differences. After a finite time, the macroscopic moments of two nearly identical flows may be nearly the same, but the large-scale turbulent structures can be out of phase in the two solutions with point-to-point differences of the order of unity. This is not pathology, just a natural consequence of the the Navier–Stokes equations. Not only can the true solution never be found: it fundamentally does not exist. In meteorology and climatology this is sometimes called the *butterfly effect* and is understood to limit the predictability of future conditions based in current understanding in a fundamental way.

Therefore, studying LES convergence becomes prohibitively expensive. If a second-order-accurate method is used on a steady problem, halving the cell size should reduce the numerical truncation errors by a factor of 8. In a turbulent (unsteady) problem, the time-integration truncation errors could accumulate systematically so the second-order-accurate method may have cumulative errors that reduce only by a factor of 4 when the cell size is halved. Therefore, doubling the grid for a three-dimensional computation to reduce the error by a factor of 4 would require 16 times as much work (with the time step also a factor of 2 shorter). If the diagnostic for convergence involves statistical averaging across physically meaningful fluctuations in the flow, as is sure to be true at least asymptotically for LES, the length of the run will also have to be increased by a factor of 16 to reduce the statistical measurement error by the same factor for 4 improvement sought in the solution. This means that the real computational cost to show second-order convergence, for a globally averaged quantity in a realistic unsteady problem, is a factor of 256 – if it can be done at all. Two grid doublings (as often recommended for steady-state problems) costs a factor of 65,536.

Consider, as a simple test of MILES using FCT, the entrainment of ambient air into a periodic jet that is initially perturbed helically. These simulations, our first tests designed to check the MILES hypothesis, were conducted in 1990 and 1991 (Boris et al. 1992a, 1992b; Oran and Boris 1993). They used the NRL Connection Machine 200, a computer powerful enough to execute turbulent simulations up to 256-cubed in a few hours. This resolution allows a wide enough range of scales in three dimensions to begin considering convergence with increasing resolution. By 1990 it was possible to resolve an analytically specified jet shear layer with at least two or three cells on the coarsest grid and with 8 to 10 cells in the fine-grid cases, ensuring that the most unstable modes were well resolved by 20–40 cells per wavelength. The goal was to have the different grid resolutions all behave the same way in their growth phases so that the beginning of the turbulent inertial casade would be similar at each resolution and to have long enough wavelengths that growth of subharmonics of the initial perturbations could be supported.

Figure 1.7 shows 13 cross sections through the center of the jet in the nonlinear mixing phase of one of these $128 \times 128 \times 256$ compressible simulations designed to expose the action of the built-in MILES subgrid model as additional short-wavelength structure is resolved in the solution. Gray-level contours of the jet fluid are shown as it mixes with the background air. Some of the fine-scale structure can be seen at this medium resolution, but fine-resolution calculations show still more structure. Viewed globally, the nonlinear Kelvin–Helmholtz instability in this geometry seems to deform the jet into a roughly helically symmetric core with a thin, stretching, helical shroud nearly encircling it. This shroud can be seen as thin strips of fluid indicated in all cross sections. Between the shroud and the core there is a layer of engulfed background fluid that is essentially vorticity free at this time and is moving only because of potential-flow effects. The primary goal of these simulations was the turbulent entrainment and mixing that follows the nonlinear saturation of the initial Kelvin–Helmholtz instability of a free jet. The shorter-wavelength Kelvin–Helmholtz modes constitute the secondary instabilities of turbulent cascade and contribute significantly to the additional entrainment in the fine-grid FCT run relative to medium- and coarse-grid runs when the flow is fully developed.

Macroscopic measures of complex turbulent interactions generally involve integrals of relatively low moments of the flow. Such measures as entrainment, heat flux, and mass transport are weighted by the relatively steep Kolmogorov spectrum toward the larger, energy-containing scales that numerical simulations are capable of resolving accurately. Measures that depend on higher derivatives may be of theoretical interest, but they more strongly emphasize the unresolved small scales in the turbulence. In this study we considered the increasing volume of gas in which at least a minimal density of the initial jet fluid can be found. This is the so-called passive scalar entrainment and allows convergence with resolution to be *estimated* by examining the temporal evolution of the total jet volume. The entrainment of one fluid into another is controlled by the speed at which the contorting interfaces of the largest scales (see Figure 1.7) move into and engulf portions of the surrounding fluid, and it depends on viscosity only at the very shortest scales.

A passive scalar ϕ, initialized with the same linear profile ($0 \leq \phi \leq 1$) through the shear layer as the X component of the velocity, was the diagnostic used to mark the fluid in the jet. The volume of fluid in the entire system that has $\phi \geq \epsilon = 0.05$ at any given time defines the entrainment volume for $0 < \epsilon < 1$. As the jet fluid spreads, the average value of the passive scalar in the jet plus mixed region drops. The low value, $\epsilon = 0.05$, used here, allows calculations to progress longer before the entrainment diagnostic loses its meaning as a result of excess dilution. Similar integral diagnostics were used previously to measure the size and cooling rate of channels generated by lasers, lightning bolts, and charged-particle beams in air (Boris et al. 1992b; Picone and Boris 1983; Picone et al. 1981), as already described in Section 1.4.

Now consider how a MILES solution of this turbulent entrainment might be expected to behave with increasing resolution and how this behavior might differ from classical LES. Figure 1.8 illustrates the expected entrainment behavior in this jet problem as a function of computational resolution for a conventional LES model and for a MILES model in a turbulent system. The figure is drawn at a particular time in the flow evolution so a range of statistically distributed values for the measured entrainment volume can be expected from the real system. The horizontal line labeled "answer" near the value 3.0 is rather broad in this schematic to indicate this variability. Note that both the *monotone algorithm* value and the corresponding notional curve for the *linearly filtered explicit subgrid* (nonmonotone) model are shown to approach this approximate answer as resolution increases to the right [larger values of $\log(r_{jet}/\delta x)$].

With MILES algorithms, the effective dissipation is nonlinear and local, and represents the unresolved mixing occurring in the subgrid scales of motion. The residual numerical (nonphysical) dissipation does not extend significantly to long wavelengths because of the strong dependence, $k^{3.3}$ to $k^{3.8}$, thus, the residual dissipation is expected to be small compared with nonmonotone methods. The curve labeled "monotone algorithm" in the figure shows the expected entrainment volume, measured at a given time in a series of different-resolution simulations. This curve illustrates the minimum entrainment that is expected at intermediate resolution where the numerical dissipation

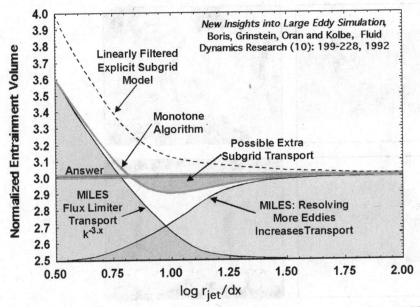

Figure 1.8. Expected convergence with increasing grid resolution according to the MILES hypothesis (modification of Figure 1 in Boris et al. 1992a).

has fallen below the turbulent entrainment that would be due to the unresolved scales. This minimum occurs when many short wavelengths, which would provide some additional entrainment, cannot be resolved but when the residual numerical diffusion present is smaller than the *eddy diffusivity* of the unresolved turbulence. This trade-off is illustrated schematically by the shaded region in the lower left showing increasing transport (mixing) due to resolved eddies as the resolution is increased. The curve in the lower left shows increasing *transport* from the flux limiter in monotone algorithms as resolution is decreased. Beyond this minimum, increasing resolution actually increases the entrainment because removing the remaining grid-scale numerical filtering has less effect than adding the corresponding eddy diffusivity from the unresolved scales.

With linearly filtered LES algorithms and an explicit eddy viscosity, the correct entrainment for a turbulent high-R_e flow appears to be approached from above as shown by the upper curve. Enough dissipation always has to be present to stabilize the computation. The computed solutions at finite resolution must be defiltered to correct some macroscopic quantities like the entrainment or the turbulent kinetic energy to their infinite numerical resolution values, and defiltering is generally unstable.

Figure 1.9 shows tracer density and the up–down (y) velocity fields at three times during an intermediate resolution simulation of the helically perturbed elliptical jet. In the figure, the small-scale secondary instabilities associated with the beginning of the turbulent cascade are increasing the entrained volume of the jet even as the large-scale, primary Kelvin–Helmholtz instability continues growing. This actual set of

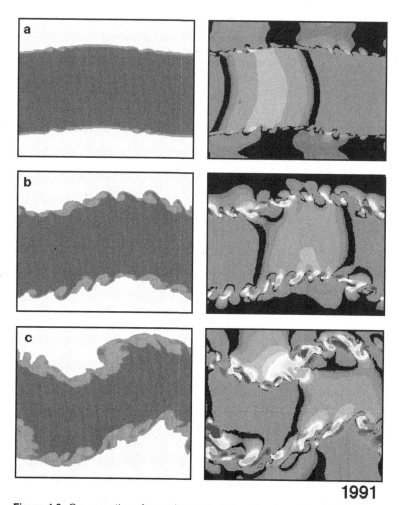

1991

Figure 1.9. Cross sections from a jet computation showing saturated short-wavelength modes superimposed on the primary perturbation. Left panels: tracer density with mixed region in lighter gray; right panels: vertical (Y) velocity from the nonlinear Kelvin–Helmholtz instability. Cross-sectional views reveal the mixing of fluid in an unstable, elliptical jet flowing from left to right at 100 m/s. The views are through the center of the jet along its major axis. The simulations were done on a 256^3 grid on the CM-200, using FAST3D. Left panels: white indicates ambient air, dark gray indicates unmixed jet gas, lighter gray indicates the mixed fluid region. Right panels: black indicates small vertical velocity; others shadings indicate alternate regions of upward and downward flow. Three times are shown: (a) 0.4 ms, (b) 0.9 ms, and (c) 1.35 ms (Boris et al. 1992a).

convergence tests on the elliptical-jet configuration is summarized in Figure 1.10. As the resolution ($r_{jet}/\delta x$) increases, the macroscopic entrainment volume quickly converges despite the existence of a wide range of unresolved small scales in the actual physical problem and the lack of an added subgrid turbulence model. These convergence tests teach us about the behavior of monotone algorithms and their nonlinear, numerical, built-in subgrid model. Unfortunately, this approach cannot be directly applied to most standard nonmonotone methods because they become unstable, or at least display a dismaying spectrum of purely numerical oscillations, as the physical viscosity is reduced to zero.

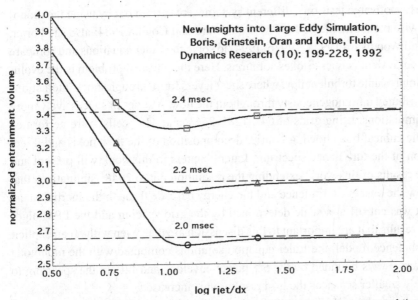

Figure 1.10. Increasing resolution (log r_{jet}/dx increasing) leads to convergence of the mixed fluid volume in helically perturbed, circular-jet simulations using FCT on the CM-200. Three different grid resolutions, $R_{jet}/\delta x = 6, 12$, and 24, are shown plotted along the horizontal axis at three times. The weak minimum in each curve results from the counteracting effects of numerical dissipation, which decreases with improving grid resolution, and physical mixing from inertial range eddies, which increases as smaller eddies are resolved (Boris et al. 1992a).

Figure 1.10 shows this entrainment-volume diagnostic computed as a function of time for *fine, medium,* and *coarse* MILES grids. The entrainment volume was normalized by dividing out the initial volume in each of the three runs, because each of these initial volumes was slightly different as a result of the resolution differences. With MILES algorithms, the effective filtering is nonlinear and thus the nonphysical diffusion does not extend significantly to long wavelengths. The three curves in Figure 1.10, corresponding to 2.0, 2.2, and 2.4 ms in the evolution of the jet at the three resolutions chosen, demonstrate that the expected minimum is indeed found at intermediate resolution.

The minimum probably occurs because short wavelengths, which provide additional entrainment, cannot be resolved. The residual numerical diffusion present in high-order monotone algorithms is smaller than the eddy diffusivity of the turbulent flow, so increasing resolution increases the entrainment. The minimum is not very deep and is resolved with only three points in these simulations, but it appears to be getting deeper as time progresses, another indication that the additional scales of vorticity in the fine-grid case are continuing to increase the entrainment volume relative to the medium and coarse grid. The fact that the minimum is so shallow, at most 2% to 3% deep, is actually beneficial. A shallow minimum, though correspondingly more difficult to measure accurately, means that virtually no additional subgrid-scale transport is needed to get the "right" answer. The intrinsic subgrid model provided by FCT, at least for these cases, is very good.

Similarly motivated tests on a different system were carried out by using the monotone piecewise parabolic method (PPM–Woodward and Colella 1984; also see Porter, Pouquet, and Woodward 1992a, 1992b, 1994) to solve the Euler equations and compare them with equivalent Navier–Stokes solutions. Here the physical problem is the evolution of a compressible turbulent flow where shear layers form through shock interactions and are distributed inhomogeneously throughout the flow. A series of successively more resolved simulations, using grids of 64^3, 128^3, 256^3, and 512^3 cells, were performed in a periodic cubical box allowing Fourier decomposition of the flow fields and direct computation of the turbulence spectrum. Later chapters in this book will present and discuss the results of this work, extending the resolution out to 2048^3 simulations that clearly show the onset of turbulence and the energy cascade through the inertial range to where it was cut off at a scale determined by the grid spacing and the PPM algorithm. The results that are important to this discussion are that, even without an explicit subgrid turbulence model, the Euler equation solutions computed with the monotone algorithm are always bounded by the 5/3 power envelope and fill out the spectrum to progressively smaller scales as the grid resolution is increased.

In accord with the MILES hypothesis, it was found that the adjustments in the numerical fluxes on the grid scale in these algorithms, carefully constructed to preserve the causal, local nature of monotone, conservative convection, constitute a numerical filter of the smallest resolved scales in the calculation. These nonlinear monotone methods filter out structures smaller than a few cell sizes by spreading out these structures on the grid (dissipating them). The remaining question for using these algorithms to describe turbulent flows is this: What are the effects of the small-scale cutoff on the large-scale structures?

1.8 Summary

The MILES paper (Boris 1990) made this argument:

... nonlinear monotone methods really have at least a minimal built-in LES filter and a matched subgrid model that do minimal damage to the longer wavelengths while still incorporating, at least qualitatively, most of the local and global effects of the unresolved turbulence expects of a large eddy simulation. When properly formulated, a wide variety of these monotone convection algorithms transform unresolvable structure in the fluid field variables, as it is pushed to shorter wavelengths by nonlinear convection effects and instabilities in the appropriate resolved fields.

This grid-scale variability is locally converted to the correct macroscopic variable averages, e.g., viscous dissipation of the unresolved scales appears as heat. Furthermore, these methods are quite capable of capturing quantitatively how much unresolved structure from the long wavelengths is actually present. Diffusion of the eddy transport type is automatically left in the flow as required but the fluctuating driving effects of random phase, unresolved eddies on the large scales is missing unless specifically included as a subgrid phenomenology. A factor of two increase in the spatial resolution of such LES models will bring more improvement in the accuracy of the well-resolved scales than all the work in the world on the subgrid model of a more coarsely resolved LES model.

In the chapters following, we will see what others have done to explain and exploit these MILES algorithms in the ensuing 15 years. Some of our contributions at the Naval Research Laboratory, treated in the chapter by Gopal Patnaik, Fernando Grinstein, and Jay Boris, have concerned reducing the numerical dissipation still further for detailed urban and building air flows where the three-dimensional turbulence is treated by MILES with stochastic backscatter in a fourth-order phase, time-accurate FCT finite-volume model for detailed building and city aerodynamics.

Acknowledgments

I thank Elaine Oran, John Gardner, Fernando Grinstein, Gopal Patnaik, Ted Young, Christer Fureby, Ronald Kolbe, Alan Kuhl, Sandy Landsberg, Charles Lind, Len Margolin, David Porter, Bill Rider, and Paul Woodward for their collaboration and interactions with me in this research, and for use of their graphics. The work discussed here was sponsored over the years by the Office of Naval Research through the U.S. Naval Research Laboratory.

REFERENCES

Batchelor, G. K. 1953. *The Theory of Homogeneous Turbulence*, 197 pp. New York: Cambridge University Press.

Belotserkovskii, O. M. 1999. *Turbulence and Instabilities*, MIPT: Moscow, p. 349.

Book, D., Boris, J., Kuhl, A., Oran, E., Picone, M., & Zalesak, S. 1981. pp. 84–90. Simulation of complex shock reflections from wedges in inert and reactive gaseous mixtures, in *Proceedings of the 7th International Conference on Numerical Methods in Fluid Dynamics*. Palo Alto, CA: Stanford University, pp. 84–90.

Book, D. L., Li, C., Patnaik, G., & Grinstein, F. F. 1991 Quantifying residual numerical diffusion in flux-corrected transport algorithms. *J. Sci. Comput.* **6**, 323–343.

Boris, J. P., 1971 A fluid transport algorithm that works, in *Computing as a Language of Physics*. Vienna: International Atomic Energy Agency, 171–189.

Boris, J. P., & Book, D. L. 1973 Flux-corrected transport I: SHASTA – a fluid transport algorithm that works. *J. Comput. Phys.* **11**, 38–69.

Boris, J. P., & Book, D. L., Hain, K. 1975 Flux-corrected transport II: Generalizations of the method. *J. Comput. Phys.* **18**, 283–284.

Boris, J. P., & Book, D. L. 1976a Solution of continuity equations by the method of flux-corrected transport, in *Methods in Computational Physics*. New York: Academic Press, 85–129.

Boris, J. P., & Book, D. L. 1976b Flux-corrected transport III. Minimal-error FCT algorithms. *J. Comput. Phys.* **20**, 397–431.

Boris, J. P., & Oran, E. S. 1981 Modeling turbulence: Physics or curve fitting?, in *Combustion in Reactive Systems: AIAA Progress in Astronautics and Aeronautics 76*, ed. J. R. Bowen et al. New York: 187–210.

Boris, J. P., 1990 On large eddy simulation using subgrid turbulence models, in *Whither Turbulence? Turbulence at the Crossroads*, ed. J. L. Lumley. New York: Springer-Verlag, 344–353.

Boris, J. P. Hubbard, R., Oran, E., & Slinker, S. 1991 Enhanced mixing from shock-generated turbulence in dusty air, in *Proceedings of the 18th International Symposium on Shock Waves*, ed. K. Takayama. Sendai, Japan: Institute of Fluid Science, Tohoku University, vol. 1, pp. 553–558.

Boris, J. P., Grinstein, F. F., Oran, E. S., & Kolbe, R. L. 1992 New insights into large eddy simulation. *J. Fluid Dyn. Res.* **10**, 199–228.

Boris, J. P., 1992 Compressibility in turbulence generation and mixing, in *Advances in Compressible Turbulent Mixing*, ed. W. P. Dannevik, A. C. Buckingham, & C. E. Leith. Washington, DC: GPO Publication 1992–687-052/P9000, 137–160.

Boris, J. P., Landsberg, A. M., Oran, E. S., & Gardner, J. H. 1993 *LCPFCT – A Flux-Corrected Transport Algorithm for Solving Generalized Continuity Equations*. U.S. Naval Research Laboratory, Washington, DC: NRL Memorandum Report 6410-93-7192.

Clark, R. A., Ferziger, J. A., & Reynolds, W. C. 1989 Evaluation of subgrid-scale models using an accurately simulated turbulent flow. *J. Fluid Mech.* **91**, 1–16.

Colella, P., & Woodward, P. R. 1984 The piecewise parabolic method (PPM) for gas-dynamical simulations. *J. Comput. Phys.* **54**, 174–201.

Colella, P. 1985 A direct Eulerian MUSCL scheme for gas dynamics, *SIAM J. Sci. Statist. Comput.* **6**, 104–17.

Colella, P., & Glaz, H. M. 1985 Efficient solution algorithms for the Riemann problem for real gases. *J. Comput. Phys.* **59**, 264–89.

Dahm, W. J. A., Southerland, K. B., & Buch, K. A. 1991 *Phys. Fluids A* **3**, 1115–1127.

Davis, S. F., 1984 *TVD Finite Difference Schemes and Artificial Viscosity*. NASA Langley Research Center, Hampton VA: ICASE Report 84-20.

DeVore, C. R., 1989 *Flux-Corrected Transport Algorithms for Two-Dimensional Compressible Magnetohydrodynamics*. U.S. Naval Research Laboratory, Washington DC: NRL Memorandum Report, 6544.

DeVore, C. R., 1991 Flux-corrected transport for multidimensional compressible magnetohydrodynamics. *J. Comput. Phys.* **92**, 142–160.

Domaradzki, J. A., 1992 Analysis of subgrid-scale eddy viscosity with the use of the results from direct numerical simulations. *Phys. Fluids A* **4**, 2037–2045.

Emery, M. H., Gardner, J. H., Boris, J. P., & Cooper, A. L. 1984 Vortex shedding due to laser ablation. *Phys. Fluids* **2**: 1338–1340.

Frisch, U., & Orszag, S. A. 1990 Turbulence: challenges for theory and experiment, *Physics Today* **43**, 24–32.

Godunov, S. K., 1959 Finite difference methods for numerical computation of discontinuous solutions of the equations of fluid dynamics. *Mat. Sb.* **47**, 271–306.

Gottlieb, D., & Orszag, S. A. 1997 *Numerical Analysis of Spectral Methods: Theory and Applications*. Philadelphia: SIAM.

Grinstein, F. F., & DeVore, C. R. 1992 Coherent structure dynamics in spatially developing square jets. Paper AIAA 92–3441.

Grinstein, F. F., & Guirguis, R. H. 1992 Effective viscosity in the simulation of spatially evolving shear flows with monotonic FCT models. *J. Comput. Phys.* **101**, 165–175.

Grinstein, F. F., Gutmark, E., & Parr, T. 1995 Near field dynamics of subsonic free square jets. A computational and experimental study. *Phys. Fluids* **7**, 1483–1497.

Grinstein, F. F. 2001 Vortex dynamics and entrainment in rectangular free jets. *J. Fluid Mech.* **437**, 69–101.

Harten, A., 1974 *The Method of Artificial Compression*. Courant Institute of Mathematical Sciences, New York University: Report C00-3077-50.

Harten, A., 1983 High resolution schemes for hyperbolic conservation laws. *J. Comput. Phys.* **49**, 357–393.

Harten, A., Engquist, B., Osher, S., & Chakravarthy, S. 1987 Uniformly high-order accurate essentially nonoscillatory schemes, III. *J. Comput. Phys.* **71**, 231–303.

Herring, J. R., & Kerr, R. M. 1982 Comparison of direct numerical simulation with prediction of two-point closures. *J. Fluid Mech.* **118**, 205–219.

Ikeda, T., & Nakagawa, T. 1979 On the SHASTA FACT algorithm. *Math. Comp.* **33**, 1157–1169.

Karniadakis, G. Em., & Sherwin, S. 1998. *Spectral/hp Element Methods for CFD*. Oxford: Clarendon Press.

Kolmogorov, A. N. 1941 *Dokl. Akad. Nauk. SSSR* **30**, 299.

Kolmogorov, A. N. 1962. A refinement of previous hypotheses concerning the local structure of turbulence in a viscous incompressible fluid at high Reynolds number. *J. Fluid Mech.* **13**, 82–85.

Kuhl, A. L., Furguson, R. E., Chien, K.-Y., Collins, J. P., & Glowacki, W. J. 1993 *Japanese National Symposium on Shock Waves*, ed. K. Takayama. Sendai, Japan: Tohoku University Press.

Kuzmin, D., Löhner, R., & Turek, S. eds. 2005 *Flux-Corrected Transport: Principles, Algorithms, and Applications*, Heidelberg: Springer-Verlag.

Landsberg, A. M., Young, T. R., & Boris, J. P. 1994 An efficient, parallel method for solving complex three-dimensional flows, presented at the 32rd Aerospace Sciences Meeting, January 10–13, Reno NV, Paper AIAA 94-0413.

Landsberg, A. M., & Boris, J. P. 1997 The virtual cell embedding gridding method: A simple approach for complex geometries, presented at the 13th CFD Conference, June 29–July 2, Snowmass CO, Paper AIAA 97-1982.

Lind, C. A., Biltoft, C. A., Oran, E. S., Boris, J. P., & Mitchell, W. J. 1996 Source characterization modeling for demil operations, *20th U.S. Army in Science Conference (June 1996): Award Winning Papers*, eds. R. Chait, C. Kominos, M. S. Schur, M. Stroscio, & J. J. Valdes. Singapore: World Scientific, 103–110.

Lind, C. A., Boris, J. P., Oran, E. S., Mitchell, W. J., & Wilcox, J. L. 1997 The response of an open air detonation facility to blast loading, in *Structures Under Extreme Loading Conditions*, Vol. 351 of ASME PVP. 109–126.

Löhner, R., Patnaik, G., Boris, J., Oran, E., & Book, D. 1986 Applications of the method of flux corrected transport to generalized meshes, in *Proceedings, Tenth International Conference on Numerical Methods in Fluid Dynamics*, ed. F. G. Zhuang, & Y. L. Zhu. New York: Springer-Verlag, vol. 264, pp. 428–434.

Löhner, R., Morgan, K., Peraire, J., & Vahdati, M. 1987b Finite element flux-corrected transport (FEM-FCT) for Euler and Navier-Stokes Equations. *Int. J. Numer Meth. in Fluids* 7, 1093–1109.

Löhner, R., Morgan, K., Vahdati, M., Boris, J. P., & Book, D. L. 1988 FEM-FCT: Combining unstructured grids with high resolution. *Commun. Appl. Numer. Meth.* 4, 717–730.

Lumley, J. L., ed. 1990 *Whither Turbulence? Turbulence at the Crossroads* Vol. 357, pp. 439–485, in Lecture Notes in Physics. New York: Springer-Verlag.

McDonald, B. E., Ambrosiano, J., & Zalesak, S. 1985 The pseudospectral flux correction (PSF) method for scalar hyperbolic problems, in *Proceedings of the Eleventh International Association for Mathematics and Computers in Simulation World Congress*, ed. R. Vichnevetsky. New Brunswick, NJ: Rutgers University Press, Vol. 1, 67–70.

Ogawa, T., & Oran, E. S. 2004 The multidimensional flux-corrected transport algorithm on the fully threaded tree, *Bull. Am. Phys. Soc.* 49, 172–173.

Oran, E. S., & Boris, J. P. 1981 Theoretical and computational approach to modelling flame ignition, in *Combustion in Reactive Systems*, Vol. 76 of AIAA Progress in Astronautics and Aeronautics, ed. J. R. Bowen et al. New York: AIAA, 154–171.

Oran, E. S., & Boris, J. P. 1987 *Numerical Simulation of Reactive Flow*. New York: Elservier.

Oran, E. S., & Boris, J. P. 1993 Computing turbulent shear flows – a convenient conspiracy. *Comput. Phys.* 7, 523–533.

Patnaik, G., Guirguis, R. H., Boris, J. P., & Oran, E. S. 1987 A barely implicit correction for flux-corrected transport. *J. Comput. Phys.* 71, 1–20.

Patnaik, G., & Kailasanath, K. 1996 A new time-dependent, three dimensional, flame model for laminar flames, in *26th Symposium (International) on Combustion*. Pittsburgh: The Combustion Institute, 899–905.

Patnaik, G., Grinstein, F. F., & Boris, J. P. 2006 Large scale urban simulations, in *Monotone Integrated Large Eddy Simulation: Computing Turbulent Fluid Dynamics*. New York: Cambridge University Press.

Picone, J. M., Boris, J. P., Greig, J. R., Raleigh, M., & Fernsler, R. F. 1981 Convective cooling of lightning channels. *J. Atmos. Sci.* 38: 2056–2062.

Picone, J. M., & Boris, J. P. 1983 Vorticity generation by asymmetric energy deposition in a gaseous medium. *Phy. Fluids* 26, 365–382.

Porter, D. H., Pouquet, A., & Woodward, P. R. 1992a Three-dimensional supersonic homogeneous turbulence: A numerical study. *Phys. Rev. Lett.* 68, 3156–3159.

Porter, D. H., Pouquet, A., & Woodward, P. R. 1992b A numerical study of supersonic turbulence. *Theor. Comput. Fluid Dyn.* 4, 13–49.

Porter, D. H., Pouquet, A., & Woodward, P. R. 1994 Kolmogorov-like spectra in decaying three-dimensional supersonic flows. *Phys. Fluids* **6**, 2133–2142.

Reynolds, W. C. 1990 On the potential and limitations of direct and large eddy simulations, in *Whither Turbulence? Turbulence at the Crossroads* ed., J. L. Lumley, New York: Springer-Verlag, 313–343.

Smolarkiewicz, P. K. 1983 A simple positive definite advection scheme with small implicit diffusion. *Monthly Weather Review* **111**, 479–486.

Smolarkiewicz, P. K. 1984 A fully multidimensional positive definite advection transport algorithm with small implicit diffusion. *J. Comput. Phys.* **54**, 325–362.

Smolarkiewicz, P. K., and L. G. Margolin, 1998. MPDATA: A finite difference solver for geophysical flows. *J. Comput. Phys.* **140**, 459–480.

Suzuki, M., & Kuwahara, K. 1989 Stratified flow past a bell-shaped hill. Paper AIAA 89–1824.

Sweby, P. K. 1984 High resolution schemes using flux limiters for hyperbolic conservation laws. *SIAM J. Numer. Anal.* **21**, 995–1011.

Tamura, T., and Kuwahara, K. 1989. Numerical analysis on aerodynamic characteristics of an inclined square cylinder. Paper AIAA 89–1805.

Tsuboi, K., Tamura, T., & Kuwahara, K. 1989 Numerical study of vortex induced vibration of a circular cylinder in high Reynolds number flow. Paper AIAA 89–0294.

van Leer, B. 1973 Towards the ultimate conservative difference scheme. I. The quest of monotonicity, in eds., H. Cabannes and R. Temam, vol. 18 of Lecture Notes in Physics. Berlin: Springer-Verlag, 163–168.

van Leer, B. 1979 Towards the ultimate conservative difference scheme. V. A second-order sequal to Godunov's method. *J. Comput. Physics* **32**, 101–136.

Vuillermoz, P., & Oran, E. S. 1992a The effects of viscosity and diffusion on a supersonic mixing layer, in *Proceedings of the Eighth Symposium on Turbulent Shear Flows*. New York: Springer-Verlag.

Vuillermoz, P., & Oran, E. S. 1992b *Mixing Regimes in a Spatially Confined, Two-Dimensional, Supersonic Shear Layer*, U.S. Naval Research Laboratory, Washington DC: Memorandum Report 4404-92-7106.

Woodward, P. R., & Colella, P. 1984 The numerical simulation of two-dimensional fluid flow with strong shocks. *J. Comput. Phys.* **54**, 115–173.

Yee, H. C., Warming, R. F., & Harten, A. 1983 Implicit total variation diminishing (TVD) schemes for steady-state calculations. Paper AIAA No. 83-1902,

Yee, H. C., Warming, R. F., & Harten, A. 1985 Implicit total variation diminishing (TVD) schemes for steady-state calculations. *J. Comput. Phys.* **57**, 327–360.

Yeung, P. K., & Brasseur, J. G., 1991 The response of isotropic turbulence to isotropic and anisotropic forcing at the large scales. *Phys. Fluids A* **3**, 884–896.

Young, T., A. Landsberg, M., & Boris, J. P. 1993 Implementation of the full 3D FAST3D (FCT) code including complex geometry on the Intel iPSC/860 parallel computer, presented at the SCS Simulation Multiconference, March 29–April 1, Arlington VA.

Youngs, D. L. 1982 Time-dependent Multi-material Flow with Large Fluid Distortion, in *Numerical Methods for Fluid Mechanics*, K. W. Morton and M. J. Baines, eds: Academic Press: London.

Zalesak, S. T. 1979 Fully multidimensional flux-corrected transport algorithms for fluids. *J. Comput. Phys.* **31**, 335–362.

Zalesak, S. T. 1981 Very high order and pseudospectral flux-corrected transport (FCT) algorithms for conservations laws, in *Advances in Computer Methods for Partial Differential Equations*, ed. R. Vichnevetsky & R. S. Stepleman. New Brunswick, NJ: Rutgers University, Vol. IV, 126–134.

Zalesak, S. T. 1987 Preliminary comparison of modern shock-capturing schemes: Linear advection, in *Advances in Computer Methods for Partial Differential Equations*, ed. R. Vichnevetsky & R.S. Stepelman. New Brunswick, NJ: Rutgers University: Vol. VI, 15–22.

Zalesak, S. T. 1997 Introduction to flux-corrected transport I: SHASTA – a fluid transport algorithm that works. *J. Comput. Phys.* **135**, 170–171.

Zohar, Y., Moser, R. D., Buell, J. C., & Ho, C. M. 1990. Length scales and dissipation of fine eddies in a mixing layer, in *Proceedings of the Summer Program*, Stanford, CA: Center for Turbulence Research.

2 A Rationale for Implicit LES

Fernando F. Grinstein, Len G. Margolin, and
William J. Rider

2.1 Introduction

High-Reynolds' number turbulent flows contain a broad range of scales of length and time. The largest length scales are related to the problem geometry and associated boundary conditions, whereas it is principally at the smallest length scales that energy is dissipated by molecular viscosity. Simulations that capture all the relevant length scales of motion through numerical solution of the Navier–Stokes equations (NSE) are termed *direct numerical simulation* (DNS). DNS is prohibitively expensive, now and for the foreseeable future, for most practical flows of moderate to high Reynolds' numbers. Such flows then require alternate strategies that reduce the computational effort. One such strategy is the Reynolds-averaged Navier–Stokes (RANS) approach, which solves equations averaged over time, over spatially homogeneous directions, or across an ensemble of equivalent flows. The RANS approach has been successfully employed for a variety of flows of industrial complexity. However, RANS has known deficiencies when applied to flows with significant unsteadiness or strong vortex-acoustic couplings.

Large eddy simulation (LES) is an effective approach that is intermediate in computational complexity while addressing some of the shortcomings of RANS at a reasonable cost. An introduction to conventional LES is given in Chapter 3. The main assumptions of LES are (1) that the transport of momentum, energy, and passive scalars is mostly governed by the unsteady features in the larger length scales, which can be resolved in space and time; and (2) that the smaller length scales are more universal in their behavior so that their effect on the large scales (e.g., in dissipating energy) can be represented by using suitable subgrid-scale (SGS) models. Many different approaches have been developed for the construction of SGS models; some of these are described in Chapter 3. It is essential to recognize that in the absence of a universal theory of turbulence, the construction of SGS models is unavoidably pragmatic, based primarily on the rational use of empirical information.

We distinguish between two general classes of SGS models. Simple *functional* SGS models focus on dissipating energy at a physically correct rate and are based on an artificial "eddy" viscosity. More sophisticated and accurate *structural* models attempt to address issues of the transfer of energy between length scales, and are based on a variety of ideas such as scale similarity and approximate deconvolution. The latter models typically do not dissipate sufficient energy to ensure computational stability, which has led to the development of mixed models that combine the positive features of the two classes of models. The results of such mixed models have been more satisfactory, but the complexity of their implementation and the computational effort required to employ them have limited their popularity. This situation has motivated the investigation of unconventional LES approaches, such as implicit LES (ILES) – the subject of this volume, adaptive flux reconstruction (e.g., Adams 2001), and variational schemes with embedded subgrid stabilization (e.g., Hughes 1995). The underlying idea of these new approaches is to represent the effects of the unresolved dynamics by regularizing the larger scales of the flow. Such regularization may be based on physical reasoning resulting from an ab initio scale separation, or on numerical constraints that enforce the preservation of monotonicity or, more generally, ensure nonoscillatory solutions. Enforcing such numerical constraints is the common thread that relates the various nonoscillatory finite-volume (NFV) numerical methods employed in ILES. The absence of explicit SGS models in the ILES approach offers many practical advantages, both of computational efficiency and ease of implementation. However, these alone are not sufficient reasons to justify ILES. At a more fundamental level, it is essential to understand why and how well this approach works in practical circumstances, while simultaneously recognizing its limits of applicability. One might argue that the more conventional LES approaches should be similarly scrutinized, though in general this is not systematically done. Nevertheless, in this chapter we will attempt to justify the ILES approach. Our basic thesis is this: *ILES works because it solves the equations that most accurately represent the dynamics of finite volumes of fluid – i.e., governing the behavior of measurable physical quantities on the computational cells.*

In general, there are approximation errors in numerical simulations even for the resolved scales of motion. One can identify the errors of a numerical algorithm by using modified equation analysis (MEA; see Chapter 5); these errors take the form of truncation terms that augment the analytic equations. It has been pointed out by Hirt (1969) and more recently by Ghosal (1996) that, in typical flow regimes, these truncation terms have the same order of magnitude as the SGS terms in LES. The purpose of those observations was to emphasize the importance of controlling the truncation errors; that is, a well-resolved LES requires accurate discretizations and adequate computational grids. However, one might naturally ask whether one could design the numerical algorithms so that the truncation terms would themselves serve as SGS models.

Why should the governing equations for numerical simulation be different from the continuum partial differential equations (PDEs)? The PDEs such as NSE that govern fluid motion are first derived in integral form by using the conservation principles of physics. The well-known PDE forms are then recovered in the limit that the integration volume shrinks to a point. The operable question then is this: What form do the equations take for finite values of the integration volume such as a computational cell? From the point of view of consistency, one would expect that the governing equations for these finite volumes would be the PDEs, augmented by additional terms that depend on the size of the volume. We will refer to these governing equations as *finite-volume equations*, and the additional terms as *finite-volume corrections*. We note that, most generally, the volume will include both space and time scales.

We begin in the next section by providing a historical perspective of ILES. We then continue by describing other theories, both analytic and numerical, that address the form of these finite-volume corrections. We shall find that all of these treatments lead to remarkably similar corrections; when a finite-volume momentum equation is derived, it contains new terms that are a nonlinear combination of first and second spatial derivatives with a dimensional coefficient that depends on the volume of integration. In addition, we recount the connection between the SGS models and NFV numerical methods, which have a common origin in the artificial viscosity of von Neumann and Richtmyer.

In Section 2.3, we will present a derivation of the finite volume equation for two-dimensional incompressible Navier–Stokes flows. The derivation closely follows the format described in Margolin and Rider (2002) for the one-dimensional Burgers' equation. However, the multidimensionality of the calculation brings out a new feature of the finite-volume equations, namely that their tensor properties depend on the details of the shape, as well as the magnitude, of volume of integration.

In Section 2.4, we exhibit the MEA derived from the approximation of the 2D NSE obtained with a particular class of NFV algorithms, known as MPDATA (see Chapter 4d). We will discuss the similarities to the finite-volume 2D NSE as well as the differences. In Section 2.5, we delve more deeply into the energy equations associated with both the finite-volume and the MPDATA approximate of 2D NSE. The purpose of these sections is to lay the groundwork for Chapter 5, where we will identify the features of a numerical method required for ILES and where we will compare the strengths and weaknesses of individual NFV methods on which ILES is based, in terms of their inherent dissipative properties.

2.2 Historical perspective

In their 1993 paper, Oran and Boris (1993) noted a "convenient conspiracy" in the numerical simulation of certain complex flows, wherein a physical model can combine with the numerical method to produce excellent results. Beyond the scope of their original discussion, the important point to note here is that the class of physical models that they considered – e.g., Burgers' equation, and the compressible and incompressible versions of NSEs – have the common feature of a quadratic nonlinearity in the advective terms. Further, on the numerical side, the advective terms discussed by Oran and Boris were formulated with monotonicity-preserving approximations. We will see in the next section that a MEA of such approximations leads to a dissipation of kinetic energy proportional to the cube of the velocity gradients. This is similar to theoretical results that are described below.

To better understand the connection of physical theory and numerical discretizations, it is useful to begin by describing the role of dissipation in ensuring numerical stability. In order for a numerical method to be stable, it must be dissipative in the sense of the energy or the L_2 norm. However, stability is not sufficient to guarantee physically realizable solutions. To ensure unique, physically meaningful solutions, a finite amount of dissipation must be present at a minimum. *This finite dissipation is referred to as an entropy condition. In principle, this dissipation may be implemented as part of the physical model (i.e., explicitly) or as part of the numerical method (i.e., implicitly).*

Historically, numerical dissipation was found to be necessary and was implemented in Lagrangian simulations of high-speed flows with shocks, where the equations were explicitly augmented by dissipative terms known as *artificial viscosity* (Richtmyer 1948; von Neumann and Richtmyer 1950). As we shall recount shortly, this idea was extended to turbulent flows, where the dissipative terms became known as SGS models. The development of explicit SGS models for turbulence has continued and grown ever more sophisticated. However, the evolution of artificial viscosity took a different turn in the early 1970s, when Jay Boris and Bram van Leer independently introduced the first nonoscillatory methods. In these methods, the entropy condition is satisfied implicitly as part of the numerical method. Over the past 30 years, many new improved nonoscillatory methods have been developed.

Nonoscillatory methods for shock flows exhibit many advantages over other approaches, including nonlinear stability, computational efficiency, ease of implementation, and, above all, accurate and realistic results. Examples of these methods used in ILES will be described in Chapter 4. For these reasons, such methods have become the preferred choice for many problems in the field of computational fluid mechanics. It would seem compelling, then, that this implicit approach should be investigated for turbulent flows. This is, in fact, ILES – the subject of this volume.

While it may not be possible to sort out the earliest efforts at simulating turbulent flows with NFV schemes, it is clear that credit for the first public documentation of the approach belongs to Jay Boris and colleagues at the U.S. Naval Research Laboratory (Chapters 1 and 8). Boris made the crucial early connection (Boris 1990), namely that the truncation errors of such algorithms could in fact serve as a SGS model in what he denoted the Monotone Integrated LES (MILES) approach. Further, he recognized that this was not a special feature of the flux-corrected transport (FCT) algorithm (Chapter 4a) on which he based MILES, but that this implicit property *could apply equally for a number of other suitably formulated monotone methods as well.* MILES applications using monotonic algorithms coupled to various physical processes in shear-flow engineering applications are extensively reviewed in this volume (Chapters 8–11, 16, and 17). The ILES work of Woodward and colleagues with the piecewise parabolic method (Chapter 4b) involved studies of homogeneous turbulence in the early 1990s (Chapter 7) and astrophysical problems in regimes of highly compressible flow with extremely high Reynolds' numbers (Chapter 15). At about the same time, David Youngs and colleagues applied van Leer methods (Chapter 4c) to modeling the growth of turbulent regions and the mixing resulting from fluid instabilities, including Raleigh–Taylor, Kevin–Helmholtz, and Richtmyer–Meshkov (Chapter 13). These applications involve adjacent regions of very high and very low Reynolds' number, illustrating a very useful feature of ILES – that the same fluid solver can be used for smooth and for turbulent flows. The vorticity confinement method (Chapter 4e) introduced by Steinhoff, also in the early 1990s, invoked ideas similar to those in shock capturing. This is an approach to ILES based directly on the discrete equations satisfied within thin, modeled vortical regions; this approach is especially well suited to treat engineering flows over blunt bodies, including attached and separating boundary layers, and resulting turbulent wakes (Chapter 12). Margolin, Smolarkiewicz, and colleagues published the first applications of ILES to geophysics using MPDATA (Chapters 4d and 14). As in the astrophysics cases, the geophysical calculations typically involve very high Reynolds' numbers (Re $\sim 10^6$), but with stratified and nearly incompressible flow.

The effectiveness of the ILES approach demonstrated in a wide range of applications in engineering, astrophysics, and geophysics does not address the question of why the approach is successful. A significant contribution was made by Fureby and Grinstein (1999, 2002), regarding the similarity between the effects of certain NFV schemes and those of explicit SGS models used in conventional LES. These authors used the MEA framework to show that a particular class of flux-limiting algorithms (Chapter 4a) with dissipative leading-order terms provide appropriate built-in (implicit) SGS models of a mixed tensorial (generalized) eddy-viscosity type. Key features in the comparisons with classical LES leading to the identification of this implicit SGS model were the MEA framework and the finite-volume formulation (also used in this chapter), which readily allowed the recasting of the leading-order truncation terms in divergence form. A similar direction was also explored by Rider and Margolin (2003), who compared the implicit SGS models resulting from a MEA of several NFV algorithms in one dimension and showed the connections to explicit SGS models. A systematic MEA of ILES is further presented in Chapter 5.

The intuitive basis for the pioneering ILES work of Boris, Woodward, Youngs, Steinhoff, and their collaborators was most largely formed as a natural follow-up to shock-capturing methods. However, a more rigorous physical basis for ILES suggested by Margolin and Rider (2002) arose from examining the correspondence of the entropy conditions themselves, as derived in various theories compared with the MEA of nonoscillatory methods. Our next step, then, is to list some of these fundamental theoretical results.

Frisch (1995) derived a formula for the dissipation of energy in a Burgers' fluid arising solely at the shock wave:

$$\left\langle \frac{\partial K}{\partial t} \right\rangle \ell = \frac{1}{12} \langle \Delta u \rangle^3, \tag{2.1}$$

where K is the kinetic energy, and angle brackets indicate spatial averaging over the length ℓ.

Bethe (1942) showed that the rate of entropy production across a shock is

$$T\frac{\partial S}{\partial t} = -\frac{\mathcal{G}}{12c_s} \langle \Delta u \rangle^3,$$ (2.2)

where S is the entropy, T is the temperature, c_s is the sound speed, and $\mathcal{G}$ is the fundamental thermodynamic derivative

$$\mathcal{G} \equiv \frac{\partial^2 P}{\partial V^2}.$$ (2.3)

Kolmogorov (1962) derived a remarkably similar form for the inviscid dissipation of kinetic energy in isotropic incompressible turbulence:

$$\left\langle \frac{\partial K}{\partial t} \right\rangle \ell = -\frac{\partial S}{\partial t} \ell = \frac{5}{4} \langle \Delta u \rangle^3.$$ (2.4)

This similarity of the forms of energy dissipation, or entropy creation, was noted by Margolin and Rider (2002), who pointed out that each case combines the features of *inviscid* dissipation of kinetic energy with finite scales of observation. Further, these theoretical results show a connection between the large-scale behavior of shocked flows and of turbulence. The authors also noted the similarity of these forms to the classic artificial viscosity of von Neumann and Richtmyer (1950), which is used to ensure sufficient entropy production in numerical simulations of shocks. They went on to recount the historic connection of artificial viscosity and the early SGS turbulence models of Smagorinsky, which is reproduced here. Smagorinsky's generalization of artificial viscosity employs a scalar (i.e., isotropic) diffusivity. However, it is now well established that the near-dissipation end region of the inertial subrange is inherently anisotropic and characterized by very thin filaments (worms) of intense vorticity with largely irrelevant internal structure, embedded in a background of weak vorticity (e.g., Jimenez et al. 1993). As previously noted by Fureby and Grinstein (1999, 2002), the implicit SGS models associated with NFV methods naturally contain a tensor diffusivity that is able to regularize the unresolved scales without losing essential directional information, while their ability to capture steep gradients can be used to emulate (near the ILES cutoff) the dissipative features of the *high end* of the physical inertial subrange region. In Section 2.4, we will exhibit the tensor diffusivity of a particular NFV scheme to illustrate this point (see also Chapters 4a and 5).

In the rest of this section, we will expand on the connection between numerical simulations of shock flows and turbulence and extend the discussion to include nonoscillatory numerical methods. Indeed, the developments of NFV methods and SGS models for turbulent flow both stem from the earlier concept of artificial viscosity for the computation of shock waves on a finite grid. This viscosity is constructed to mimic the physical production of entropy across a shock – such as shown in Eq. (2.2) – without resolving the viscous processes that are responsible, and to reproduce the correct jump conditions. The strategy is often referred to as *shock capturing* or *regularization*. One important and noticeable result of artificial viscosity is the suppression of Gibbs phenomena (unphysical oscillations) associated with the discrete jump. Further, the artificial viscosity guarantees the nonlinear stability of the simulation (under a proper time-step limit).

Artificial viscosity was conceived for Lagrangian simulations of shocks; however, Eulerian simulations of shocks and also of turbulence also exhibit unphysical oscillations, and it is not difficult to imagine that one might try to extend the concept to these simulations as well. The connection to turbulent flows and SGS models appeared first, at the very dawn of numerical weather and climate prediction.

After World War II, John von Neumann worked to expand the role of simulation in science. One of his efforts was in the area of numerical weather prediction, where he worked with Jules Charney at the Institute for Advanced Study during the early 1950s. In 1956, von Neumann and Charney were present at a conference where Norman Phillips presented his two-dimensional simulation of a month of weather of the Eastern North America area. Also present was a graduate student, Joseph Smagorinsky. It was observed that Phillip's calculation was polluted by ringing late in the simulated month, and Charney made the suggestion that von Neumann's viscosity could be used to eliminate that ringing. Smagorinsky was given the task of extending Phillip's results to three dimensions, including artificial viscosity.

Smagorinsky's implementation (Smagorinsky 1963, 1983) resulted in the first SGS model, and it formed the basis for much future work. After the fact, a more rigorous connection of the Smagorinsky eddy viscosity to turbulence theory was made by Lilly (1966). SGS modeling has since grown and evolved. However, the energy dissipation associated with the original Smagorinsky form persists in many more sophisticated models such as the popular dynamic Smagorinsky models and mixed models.

The path to nonoscillatory methods was a little less direct, and began with the early work of Peter Lax (Lax 1954, 1972). In particular, a paper by Lax and Wendroff (1960) first emphasized the importance of conservative methods (cf. finite-volume methods; flux methods). A later paper (Lax and Wendroff 1964) described a second-order-accurate numerical approximation for the advective terms in which first-order diffusive errors are directly compensated. That Lax–Wendroff scheme produces oscillatory fields behind a shock wave, in contrast to the Lax–Friedrichs method, which produces monotone shock transitions but is overly diffusive.* Both methods differ in an essential way from the artificial viscosity methods in that they are "linear." Specifically, a linear method uses the same stencil everywhere, whereas artificial viscosity is nonlinear in the sense that its magnitude depends on the flow variables. However, it took almost 20 more years for the importance of nonlinearity to be recognized.

Sergei Godunov was a graduate student in the Soviet Union in the early 1950s. As part of his doctoral thesis, he was assigned to calculate shock-wave propagation. At the time, existing algorithms in the Soviet Union were not sufficiently accurate, and

---

* Lax–Wendroff methods usually employ an artificial viscosity to control oscillations behind shocks.

Godunov was not familiar with the work of Lax and Friedrichs (not easily available to him because of the Cold War; see Godunov 1999). Instead, he developed a new methodology, based on solving local Riemann problems, that not only satisfied his degree requirements but sowed the seeds of a computational revolution 20 years later. Perhaps of equal importance, Godunov proved a fundamental theorem, which states that no numerical method can be simultaneously linear, second-order accurate, and monotonicity preserving (see, e.g., LeVeque 1999).

Godunov's method and theorem were published in 1959 (the manuscript was completed in 1956). However, it was largely ignored and lay dormant for over a decade. Then in 1971, two scientists, Jay Boris in the United States and Bram van Leer in the Netherlands, overcame the barrier of Godunov's theorem by recognizing the feasibility of giving up linearity. The FCT method (Boris and Book 1973) focused on eliminating unphysical oscillations. It is a hybrid scheme that mixes first-order and second-order accuracy in a nonlinear manner. The MUSCL schemes of van Leer (1979) are a more direct generalization of Godunov's method in which the initial states of the underlying Riemann problem are modified by limiting the magnitude of gradients.

(Aside: As fate would have it, a third scientist, Kolgan, produced an alternate generalization of Godunov's method that also overcame the barrier of Godunov's theorem. In today's parlance, his method would take the label of a second-order essentially nonoscillatory, or ENO, method. Unfortunately, this work went unnoticed and Kolgan died before receiving any recognition.)

By the 1980s, the nonoscillatory approach was widely accepted and produced a plethora of ideas and implementations. Some of these are described in more detail in Chapter 4 of this volume. Many more may be expected to work for ILES, but perhaps not all. In Section 2.4 of this chapter and later in Chapter 5, we will investigate the requirements for successful ILES in more detail.

To summarize this section, sufficient dissipation to satisfy entropy conditions is necessary to achieve physically realizable simulations. The entropy conditions can be seen to be dependent on the mesh resolution, based on the successful form of artificial viscosity and more simply on dimensional analysis. This is consistent with our observation in the previous section, that the governing equations for volume-averaged quantities should depend on the size of the volume elements. Furthermore, we note that it is the nonlinearity of the advective terms that gives rise to these extra terms, and that it is the use of nonlinear approximations that allows the formulation of ILES methods. In the next section, we will explore these ideas in more mathematical detail.

2.3 A Physical perspective

In this section, we derive the finite-volume equation for the 2D incompressible NSE. Our derivation generally follows that in Margolin and Rider (2002), which was applied to the 1D Burgers' equation; however, we will ignore the averaging in time here. First, we will consider the case of smooth (laminar) flow. Then we will show that the same results apply to turbulent flow.

2.3.1 Laminar flows

The Navier–Stokes equations for the two-component velocity vector, $U = (u, v)$, are

$$\frac{\partial u}{\partial t} = -(u\,u)_x - (u\,v)_y - P_x + \nu\,(u_{xx} + u_{yy}),$$

$$\frac{\partial v}{\partial t} = -(u\,v)_x - (v\,v)_y - P_y + \nu\,(v_{xx} + v_{yy}),$$

(2.5)

plus the equation of incompressibility

$$u_x + v_y = 0,$$

(2.6)

where the shorthand notation, $u_x \equiv \frac{\partial u}{\partial x}$, is used for spatial differentiation. Here, P is the pressure and ν is the coefficient of physical viscosity. We note that the pressure is a diagnostic variable and can be found by solving an elliptic equation that enforces incompressibility.

We define volume-averaged velocities

$$\bar{u}(x, y) \equiv \frac{1}{\Delta x\,\Delta y} \int_{x-\frac{\Delta x}{2}}^{x+\frac{\Delta x}{2}} \int_{y-\frac{\Delta y}{2}}^{y+\frac{\Delta y}{2}} u(x', y')\,dx'\,dy'$$

(2.7)

and

$$\bar{v}(x, y) \equiv \frac{1}{\Delta x\,\Delta y} \int_{x-\frac{\Delta x}{2}}^{x+\frac{\Delta x}{2}} \int_{y-\frac{\Delta y}{2}}^{y+\frac{\Delta y}{2}} v(x', y')\,dx'\,dy'.$$

(2.8)

That is, here we have chosen the volume of integration to be a rectangle, mimicking a computational cell in a regular mesh.

Our goal is to find evolution equations for $\bar{U} = (\bar{u}, \bar{v})$. To begin, we note that (2.6) is linear. Hence the spatial differentiation and the volume averaging commute, and it immediately follows that

$$\bar{u}_x + \bar{v}_y = 0.$$

(2.9)

Similar arguments apply to the time derivatives and the viscous terms in (2.5). However, the nonlinearity of the advective terms requires more care. Here we will generalize the calculation of Margolin and Rider (2002) to two dimensions. To evaluate terms such as

$$\mathcal{I}_1 = \frac{1}{\Delta x\,\Delta y} \int_{x-\frac{\Delta x}{2}}^{x+\frac{\Delta x}{2}} \int_{y-\frac{\Delta y}{2}}^{y+\frac{\Delta y}{2}} u\,u_x\,dx'\,dy',$$

the basic idea is to expand the integrand in a Taylor series. For this series to converge, the velocity field has to be smooth on the length scales Δx and Δy. This is true for low-Reynolds' number flows, which are amenable to DNS. It is unlikely to be true for flows that we consider for LES, where the Reynolds' number is large and the dissipative scales are not resolved on the mesh. In this subsection, we will consider the case of smooth flows and indicate the steps to evaluate integrals like $\mathcal{I}_1$. In the next section, we will show how one can extend these results to LES regimes.

We begin by assuming that the velocity field can be expanded in a convergent Taylor series on the scales Δx and Δy:

$$u(x + x', y + y') \approx u(x, y) + u_x\, x' + u_y\, y' + u_{xx}\frac{(x')^2}{2}$$
$$+ u_{xy}\, x'y' + u_{yy}\frac{(y')^2}{2} + \text{HOT},$$
$$v(x + x', y + y') \approx v(x, y) + v_x\, x' + v_y\, y' + v_{xx}\frac{(x')^2}{2}$$
$$+ v_{xy}\, x'y' + v_{yy}\frac{(y')^2}{2} + \text{HOT}, \tag{2.10}$$

where HOT indicates terms of higher order. Substituting these expansions into definitions (2.7) and (2.8) immediately yields

$$\bar{u}(x, y) \approx u(x, y) + \frac{1}{6}\left(\frac{\Delta x}{2}\right)^2 u_{xx} + \frac{1}{6}\left(\frac{\Delta y}{2}\right)^2 u_{yy} + \text{HOT} \tag{2.11}$$

and

$$\bar{v}(x, y) \approx v(x, y) + \frac{1}{6}\left(\frac{\Delta x}{2}\right)^2 v_{xx} + \frac{1}{6}\left(\frac{\Delta y}{2}\right)^2 u_{yy} + \text{HOT}. \tag{2.12}$$

These volume-averaged velocities are continuous functions of space and time. Note that, by symmetry, the averaged functions are even in Δx and Δy. Now the higher-order derivatives like $\bar{u}_x$ can be derived by differentiating (2.11) and (2.12). For example,

$$\bar{u}_x \approx u(x, y) + \frac{1}{6}\left(\frac{\Delta x}{2}\right)^2 u_{xxx} + \frac{1}{6}\left(\frac{\Delta y}{2}\right)^2 u_{xyy} + \text{HOT}. \tag{2.13}$$

We will have the need for the inverse relations corresponding to (2.11) and (2.12). These are easily found to be

$$u(x, y) \approx \bar{u} - \frac{1}{6}\left(\frac{\Delta x}{2}\right)^2 \bar{u}_{xx} - \frac{1}{6}\left(\frac{\Delta y}{2}\right)^2 \bar{u}_{yy} + \text{HOT} \tag{2.14}$$

and

$$v(x, y) \approx \bar{v} - \frac{1}{6}\left(\frac{\Delta x}{2}\right)^2 \bar{v}_{xx} - \frac{1}{6}\left(\frac{\Delta y}{2}\right)^2 \bar{v}_{yy} + \text{HOT}. \tag{2.15}$$

There are four quadratic terms to evaluate:

$$\mathcal{I}_1 = \frac{1}{\Delta x \, \Delta y} \int_{x-\frac{\Delta x}{2}}^{x+\frac{\Delta x}{2}} \int_{y-\frac{\Delta y}{2}}^{y+\frac{\Delta y}{2}} 2 u \, u_x \, dx' \, dy',$$

$$\mathcal{I}_2 = \frac{1}{\Delta x \, \Delta y} \int_{x-\frac{\Delta x}{2}}^{x+\frac{\Delta x}{2}} \int_{y-\frac{\Delta y}{2}}^{y+\frac{\Delta y}{2}} (v \, u_y + v_y \, u) \, dx' \, dy',$$

$$\mathcal{I}_3 = \frac{1}{\Delta x \, \Delta y} \int_{x-\frac{\Delta x}{2}}^{x+\frac{\Delta x}{2}} \int_{y-\frac{\Delta y}{2}}^{y+\frac{\Delta y}{2}} (u \, v_x + u_x \, v) \, dx' \, dy',$$

$$\mathcal{I}_4 = \frac{1}{\Delta x \, \Delta y} \int_{x-\frac{\Delta x}{2}}^{x+\frac{\Delta x}{2}} \int_{y-\frac{\Delta y}{2}}^{y+\frac{\Delta y}{2}} 2 v \, v_y \, dx' \, dy'. \tag{2.16}$$

The general strategy to evaluate each of these terms is to insert the Taylor expansions into the integrand and multiply. We note that only terms that are even in the integration variables x' and y' will contribute. Then

$$\mathcal{I}_1 = 2uu_x + \frac{3 \, u_x \, u_{xx} + u \, u_{xxx}}{\Delta x \, \Delta y} \int_{-1/2\Delta x}^{1/2\Delta x} \int_{-1/2\Delta y}^{1/2\Delta y} (x')^2 \, dx' \, dy'$$

$$+ \frac{2 \, u_y \, u_{xy} + u \, u_{xyy} + u_x \, u_{yy}}{\Delta x \, \Delta y} \int_{-1/2\Delta x}^{1/2\Delta x} \int_{-1/2\Delta y}^{1/2\Delta y} (y')^2 \, dx' \, dy'. \tag{2.17}$$

Evaluating the integrals leads to

$$\mathcal{I}_1 = 2uu_x + \frac{1}{3} \left(\frac{\Delta x}{2} \right)^2 (3u_x u_{xx} + u u_{xxx}) + \frac{1}{3} \left(\frac{\Delta y}{2} \right)^2 (2u_y u_{xy} + u_x u_{yy} + u u_{xyy}). \tag{2.18}$$

Equation (2.18) is not yet in useful form, since $\mathcal{I}_1$ is written in terms of u rather than $\bar{u}$. We can rewrite the equation in the desired form by using the inverse relations (2.14) and (2.15). This approach is similar to approximate deconvolution described in more detail in Chapter 6. For example,

$$uu_x \approx \left[\bar{u} - \left(\frac{\Delta x^2}{24} \right) \bar{u}_{xx} - \left(\frac{\Delta y^2}{24} \right) \bar{u}_{yy} \right] \left[\bar{u}_x - \left(\frac{\Delta x^2}{24} \right) \bar{u}_{xxx} - \left(\frac{\Delta y^2}{24} \right) \bar{u}_{xyy} \right]$$

$$\approx \bar{u}\bar{u}_x - \left(\frac{\Delta x^2}{24} \right) (\bar{u}_x \bar{u}_{xx} + \bar{u}\bar{u}_{xxx}) - \left(\frac{\Delta y^2}{24} \right) (\bar{u}_x \bar{u}_{yy} + \bar{u}\bar{u}_{xyy}). \tag{2.19}$$

Putting all the terms together, the result for $\mathcal{I}_1$ is

$$\mathcal{I}_1 = 2\bar{u}\bar{u}_x + \frac{2}{3} \left(\frac{\Delta x}{2} \right)^2 \bar{u}_x \bar{u}_{xx} + \frac{2}{3} \left(\frac{\Delta y}{2} \right)^2 \bar{u}_y \bar{u}_{xy}$$

$$= (\bar{u}^2)_x + \frac{1}{3} \left(\frac{\Delta x}{2} \right)^2 [(\bar{u}_x)^2]_x + \frac{1}{3} \left(\frac{\Delta y}{2} \right)^2 [(\bar{u}_y)^2]_x. \tag{2.20}$$

The same procedure can be applied for $\mathcal{I}_2$, $\mathcal{I}_3$, and $\mathcal{I}_4$. The final result for the volume-averaged momentum equations is, to $\mathcal{O}(\Delta x^2, \Delta y^2)$,

$$\frac{\partial \bar{u}}{\partial t} = -(\bar{u}^2)_x - (\bar{v}\bar{u})_y - \frac{1}{3}\left(\frac{\Delta x}{2}\right)^2 \left[(\bar{u}_x\bar{u}_x)_x + (\bar{v}_x\bar{u}_x)_y\right]$$

$$- \frac{1}{3}\left(\frac{\Delta y}{2}\right)^2 \left[(\bar{u}_y\bar{u}_y)_x + (\bar{u}_y\bar{v}_y)_y\right] - \bar{P}_x + \nu(\bar{u}_{xx} + \bar{u}_{yy}), \qquad (2.21)$$

$$\frac{\partial \bar{v}}{\partial t} = -(\bar{u}\bar{v})_x - (\bar{v}^2)_y - \frac{1}{3}\left(\frac{\Delta x}{2}\right)^2 \left[(\bar{u}_x\bar{v}_x)_x + (\bar{v}_x\bar{v}_x)_y)\right]$$

$$- \frac{1}{3}\left(\frac{\Delta y}{2}\right)^2 \left[(\bar{u}_y\bar{v}_y)_x + (\bar{v}_y\bar{v}_y)_y\right] - \bar{P}_y + \nu(\bar{v}_{xx} + \bar{v}_{yy}). \qquad (2.22)$$

We emphasize that this result is specific for the rectangular volume of integration.

2.3.2 Renormalization

When the velocity field is not smooth on the scales of the integration volume, its truncated Taylor series is not an accurate approximation throughout the volume and the derivation of the previous section is not justifiable. In this section, we will consider a more general approach that extends our results to high-Reynolds' number flows. Remarkably, we will find that the results of the previous section remain valid, indicating the renormalizability of the averaging of the advective terms.

The fact that the averaging process is also a smoothing process leads naturally to a new question: Will the averaged velocity always be "smooth enough" on the scale of the averaging? We will simply assume this to be true, seeing it as a necessary prerequisite to practical computer simulation of turbulence. More precisely, let us assume that the finite-scale equations, (2.21) and (2.22), are valid on the length scales Δx and Δy. We will show that this implies their validity on the scales $2\Delta x$ and $2\Delta y$. Since we know these equations are valid on some scales, such as in the DNS range, we can use induction to conclude that the finite-scale equations derived in the previous section are valid at all scales. This implies that the entropy condition is ensured through this process.

Let us consider a rectangle $2\Delta x$ by $2\Delta y$. This can be thought of as four Δx by Δy rectangles. A volume-averaged velocity $\bar{U} = (\bar{u}, \bar{v})$ is defined at the center of each of these smaller rectangles. Because the integrals that define $\bar{U}$ are simply additive – see (2.7) – we can define the volume average over the larger rectangle:

$$\widehat{u}(x, y) = \frac{1}{4\Delta x \Delta y} \int_{-\Delta x}^{\Delta x} \int_{-\Delta y}^{\Delta y} u(x + x', y + y')\, dx'\, dy'$$

$$= \frac{1}{4}\left[\bar{u}(+, +) + \bar{u}(+, -) + \bar{u}(-, -) + \bar{u}(-, +)\right], \qquad (2.23)$$

where for brevity we have written $\bar{u}(+, +) \equiv \bar{u}\left(x + \frac{\Delta x}{2}, y + \frac{\Delta y}{2}\right)$. Taylor expanding the four terms inside the brackets in this equation leads (to second order) to

$$\widehat{u} = \bar{u} + 1/2\bar{u}_{xx}\left(\frac{\Delta x}{2}\right)^2 + 1/2\bar{u}_{yy}\left(\frac{\Delta y}{2}\right)^2, \tag{2.24}$$

where now all functions are centered at coordinates (x, y). This implies the inverse relation,

$$\bar{u} = \widehat{u} - 1/2\widehat{u}_{xx}\left(\frac{\Delta x}{2}\right)^2 - 1/2\widehat{u}_{yy}\left(\frac{\Delta y}{2}\right)^2. \tag{2.25}$$

Now let us calculate

$$\widehat{\mathcal{I}}_1 = \frac{1}{4\Delta x\,\Delta y} \int_{-\Delta x}^{\Delta x} \int_{-\Delta y}^{\Delta y} 2uu_x dx'\, dy'$$

$$= \frac{1}{4}\left[\mathcal{I}(+, +)_1 + \mathcal{I}(+, -)_+ \mathcal{I}(-, -)_1 + \mathcal{I}(-, +)_1\right]. \tag{2.26}$$

From the symmetry of (2.26), it is clear that linear terms in Δx and Δy will cancel, and the quadratic terms are identical for all four terms. So it is sufficient to consider $\mathcal{I}(+, +)$, keeping only the even terms. Then

$$\bar{u}(+, +) = \bar{u} + \bar{u}_x\frac{\Delta x}{2} + \bar{u}_y\frac{\Delta y}{2} + u_{xx}\frac{\Delta x^2}{4} + u_{yy}\frac{\Delta y^2}{4}\cdots$$

$$\bar{u}_x(+, +) = \bar{u}_{xx} + \bar{u}_{xx}\frac{\Delta x}{2} + \bar{u}_{xy}\frac{\Delta y}{2} + u_{xxx}\frac{\Delta x^2}{4} + u_{xyy}\frac{\Delta y^2}{4}\cdots \tag{2.27}$$

and

$$\widehat{\mathcal{I}}_1 = 2\bar{u}\bar{u}_x + 2\left(\frac{\Delta x}{2}\right)^2\left[\frac{11}{6}\bar{u}_x\bar{u}_{xx} + 1/2\bar{u}\bar{u}_{xxx}\right]$$

$$+ 2\left(\frac{\Delta y}{2}\right)^2\left[\frac{8}{6}\bar{u}_y\bar{u}_{xy} + 1/2\bar{u}\bar{u}_{xyy} + 1/2\bar{u}_x\bar{u}_{yy}\right]. \tag{2.28}$$

Finally, we use the inverse relations, (2.25), to rewrite this expression in terms of $\widehat{u}$:

$$2\bar{u}\bar{u}_x = 2\widehat{u}\widehat{u}_x - \left(\frac{\Delta x}{2}\right)^2\left[\widehat{u}_x\widehat{u}_{xx} + \widehat{u}\widehat{u}_{xxx}\right] - \left(\frac{\Delta y}{2}\right)^2\left[\widehat{u}_x\widehat{u}_{yy} + \widehat{u}\widehat{u}_{xyy}\right],$$

leading to

$$\widehat{\mathcal{I}}_1 = 2\widehat{u}\widehat{u}_x + \frac{2\Delta x^2}{3}\widehat{u}_x\widehat{u}_{xx} + \frac{2\Delta y^2}{3}\Delta y^2\widehat{u}_y\widehat{u}_{xy}$$

$$= (\widehat{u^2})_x + \frac{\Delta x^2}{3}\left[(\widehat{u}_x)^2\right]_x + \frac{\Delta y^2}{3}\left[(\widehat{u}_y)^2\right]_x. \tag{2.29}$$

That is, we have reproduced (2.20) when the averaging volume now is $2\Delta x$ by $2\Delta y$. The same calculation is readily repeated for $\mathcal{I}_2$, $\mathcal{I}_3$, and $\mathcal{I}_4$, and each of these terms is unchanged except for the doubling of the length scales. Thus, we conclude that the finite-scale equations, (2.21) and (2.22), are valid for all averaging volumes, whether the Reynolds' number associated with those scales is small (DNS) or large (LES).

2.3.3 Discussion of the finite-scale equations

Here, we will make several observations about the finite-scale equations, (2.21) and (2.22). First, we repeat that the derivation is a straightforward extension of that in Margolin and Rider (2002), and that there is no obstacle to further extending the results to three spatial dimensions and to include time averaging as well.

Second, we note that the lowest-order finite-scale corrections are quadratic in the mesh spacings Δx and Δy (at lowest order) and that there are no terms of order $\Delta x \, \Delta y$. The result does depend on both the size and shape of the volume chosen for averaging. We have chosen rectangles in two dimensions to emphasize the comparisons with our numerical results in the next two sections.

Third, we note that the finite-scale corrections have the form of the divergence of a symmetric tensor, which implies the conservation of the finite-scale momentum. In the language of LES, the finite-scale corrections correspond to the divergence of a SGS tensor, $\nabla \cdot \tau$, where

$$\tau^{xx} = -\frac{1}{3}\left[\left(\frac{\Delta x}{2}\right)^2 \bar{u}_x^2 + \left(\frac{\Delta y}{2}\right)^2 \bar{u}_y^2\right] \tag{2.30}$$

$$\tau^{xy} = \tau^{yx} = -\frac{1}{3}\left[\left(\frac{\Delta x}{2}\right)^2 \bar{u}_x \, \bar{v}_x + \left(\frac{\Delta y}{2}\right)^2 \bar{u}_y \, \bar{v}_y\right] \tag{2.31}$$

$$\tau^{yy} = -\frac{1}{3}\left[\left(\frac{\Delta x}{2}\right)^2 \bar{v}_x^2 + \left(\frac{\Delta y}{2}\right)^2 \bar{v}_y^2\right]. \tag{2.32}$$

Here the (Cartesian) tensor indices are shown as superscripts to distinguish them from the subscripts that denote spatial differentiation.

Fourth, we emphasize the similarity of this SGS tensor to the LES model of Clark – see, for example, page 628 in Pope (2000). This model belongs to the class of similarity models, which are found to accurately represent the nonlinear dynamics and energy transfer of turbulence in simulations, but which are generally not sufficiently dissipative. The Clark model is often used in conjunction with Smagorinsky to form mixed models (cf. Pope 2000 or Meneveau and Katz 2000).

Finally, we call attention to the fact that the finite-scale equations depend sensitively on the length scales of the averaging. We emphasize that these are not intrinsic scales of the flow, but in fact *represent the length scales of the observer*. From a physics point of view, an observer measures aspects of a flow using experimental apparatus that itself has finite scales – such as the diameter of a wire, the response time of a detector, and so on. The fact that the equations and their solutions change as the length scales change is not a flaw, but a necessary consequence of their interpretation as a model of reality, as measured by a particular observer.

2.4 NFV modified equation

In this section, we will compare the finite-scale equation derived in the previous section with the MEA of a particular NFV scheme, MPDATA (Chapter 4d). MPDATA has been used successfully in ILES simulations of the atmosphere, both on mesoscale and global-scale problems; some examples are described in Chapter 14.

MPDATA is constructed directly by using the properties of iterated upwinding, in contrast to the majority of NFV schemes, which are based on the idea of flux limiting. Nevertheless, MPDATA's properties, as exposed by MEA, are typical of many NFV schemes. In the next section, we will delineate the common features that make for successful ILES as well as important distinctions that relate to the dissipative process.

2.4.1 Implicit SGS stresses

Modified equation analysis is a technique for generating a PDE whose solution closely approximates the solution of a numerical algorithm. Comparison of the MEA of an algorithm with its model PDE gives useful information about the accuracy and the stability of the algorithm. A description of this important technique can be found in Chapter 5 of this volume and associated references.

We will assume a simple data structure where both components of velocity are located at the cell centers. A straightforward, if somewhat tedious, Taylor analysis of the MPDATA algorithm applied to the 2D Navier–Stokes equations, (2.5), leads to the modified equations. Here, we focus only on the semidiscrete equations by letting the time step $\Delta t \to 0$, to correspond to the finite-volume equations (2.21) and (2.22) where we did only spatial averaging. Also, we exhibit only the truncation terms originating in the advective terms, to allow direct comparison with the finite-scale "subgrid stress" terms in (2.30) through (2.32). It turns out that these truncation terms also can be written as the divergence of a tensor. To lowest order, the implicit subgrid stress $\mathcal{T}$ of the MPDATA algorithm is

$$\mathcal{T}_{xx} = \left[\frac{1}{4}u_x\,|u_x| + \frac{1}{12}u_x\,u_x + \frac{1}{3}u\,u_{xx}\right]_x \Delta x^2 \qquad (2.33)$$

$$\mathcal{T}_{xy} = \left[\frac{1}{4}u_y\,|v_y| + \frac{1}{12}u_y\,v_y + \frac{1}{6}(u\,v_{yy} + v\,u_{yy})\right]_y \Delta y^2 \qquad (2.34)$$

$$\mathcal{T}_{yx} = \left[\frac{1}{4}v_x\,|u_x| + \frac{1}{12}u_x\,v_x + \frac{1}{6}(u\,v_{xx} + v\,u_{xx})\right]_x \Delta x^2 \qquad (2.35)$$

and

$$\mathcal{T}_{yy} = \left[\frac{1}{4}v_y\,|v_y| + \frac{1}{12}v_y\,v_y + \frac{1}{3}v\,v_{yy}\right]_y \Delta y^2. \qquad (2.36)$$

Let us now do a detailed comparison between the finite-scale SGS stress of equations (2.30) through (2.32) and the implicit SGS stress of equations (2.33) through (2.36).

First of all, we note that the truncation terms can be written as a second-order tensor. This is a direct consequence of the finite-volume nature of the approximation and underlines the importance of the "FV" in NFV methods.

Second, we note that each of the components in $\mathcal{T}$ is quadratic in Δx or Δy, similar to the properties of τ. This is a direct consequence of the second-order accuracy of MPDATA and explains why first-order schemes such as donor cell are not suitable for ILES, even though they are nonoscillatory. Perhaps of equal importance is the implication that higher-order (than second) schemes will not have the proper dimensional dependence and are also unsuitable for ILES.

Third, we note that $\mathcal{T}$ is not symmetric in its off-diagonal components, whereas τ is symmetric. More generally, we may note the lack of certain terms in $\mathcal{T}$ that are present in τ. For example, there are no terms of order Δy^2 in $\mathcal{T}_{xx}$. This source of this deficit is easy to uncover, and in fact results from the particular form used by MPDATA to estimate the velocity at the center of the edge of a computation cell, specifically the average of the two values of the adjacent cells. This ignores the perpendicular variation of these values. This is also a relatively easy deficiency to fix. In computational experiments, however, we have seen little difference when "fuller" stencils are used for this averaging, indicating a relative lack of importance of these terms.

2.4.2 Energy analysis and computational stability

Both the finite-scale subgrid stresses, (2.30) through (2.32), and the implicit subgrid stresses, (2.33) through (2.36), are just the lowest order in an infinite series of terms with higher-order derivatives and larger (even) powers of Δx and Δy. Our assumptions about the smoothness of the averaged flow are designed to imply that these higher-order terms can be ignored *from the point of view of accuracy*. Stability of the equations is an independent issue. Stability can be studied through the energy equation.

The total rate of inviscid energy dissipation – that is, independent of the physical viscosity ν – is

$$\frac{dE}{dt} = \frac{1}{2} \int_D \left[u\,\tau_x^{xx} + u\,\tau_y^{xy} + v\,\tau_x^{xy} + v\,\tau_y^{yy} \right] dx\,dy, \tag{2.37}$$

where D is the two-dimensional domain. Integrating by parts and neglecting surface terms (work done by external forces) yields

$$\frac{dE}{dt} = -\frac{1}{2} \int_D \left[u_x\,\tau^{xx} + u_y\,\tau^{xy} + v_x\,\tau^{yx} + v_y\,\tau^{yy} \right] dx\,dy. \tag{2.38}$$

Substituting the finite-scale subgrid stresses into (2.38) yields:

$$\frac{dE_{FS}}{dt} = \frac{1}{6}\left(\frac{\Delta x}{2}\right)^2 \langle \bar{u}_x^3 \rangle + \frac{1}{6}\left(\frac{\Delta y}{2}\right)^2 \langle \bar{v}_y^3 \rangle + \frac{1}{6}\left[\left(\frac{\Delta x}{2}\right)^2 - \left(\frac{\Delta y}{2}\right)^2\right]\langle \bar{u}_x \bar{u}_y \bar{v}_x \rangle.$$

$$\tag{2.39}$$

Here the brackets indicate spatial integration over the domain. Note that for solutions (u, v) of NSE, $\langle u_x^3 \rangle < 0$ and $\langle v_y^3 \rangle < 0$ by Kolmogorov's 4/5 law; these inequalities are verified computationally in Chapter 14 for MPDATA solutions of decaying turbulence. In an isotropic flow $\langle \bar{u}_x \bar{u}_y \bar{v}_x \rangle$ would vanish. Thus, in the absence of forces, kinetic energy is absolutely decreasing and the (truncated) finite-volume equations are globally stable.

Next, substituting the implicit subgrid stresses of the modified equations into (2.38) yields

$$
\begin{aligned}
\frac{dE_{ME}}{dt} = &\left(\frac{\Delta x^2}{2} \right) \left[\frac{1}{3} \langle \bar{u}_x^3 \rangle - \langle |\bar{u}_x^3| \rangle + \frac{1}{3} \langle \bar{u}_x \bar{v}_x^2 \rangle - \langle |\bar{u}_x| \bar{v}_x^2 \rangle \right] \\
&+ \left(\frac{\Delta y^2}{2} \right) \left[\frac{1}{3} \langle \bar{v}_y^3 \rangle - \langle |\bar{v}_y^3| \rangle + \frac{1}{3} \langle \bar{v}_y \bar{u}_y^2 \rangle - \langle |\bar{v}_y| \bar{u}_y^2 \rangle \right] \\
&- \left(\frac{\Delta x^2}{3} \right) \left[\langle \bar{v} \bar{v}_x \bar{u}_{xx} \rangle - \langle \bar{u} \bar{v}_x \bar{v}_{xx} \rangle \right] \\
&- \left(\frac{\Delta y^2}{3} \right) \left[\langle \bar{u} \bar{u}_y \bar{v}_{yy} \rangle - \langle \bar{v} \bar{u}_y \bar{u}_{yy} \rangle \right].
\end{aligned}
\tag{2.40}
$$

We note the similarities to (2.39), and proceed to discuss the differences next.

The NFV methods are nonlinearly stable by construction. This can be seen in (2.40), where several of the terms can be grouped to ensure that the integrands are negative definite – for example, $\langle \frac{1}{3} \bar{u}_x^3 - |\bar{u}_x^3| \rangle < 0$. This is a different kind of stability from that of the finite-scale equations, as it does not depend on the solution. In the language of numerical analysis, the MPDATA modified equations for the NSE are locally stable, whereas the finite-scale equations are globally stable.

Based on the energy analysis, the MPDATA implicit subgrid stress tensor can be written as the sum of two parts, one of which is absolutely dissipative and one of which corresponds to the corrective terms of the finite-scale equation. In the case of MP-DATA, the dissipative stress is similar to a tensor version of the common Smagorinsky model. As we remarked in Section 2.4, the nonlinear terms are the same as those of the self-similar Clark model (Pope 2000). Thus, *MPDATA has the form of a mixed LES model.*

2.5 A Discussion of energy dissipation

Successful ILES simulations have been found to be a property of most, but not all, NFV algorithms. This does not imply that the results of using different NFV algorithms are entirely equivalent. In this section, we discuss the differences among these results. The conceptual framework of this discussion requires us to delve more deeply into the details of the numerical regularization. A computational validation of our discussion and conclusions in this section is presented in Chapter 5.

The theoretical analysis of Section 2.3 and the comparisons with the MPDATA-modified equation in Section 2.4 suggest that it is possible to identify the essential algorithmic elements required for ILES. We begin by summarizing these elements:

- The first element is the appearance of the self-similar term (i.e., Clark model) in the MEA of the algorithm. It is not difficult to show that the presence of this term is a direct consequence of finite-volume differencing. This term scales generically like the square of a characteristic length scale, h, of the averaging volume, such as Δx^2 or Δy^2 in two-dimensional simulations.
- The second element is the presence of sufficient dissipation to "regularize" the equations. A corollary to this is that there should not be too much dissipation – that is, should not dominate the self-similar term. In our analysis of MPDATA, we found the dissipation also scales like h^2; in fact, is comparable in size in regions of compression, but vanishes in regions of expansion.

Although most NFV schemes can be used for ILES, there are differences among the results. In particular, one might suppose that there are advantages to NFV algorithms whose dissipation scales with a higher power of h, which may minimize the interaction with the self-similar term. It is possible to test this hypothesis within the context of MPDATA, which has an option to specify the number of corrective iterations used. This option, termed IORD, is described in Chapter 4d and involves iterating the basic scheme to further reduce the simulation error. When IORD = 1, the algorithm is a simple donor cell, which is found to be too diffusive for ILES. When IORD = 2, the scheme described in Section 2.5 results. When IORD = 3, dissipation is further reduced; for comparison with equation (2.33), the particular subgrid stress component $\mathcal{T}_{xx}$ has the modified form

$$\mathcal{T}_{xx} = \left(\frac{1}{12} u_x u_x + \frac{1}{3} u u_{xx} \right)_x \Delta x^2 + \mathcal{O}(\Delta x^3) + \frac{1}{6} \operatorname{sgn}(u_x) u_{xx}^2 \Delta x^4, \qquad (2.41)$$

where $\operatorname{sgn}(u_x) \equiv \frac{|u_x|}{u_x}$. The terms $\mathcal{O}(\Delta x^3)$ are not dissipative; the first locally dissipative term shows up at fourth order (other fourth-order terms also are present but are not dissipative). In general, incrementing IORD by 1 preserves the self-similar term while increasing the order of dissipation by 2.

The effect of increasing the order of dissipation in MPDATA simulations is studied in Chapter 5. To summarize these results, the difference between IORD = 2 and IORD = 3 is substantial, both in the mean flow characteristics and in the intermittency. The further increase to IORD = 4 and above has minimal effect on the results. This latter result is a little surprising, and is important to understand as a limitation of the MEA approach. In the simulation, energy is dissipated in regions of steep compressive gradients and also at local extrema. Taylor expansion, which underlies MEA, is not sufficiently convergent to allow an accurate representation of the equations at sharp extrema.

To demonstrate the importance of dissipation at extrema, we have also included simulations in Chapter 5 using a monotonicity-preserving scheme with fourth-order dissipation equivalent to MPDATA with IORD = 3. To summarize these comparisons, the results of the monotonicity-preserving scheme results show a better agreement with mean flow characteristics than MPDATA, but show substantially less intermittency. The latter results give some insight into a question that will also surface elsewhere in this volume (Chapter 8) – namely, is the strict preservation of monotonicity an ingredient for optimal ILES, or are weaker conditions such as sign preservation or positivity preferable? In particular, it is clear that different NFV schemes provide qualitatively different implicit models by which SGS fluctuations affect the resolved flow features. The apparent conclusion is that each approach has potential advantages for particular problems, and so there is no optimal scheme for all problems; the choice of NFV scheme should depend on the specific questions that a simulation is meant to address.

2.6 Summary

Our goal in this chapter has been to provide a rationale for the ILES approach. Our strategy has been to argue that the finite-volume Navier–Stokes equations are the most appropriate model for simulating turbulent flows, to derive these for the case of the 2D Navier–Stokes equation, and to show that a particular numerical algorithm, MPDATA, is effectively solving these equations. We also exposed the connection of the implicit subgrid model of MPDATA to the class of mixed models in explicit LES. We will extend these ideas more generally to other methods of the class of NFV schemes in the next three chapters. In Chapter 3, we will provide more detailed background on the explicit LES approach, the ideas that underpin it, and some of the methodologies it includes. In Chapter 4, we will provide more detailed descriptions of the particular NFV schemes that are used in the body of the volume. Although all belong to the general class of NFV methods, we will see that the underlying concepts are quite varied. Then, in Chapter 5, we will draw out the common threads of these schemes and elucidate the features necessary for successful ILES. In closing and further integrating this section on *Capturing Physics*, Chapter 6, by Adams, Hickel, and Domaradzki, describes how the approximate deconvolution method can be used as a powerful bridge connecting LES and ILES.

REFERENCES

Adams, N. 2001 The role of deconvolution and numerical discretization in subgrid scale modeling, in *Direct and Large Eddy Simulation IV*, ed. Geurts, B. J., Friedrich, R. & Metais, O. Dordrecht: Kluwer, 311–320.

Bethe, H. 1998 On the theoy of shock waves for an arbitrary equation of state, in *Classic Papers in Shock Compression Science*, New York: Springer-Verlag. (Original report NDRC-B-237, 1942)

Boris, J. P., & Book, D. L. 1973 Flux-corrected transport: 1. SHASTA, a fluid transport algorithm that works, *J. Comput. Phys.* **11**, 38–69.

Boris, J. P. 1990 On large eddy simulation using subgrid turbulence models, in *Whither Turbulence? Turbulence at the Crossroads*, ed. J. L. Lumley. New York: Springer-Verlag, 344–353.

Frisch, U. 1995 *Turbulence: The Legacy of A. N. Kolmogorov*. New York: Cambridge Universiy Press.

Fureby, C., & Grinstein, F. F. 1999 Monotonically integrated large eddy simulation of free shear flows. *AIAA J.* **37**, 544–556.

Fureby, C., & Grinstein, F. F. 2002 Large eddy simulation of high Reynolds-number free and wall bounded flows. *J. Comput. Phys.* **181**, 68.

Ghosal, S. 1996 An analysis of numerical errors in large-eddy simulations of turbulence. *J. Comput. Phys.* **125**, 187–206.

Godunov, S. K. 1999 Reminiscences about difference schemes. *J. Comput. Phys.* **153**, 6–25.

Hirt, C. W. 1969 Computer studies of time-dependent turbulent flows. *Phys. Fluids* **II**, 219–227.

Hughes, T. J. R. 1995 Multiscale phenomena: Green's function, the Dirichlet-to-Neumann formulation, subgrid scale models, bubbles, and the origin of stabilized methods. *Comput. Methods Appl. Mech. Eng.* **127**, 387–401.

Jimenez, J., Wray, A. A., Saffman, P. G., & Rogallo, R. S. 1993 The structure of intense vorticity in isotropic turbulence. *J. Fluid Mech.* **255**, 65–90.

Kolmogorov, A. N. 1962 A refinement of previous hypothesis concerning the local structure of turbulence in viscous incompressible fluid at high Reynolds number. *J. Fluid Mech.* **13**, 82–85.

Lax, P. D. 1954 Weak solutions of nonlinear hyperbolic equations and their numerical computation. *Commun. Pure Appl. Math.* **7**, 159–193.

Lax, P. D. 1972 *Hyperbolic Systems of Conservation Laws and the Mathematical Theory of Shock Waves*. Philadelphia: SIAM.

Lax, P. D. & Wendroff, B. B. 1960 Systems of Conservation Laws. *Commun. Pure Appl. Math.* **13**, 217–237.

Lax, P. D. & Wendroff, B. B. 1964 Difference schemes for hyperbolic equations with high order of accuracy. *Commun. Pure Appl. Math.* **17**, 381.

LeVeque, R. J. 1999 *Numerical Methods for Conservation Laws*. Boston: Birkhauser Verlag, 2nd ed.

Lilly, D. K. 1966 *On the Application of the Eddy Viscosity Concept in the Inertial Subrange of Turbulence* Boulder, CO: National Center for Atmospheric Research, Manuscript 123.

Margolin, L. G., & Rider, W. J. 2002 A rationale for implicit turbulence modeling. *Int. J. Num. Meth. Fluids.* **39**, 821–841.

Margolin, L. G., Smolarkiewicz, P. K. & Sorbjan, Z. 1999 Large eddy simulations of convective boundary layers using nonoscillatory differencing. *Physica D* **133**, 390–397.

Margolin, L. G., Smolarkiewicz, P. K., & Wyszogrodzki, A. A. 2002 Implicit turbulence modeling for high Reynolds number flows. *J. Fluids Eng.* **24**, 862–867.

Meneveau, C., & J. Katz, J. 2000 Scale-invariant and turblence models for large-eddy simulation. *Ann. Rev. Fluid Mech.* **32**, 1–32.

Oran, E. S., & Boris, J. P. 1993 Computing turbulent shear flows – a convenient conspiracy. *Comput. Phys.* **7**, 523–533.

Pope, S. 2000 *Turbulent Flows*, New York: Cambridge University Press.

Richtmyer, R. D. 1948 *Proposed Numerical Method for Calculation of Shocks*. Los Alamos, NM: Los Alamos Scientific Laboratory LA-671.

Rider, W. J., & Margolin, L. G. 2003 From numerical analysis to implicit subgrid turbulence modeling, presented at the 16th AIAA CFD Conference, June 23–26, Orlando, FL, paper AIAA 2003–4101.

Smagorinsky, J. 1963 General circulation experiments with the primitive equations. I. The basic experiment. *Mon. Weather Rev.* **101**, 99–164.

Smagorinsky, J. 1983 The beginnings of numerical weather prediction and general circulation modeling: Early recollections. *Adv. Geophys.* **25**, 3–37.

van Leer, B. 1979 Towards the ultimate conservative difference scheme. V. A second-order sequel to Godunov's method. *J. Comput. Phys.* **32**, 101–136; reprinted 1997, **135**, 229–248.

von Neumann, J., & Richtmyer, R. 1950 A method for the numerical calculation of hydrodynamic shocks. *J. Appl. Phys.* **21**, 532–537.

CAPTURING PHYSICS WITH NUMERICS

3 Subgrid-Scale Modeling: Issues and Approaches

Pierre Sagaut

3.1 Large eddy simulation: From practice to theory

3.1.1 LES: Statement of the problem

It is well known that the direct numerical simulation (DNS) of fully developed turbulent flows is far beyond the range of available supercomputers. Indeed, the computational effort scales like the cube of the Reynolds number for the very simple case of incompressible isotropic turbulence, showing that an increase by a factor of 1,000 in the computational cost will only permit a gain of a factor of about 10 in the Reynolds number. The actual possibilities are illustrated by the results obtained on a grid of 4096^3 points by Kaneda et al. (2003) simulating incompressible isotropic turbulence at $Re_\lambda = 1201$ (where Re_λ is the Reynolds number based on the Taylor microscale).

The main consequence is that to obtain results at high Reynolds number, all the dynamically active turbulent scales cannot be simulated at the same time: some must be discarded. But, because of the intrinsically nonlinear nature of the Navier–Stokes equations, all turbulent scales are coupled in a dynamic way so that the effects of the discarded scales on the resolved scales must be taken into account to ensure the reliability of the results. This is achieved by augmenting the governing equations for the resolved scales to include new terms that represent the effects of the unresolved scales. The large eddy simulation (LES) technique computes the large scales (where the notion of "large" will be defined) of the flow, while modeling their interactions with small unresolved scales (referred to as *subgrid scales*) through a subgrid model. It is worth noting that the dual approach (i.e., computing the smallest scales while modeling the largest ones) is commonly used in meteorology, where mesoscale models are used to represent the large-scale motion.

The intuitive approach to LES presented above must be refined and recast in an appropriate mathematical framework to derive a set of governing equations for LES and to gain useful insight into its reliability. Let us consider the vector field u defined as the exact solution of the conservation law system:

$$\frac{\partial u}{\partial t} + \nabla \cdot F(u, u) = 0. \tag{3.1}$$

Here, $F(u, u)$ is a nonlinear flux function, which may exhibit any kind of dependency (polynomial, exponential, ...) on u. This equation is to be solved on a domain, Ω, with boundary conditions

$$u = f \text{ on } \partial\Omega_1, \quad \frac{\partial u}{\partial n} = g \text{ on } \partial\Omega_2, \qquad (3.2)$$

where $\partial\Omega_1 \bigcup \partial\Omega_2 = \partial\Omega$ is the boundary of the computational domain and n is the associated outward-pointing unit normal vector. The initial condition is given as $u(x, t = 0) = u_0(x)$. We also assume that u is a turbulent solution, whose smallest active scales have a characteristic size η (the Kolmogorov scale in usual turbulence theories). Now let u_h be the computed discrete approximation of u. The computational grid has a characteristic mesh size h, and the numerical method used to compute u_h has a characteristic resolution length h_r. The cutoff frequency of the simulation, conceptually related to the Nyquist frequency, is associated with h_r. The LES case corresponds to $h_r \gg \eta$, while the DNS is defined as $h_r \approx \eta$. The length scale, h_r, can be larger than h (e.g., as in usual finite-difference schemes) or smaller than h (e.g., as in spectral element methods), and it is the relevant length to define the cutoff frequency of the simulation.

The discrete problem is written as follows:

$$\frac{\delta u_h}{\delta t} + \nabla_h \cdot F_h(u_h, u_h) = 0, \qquad (3.3)$$

where $\delta/\delta t$, ∇_h, and F_h are the discrete approximations of $\partial/\partial t$, ∇, and F, respectively. This discrete equation is supplemented by discretized boundary and initial conditions:

$$u_h = f_h \text{ on } \partial\Omega_1^h, \quad \frac{\partial u_h}{\partial n_h} = g_h \text{ on } \partial\Omega_2^h. \qquad (3.4)$$

Here, n_h is the normal outward vector to $\partial\Omega^h = \partial\Omega_1^h \bigcup \partial\Omega_2^h$, the boundary of discretized physical domain Ω^h. The corresponding initial condition is $u_h(x, t = 0) = u_{h,0}(x)$. The approximated flux, F_h, is assumed to satisfy consistency condition $F_h(u, u) = F(u, u)$.

The key problem in ensuring the reliability of results is to control the error $e(h_r, \eta) = \|u - u_h\|_\mathcal{M}$, where $\|\cdot\|_\mathcal{M}$ is an arbitrary norm. An important remark is that, since the solution is assumed to be turbulent, it exhibits a chaotic behavior and all perturbations in the boundary and initial conditions will be continuously amplified. Consequently, only control of the statistical moments (or in an equivalent way of the probability density function) of the solution can be expected over long times. Therefore, a useful definition of the error must rely on statistical moments of u and u_h and must be consistent with the definition of DNS; in other words,

$$\lim_{h_r \to \eta} e(h_r, \eta) = 0. \qquad (3.5)$$

A deeper insight in the nature of the committed error can be gained, at least symbolically, by identifying the various sources of error:

$$e(h_r, \eta) = e_\Pi(h_r, \eta) + e_h(h_r, \eta) + e_r(h_r, \eta), \tag{3.6}$$

where the following are true.

First, $e_\Pi(h_r, \eta)$ is the projection error, originating in the fact that the exact continuous solution is approximated by using a finite number of degrees of freedom (even though some numerical approximations, e.g., finite elements, provide continuous approximations of the exact solution). This is an intrisic error that cannot be eliminated in general. A constraint based on consistency is

$$\lim_{h_r \to 0} e_\Pi(h_r, \eta) = 0. \tag{3.7}$$

Second, $e_h(h_r, \eta)$ represents the numerical error, which arises from the fact that space and time derivatives are not evaluated exactly (with the well-known exception of the Fourier spectral methods):

$$\nabla_h(u) - \nabla(u) \neq 0. \tag{3.8}$$

Because the numerical method is consistent, we have

$$\lim_{h_r \to 0} e_r(h_r, \eta) = 0. \tag{3.9}$$

Third, $e_r(h_r, \eta)$ is the resolution error, not equivalent to the discretization error. It accounts for the fact that, even if the derivatives are exactly evaluated, the flux function is not exact since all scales of the exact solution are not resolved:

$$F_h(u_h, u_h) \neq F_h(u, u) = F(u, u). \tag{3.10}$$

According to the definition of DNS, $e_r(h_r, \eta)$ vanishes when $h_r \approx \eta$.

The problem in LES is to minimize both the numerical and resolution error; the best (or ideal) LES solution is defined as

$$e(h_r, \eta) = e_\Pi(h_r, \eta) \quad \text{(ideal LES)}. \tag{3.11}$$

A major difficulty in LES is that both the topology of the computational grid (parameter h) and the details of the numerical method enter the error definitions, with the result that each case is *a priori* unique. To deal with this problem, several ways to minimize the error have been proposed that rely on mathematical models of the ideal LES problem just discussed. The most popular mathematical representation of LES will be presented in Subsection 3.1.2; existing results treating errors in explicit LES are surveyed later. Explicit LES relies on the addition of a new forcing term, F_{LES}, in Eq. (3.3), leading to a new problem:

$$\frac{\delta u_h}{\delta t} + \nabla_h \cdot F_h(u_h, u_h) = F_{\text{LES}}. \tag{3.12}$$

The consistency constraint for this subgrid-scale term is $F_{\text{LES}} \longrightarrow 0$ when $h_r \longrightarrow \eta$. As will be shown, in the most common explicit LES approach, F_{LES} is designed to cancel the resolution error – that is, to approximate $\nabla_h \cdot [F_h(u_h, u_h) - F_h(u, u)]$. Consequently, the numerical method must be sufficiently accurate to have $e_h(h_r, \eta) \approx 0$. This is the main reason why several authors recommend the use of high-order numerical methods for LES. However, an important point is that the numerical scheme does not need to be perfect: It is only required that it be *neutral* with respect to the selected error norm. As will be discussed, most explicit subgrid modeling strategies are based on the budget of the resolved kinetic energy, and they are designed in order to enforce the correct source–sink term in the corresponding equation. Therefore, the corresponding models do not aim at reproducing either $F_h(u_h, u_h) - F_h(u, u)$ or u (or $u - u_h$); they are referred to as *functional models*. The alternative is to derive *structural models*, which are built to directly approximate u.

Each closure strategy can be classified according to its properties (explicit or implicit, functional or structural). Explicit subgrid techniques are emphasized in this chapter, whereas the balance of this book is devoted to the implicit approach. Because subgrid models are not exact, one can introduce a new error term: the modeling error, which accounts for the discrepancies that exist between the ideal subgrid model (i.e., the one that drives the resolution error to zero) and the subgrid model effectively used during the simulation.

3.1.2 LES: Mathematical models

As already mentioned, the true LES problem (i.e., what is really done on a computer) cannot be analyzed without explicit reference to the computational grid and to the numerical method. This lack of generality renders the theoretical analysis very difficult, and so some idealized mathematical models for the general LES problem are often used to derive information on the nature and the properties of the LES solution.

3.1.2.1 The filtered Navier–Stokes equations model

The most popular theoretical model for the LES problem is the system of filtered Navier–Stokes equations, as proposed by Leonard in the 1970s (Leonard 1974). The underlying idea is that the main property of the LES solution is that it represents a regularized, smoothed solution of the Navier–Stokes equations because the small scales have been eliminated. The gradients of the LES solution are assumed to be less sharp and the correlation length of the resolved fluctuations to be greater than, or equal to, that of the exact solution. The true, discrete LES solution u_h is then approximated by $\overline{u}$, which is the filtered Navier–Stokes solution defined as

$$\overline{u}(x, t) \equiv G(\Delta) * u(x, t) = \int_{\Omega} G(\Delta, x, y)u(y, t)dy, \qquad (3.13)$$

where $G(\Delta, x)$ and Δ are the filter kernel and the filter characteristic length, respectively. Here, Δ will play the role of h_r and all the properties of the numerical method are included in the definition of G. The filter kernel in Eq. (3.13) is expressed as a function of space coordinates only, since almost all explicit LES methods published at the present time rely on spatial filtering. Common filter kernels are discussed in Chapter 6. Some developments dealing with Eulerian domain time-filtering LES were recently published by Pruett et al. (2003), but they are not discussed here.

The use of the convolution filter model for LES is a convenient simplification of the true LES problem, but it has some drawbacks and specifics that are now discussed. The first issue is that it does not account for the projection error: The filtered solution is a continuous solution involving an infinite number of degrees of freedom. It must still be projected onto a discrete, finite-dimension basis to fully mimic the properties of the true LES solution. Accordingly, and observing that no scale smaller than h_r can be captured, several authors, such as Carati, Winckelmans, and Jeanmart (2001), have advocated that the LES approach requires a double-filtering model: a first sharp cutoff filter with cutoff length 2Δ to account for the grid Nyquist cutoff and a second filter with a smoother transfer function and the same cutoff length to account for the properties of the numerical method. This doubly-filtered model is discussed in the next section.

The main drawback of the filtered Navier–Stokes model for LES appears when the corresponding governing equations are sought. For the sake of convenience, my presentation uses the generic conservation law system (3.1). The full system of filtered Navier–Stokes equations for both compressible and incompressible flows is presented in Section 3.6. The filtered equation is derived by applying the convolution filter to the original equation (3.1):

$$\frac{\overline{\partial u}}{\partial t} + \overline{\nabla \cdot F(u, u)} = 0. \tag{3.14}$$

The next step consists of restricting the possible choices for the filter kernel and retaining filters that comply with the following constraints.

First, uniform fields are not modified by the filtering process, leading to

$$\int_{\Omega} G(\Delta, y)dy = 1. \tag{3.15}$$

Second, the filter commutes with both time and space derivatives:

$$\left[\frac{\partial}{\partial s}, G*\right](u) \equiv 0, \quad s = x, t, \tag{3.16}$$

where the commutator is defined as $[f, g](u) = f[g(u)] - g[f(u)]$.

When such a filter is used, the filtered LES model simplifies to an evolution equation for filtered field $\overline{u}$:

$$\frac{\partial \overline{u}}{\partial t} + \nabla \cdot \overline{F(u, u)} = 0. \tag{3.17}$$

The closure problem is formally similar to the one we encounter when we deal with Reynolds-averaged numerical simulations (RANS): The unknown term $\overline{F(u, u)}$ must be expressed as a function of $\overline{u}$, leading to

$$\frac{\partial \overline{u}}{\partial t} + \nabla \cdot F[\overline{u}, \overline{u}] = \nabla \cdot \left[F(\overline{u}, \overline{u}) - \overline{F(u, u)}\right] = \mathcal{F}_{\text{LES}}, \tag{3.18}$$

where subgrid-scale acceleration $\mathcal{F}_{LES}$ is the counterpart of the F_{LES} term appearing in Eq. (3.12) within the filtered model framework. This term cannot be exactly evaluated and must then be *modeled*.

The simple derivation process detailed here is based on the very stringent constraint of Eq. (3.16), which is unrealistic when bounded domains are considered (as is the case in most practical LES problems!). Bounded domains imply that the support of the filter kernel (or equivalently the cutoff length) must vary when approaching domain boundary $\partial\Omega$ to maintain a well-posed problem. The use of a position-dependent (i.e., nonhomogeneous) filter kernel automatically introduces commutation errors, whose compact expression was given by Fureby and Tabor (1997):

$$[\nabla, G*](u) = \nabla\Delta\left(\frac{\partial G}{\partial\Delta} * u\right) + \int_{\partial\Omega} G[x - y, \Delta(x)]u(y)dy. \tag{3.19}$$

The first term accounts for the spatial gradient of the cutoff length while the second term arises from the trace of G on the domain boundary. It is also very important to note that the use of nonhomogeneous filters is also a way to model the effect of nonuniform grids in true LES computations. Equation (3.19) reveals that the set of governing equations obtained by using a nonhomogeneous filter kernel are much more complicated than Eq. (3.18), and Ghosal and Moin (1995) have shown that it can yield inconsistent equations if no additional constraint is imposed on G. This analysis was extended by Vasilyev, Lund, and Moin (1998), who showed that the commutation error scales as Δ^p if the moments of order 1 to $(p - 1)$ of the convolution kernel are identically zero and all higher-order moments are bounded. The main remaining problem here is to find kernels G fulfilling these conditions.

Another way to derive filtered Navier–Stokes equations that mimic the use of a structured nonuniform computational grid was proposed by Jordan (1999, 2001) who observed that most of the difficulties arise from the fact that the Navier–Stokes equations, written in Cartesian coordinates, are first filtered and then rewritten in general coordinates. Inverting these two operations, that is, first writing the equations in general coordinates and then filtering them, makes it possible to use constant-cutoff-length filters (but the problem of the definition of $\overline{u}$ on $\partial\Omega$ remains). This is illustrated by writing the model conservation law system (3.1) in general coordinates:

$$\frac{\partial J^{-1}u}{\partial t} + \nabla_\xi \cdot F_\xi(U^\xi, u) = 0, \tag{3.20}$$

where J^{-1} is the Jacobian of the transformation, and ∇_ξ, F_ξ, and U^ξ are the gradient in the reference space, the flux in the reference space, and the contravariant unknowns, respectively. As the new coordinate system is associated with a uniform grid distribution, a homogeneous filter is used, ensuring that the filtering operator now commutes with the space derivatives and yielding

$$\frac{\partial \overline{J^{-1}u}}{\partial t} + \nabla_\xi \cdot F_\xi(\overline{U^\xi}, \overline{u}) = \nabla_\xi \cdot \left[F_\xi(\overline{U^\xi}, \overline{u}) - \overline{F_\xi(U^\xi, u)}\right]. \tag{3.21}$$

The difficulty is now that the subgrid terms appearing on the right-hand side of Eq. (3.21) involve metric terms and contravariant quantities, whose behavior cannot be directly analyzed in the framework of the dynamics of turbulent flows. We can partially simplify this problem by assuming that the geometric parameters are invariants of the filtering operation, since they are computed by using discrete schemes and can be considered as being already filtered. Explicit model derivation becomes more complicated than in the previous case, since the unknowns cannot be directly tied to physical quantities (contravariant velocity cannot be measured in a wind tunnel!). As a consequence, functional models used within this framework are most often obtained by translating the usual models derived in the previous approach that rely on physical unknowns.

3.1.2.2 A more realistic model: The twice-filtered Navier–Stokes equations

The filtered Navier–Stokes equations model just discussed cannot account for all the error sources present in the true LES problem when simple filter kernels are considered. To recover a more realistic mathematical model still based on the convolution filter approach, several authors use a double-filtering technique:

- The first filter (usually a smooth filter such as the Gaussian filter or the box filter) accounts for the smoothing properties of the LES approach. It is exactly the same filtering step as in the preceding section. The cutoff length is usually related to the resolution length, h_r, and thus accounts for some features of the numerical method used in the true LES problem. The scales removed by this filter are referred to as the *subfilter scales*. The associated field convolution is written as $\overline{u} = \overline{G} * u$.
- The second filter is a spectral sharp cutoff filter related to the Nyquist cutoff frequency of the computational grid, h. The scales filtered out at this stage are called the *subgrid scales*. This filter level is noted as $\widetilde{\overline{u}} = \widetilde{G} * \overline{u} = \widetilde{G} * \overline{G} * u$.

The introduction of the new second filtering step aims to take into account the fact that scales of motion smaller that the Nyquist cutoff length are irremediably lost in the true LES problem. The use of a single smooth filter does not account for this loss of information since smooth filters can be inverted, allowing a theoretical perfect reconstruction of the exact solution. Therefore, in the case of the single-filter approach with a smooth filter, the best LES solution is

$$e(\Delta, \eta) = 0, \tag{3.22}$$

while, in the case of the double-filter approach,

$$e(\Delta, \eta) = e_\Pi(\Delta, \eta), \tag{3.23}$$

which is consistent with the analysis of the true LES problem already given using the formal equivalence $\Delta \longleftrightarrow h_r$. The special case of the single-filter approach based on the sharp cutoff filter is also consistent with the true LES problem, but it represents only the special case of LES computations carried out with Fourier spectral methods.

Within the double-filter approach, the true LES solution is approximated as $\widetilde{\overline{u}}$ (where the bar and tilde symbols are related to the first and second filtering step, respectively), which is the solution of

$$\frac{\partial \widetilde{\overline{u}}}{\partial t} + \nabla \cdot F(\widetilde{\overline{u}}, \widetilde{\overline{u}}) = \nabla \cdot \left[F(\widetilde{\overline{u}}, \widetilde{\overline{u}}) - \widetilde{\overline{F(u, u)}} \right]. \tag{3.24}$$

The unknown term on the right-hand side can be further decomposed by introducing the triple decomposition

$$u = \widetilde{\overline{u}} + u' + u'', \tag{3.25}$$

where

$\widetilde{\overline{u}}$ represents the scales that are captured on the grid and resolved in the sense that they are not destroyed by the smooth filter;

u' are the subfilter scales that are captured on the computational grid (these scales are not part of the approximate LES solution but could be computed on the computational grid); and

u'' are the "true" subgrid scales, that is, the scales that cannot be computed on the grid and are definitively missing.

The field that can be captured on the grid is thus equal to $\widetilde{u} = \widetilde{\overline{u}} + u'$. According to this decomposition, one obtains

$$\widetilde{\overline{F(u, u)}} = \underbrace{\widetilde{\overline{F(\widetilde{\overline{u}}, \widetilde{\overline{u}})}} + \widetilde{\overline{F(u', \widetilde{\overline{u}})}} + \widetilde{\overline{F(\widetilde{\overline{u}}, u')}}}_{F_0} + \underbrace{\widetilde{\overline{F(u', u')}}}_{F_1}$$

$$+ \underbrace{\widetilde{\overline{F(\widetilde{\overline{u}} + u', u'')}} + \widetilde{\overline{F(u'', \widetilde{\overline{u}} + u')}}}_{F_2} + \underbrace{\widetilde{\overline{F(u'', u'')}}}_{F_3}. \tag{3.26}$$

The two first terms, F_0 and F_1, involve only scales that can be captured on the computational grid, and thus might be exactly computed if u' was known. The other terms include u'' and must be modeled.

3.1.2.3 Additional mathematical models

The filtering approach is the most popular mathematical model of LES, but it was shown in the preceding section that it introduces some new problems (the main one being its extension to nonuniform grids on bounded domains) which are *a priori* not present in the true LES problem. To preclude these *artifacts* of the filtering approach, a few authors recommend using statistical operators or projection operators rather than a convolution filter.

A procedure based on conditional elimination of the small scales was advocated independently by McComb, Hunter, and Johnston, (2001) and Yoshizawa (1991). The underlying idea is that the random chaotic character of the turbulent motion is more pronounced at the very small scales than at the larger one. The rationale behind this hypothesis, referred to as the *local chaos hypothesis* (locality being in terms of wave number), is close to Kolmogorov's local isotropy assumption: Very small scales are isotropic and decorrelated from the large ones, which are deterministically affected by the boundary conditions and therefore are more uncertain than the latter. That loss of memory is associated with the picture of the kinetic energy cascade. The associated scale-separation procedure is a statistical average over scales smaller than Δ, while larger scales are left unchanged.

The main advantage of this procedure is that the artifacts of the convolution filter are now avoided, but since it relies on a statistical average, several realizations of the flow are required, while the convolution filter procedure requires only one. This requirement is not important from a purely theoretical viewpoint, but it makes the statistical procedure very difficult to implement in practice.

The use of a projection operator to model the LES problem has been advocated by several authors. The most popular approach is the variational multiscale method developed by Hughes and coworkers (Hughes 1995; Hughes et al. 1998; Hughes, Mazzei, and Jansen 2000; Hughes, Mazzel, and Oberai 2001). Writing the full solution as the sum of orthogonal components (or, equivalently, decomposing the solution space as a sum of orthogonal subspaces),

$$u \in \mathcal{W}, \quad \mathcal{W} = \mathcal{W}_1 \oplus \mathcal{W}_2 \oplus \cdots, \tag{3.27}$$

we obtain an elegant model for the LES problem by truncating the sum and discarding some components (i.e., operating by projection onto an arbitrary subspace $\mathcal{W}_h$ spanned by an arbitrary set of $\mathcal{W}_i$), yielding

$$u_h \in \mathcal{W}_h, \quad u' \equiv (u - u_h) \in \mathcal{W}' = \mathcal{W} - \mathcal{W}_h. \tag{3.28}$$

This approach is a natural one to analyze the properties of LES simulations relying on variational numerical methods such as finite elements.

3.2 Explicit subgrid-scale models

Explicit LES approaches are based on the introduction of a forcing term in the governing equations, such as F_{LES} in Eq. (3.12) or $\mathcal{F}_{\text{LES}}$ in Eq. (3.20), which cannot be directly computed and must be modeled as a function of the available unknowns. As already mentioned, two classes of models can be distinguished and are discussed in the following subsections. The present survey is restricted to models developed for Newtonian, nonreactive, single-phase flows without external forcing or mutiphysics coupling such as in magnetohydrodynamics. Therefore, the governing equations presented in Section 3.6 are used as relevant models to describe the LES problem and to present the models. It is important to emphasize that almost all existing explicit subgrid-scale models have been developed within the framework of the filtered Navier–Stokes equations.

3.2.1 Functional subgrid-scale models

3.2.1.1 The basic model

Functional subgrid-scale models are designed to reproduce the effects of the small unresolved scales on the resolved ones. They are built on physical considerations on the nature of this interaction, and they do not aim at producing a good approximation of either the subgrid scales or the subgrid-scale tensor. An examination of the existing literature shows that almost all functional models have been built to enforce the correct resolved kinetic energy balance. The reason for this is that they are designed to mimic the kinetic energy cascade from large to small scales, which is known to be a universal and dominant physical mechanism in fully developed turbulent flows. More complex dynamical effects exhibit less generality, are much more difficult to take into account by means of a model, and so are usually neglected in the functional modeling approach. Nevertheless, some modified functional models for stratified flows have been proposed (see Sagaut 2005 for a review). It is also worth noticing that some effects on turbulence, such as those due to rotation, are very difficult to take into account in a physical model. An important consequence is that all scales at which important physical mechanisms other than the energy cascade are at play must be directly captured during the computation to get reliable results. This results in a *physical guide* for the definition of resolved–subgrid scales in explicit LES with functional subgrid-scale models.

The main effect of the kinetic energy cascade is a gross drain of the energy of the large scales (i.e., resolved scales in the present LES framework) by the small ones (i.e., subgrid scales). This drain is expressed within the filtered Navier–Stokes framework as

$$\varepsilon_t = -\tau : \nabla \overline{u}, \tag{3.29}$$

where ε_t and τ are the kinetic energy cascade rate and the subgrid tensor, respectively. A simple empirical mathematical model for this kinetic energy loss is a *subgrid viscosity* ν_t, whose amplitude is calibrated to enforce the desired mean energy loss. The resulting term in the filtered Navier–Stokes equations (in the incompressible case) is

$$\mathcal{F}_{\text{LES}} = \nabla \cdot (-2\nu_t \overline{S}), \quad \overline{S} \equiv \frac{1}{2} \left(\nabla \overline{u} + \nabla^t \overline{u} \right). \tag{3.30}$$

This simple approach is supported by the analogy with kinetic gas theory, in which the macroscopic viscosity and diffusivity of a gas originate in the small-scale molecular motion. However, it is worth noting that this is nothing but an analogy: Viscosity and diffusivity characterize the fluid and can be considered as constants because there is a clear scale separation between the molecular motion and the velocity fluctuations at the macroscopic level, while there is no spectral gap in the turbulent spectrum, resulting in a flow-dependent definition of the subgrid viscosity. The idea of representing the small turbulent scale effects through the definition of a turbulent viscosity (a subgrid viscosity in the present parlance) is not new. Parameterizing small scales of turbulence through the use of an effective viscosity can be traced back to the early 19th century: Saint-Venant in 1834 clearly distinguished two scales of motion in a turbulent flow, namely a larger scale at which the average (over a fluid element) velocity varied smoothly in space and time, and a smaller scale at which the motion could be very irregular. In his description, the effective viscosity parameter defined at larger scales depended on the irregular motion at the small scales. Thus it could vary from one point to another and from one kind of flow to another. A consequence[*] of Eq. (3.30) is the following approximation for subgrid tensor τ:

$$\tau = -2\nu_t \overline{S}. \tag{3.31}$$

In the incompressible flow case, tensor $\overline{S}$ is traceless and the relationship (3.31) holds for the deviatoric part of τ only:

$$\tau - \frac{\tau_{kk}}{3}I = -2\nu_t \overline{S}. \tag{3.32}$$

The most popular subgrid-scale viscosity model (and the oldest one in the history of modern LES) was published by Smagorinsky (1963). The dimension of ν_t is $\nu_t = [L]^2[T]^{-1}$; the problem is to identify the characteristic space and time scales for the subgrid-scale motion. The space scale is defined as $C_S\Delta$ (where C_S is a constant to be adjusted to tune the model) and the time scale is evaluated through an evaluation of the local shear, $|\overline{S}| \equiv \sqrt{2\overline{S}:\overline{S}}$, yielding

$$\nu_t = (C_S\Delta)^2|\overline{S}|. \tag{3.33}$$

We can evaluate the constant, C_S, in the very simple case of isotropic incompressible turbulence at a very high Reynolds number: Assuming that (i) the Kolmogorov spectrum shape $E(k) = C_K\varepsilon^{2/3}k^{-5/3}$ (with $C_K = 1.4$, ε, and k the Kolmogorov constant, the turbulent dissipation rate, and the wave number, respectively) holds at all scales, and (ii) that turbulent scales are in local equilibrium (the turbulent kinetic energy production rate, kinetic energy cascade rate, and viscous dissipation rate are equal), we find the value $C_S = 0.18$, which yields good LES results for such flows. But numerical experiments show that this model is too dissipative in shear flows. We can understand this easily by writing the associated subgrid-scale dissipation ε_t (which appears as a sink term in the resolved kinetic energy equation and a source term in the subgrid kinetic energy equation):

$$\varepsilon_t \equiv -\tau:\overline{S} = 2\nu_t\overline{S}:\overline{S} = (C_S\Delta)^2|\overline{S}|^3. \tag{3.34}$$

Since the Smagorinsky model is based on the local isotropy hypothesis, its use for shear-flow simulation will result in an overestimation of $|\overline{S}|$ since the mean shear will be incorporated in the evaluation of $\overline{S}$ but should not be. That flaw is directly tied to the Gabor–Heisenberg uncertainty principle: The Smagorinsky model is local in space (it is computed by using a local evaluation of the gradient) and nonlocal in wave number (all resolved scales contribute to the evaluation of $\overline{S}$). A direct consequence is that the Smagorinsky model or similar subgrid-viscosity models based on the first-order

---

[*] Note that this parameterization of the subgrid tensor is deduced from the subgrid viscosity assumption, since the functional modeling strategy does not aim at reproducing it at all.

derivatives of the resolved velocity field, such as the structure function model of Métais and Lesieur (1996) or the WALE model of Nicoud and Ducros (1999), are not consistent from the spectral point of view; that is, they will not automatically vanish when all the scales are resolved. The limiting DNS case will not be recovered in the proper way. A good example is that the Smagorinsky subgrid viscosity does not return a null subgrid viscosity when applied to a laminar Poiseuille or Couette flow (where there is no subgrid scale at all!). Since the subgrid viscosity depends directly upon the cutoff length Δ, tuning this length makes it possible to improve the results. This is what is done in the very near wall region, where the subgrid viscosity is supposed to scale as the cube of the distance to the wall. In this region, a new definition of Δ is employed to enforce the correct asymptotic behavior. A classic example is the van Driest damping function, which is based on the following modification of the characteristic length Δ in the near wall region:

$$\Delta \longleftarrow \Delta f_w(z), \quad f_w(z) = 1 - e^{-z u_w / 25 \nu}, \tag{3.35}$$

where z, u_w, and ν are the distance to the wall, the friction velocity, and the molecular viscosity, respectively.

In high-Reynolds-number turbulent shear flows, a heuristic criterion to obtain satisfactory results is to chose the cutoff length Δ such that (Bagget, Jimenez, and Kravchenko 1997)

$$\frac{L_{\varepsilon_t}}{\Delta} = 10 - 20, \quad L_{\varepsilon_t} \propto \sqrt{\varepsilon / |S|^3}, \tag{3.36}$$

where ε_t is the turbulent dissipation and S is the mean shear.

All recent improvements of the Smagorinsky model (or similar subgrid-viscosity models based on the first-order derivatives of the resolved velocity field) share the same key idea of increasing the locality of the subgrid viscosity in terms of wave number and to evaluate the subgrid term by using the smallest resolved scales only. The expected benefit is twofold: first, the resulting model is expected to be self-adaptive in the sense that it will automatically vanish when all scales are resolved (i.e., when the energy of the smallest scales captured on the computational grid is zero or nearly zero); second, the rate of energy transfer is expected to be predicted more accurately since the physical analysis reveals that the energy transfer is local in wave number.[†] Being more local in wave number, these improved models are less local in space and the width of their discrete stencil is larger, yielding possible nontrivial problems when one is dealing with their definition near the boundaries of the computational domain.

3.2.1.2 A few improvement strategies

The main strategies for improving the basic subgrid-viscosity models are now surveyed.

The first strategy is the removal of the mean flow contribution. The key idea here is for us to split the instantaneous resolved field, $\overline{u}(x, t)$, into the sum of its statistical

---

[†] At least 75% of the energy transfer of mode k involves modes in the range $[k/2, 2k]$.

mean value $\langle \overline{u} \rangle (x, t)$ and its statistical fluctuation $\overline{u}''(x, t)$. We then evaluate the subgrid viscosity by using $\overline{u}''(x, t)$, resulting in a cancellation of the influence of the mean shear. This technique, first proposed by Schumann (1975), has been shown to yield good results. The main underlying problem is in computing $\langle \overline{u} \rangle (x, t)$. In his original work, Schumann proposed to compute it by averaging the resolved field in directions of spatial periodicity (resulting in a fully nonlocal model in these directions); Carati, Rogers, and Wray (2002) recently proposed the use of several statistically equivalent simulations (performed on a parallel computer) to carry out the statistical average.

The second strategy is the use of the kinetic energy at the cutoff or the subgrid kinetic energy. The basic idea here is that the kinetic energy of the smallest resolved scales and the subgrid kinetic energy are very relevant parameters for subgrid-scale detection and parameterization. The subgrid viscosity is then rewritten as

$$\nu_t = C \sqrt{q} \Delta, \tag{3.37}$$

where $C \simeq 0.1$ and q are a constant and the kinetic energy under consideration, respectively. In the case where the smallest resolved scale energy is looked at, we compute q by applying a discrete low-pass filter (referred to as the *test filter*) to the instantaneous resolved field and computing the kinetic energy of the small scales educed this way. The locality of the resulting subgrid viscosity is then directly governed by that of the test filter, and equivalently by its stencil width. This approach was first proposed by Bardina, Ferziger, and Reynolds (1983). A common example is the following one-dimensional discrete test filter (where i is the grid point index and ϕ is a dummy variable):

$$\overline{\phi}_i = \frac{1}{4} (\phi_{i+1} + 2\phi_i + \phi_{i-1}). \tag{3.38}$$

The kinetic energy of the subgrid scales cannot be accurately extracted this way, and an additional evolution equation is solved to compute it. The most common closed equation is very similar to the Kolmogorov–Prandtl model:

$$\frac{\partial q}{\partial t} + \overline{u} \nabla q = 2\nu_t \overline{S} : \overline{S} - \frac{q^{3/2}}{\Delta} + \nabla \cdot (\nu_t \nabla q) + \nu \nabla^2 q, \tag{3.39}$$

where terms appearing on the right-hand side represent the production by the resolved scales, the turbulent dissipation, the turbulent diffusion, and the viscous dissipation, respectively. The locality in space of the original model is not modified, strictly speaking, since only local quantities are used, but the equation for q introduces a nonlocal history effect that is equivalent. Some algebraic estimates for the subgrid kinetic energy have been proposed by several authors, which have been observed to be accurate in isotropic turbulence (Knaepen, Debliquy, and Carati 2002; Voekl, Pullin, and Chan 2000).

We can define more complex forms of the subgrid viscosity by using nonlinear combinations of (3.33) and (3.37), yielding the one-parameter mixed-scale model family proposed by Sagaut and Loc (Sagaut 2005):

$$\nu_t = C |\overline{S}|^\alpha q^{\frac{1-\alpha}{2}} \Delta^{1+\alpha}, \tag{3.40}$$

where C and α are two real parameters. These models yield better results than rough subgrid-viscosity models, but their efficiency is very sensitive to the test filter or the way complex mechanisms such as transition are taken into account in Eq. (3.39).

The third strategy is to find the best constant value in the least-squares sense (dynamic models). The idea here, proposed by Germano et al. (1991) and Lilly (1992), is to find the optimal value (local in space and time) of the constant C, that is, the value of the constant that will minimize a given error estimate for the considered formulation of the subgrid-viscosity model under consideration: $v_t = Cf(\overline{u}, \Delta)$. The locality in wave number is increased by using an ad hoc error estimate based on the following exact relationship, referred to as the *Germano relationship*, which ties the subgrid tensor at two different filtering levels (the first one being related to the bar symbol and the second, the test filter with cutoff length $\beta\Delta$, $\beta > 1$, to the tilde symbol):

$$\underbrace{(\widetilde{\overline{u} \otimes \overline{u}} - \widetilde{\overline{u}} \otimes \widetilde{\overline{u}})}_{L} - \underbrace{(\widetilde{u \otimes u} - \widetilde{\overline{u}} \otimes \widetilde{\overline{u}})}_{T} + \underbrace{(\widetilde{u \otimes u} - \overline{u} \otimes \overline{u})}_{\widetilde{\tau}} = 0, \qquad (3.41)$$

where τ, T, and $\widetilde{\tau}$ are the subgrid tensor at the first filter level, the subgrid tensor at the second filter level, and the subgrid tensor at the first filter level filtered at the second level, respectively. It is important to note that we can compute the tensor L directly from the resolved LES field $\overline{u}$ by using the test filter. The next step consists of defining an error estimate. We achieve this by replacing exact subgrid tensors τ and T by the corresponding models

$$\tau = -2Cf(\overline{u}, \Delta), \quad T = -2Cf(\widetilde{\overline{u}}, \beta\Delta)$$

in Eq. (3.41), yielding the definition of the tensorial error estimate, E:

$$E = L + 2Cf(\widetilde{\overline{u}}, \beta\Delta) - 2\widetilde{Cf(\overline{u}, \Delta)}. \qquad (3.42)$$

Assuming that constant C is nearly constant over intervals of length $\beta\Delta$, we obtain the simplified formula

$$E = L + 2C[f(\widetilde{\overline{u}}, \beta\Delta) - \widetilde{f(\overline{u}, \Delta)}] = L + 2CM. \qquad (3.43)$$

To get a scalar subgrid viscosity, Lilly (1992) proposed to compute C to minimize E in the least-squares sense, leading to

$$C = \frac{1}{2}\frac{L : M}{M : M}. \qquad (3.44)$$

This method is known to yield the expected properties: The subgrid viscosity vanishes in fully resolved regions and also exhibits correct asymptotic behavior in the near wall region on fine grids. It successfully decreases the dissipation of the Smagorinsky viscosity during the transition to turbulence, allowing an accurate description of the transition phase. Unfortunately, the method has some severe drawbacks in the area of numerical stability: Constant C defined as in Eq. (3.44) is observed to take very large values (corresponding to mathematically ill-posed problems) and negative values can occur over long times, leading to an exponential growth of perturbations. To deal with these problems, authors have proposed many techniques (see Sagaut 2005 for a review): to clip the constant between 0 and an arbitrary upper bound; to perform averages over homogeneous directions, closest neighbors, or streamlines; or combinations of these methods. Iterative methods have also been proposed. The resulting subgrid viscosity is obviously less local in space than the original one. Additional difficulties are known to arise when the two cutoffs (of the primary LES filter and of the test filter) are not located within the inertial range of the spectrum, and the procedure must be modified to recover consistent evaluations of the subgrid viscosities.

The fourth strategy involves filtered models, variational multiscale methods, and hyperviscosities. This strategy is very close to those based on the removal of the mean flow contribution already described, the main difference being that a low-pass filter in space is used instead of a statistical average, adding flexibility to its use. Several implementations of the same basic idea have been proposed indepently, which are subsequently described by using the framework developed by Hughes et al. (2001). The instantaneous resolved velocity field is split into a large-scale component, $u^<(x, t)$, and a small-scale component, $u^>(x, t)$. The key idea underlying all these models consists of using $\bar{u}$ or $u^>$ to evaluate each term appearing in the model for the full subgrid tensor, namely the subgrid viscosity and the shear stress tensor. In practice, this splitting is achieved at each time step of an LES computation by applying a discrete test filter, G_d, to the LES resolved field $\bar{u}$:

$$u^< = G_d(\bar{u}), \quad u^> = \bar{u} - u^< = (1 - G_d)(\bar{u}). \tag{3.45}$$

The four possible combinations are illustrated here; the Smagorinsky model (3.33) is taken as an example.

- The large-large model, which corresponds to the original model (3.33) in which both parts are computed by using the full resolved field, leading to very poor local properties in terms of wave number;
- The large-small model proposed by Hughes, in which the subgrid viscosity is evaluated by using the full resolved field, while the velocity gradient tensor is restricted to $u^>(x, t)$:

$$\tau = -2(C_S\Delta)^2|\bar{S}|S^>, \quad S^> \equiv \frac{1}{2}\left(\nabla u^> + \nabla^t u^>\right); \tag{3.46}$$

- The small-small model of Hughes, in which both parts are computed by using the small resolved scales only:

$$\tau = -2(C_S\Delta)^2|S^>|S^>; \tag{3.47}$$

- The small-large model, which is strictly equivalent to the filtered model concept developed by Ducros on the grounds of the structure function model and further extended by Sagaut et al.,

$$\tau = -2(C_S\Delta)^2|S^>|\bar{S}. \tag{3.48}$$

The small-small and large-small models share the property that, if the $u^>$ and $u^<$ are orthogonal, the subgrid energy drain will act on the $u^>$ component only. In the case of a Fourier spectral implementation, this property of orthogonality was shown to yield spurious energy pileup in the $u^<$ component because nonlocal energy transfers toward subgrid modes were not taken into account. These two models can also be easily recast as hyperviscosity models assuming the following differential approximation of test filter G_d is valid:

$$G_d(\overline{u})(x) = \overline{u} - \alpha \Delta^{2p} \nabla^{2p} \overline{u} + o(\Delta^{2p}). \tag{3.49}$$

Here α and p are filter-dependent parameters, which leads to

$$u^> \simeq \alpha \Delta^{2p} \nabla^{2p} \overline{u}. \tag{3.50}$$

This yields the following approximations:

$$\tau = -2\alpha C_S^2 \Delta^{2(p+1)} |\overline{S}| \nabla^{2p}(\overline{S}), \quad \text{large-small model}, \tag{3.51}$$

$$\tau = -2\alpha^2 C_S^2 \Delta^{2(2p+1)} |\nabla^{2p}(\overline{S})| \nabla^{2p}(\overline{S}), \quad \text{large-small model}. \tag{3.52}$$

These new formulations make the improved spectral locality properties clearer and show how they are governed by those of the test filter. Generalized versions based on a filtering approach have been proposed by Vreman (2003).

3.2.2 Structural subgrid-scale models

The structural models aim at predicting the subgrid scales (or the subgrid tensor) directly, rather than recovering their effects on the resolved scales through the use of a forcing term. Obviously, an accurate prediction of the subgrid scales would lead to a satisfactory estimation of the forcing term, $\mathcal{F}_{\text{LES}}$. While the distinction between the single- and the twice-filtered Navier–Stokes model was not important in the functional modeling case, it is a central feature in the structural model case. Looking at Eq. (3.25), we see that it is clear that only the u' field can be reconstructed on the computational grid. As a consequence, we can distinguish between two general classes of structural models.

The first class is *models and methods for reconstructing u' only*. This approach was coined the *soft-deconvolution problem* by Adams, and it makes it possible to evaluate the term F_1 in Eq. (3.26). Since interactions with the field u'' are not taken into account, these models must be supplemented by a secondary model of the structural type that will acount for F_2 and F_3 in Eq. (3.26), leading to the definition of *mixed models*, also called *linear combination models*. This combination is also supported by physical arguments developed by Shao et al. (1999), who found in a priori tests that F_0 and F_1 are rapid terms (i.e., terms that react quickly to changes in the resolved field) that carry most of the information tied to the anisotropy of resolved scales, while F_2 and F_3 are slow terms mostly related to an isotropic energy cascade.

The second class is *models and methods for estimating both u' and u''*. An auxiliary computational grid is now introduced to capture u'' or its surrogate. This approach can be referred to as a *full reconstruction approach*.

3.2.2.1 Soft-deconvolution models

All models belonging to this family rely on a partial inversion of the first filtering step introduced within the twice-filtered Navier–Stokes equations LES model framework. The purpose is to evaluate the field $u^* \simeq \widetilde{u} \equiv \widetilde{\overline{u}} + u'$, using $\widetilde{\overline{u}}$ as a starting point, and then to use it as a predictor of the whole resolvable field $\widetilde{u}$, yielding the following expression for generalized soft-deconvolution models:

$$\tau \approx \widetilde{u^* \otimes u^*} - \widetilde{u}^* \otimes \widetilde{u}^*. \tag{3.53}$$

Many strategies have been developed, which are extensively discussed in Chapter 6.

All soft-deconvolution models are underdissipative in the sense that they do not prevent energy pileup in the resolved scales. This is consistent with the fact that they do not account for interaction with "true" subgrid scales represented by the u'' field. In the case of the tensor diffusivity model, this lack of dissipation is ascribed to antidiffusive properties along proper axes of $\widetilde{\overline{S}}$ associated with negative eigenvalues.

3.2.2.2 Full reconstruction of subgrid scales

These models require the definition of an auxiliary fine grid with characteristic mesh size $h' < h$ on which the field u'' will be reconstructed. They can be recast within the general framework of multilevel techniques. The original LES grid with mesh size h is then referred to as the *coarse grid* (or *coarse-resolution level*). This new resolution level is represented by using another filtering step within the filtered Navier–Stokes equations framework. The common structure of all these methods is the following:

1. Extend the coarse grid solution $\widetilde{\overline{u}}$ to the fine grid. One can achieve this by using an interpolation step, which can be supplemented by a deconvolution step to recover $\widetilde{u}$.
2. Reconstruct subgrid-scale field u'' on the fine grid, and add it to $\overline{u}$ to get u.
3. Compute nonlinear term $F(u, u)$ on the fine grid.
4. Restrict it at the coarse-resolution level to obtain $\widetilde{F(u, u)}$. This step can be implemented as a projection or a combination of a filtering step and a projection step.

One achieves the gain in terms of computational time with respect to a usual LES computation performed on the fine grid by using at least one of the two techniques listed here (see Sagaut 2005, and Domaradzki and Adams 2002, for reviews):

1. Use a cheap method (i.e., cheaper than solving the Navier–Stokes equations) to evaluate u'' on the fine grid. One can do this by using many methods, such as (i) fractal interpolation of the resolved field or (ii) kinematic extrapolation based on an estimate of the production term.
2. Define a cycling strategy between the two resolution levels; that is, freeze the u'' on the fine grid for some time. One can prescribe this time by using phenomenological arguments or compute it by using a priori error estimates. One can achieve

further cost reduction by using a set with an embedded resolution level (Terracol, Sagaut, and Basdevant 2001 use up to four resolution levels in the case of a compressible mixing layer).

3.2.3 Extension for compressible flows

The extension of the subgrid models to compressible flows can mostly be interpreted as a variable-density extension of the incompressible models, rather than as the development of true compressible subgrid models relying on the physics of compressible turbulence. The main reasons are as follows.

First, the subgrid Mach number M_{SGS} is generally small, and so compressibility effects at the subgrid motion level are expected to be small. This is consistent with the observation that the velocity divergence spectrum exhibits very small values at large wave numbers in compressible isotropic turbulence (Samtaney, Pullin, and Kosovic 2001). Erlerbacher and coworkers (1992) observe that subgrid compressiblity effects can be neglected for subgrid Mach numbers less than or equal to 0.4, and, because the subgrid Mach number is smaller than the turbulent Mach number, M_t, they propose to neglect these effects in flows with a turbulent Mach number smaller than 0.6. It is also observed in most engineering applications at a low turbulent Mach number that the ratio of the compressible to the solenoidal kinetic energy scales as M_t^α (with $\alpha = 2$ or $\alpha = 4$), resulting in a very small contribution of the compressible part in the total energy transfer.

Another point is that a priori evaluation of subgrid terms using DNS data reveal that some subgrid terms can be neglected for a large class of flows. An examination of DNS data shows that the subgrid term A_2 in Eq. (3.74) in the momentum equations can be neglected in all cases. The most significant terms in the energy equation as defined in Eq. (3.74) are B_1 through B_5. The magnitude of B_6 and B_7 is about 1 to 2 that orders of magnitude smaller.

A third point is that the main compressibility effects are expected to be taken into account by the resolved modes. In addition, interscale interactions and energy transfers in the compressible case exhibit very different regimes, depending on many parameters (e.g., subgrid Mach number, turbulent Mach number, initial distribution of energy between compressive and solenoidal modes), which have not been modeled at the present time.

As a consequence, functional models are extended through variable-density formulations of models for the incompressible momentum equations, while models for the passive scalar problem are used as a basis for the energy equation.

Structural models will also be extended, but are not so restrictive, because they do not rely on assumptions about the nature of the interscale couplings. Starting from the basic relationship (3.32), we can parameterize the deviatoric part of the subgrid tensor as

$$\tau - \frac{\tau_{kk}}{3} I = -2\overline{\rho} \nu_t \left(\widetilde{S} - \frac{1}{3} \widetilde{S}_{kk} I_d \right), \tag{3.54}$$

where we evaluate ν_t by using one of the previously discussed subgrid-viscosity models. It is important to remember that $\widetilde{S}$ is related here to the density-weighted resolved shear stress tensor (see Section 3.6.2 for a definition). A few subgrid models for the isotropic part τ_{kk} have been proposed (Erlebacher et al. 1990, 1992; Salvetti and Banerjee 1995), but it appears that in most of the existing computations it is very small compared with the filtered pressure, and then is often neglected.

In the filtered energy equation discussed in Section 3.6.2, terms B_1 and B_2 still need to be parameterized. Vreman (2003) and almost all authors who have dealt with similar formulations have used a simple Boussinesq-type model:

$$B_1 + B_2 = -\nabla \cdot \left(\kappa_t \nabla \overline{T} \right), \tag{3.55}$$

where the eddy diffusivity, κ_t, is tied to the eddy viscosity through a turbulent Prandtl number, $\mathrm{Pr}_t = \nu_t / \kappa_t$. Typical values for this Prandtl number are $0.3 \leq \mathrm{Pr}_t \leq 0.9$. One can also compute it by using a dynamic procedure.

Reformulating structural subgrid models is straightforward and will not be detailed here. The procedure is the same as in compressible flows considering density-weighted filtered variables. An important remark is that no subgrid models for true compressible effects such as shock–turbulence interaction have been derived up to now, and that existing results have been obtained on grids fine enough to ensure the direct capture of compressible mechanisms (such as turbulence anisotropization and shock corrugation in shock–turbulence interaction). The consequence is the need for very fine grids in regions where compressiblity effects are important at small scales.

3.3 The boundary condition issue

3.3.1 General statement of the problem

The issue of adequate boundary conditions is a classic problem in computational mechanics. Results in this section are not discussed in the most general way, but rather put the emphasis on two very important specific problems: the development of unsteady boundary conditions representing a turbulent inflow and the representation of solid walls on coarse grids. In Section 3.1.2 we saw that the filtered Navier–Stokes equations model introduces additional problems dealing with solid boundaries. Since these are *artifacts* of the mathematical model and only a very few works deal with this problem, they are not discussed here.

We can analyze the problem of the definition of adequate boundary conditions for the true LES problem by looking at Eqs. (3.2) and (3.4). The problem is twofold. First, since the computed solution u_h is not equal to the exact solution u, even in the ideal case where both the subgrid model and the numerical scheme are perfect ($u_h = u_\Pi$), there is no a priori reason why the boundary conditions for these two fields should be the same, leading to the statement that

$$f \neq_h, \quad g \neq g_h. \tag{3.56}$$

The second part of the problem is that, in the general case, the geometry of the boundary of the domain is not exactly reproduced in the discretization, leading to $n \neq n_h$.

The most suitable boundary conditions for LES can be defined as those boundary conditions such that the growth of the error $e(h_r, \eta)$ will be minimized. From a mathematical standpoint, this issue is close to the definition of a control problem, in which values of the solution and its derivatives on the boundaries of the domain are defined to minimize a given cost function (the LES error in the present case). Since the exact solution is not known in most practical cases, the LES boundary conditions cannot be sought that way in the general case.[‡]

The rigorous way to find LES boundary conditions should be to apply the mathematical model used to represent the LES problem for exact boundary conditions. Since the filtered Navier–Stokes model involves many commutation errors on bounded domains, this is generally not done; most inflow–outflow conditions used in practice are no more than discretized versions of the boundary conditions of the continuous problem. A physical requirement for obtaining good results is that all resolved spatiotemporal modes must be taken into account in the definition of the boundary conditions.

The problem becomes more complex when important physical mechanisms are governed by the boundary condition (e.g., the boundary layer and the no-slip boundary conditions). If the mesh is too coarse on the boundary, all critical scales of motions cannot be captured, and the boundary conditions should account for them, leading to a very complex problem. Two critical cases are presented here: turbulent inflow conditions and solid-wall modeling.

3.3.2 Turbulent inflow conditions

Prediction of the turbulent flow in full systems (engines, aircraft, etc.) is very often out of range of available computing facilities, and the computational domain is excised from the full geometry. As a consequence, the flow at the inlet plane of the LES domain is already turbulent, and the problem arises of prescribing turbulent fluctuations at the inflow. A first important point is that recent analyses by George and Davidson (2004) show that a memory effect associated with the inlet conditions exists in turbulent flows. A very important consequence is that the efficiency of LES as a predictive tool is partly governed by the quality of the inflow conditions. Numerical experiments have shown that the use of steady flow conditions at the inlet plane (i.e., prescribing the mean flow profile only) has deleterious effects on the solutions: A (very large) buffer region develops behind the inflow plane in which turbulent fluctuations are rebuilt by nonlinear interactions. The LES solution is then corrupted in two different ways. First, the fluctuations do not correspond to turbulent fluctuations but are associated with the growth of unstable modes. Second, as the fluctuations are not correctly captured, their effect on the mean flow is poorly predicted, yielding errors on the mean flow profile itself. A consequence is that results need to be recalibrated to match experimental data.

[‡] However, note that some attempts have been made in the plane channel case, where the mean flow profile and the Reynolds stress profiles are known.

A well-known example is the turbulent jet case, in which the axial coordinate must be translated a posteriori to recover the correct location of the end of the potential core. The field at the inlet plane is usually written as (the inlet plane is defined as $x = x_0$)

$$u_h(x_0, y, zt) = U(y, z) + v(y, z, t),\qquad(3.57)$$

where U and v are the mean flow and the fluctuating field, respectively. The mean flow profile is assumed to be known for all variables, and it can be obtained by use of analytical laws, experimental data, or RANS computations. Existing methods for prescribing the fluctuationg part can be grouped into two classes: stochastic methods and deterministic methods.

3.3.2.1 Stochastic turbulence generation methods

These methods are related to the so-called issue of "synthetic turbulence" models, and they rely on the definition of random fields with prescribed statistical properties. All these methods share two basic features: the induced extra cost is very low compared with the cost of the full LES method, and they require a complex foreknowledge of the statistical properties of solutions at the inlet plane.

The simplest solution consists of generating independent random series for each variable at each space location with prescribed kinetic energy. This results in uncorrelated fluctuations and is known to yield very poor results, since the fluctuations are not solutions of the Navier–Stokes equations and have no common features with true turbulent fluctuations. In most known numerical examples, the fluctuations obtained in this way exhibit a very severe damping near the inflow plane, followed by an amplification phase once again associated with the growth of partially scrambled linearly unstable modes. The initial damping is associated with the fact that the random field does not belong to the space of solutions spanned by the numerical method, and that the projection of the imposed random field onto this space results in a loss of kinetic energy.

Some very significant improvements (reduction of the spurious transition region) can be realized if two-point correlations are known at the inlet. In this case, one can obtain coherent fluctuations by using the linear stochastic estimation method or a hybrid proper orthogonal decomposition–linear stochastic estimation approach (Druault et al. 2004). Bonnet and coworkers showed that this method can produce very good results for the plane mixing layer case. They also proposed a method to interpolate the two-point correlation tensor based on Gram–Charlier polynomials, leading to a drastic reduction of the required information: three points were enough in the incompressible mixing layer case.

3.3.2.2 Deterministic methods

The need for fluctuations with correct space and time correlations led several researchers to propose deterministic models for the turbulent fluctuations. The most accurate method (and also the most expensive one) consists of computing the flow at the inlet plane as an auxiliary LES problem. This new LES simulation, referred to as the *driver* or the *precursor simulation*, is carried out in a small computational domain in canonical configurations (zero-pressure gradient flat plate boundary layer, plane channel flow, time-developing mixing layer, etc). The results are very good in the incompressible case, but nonreflective conditions must be developed in the compressible case to prevent spurious acoustic resonance. The main restriction of this approach is that it is efficient for very simple predictor simulations only. One can achieve further gain by storing the precursor solution over a short time only (or using the time-developing approximation) and prescribing it cyclically at the inflow. This method can yield spurious periodic forcing in the LES solution and must be used with care. Partial scrambling by the addition of a random-noise component may help in preventing this drawback.

Another solution was proposed by Lund, Wu, and Squires (1998) to avoid the computation of the auxiliary LES problem. It relies on the use of universal scaling laws for turbulent fluctuations in some fully developed turbulent flows that exhibit a self-similar regime. The key idea is to extract the fluctuations within the computational domain in a plane located far downstream of the inlet plane, to rescale them, and to impose the resulting fluctuating field at the inflow plane. Numerical experiments show that the results are very satisfactory if the extraction plane is taken far enough downstream to prevent spurious coupling with the inflow. That method was originally developed to compute incompressible canonical boundary layers. Its extension to compressible flows is more problematic, since all the variables (velocity, temperature pressure, etc.) do not obey the same scaling laws, yielding some lack of consistency between the fluctuations of different variables.

The last class of solutions encompasses deterministic models of coherent structures in turbulent flows. The simplest solution is to use linearly unstable modes associated with the mean flow profile, U. More relevant ad hoc models can be built in some cases. This approach is illustrated by the five-mode model proposed by Sandham, Yao, and Lawal (2003) for reconstructing turbulent fluctuations in a compressible boundary layer.

3.3.3 Solid walls on coarse grids: Wall models

3.3.3.1 Statement of the problem

It was already mentioned that, in the LES approach, all events associated with turbulence production escape traditional subgrid-scale modeling and must be directly captured on the computational grid. In the turbulent boundary layer that develops at solid boundaries, this would generate very drastic mesh-resolution requirements. Indeed, the maximum of the turbulence production is known to take place in the inner region of the boundary layer (at a distance nearly equal to 15 wall units from the wall), and turbulence production events (mainly the sweeps and the ejections) are known to

Table 3.1. *Typical mesh size (in wall units) for DNS and LES of the boundary layer flow*

	DNS	Wall-resolving LES	LES with wall model
Δx^+ (streamwise)	10–15	50–100	100–600
Δy^+ (spanwise)	5	10–20	100–300
Min(Δz^+) (wall normal)	1	1	30–150
No. of points in $0 < z^+ < 10$	3	3	–

have nearly universal characteristic scales in wall units. As a consequence, the grid must be fine enough to capture these events. Typical mesh sizes in attached boundary layer computations are displayed in Table 3.1. Numerical experiments show that the resolution in the streamwise and the spanwise directions are very important parameters that govern the quality of the solution. A large effect on the numerical error is also observed, whose analysis is not be discussed here.

In the case that coarser grids are used, the no-slip condition appropriate to the rigid wall is not relevant, because scales of motion in the very near wall region are not captured. As a consequence, a specific subgrid model for the inner layer, referred to as a *wall model*, must be defined and must take into account the wall dynamics of the inner region of the boundary layer, including turbulence production. This will provide the simulation with adequate conditions on the variables, the fluxes, or both.

3.3.3.2 A few wall models

This section describes the two main groups of wall models for LES in the incompressible regime. Their extension to compressible flows is straightforward with the use of the van Driest scaling (if no compressibility effects are present, such as very cold wall effects).

The first group is the algebraic equilibrium boundary layer model. This model assumes that the instantaneous streamwise velocity component at the first grid point and the instantaneous wall stress are in phase. In a staggered finite-volume implementation, the following conditions are imposed at the first grid point, which is located at distance Δz to the wall (here assumed to be located at $z = 0$):

$$\tau_{p,13} = \frac{\bar{u}_1(x, y, \Delta z)}{\langle \bar{u}_1(x, y, \Delta z) \rangle} \langle \tau_p \rangle, \quad \bar{u}_3 = 0, \quad \tau_{p,23} = \frac{2}{\text{Re}_\tau} \left[\frac{\bar{u}_3(x, y, \Delta z)}{\Delta z} \right], \quad (3.58)$$

where
   streamwise, spanwise, and wall-normal velocity components are respectively u_1, u_2, and u_3;
   $\langle \rangle$ denote the statistical average (over homogeneous direction or time);
   $\tau_{p,i3}(x, y) = \nu \bar{S}_{i3}(x, y, 0)$;
   $\tau_p = \sqrt{\tau_{p,13}^2 + \tau_{p,23}^2}$ is the skin friction; and
   $\text{Re}_\tau = \delta u_\tau / \nu$ is the friction Reynolds number (δ being the boundary layer thickness and $u_\tau = \sqrt{\tau_p}$ the friction velocity).

The skin friction is computed at each point and each time step from the classic logarithmic law for the mean velocity profile (using a Newton-like algorithm):

$$u_1^+(x, y, \Delta z) = \frac{u_1(x, y, \Delta z)}{u_\tau(x, y)} = 2.5 \ln[\Delta z u_\tau(x, y)/v] + 5.5. \qquad (3.59)$$

The second group consists of two-layer models based on thin-boundary-layer equations (TBLEs; see Balaras, Benocci, and Piomelli 1996; Cabot and Moin 1999), which are the most complex. They rely on the simplified equations for thin boundary layers with a zero wall-normal pressure gradient assumption, which are to be solved on an auxiliary grid, located inside the first cell of the LES grid. TBLEs can be recast as an equation for the wall-normal stresses:

$$\frac{\partial}{\partial z}\left[(v + v_t)\frac{\partial \overline{u}_i}{\partial z}\right] = \underbrace{\frac{\partial \overline{u}_i}{\partial t} + \frac{\partial}{\partial x}(\overline{u}_1\overline{u}_i) + \frac{\partial}{\partial z}(\overline{u}_3\overline{u}_i) + \frac{\partial \overline{p}}{\partial x_i}}_{F_i}, \quad i \neq 3, \qquad (3.60)$$

$$\overline{u}_3(x, y, \Delta z) = \int_0^{\Delta z} \left[\frac{\partial \overline{u}_1}{\partial x}(x, y, \zeta) + \frac{\partial \overline{u}_2}{\partial y}(x, y, \zeta)\right] d\zeta. \qquad (3.61)$$

The first equation is derived from the incompressible Navier–Stokes equations using thin layer approximations, while the wall-normal velocity component, $\overline{u}_3$, is computed by use of the incompressibility constraint. In practice, the right-hand side of Eq. (3.60), which is either computed on the auxiliary mesh or evaluated by using an LES solution on the primary grid, is often simplified as $F_i = 0$ or $F_i = \frac{\partial \overline{p}}{\partial x_i}$. The latter form was shown to yield satisfactory results for separated flows. The quality of the results also depends strongly on the choice of the model for v_t. The most commonly used model for v_t is a simple mixing length model with near wall damping:

$$v_t = v C_t z^+(1 - e^{-z^+/19}), \qquad (3.62)$$

where z^+ is the distance to the wall in wall units and C_t is a constant. Wang and Moin (2000) indicate that the best results are obtained by using a dynamically adapted algebraic RANS model. This is accomplished by locally adjusting C_t so that the RANS eddy viscosity (in the TBLE region) and the subgrid viscosity (in the outer LES domain) are equal at the interface. Boundary conditions are derived by requiring that the TBLE solution satisfy the no-slip condition at the wall and match the outer LES solution on the interface. The main advantage of the TBLE model is that, since it does not require the solution of a Poisson problem to compute the pressure, its relative cost is small. In the compressible flow case, this advantage disappears and its efficiency remains to be investigated.

These models were originally developed for plane channel flow simulations and yield acceptable results for this configuration at low or medium Reynolds numbers. At high Reynolds numbers, noticeable errors on the predicted mean velocity field are observed. Another observed flaw is the existence of spurious resolved Reynolds stresses peaks just near the boundary, corresponding to the existence of unphysical, mesh-governed streaks in that region.

Using suboptimal control theory, Nicoud et al. (2001) proved that the main sources of error are the linear relationship assumed between the instantaneous velocity and the wall stresses and the zero wall-normal velocity. The existence of these spurious streamwise vortical structures can be linked to the existence of an autonomous self-regenerating cycle in shear flows, which was analyzed by use of numerical experiments by Jimenez and his coworkers and described from a theoretical viewpoint by Waleffe. An alternate explanation is that they are related to secondary vortices created when vortical structures impinge on the boundary. No cure for this problem is known at present time, but the use of very coarse grid and very dissipative models like the Smagorinsky model is observed to yield a decrease in the intensity of these parasitic fluctuations.

3.4 LES validation

3.4.1 Sensitivity and efficiency

Validation of LES is an important and complex issue: important, because it is a mandatory step in the development of an efficient LES technique (based on either explicit or implicit modeling), and complex because the very meaning of the term *validation* must be carefully defined to prevent controversies. The basic idea underlying validation is to assess the capability of a given LES approach to give satisfactory results on a class of flows. The notion of satisfactory results can be formulated in a mathematical framework as follows: An efficient LES method for a class of flows is a method that leads to an explicit control of error $e(h_r, \eta) = \|u - u_h\|_{\mathcal{M}}$ and thus makes it possible to reach an arbitrary level of accuracy [the ideal limit at a given resolution level being $e_\Pi(h_r, \eta)$].

Two related issues are the sensitivity of the LES method and its efficiency. Sensitivity accounts for the robustness of the method, that is, the variability of the results with respect to model parameters (e.g., the mesh size, explicit constants). Robust methods are usually sought, since they allow the definition of empirical rules that summarize the practitioner know-how. Efficiency measures the cost that should be paid to reach a given level of accuracy for the class of flows under consideration. The "LES Grail" is therefore a method that should be robust and very efficient for a very wide class of flows.

These considerations show that the concept of "good" LES methods depends strongly on the definition of the error. Thus, a method can be good for one given error measure and bad for another. For example, a method could yield satisfactory results for practical engineering studies (let us say to predict the mean flow within a 1% error) and do a very poor job looking at higher-level statistics (e.g., two-points – two-times correlations). Such a method will certainly be well suited for basic aerodynamics but poorly suited for aeroacoustics. As a consequence, the validation step should be explicitly associated with a clear purpose in terms of future use.

The properties of the ideal LES solution for a given set (h_r, η) are still unknown. The existing literature shows that the mean flow can be accurately recovered, and the

Reynolds stresses can be relevantly approximated on grids that are not too coarse. Researchers have yet to investigate the accuracy of simulating higher-order statistics systematically. LES results have a general trend to "Gaussianize" the fluctuations; that is, the tails of the velocity probability density function become close to those of a Gaussian variable.

3.4.2 Validation procedures

The difficult question of what procedure to use for validation has received two answers.

One answer is to use a priori tests. A direct numerical simulation database (using a given mathematical model for LES) is used to extract pseudo-resolved and pseudo-subgrid modes, and hypotheses or models are assessed in a purely static way. This approach yields interesting results concerning the fundamental nature of nonlinear interactions, but it fails at predicting the true properties of subgrid closures.

Another answer is to use a posteriori tests. These tests consist of running LES simulations by using the LES method to be assessed and comparing the results with targeted ones [i.e., computing $e(h_r, \eta)$]. These are the only dynamic tests. They enable an evaluation of both the sensitivity and the efficiency of a given method.

A last point is that the full validation problem (for a given flow) is equivalent to finding the space of the solutions spanned by a given LES method and finding the solution(s) with the minimum error. The theoretical analysis of this space of possible solutions is still out of the range of mathematical analysis, while systematic a priori tests give an insight only into sensitivity and efficiency. The question of discovering whether a given LES method is unable to yield an arbitrary level of accuracy for a class of flows is still open.

3.5 Open problems in the explicit LES approach

The explicit LES approach raises problems in many different areas – the use of numerical methods, the design of subgrid-scale models, the development of boundary conditions, and so on. Two general problems underlying the very definition of this approach are discussed here.

The first one deals with the regularity properties of the LES solution that is computed by using explicit LES. While no major theoretical problem appears in the simple case of the incompressible velocity field, the case of coupling with thermodynamics raises the question of the realizability of the computed solution. As a matter of fact, numerical experiments show that the explicit LES approach does not guarantee that essential physical constraints will be fulfilled. Well-known examples are overshoots or undershoots of the temperature, density, or pressure, which violate the second law of thermodynamics. In the incompressible flow case, some LES solutions have been observed to develop spurious wiggles in the velocity field.

The associated corresponding problem is twofold. The first one deals with the identification of the true theoretical properties that the LES solution must fulfill. As an example, let us recall that within the filtered Navier–Stokes model framework, the local instantaneous kinetic energy of the resolved scales can be higher than its unfiltered counterpart if a nonpositive filter is employed. Another example is the nature, in terms of total variation, of the LES solution of forced incompressible turbulence at an infinite Reynolds number. Since the instantaneous spectrum can exhibit small oscillations near the cutoff that are due to modulations in the energy cascade in the time–wave number domain Kida and Okhitani (1992), the instantaneous total variation of the LES solution is bounded but unsteady. Therefore, the ad hoc numerical analysis remains to be developed.

The properties of solutions governed by both the physics of the Navier–Stokes equations and the elements of the true LES problem are case dependent and remain to be understood. However, for the sake of usability of LES results for applications, it is convenient to ensure that the LES solutions will share some (if not all) realizability properties of the exact Navier–Stokes equations.

This leads to the second problem, which is related to the use of numerical methods that enforce some desired conditions (wiggle-free solution, no unphysical extrema, etc.). In this case, the numerical method can no longer be considered as "neutral" in the sense defined at the beginning of this chapter, and the problem of the competition between the explicit LES model and the numerical stabilization becomes important.

Since fully satisfactory solutions have yet to be found in the general case, the analysis of the respective influence of these two forcing terms in the discrete problem has been addressed by several authors. Ghosal (1996) performed a theoretical analysis in the incompressible isotropic turbulence case by using the quasi-normal approximation to close the equations. He showed that the amplitude of the dispersive error committed by centered high-order finite-difference schemes (up to the eighth order of accuracy) is larger than the subgrid-scale source term. He also showed that the numerical error dominates the subgrid-scale contribution at all resoved scales. A possible theoretical cure for that is to try to reach grid convergency, which must be understood in this framework as the solution obtained for a fixed-resolution cutoff length h_r in the limit $h \longrightarrow 0$. This limit corresponds to a numerical-error-free solution. Theoretical results indicate that the numerical error will be dominated by subgrid forces using fourth-order-accurate centered schemes and $h_r / h \geq 2$. Numerical experiments confirmed that the subgrid-scale force is overwhelmed by the numerical error. In practice, grid convergence is achieved by ensuring an explicit control of the *effective filter* that governs discrete solution u_h. A common solution is to apply a discrete filter to the solution during the computation to mimic the filtered Navier–Stokes mathematical model. This option gives an explicit control of h_r through the definition of the filter cutoff length, Δ. Another solution was proposed by Magnient, Sagaut, and Delville (2001), who obtained an implicit control of the effective filter by manipulating the value Δ of the characteristic length that must be prescribed in subgrid-viscosity models. For sufficiently large values of the ratio Δ/h, numerical-error-free solutions were obtained. A last solution is to rely on the numerical error to govern the simulation, corresponding to the implicit LES approach.

Some comments must be made at this point.

1. The concept of grid-converged LES was developed for incompressible flows with the purpose of minimizing the influence of the numerical scheme. Therefore, it is not fully adequate to evaluate the problem associated with the enforcement of desired realizability properties in the LES solution.

2. It was reported that, in some cases, some partial error cancellation occurs, minimizing total error $e(h_r, \eta)$. In this case, the use of higher-order schemes yields worse results.

3. The theoretical predictions dealing with the improvement of the results are not automatically observed in numerical experiments, and it appears in the few available works that the subgrid-scale dissipation must be diminished to improve the results. The most probable explanation is that usual subgrid-scale models cannot account for interactions (energy cascade) between scales larger than h_r and scales smaller than h_r but larger than h. The theoretical results may be recovered if all scales smaller than h_r are set to zero, corresponding to the application of the sharp cutoff filter with cutoff wave number π/h_r. Such a filter cannot be implemented in the general case.

4. The theoretical analysis deals with the amplitude of the numerical error, but it does not take its nature (dispersive, dissipative) into account, and numerical simulations show that a very satisfactory prediction of the energy spectrum in forced isotropic turbulence can be obtained with a centered second-order scheme. We can explain this by recalling that the dispersive error is neutral with respect to the kinetic energy evolution. Therefore, the concept of a neutral method seems more appropriate than the error-free one to evaluate the need within the LES framework.

The related issue of defining a quantitative measure for the numerical error that would provide an accurate insight into its competition with the subgrid model action is also an open problem. Garnier et al. (1999) introduced the generalized Smagorinsky constant concept to measure the relative dissipation. This constant, C_g, is defined as the value of the constant in the Smagorinksy constant for which subgrid dissipation ε_t given by (3.34) is equal to the numerical dissipation, ε_h. Values larger than 0.18 (i.e., the theoretical value of the constant in isotropic turbulence) indicate that the numerical dissipation is dominant. Results reported by these authors show that shock-capturing-scheme dissipation masks the dissipation typical of the explicit subgrid model, even when fifth-order-accurate WENO schemes are considered. Extensive analyses of the subgrid modeling error and its sensitivity with respect to numerical error have been carried out by Geurts and coworkers (Geurts and Froelich 2001, 2002; Meyers, Geurts, and Baelmans 2003) and show that a strong nonlinear interaction exists. An important consequence is that the improvement of the reliability of explicit LES results requires work on both the numerical method and the subgrid model.

3.6 Appendix: The filtered Navier–Stokes equations

This appendix is devoted to the presentation of the filtered Navier–Stokes equations model for LES. The procedure used here is exactly the single-filter-level procedure described in Section 3.1.2. The equations of the twice-filtered models can be obtained in the same way. The convolution filter is supposed to commute with all space and time derivatives.

3.6.1 Incompressible flows

The filtered Navier–Stokes equations read as

$$\frac{\partial \overline{u}}{\partial t} + \nabla \cdot (\overline{u \otimes u}) = -\nabla \overline{p} + \nu \nabla^2 \overline{u}, \tag{3.63}$$

$$\nabla \cdot \overline{u} = 0. \tag{3.64}$$

The nonlinear term $\overline{u \otimes u}$ must be decomposed as a function of the acceptable unknowns, namely $\overline{u}$ and u'. The following decomposition was proposed by Leonard (1974). One obtains it by inserting decomposition $u = \overline{u} + u'$ into the nonlinear term, yielding

$$\overline{u \otimes u} = \overline{(\overline{u} + u') \otimes (\overline{u} + u')} \tag{3.65}$$

$$= \underbrace{\overline{\overline{u} \otimes \overline{u}}}_{\text{resolved}} + \underbrace{\overline{\overline{u} \otimes u'} + \overline{u' \otimes \overline{u}}}_{C:\text{Cross terms}} + \underbrace{\overline{u' \otimes u'}}_{R:\text{Reynolds stresses}}. \tag{3.66}$$

The resolved term can be expressed as

$$\overline{\overline{u} \otimes \overline{u}} = \underbrace{\overline{u} \otimes \overline{u}}_{\text{new resolved}} + \underbrace{(\overline{\overline{u} \otimes \overline{u}} - \overline{u} \otimes \overline{u})}_{L:\text{Leonard stress tensor}}, \tag{3.67}$$

showing the appearance of three tensors:

- the Leonard tensor, L, which corresponds to the fluctuations of the interactions between resolved scales (zero for RANS);
- the cross stress tensor, C, which accounts for direct interactions between resolved and unresolved scales (zero for RANS); and
- the subgrid Reynolds stress tensor R, which is associated with the action of subgrid scales or the resolved field (Reynolds tensor for RANS).

Two possibilities arise for the definition of the subgrid-scale tensor, τ, which depend on the choice of the formulation of the resolved convection term. The first one is

$$\tau = C + R \tag{3.68}$$

and corresponds to a resolved nonlinear term of the form $\overline{\overline{u} \otimes \overline{u}}$. The second is

$$\tau = L + C + R, \tag{3.69}$$

with a resolved convection term of the form $\overline{u} \otimes \overline{u}$.

These two decompositions can be used, but they introduce some interesting conceptual problems. Consider the philosophy of LES: Filtered equations are derived, and all the terms appearing in the equations must be filtered terms (i.e., appear as the filtered part of something). Only the first decomposition satisfies that condition. This is particularly true when the filtering operator is associated with the definition of a computational grid (and *bar* just means "defined on the grid"): The convection term is computed on the same grid as the filtered variables, and then should be written as $\overline{\overline{u} \otimes \overline{u}}$. Another point is that, in the first decomposition, neither the subgrid tensor nor the resolved convection term is invariant under Galilean transformations (but their sum is invariant), while in the second decomposition both terms are invariant. A more accurate analysis reveals that R and $L + C$ are invariant for this class of transfomation.

3.6.2 Compressible flows

3.6.2.1 Definition of the filtered variables

The definition of the filtered set of equations results from three preliminary choices:

- the original set of unfiltered variables or equations;
- the filter (same low-pass filter as in the incompressible case); and
- the set of filtered variables.

A first question arises from the fact that we can find a very large number of formulations of the compressible Navier–Stokes equations, which rely on the choice of basic variables for describing the flow. While the momentum and density equations are usually the same, several forms of the energy equations are encountered: entropy, total energy, internal energy, enthalpy, and so on. This point, which is the most difficult one when we are dealing with compressible flows, is discussed now. Consider the set of conservative variables $(\rho, \rho u, \rho E)$. A straightforward use of the filtering procedure for incompressible flows leads to the definition of the filtered variables $(\overline{\rho}, \overline{\rho u}, \overline{\rho E})$. The term $\overline{\rho u}$ can be rewritten as

$$\overline{\rho u} \equiv \overline{\rho} \widetilde{u}, \tag{3.70}$$

where $\widetilde{u} = \overline{\rho u}/\overline{\rho}$ is the mass-weighted filtered velocity. It is worthwhile noting that the tilde operator cannot be interpreted as a convolution-like filtering operator, and it does not commute with derivatives. It corresponds only to a change of variable.[§]

As the conserved variables, $\overline{\rho u}$, $\overline{\rho E}$, and $\overline{p}$, are nonlinear functions of the other variables that can be decomposed as functions of the other variables, the choice of the basic variables for the LES problem is not unique. Depending on that choice, different subgrid terms will arise from the filtered equations, inducing a need for specific subgrid modeling. There is an infinite number of possibilities for the definition of the variables of the filtered problem, starting from $(\overline{\rho}, \overline{\rho u}, \overline{\rho E})$.

[§] It is worth noting that this mass-weighting approach was proposed by Osborne Reynolds in his 1884 paper.

Here the solution proposed in certain references is presented (Sreedhar and Ragab 1998; Vreman 1995; Vreman et al. 1997). The selected variables are $(\overline{\rho}, \widetilde{u}, \widehat{E})$, supplemented by filtered pressure $\overline{p}$ and mass-weighted filtered temperature $\widetilde{T}$. The specific total energy, $\widehat{E}$, is defined as

$$\widehat{E} = \frac{\overline{p}}{\gamma - 1} + \overline{\rho}\widetilde{u}^2$$

and corresponds to the computable part of the total energy. Using this set of variables, we find that no subgrid contribution appears in the filtered equation of state.

The corresponding set of equations is

$$\frac{\partial \overline{\rho}}{\partial t} + \nabla \cdot (\overline{\rho}\widetilde{u}) = 0, \tag{3.71}$$

$$\frac{\partial \overline{\rho}\widetilde{u}}{\partial t} + \nabla \cdot (\overline{\rho}\widetilde{u}\widetilde{u}) + \nabla \overline{p} - \nabla \cdot \widehat{\tau}_v = A_1 + A_2, \tag{3.72}$$

$$\frac{\partial \overline{\rho}\widehat{E}}{\partial t} + \nabla \cdot \left[\widetilde{u} \left(\overline{\rho}\widehat{E} + \overline{p} \right) \right] - \nabla \cdot (\widetilde{u}\widehat{\tau}_v) + \nabla \cdot \widehat{q}_T$$
$$= -B_1 - B_2 - B_3 + B_4 + B_5 + B_6 - B_7, \tag{3.73}$$

where $\widehat{\tau}_v$ and $\widehat{q}_T$ are the viscous stresses and the heat flux computed from the resolved variables and are equal to their respective filtered counterparts if the viscosity and the diffusivity are assumed to be constant. The subgrid terms, $A_1, A_2, B_1, \ldots, B_7$, are defined as

$$\tau \equiv \overline{\rho}(\widetilde{u \otimes u} - \widetilde{u} \otimes \widetilde{u}), \quad A_1 = -\nabla \cdot \tau, \quad A_2 = \nabla \cdot (\overline{\tau}_v - \widehat{\tau}_v), \tag{3.74}$$

$$B_1 = \frac{1}{\gamma - 1}\nabla \cdot (\overline{pu} - \overline{p}\widetilde{u}), \quad B_2 = \overline{p\nabla \cdot u} - \overline{p}\nabla \cdot \widetilde{u}, \quad B_3 = \nabla \cdot (\tau\widetilde{u}), \quad B_4 = \tau\nabla \cdot \widetilde{u}, \tag{3.75}$$

$$B_5 = \overline{\tau_v \nabla u} - \overline{\tau}_v \nabla \widetilde{u}, \quad B_6 = \nabla \cdot (\overline{\tau_v u} - \widehat{\tau}_v \widetilde{u}), \quad B_7 = \nabla \cdot (\overline{q}_T - \widehat{q}_T). \tag{3.76}$$

Terms A_2, B_6, and B_7 arise from the possible dependency of molecular viscosity and diffusivity on the temperature, and they correspond to new nonconvective sources of nonlinearity. The use of mass-weighted filtered variable $\widetilde{u}$ instead of filtered velocity $\overline{u}$ prevents the appearance of subgrid terms in the continuity equation.

REFERENCES

Bagget, J. S., Jimenez, J., & Kravchenko, A. G. 1997 Resolution requirements in large-eddy simulations of shear flows, in *CTR Annual Research Briefs*. Stanford, CA: NASA Ames/Stanford University, 51–66.

Balaras, E., Benocci, C., & Piomelli, U. 1996 Two-layer approximate boundary conditions for large-eddy simulations. *AIAA* **34**, 1111–1119.

Bardina, J., Ferziger, J. H., & Reynolds, W. C. 1983 *Improved turbulence models based on large eddy simulation of homogeneous, incompressible, turbulent flows*. Thermosciences Division, Stanford University: Report TF-19.

Cabot, W., & Moin, P. 2000 Approximate wall boundary conditions in the large-eddy simulation of high Reynolds number flow *Flow Turb. Combust.* **63**, 269–291.

Carati, D., Rogers, M., & Wray, A. 2002 Statistical ensemble of large-eddy simulations. *J. Fluid Mech.* **455**, 195–212.

Carati, D., Winckelmans, G., & Jeanmart, H. 2001 On the modeling of the subgrid-scale and filtered-scale stress tensors in large-eddy simulation. *J. Fluid Mech.* **441**, 119–138.

Domaradzki, J. A., & Adams, N. A. 2002 Direct modeling of subgrid scales of turbulence in large eddy simulations. *J. Turb.* **3**, 1–19.

Druault, P., Lardeau, S., Bonnet, J. P., Coiffet, F., Delville, J., Lamballis, E., Largeau, J. F., & Perret, L. 2004 Generation of three-dimensional turbulent inlet conditions for large-eddy simulation. *AIAA J.* **42**, 447–456.

Erlebacher, G., Hussaini, M. Y., Kreiss, M., & Sarkar, S. 1990 The analysis and simulation of compressible turbulence. *Theor. Comput. Fluid Dyn.* **2**, 73–95.

Erlebacher, G., Hussaini, M. Y., Speziale, C., & Zang, T. A. 1992 Toward the large-eddy simulation of compressible turbulent flows. *J. Fluid Mech.* **238**, 155–185.

Fureby, C., & Tabor, G. 1997 Mathematical and physical constraints on large-eddy simulations. *Theor. Comput. Fluid Dyn.* **9**, 85–102.

Garnier, E., Mossi, M., Sagaut, P., Deville, M., & Comte, P. 1999 On the use of shock-capturing schemes for large-eddy simulation. *J. Comput. Phys.* **153**, 273–311.

George, W. K., & Davidson, L. 2004 Role of initial conditions in establishing asymptotic flow behavior. *AIAA J.* **42**, 438–446.

Germano, M., Piomelli, U., Moin, P., & Cabot, W. H. 1991 A dynamic subgrid-scale eddy viscosity model. *Phys. Fluids A* **3**, 1760–1765.

Geurts, B. J., & Froelich, J. 2001 Numerical effects contaminating LES; a mixed story, in *Modern Simulation Strategies for Turbulent Flow*, ed. B. Geurts. Philadelphia: Edwads, 309–327.

Geurts, B. J., & Froehlich, J. 2002 A framework for predicting accuracy limitations in large-eddy simulation. *Phys. Fluids* **14**, L41–L44.

Ghosal, S. 1996 An analysis of numerical errors in large-eddy simulations of turbulence. *J. Comput. Phys.* **125**, 187–206.

Ghosal, S., & Moin, P. (1995). The basic equations for the large-eddy simulation of turbulent flows in complex geometry. *J. Comput. Phys.* **118**, 24–37.

Hughes, T. J. R. 1995 Multiscale phenomena: Green's function, the Dirichlet-to-Neumann formulation, subgrid scale models, bubbles and the origin of stabilized methods. *Comput. Methods Appl. Mech. Eng.* **127**, 387–401.

Hughes, T. J. R., Feijoo, G. R., Mazzei, L., & Quincy, J. B. 1998 The variational multiscale method – a paradigm for computational mechanics. *Comput. Methods Appl. Mech. Eng.* **166**, 2–24.

Hughes, T. J. R., Mazzei, L., & Jansen, K. E. 2000 Large eddy simulation and the variational multiscale method. *Comput. Visual Sci.* **3**, 47–59.

Hughes, T. J. R., Mazzei, L., Oberai, A. A., & Wray, A. A. 2001 The multiscale formulation of large-eddy simulation: Decay of homogeneous isotropic turbulence. *Phys. Fluids* **13**, 505–512.

Jordan, S.A. 1999 A large-eddy simulation methodology in generalized curvilinear coordinates. *J. Comput. Phys.* **148**, 322–340.

Jordan, S. A. 2001 Dynamic subgrid-scale modeling for large-eddy simulations in complex topologies. *J. Fluids Eng.* **123**, 619–627.

Kaneda, Y., Ishihara, T., Yokokawa, M., Itakura, K., & Uno, A. 2003 Energy dissipation rate and energy spectrum in high resolution direct numerical simulations of turbulence in a periodic box. *Phys. Fluids* **15**, L21–L24.

Kida, S., & Okhitani, K. 1992 Fine structure of energy transfer in turbulence. *Phys. Fluids A* **4**, 1602–1604.

Knaepen, B., Debliquy, O., & Carati, D. 2002 Subgrid-scale energy and pseudo pressure in large-eddy simulation. *Phys. Fluids* **14**, 4235–4241.

Leonard, A. 1974 Energy cascade in large-eddy simulations of turbulent fluid flows. *Adv. Geophys. A* **18**, 237–248.

Lesieur, M., & Métais, O. 1996 New trends in large-eddy simulations of turbulence. *Annu. Rev. Fluid Mech.* **28**, 45–82.

Lilly, D. K. 1992 A proposed modification of the Germano subgrid-scale closure method. *Phys. Fluids A* **4**, 633–635.

Lund, T. S., Wu, X., & Squires, K. D. 1998 On the generation of turbulent inflow conditions for boundary-layer simulations. *J. Comput. Phys.* **140**, 233–258.

Magnient, J. C., Sagaut, P., & Deville, M. 2001 A study of built-in filter for some eddy viscosity models in large-eddy simulation. *Phys. Fluids* **13**, 1440–1449.

McComb, W. D., Hunter, A., & Johnston, C. 2001 Conditional mode-elimination and the subgrid-modeling problem for isotropic turbulence. *Phys. Fluids* **13**, 2030–2044.

Meyers, J., Geurts, B., & Baelmans, M. 2003 Database analysis of errors in large-eddy simulation. *Phys. Fluids* **15**, 2740–2755.

Nicoud, F., Bagget, J. S., Moin, P., & Cabot, W. 2001 Large-eddy simulation wall-modeling based on sub-optimal control theory and linear stochastic estimation. *Phys. Fluids* **13**, 2968–2984.

Nicoud, F., & Ducros, F. 1999 Subgrid stress modeling based on the square of the velocity gradient tensor. *Flow Turb. Combust.* **62**, 183–200.

Pruett, C. D. 2000 Eulerian time-domain filtering for spatial large-eddy simulation. *AIAA J.* **38**, 1634–1642.

Pruett, C. D., Gatski, T. B., Grosch, C. E., & Thacker, W. D. 2003 The temporally filtered Navier–Stokes equations: Properties of the residual stress. *Phys. Fluids* **15**, 2127–2140.

Sagaut, P. 2005 *Large-Eddy Simulation for Incompressible Flows*. Berlin: Springer, 3rd ed.

Salvetti, M. V., & Banerjee, S. 1995 A priori tests of a new dynamic subgrid-scale model for finite-difference large-eddy simulations. *Phys. Fluids* **7**, 2831–2847.

Samtaney, R., Pullin, D., & Kosovic, B. 2001 Direct numerical simulation of decaying compressible turbulence and shocklets statistics. *Phys. Fluids* **13**, 1415–1430.

Sandham, N., Yao, Y., & Lawal, A. 2003 Large-eddy simulation of the flow over a bump. *Int. J. Heat Fluid Flow* **24**, 584–595.

Schumann, U. 1975 Subgrid scale model for finite difference simulations of turbulent flows in plane channels and annuli. *J. Comput. Phys.* **18**, 376–404.

Scotti, A., & Meneveau, C. 1997 Fractal model for coarse-grained nonlinear partial differential equations. *Phys. Rev. Lett.* **78**, 867–870.

Shao, L., Sarker, S., & Patano, C. 1999 On the relationship between the mean flow and subgrid stresses in large eddy simulation of turbulent shear flow. *Physics of Fluids* **11**, 1229–1248

Smagorinsky, J. 1963 General circulation experiments with the primitive equations. I: The basic experiment. *Mon. Weather Rev.* **91**, 99–165.

Sreedar, M., & Stern, F. 1998 Larg eddy simulation of temporally developing juncture flows. *Int. J. Numer. Fluids* **28**, 47–72.

Terracol, M., Sagaut, P., & Basdevant, C. 2001 A multilevel algorithm for large-eddy simulation of turbulent compressible flows. *J. Comput. Phys.* **167**, 439–474.

Vasilyev, O., Lund, T. S., & Moin, P. 1998 A general class of commutative filters for LES in complex geometries. *J. Comput. Phys.* **146**, 82–104.

Voelkl, T., Pullin, D. I., & Chan, D. C. 2000 A physical-space version of the stretched-vortex subgrid-stress model for large-eddy simulation. *Phys. Fluids* **12**, 1810–1825.

Vreman, A. 2003 The filtering analog of the variational multiscale method in large-eddy simulation. *Phys. Fluids* **15**, L61–L64.

Vreman, B., Geurts, B., & Kuerten, H. 1995 Subgrid modeling in LES of compressible flow. *Applied Scientific Research* **54**, 191–203.

Vreman, B., Geurts, B., & Kuerten, H. 1997 Large eddy simulation of the turbulent mixing layer. *J. Fluid Mechanics* **339**, 357–390.

Wang, M., & Moin, P. 2000 Computation of trailing-edge flow and noise using large eddy simulation, *AIAA J.* **38**, 2201–2209.

Yoshizawa, A. 1991 A statistically-derived subgrid model for the large-eddy simulation of turbulence. *Phys. Fluids A* **3**, 2007–2009.

4 Numerics for ILES

4a Limiting Algorithms

Dimitris Drikakis and Marco Hahn,
Fernando F. Grinstein and Carl R. DeVore,
Christer Fureby and Mattias Liefvendahl,
and David L. Youngs

4a.1 Introduction

Large eddy simulation (LES) has emerged as the next-generation simulation tool for handling complex engineering, geophysical, astrophysical, and chemically reactive flows. As LES moves from being an academic tool to being a practical simulation strategy, the robustness of the LES solvers becomes a key issue to be concerned with, in conjunction with the classical and well-known issue of accuracy. For LES to be attractive for complex flows, the computational codes must be readily capable of handling complex geometries. Today, most LES codes use hexahedral elements; the grid-generation process is therefore cumbersome and time consuming. In the future, the use of unstructured grids, as used in Reynolds-averaged Navier–Stokes (RANS) approaches, will also be necessary for LES. This will particularly challenge the development of high-order unstructured LES solvers. Because it does not require explicit filtering, Implicit LES (ILES) has some advantages over conventional LES; however, numerical requirements and issues are otherwise virtually the same for LES and ILES. In this chapter we discuss an unstructured finite-volume methodology for both conventional LES and ILES, that is particularly suited for ILES. We believe that the next generation of practical computational fluid dynamics (CFD) models will involve structured and unstructured LES, using high-order flux-reconstruction algorithms and taking advantage of their built-in subgrid-scale (SGS) models.

ILES based on functional reconstruction of the convective fluxes by use of high-resolution hybrid methods is the subject of this chapter. We use modified equation analysis (MEA) to show that the leading-order truncation error terms introduced by such methods provide implicit SGS models similar in form to those of conventional mixed SGS models. Major properties of the implicit SGS model are related to the following: (i) the choice of high- and low-order schemes – where the former is well behaved in smooth flow regions and the latter is well behaved near sharp gradients; (ii) the choice of flux limiter, which determines how these schemes are blended locally, depending on the flow; and (iii) the balance of the dissipation and dispersion contributions to the numerical

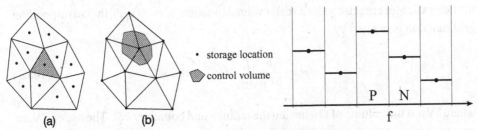

Figure 4a.1. Control volume variants of the FVM (left): cell-centered (a) and vertex-centered (b) CV tessellation. Variations of the dependent variables in the FVM (right). Legend: (· · · · · ·) exact (continuous) variation, (- - -) piecewise linear variation and (—) piecewise constant variation.

solution, which depend on the design details of each numerical method. Finally, we assess the performance of different hybrid algorithms suitable for ILES in the case of the Taylor–Green vortex problem (Brachet et al. 1983), comparing them with conventional LES and previously reported direct numerical simulation (DNS) data. We discuss flow transition and turbulence decay as well as the ability of various different ILES models to capture the correct physics, using comparisons with established theoretical findings for the kinetic energy dissipation, energy spectra, enstrophy, and kinetic energy decay.

4a.2 Finite-volume discretization of the Navier–Stokes equation

In the finite-volume method (FVM), the integral formulation of the governing equations is discretized directly in physical space. The computational domain, $D \subset R^d$, is first tessellated into a collection of nonoverlapping control volumes (CVs) covering the domain and defining the grid. Let T denote the tessellation of D, with control volumes $\Omega_P \in T$ such that $\cup_{\Omega_P \in T} \overline{\Omega}_P = \overline{D}$. The choice of CV tessellation is very flexible in the FVM, as shown in Figure 4a.1. The tessellation can be structured or unstructured: in a structured grid the CVs are arranged in an array structure. Location of neighboring CVs is implicit in the array indices, which thus allows for efficient storage and book-keeping of CV data. An unstructured grid has no inherent ordering of the CVs, and hence the arrangement of the CVs must be specified explicitly through the connectivity. Unstructured grids typically allow for greater flexibility in generating and adapting grids but at the expense of increased storage of cell information. Furthermore, we usually distinguish between cell-centered and vertex-centered FVMs. In the cell-centered FVM, the tessellation elements themselves serve as CVs, with unknowns stored on a per-tessellation-element basis. In the vertex-centered FVM, CVs are formed as a geometrical dual to the tessellation element complex, and solution unknowns are therefore stored on a per-dual-element basis.

Central to the FVM is that the values of the dependent variables (e.g., ρ, $\mathbf{v}$, E), usually considered to be continuous (or smooth) functions of the independent variables $(\mathbf{x}, t)$, will be represented by CV averages for each $\Omega_P \in T$, with Ω_P centered around the point $\mathbf{x}_P$. More precisely, for a scalar valued function $f = f(\mathbf{x}, t)$, the corresponding grid function is defined by

$$f_P = \frac{1}{\delta V_P} \int_{\Omega_P} f(\mathbf{x}, t)\, dV, \tag{4a.1}$$

where δV_P is the volume of Ω_P having the radius r and boundary $\partial \Omega$. The use of CV averages ensures conservation in a way that mimics the exact solution, since $\sum_{\forall P \in D} (f_P)$ approximates the integral of f over D, and if a conservative scheme is used, this sum will only change as a result of the boundary fluxes. Accordingly, the total mass will be preserved, or at least vary correctly, provided that the boundary conditions are properly imposed. Godunov (1959) pursued this in the discretization of the Euler equations by assuming a piecewise constant solution in each CV, with values equal to the CV average. However, the use of a piecewise constant representation renders the solution multivalued at CV interfaces, making the calculation of a single flux at these interfaces difficult. Instead, as we elaborate on later, and as suggested by Godunov (1959), we introduce a (Lipschitz continuous) numerical flux function for this purpose.

We can derive discrete FVM approximations to the differential operators by combining the divergence theorem (Gurtin 1981) with a localization theorem (Kellogg 1929); for example, the gradient of f can be expressed as $\nabla f = \lim_{r \to 0} \frac{1}{\delta V} \int_{\partial \Omega} (f) d\mathbf{A}$, where $d\mathbf{A} = \mathbf{n} dA$ is the area element of integration and $\mathbf{n}$ is the outward pointing unit normal to Ω. If we identify Ω with control volume Ω_P, located around point $\mathbf{x}_P$, and replace the surface integrals with a discrete summation over the cell faces enclosing control volume Ω_P, then it follows that $\nabla f \approx \frac{1}{\delta V_P} \sum_f [f d\mathbf{A}]_f$. Similarly, if $\mathbf{f}$ is a vector field and $\mathbf{F}$ a tensor field, then $\nabla \mathbf{f} \approx \frac{1}{\delta V_P} \sum_f [\mathbf{f} \otimes d\mathbf{A}]_f$, $\nabla \cdot \mathbf{f} \approx \frac{1}{\delta V_P} \sum_f [\mathbf{f} \cdot d\mathbf{A}]_f$, $\nabla \cdot \mathbf{F} \approx \frac{1}{\delta V_P} \sum_f [\mathbf{F} d\mathbf{A}]_f$, and $\nabla \times \mathbf{f} \approx \frac{1}{\delta V_P} \sum_f [\mathbf{f} \times d\mathbf{A}]_f$, respectively. Here, $\cdot$ denotes the scalar product, $\otimes$ the tensor product, and $\times$ the cross product. Given the vectors $\mathbf{a}$, $\mathbf{b}$, and $\mathbf{c}$, expressed as $\mathbf{a} = a_i \mathbf{e}_i$, $\mathbf{b} = b_i \mathbf{e}_i$, and $\mathbf{c} = c_i \mathbf{e}_i$ in a Cartesian basis $\{\mathbf{e}_i\}$, the scalar product is defined as $\mathbf{a} \cdot \mathbf{b} = a_i b_i$; the tensor product is defined through the transformation $(\mathbf{a} \otimes \mathbf{b})\mathbf{c} = (\mathbf{b} \cdot \mathbf{c})\mathbf{a}$, which implies that $(\mathbf{a} \otimes \mathbf{b})_{ij} = a_i b_j$; and the vector product is defined as $\mathbf{a} \times \mathbf{b} = a_i b_j \varepsilon_{ijk} \mathbf{e}_k$, where ε_{ijk} is the Levi–Civita ε tensor (Gurtin 1981).

By making use of the aforementioned approximations to the spatial differential operators, we can transform the compressible Navier–Stokes equation (NSE) into the semidiscretized compressible NSE,

$$\partial_t(\rho_P) + \frac{1}{\delta V_P} \sum_f \left[F_f^C \right] = 0,$$

$$\partial_t(\rho_P \mathbf{v}_P) + \frac{1}{\delta V_P} \sum_f \left[\mathbf{F}_f^{C,v} - \mathbf{F}_f^{D1,v} + \mathbf{F}_f^{D2,v} \right] = -(\nabla p)_P + (\rho \mathbf{f})_P, \tag{4a.2}$$

$$\partial_t(\rho_P E_P) + \frac{1}{\delta V_P} \sum_f \left[\mathbf{F}_f^{C,E} - \mathbf{F}_f^{D1,E} + \mathbf{F}_f^{D2,E} + \mathbf{F}_f^{D3,E} \right] = (\rho \sigma)_P,$$

where the convective, diffusive, and additional fluxes are defined by

$$F_f^C = (\rho \mathbf{v} \cdot d\mathbf{A})_f, \quad \mathbf{F}_f^{C,v} = (\rho \mathbf{v} \otimes \mathbf{v})_f d\mathbf{A}_f = (\rho \mathbf{v} \cdot d\mathbf{A})_f \mathbf{v}_f = F_f^C \mathbf{v}_f,$$

$$\mathbf{F}_f^{D1,v} = (\mu \nabla \mathbf{v})_f d\mathbf{A}_f, \tag{4a.3a}$$

$$\mathbf{F}_f^{D2,v} = \left[\frac{1}{3}\mu(\nabla \cdot \mathbf{v})\right]_f d\mathbf{A}_f, \quad \mathbf{F}_f^{C,E} = (\rho \mathbf{v} \cdot d\mathbf{A})_f E_f = F_f^C E_f,$$

$$\mathbf{F}_f^{D1,E} = \left(\frac{\kappa}{c_V}\nabla E\right)_f d\mathbf{A}_f, \tag{4a.3b}$$

$$\mathbf{F}_f^{D2,E} = (-p\mathbf{v})_f d\mathbf{A}_f, \quad \mathbf{F}_f^{D3,E} = \left[\frac{1}{3}(\nabla \cdot \mathbf{v})\mathbf{I} - \mu(\nabla \mathbf{v} + \nabla \mathbf{v}^T)\right]_f d\mathbf{A}_f. \tag{4a.3c}$$

Similarly, we can spatially discretize the incompressible NSE as

$$\frac{1}{\delta V_P}\sum_f [F_f^C] = 0,$$

$$\partial_t(\mathbf{v}_P) + \frac{1}{\delta V_P}\sum_f \left[\mathbf{F}_f^{C,v} - \mathbf{F}_f^{D,v}\right] = (\nabla p)_P + \mathbf{f}_P, \tag{4a.4}$$

where the convective and diffusive fluxes are defined by

$$F_f^C = (\mathbf{v} \cdot d\mathbf{A})_f, \quad \mathbf{F}_f^{C,v} = (\mathbf{v} \otimes \mathbf{v})_f d\mathbf{A}_f = (\mathbf{v} \cdot d\mathbf{A})_f \mathbf{v}_f = F_f^C \mathbf{v}_f, \quad \mathbf{F}_f^{D,v} = (v\nabla \mathbf{v})_f d\mathbf{A}_f.$$

$$\tag{4a.5}$$

The semidiscretized equations, (4a.2) and (4a.4), have to be integrated in time, and rules must be prescribed for the fluxes in Eqs. (4a.3) and (4e.6) to obtain a fully discrete set of equations that can be solved numerically to give an approximation to the solution of the NSE.

4a.3 Time integration and solution algorithms

Since the objective of LES is to find time-accurate solutions to the LES equations, we need a balanced time-accurate method in which the errors caused by the time advancement approximately equal those introduced by the finite-volume discretization. In principle, any method for integrating ordinary differential equations can be used to integrate the semidiscretized LES equations. However, here we limit the discussion to multistep and Runge–Kutta methods (Lambert 1973; Gottlieb and Shu 1998; Jameson, Schmidt, and Turkel 1981; van der Houwen 1977), since they are found to behave satisfactorily for unsteady flows, and for LES in particular. Besides introducing these schemes we consider solution algorithms for the compressible and incompressible NSE based on the FVM and the aforementioned time-integration methods.

Applying a multistep method (Lambert 1973) to the semidiscretized compressible NSE, Eq. (4a.2), yields

Table 4a.1. *Multistep time-integration parameters*

Scheme	m	α_0	α_1	α_2	β_0	β_1	β_2	Order	Stability
Euler explicit	1	-1	1	–	1	0	–	1	A stable
Euler backward	1	-1	1	–	0	1	–	1	A stable
Crank–Nicholson	1	$-1/2$	$1/2$	–	$1/2$	$1/2$	–	2	A stable
Two-point backward differencing	2	$1/2$	-2	$3/2$	0	0	1	2	A stable

$$\sum_k \left(\alpha_k (\rho)_P^{n+k} + \beta_k \frac{\Delta t}{\delta V_P} \sum_f [F_f^C]_f^{n+k} \right) = 0, \tag{4a.6a}$$

$$\sum_{i=0}^{m} \left(\alpha_i (\rho \mathbf{v})_P^{n+i} + \frac{\beta_i \Delta t}{\delta V_P} \sum_f \left[\mathbf{F}_f^{C,v} + \mathbf{F}_f^{D1,v} + \mathbf{F}_f^{D2,v} \right]^{n+i} \right)$$
$$= \Delta t \sum_{i=0}^{m} \left(\beta_i \left(-(\nabla p)_P^{n+i} + (\rho \mathbf{f})_P^{n+i} \right) \right), \tag{4a.6b}$$

$$\sum_{i=0}^{m} \left(\alpha_i (\rho E)_P^{n+i} + \frac{\beta_i \Delta t}{\delta V_P} \sum_f \left[\mathbf{F}_f^{C,E} + \mathbf{F}_f^{D1,E} + \mathbf{F}_f^{D2,E} + \mathbf{F}_f^{D3,E} \right]^{n+i} \right)$$
$$= \Delta t \sum_{i=0}^{m} \left(\beta_i (\rho \sigma)_P^{n+i} \right), \tag{4a.6c}$$

where α_i and β_i are parameters satisfying $\sum_{i=0}^{m} \alpha_i = 0$ and $\sum_{i=0}^{m} i \cdot \alpha_i = \sum_{i=0}^{m} \beta_i$. Note that for the multistep method, the resulting scheme is explicit if $\beta_i = 0$ for all $i \neq 0$; otherwise, it is implicit. Similarly, applying the multistep method to the semidiscretized incompressible NSE, Eq. (4a.4), gives

$$\frac{\beta_i \Delta t}{\delta V_P} \sum_f [F_f^C]^{n+i} = 0, \tag{4a.7a}$$

$$\sum_{i=0}^{m} \left(\alpha_i \mathbf{v}_P^{n+i} + \frac{\beta_i \Delta t}{\delta V_P} \sum_f \left[\mathbf{F}_f^{C,v} - \mathbf{F}_f^{D,v} \right]^{n+i} \right) = \Delta t \sum_{i=0}^{m} \left(\beta_i \left(-(\nabla p)_P^{n+i} + \mathbf{f}_P^{n+i} \right) \right). \tag{4a.7b}$$

Table 4a.1 presents numerical parameters m, α_i, and β_i for some selected multistep methods.

The basic idea of the Runge–Kutta methods is to evaluate the right-hand side of the equation at hand at several intermediate values in $[t, \ t + \Delta t]$, and finally to combine them in order to obtain a high-order approximation at $t + \Delta t$ (Gear 1971). For example,

$$\mathbf{U}_P^{(1)} = \mathbf{U}_P^n,$$
$$\mathbf{U}_P^{(2)} = \mathbf{U}_P^n + \alpha_2 \left\{ \frac{1}{\delta V_P} \sum_f \left[\mathbf{F}_f^C(\mathbf{U}^{(1)}) + \mathbf{F}_f^{D1}(\mathbf{U}^{(1)}) + \cdots \right] - \mathbf{S}(\mathbf{U}_P^{(1)}) \right\} \Delta t,$$
$$\cdots$$
$$\mathbf{U}_P^{(K)} = \mathbf{U}_P^n + \alpha_K \left\{ \frac{1}{\delta V_P} \sum_f \left[\mathbf{F}_f^C(\mathbf{U}^{(K-1)}) + \mathbf{F}_f^{D1}(\mathbf{U}^{(K-1)}) + \cdots \right] - \mathbf{S}\left(\mathbf{U}_P^{(K-1)} \right) \right\} \Delta t,$$
$$\mathbf{U}_P^{n+1} = \mathbf{U}_P^n + \sum_{k=1}^{K} \left\{ \beta_k \frac{1}{\delta V_P} \sum_f \left[\mathbf{F}_f^C(\mathbf{U}^{(k)}) + \mathbf{F}_f^{D1}(\mathbf{U}^{(k)}) + \cdots \right] - \mathbf{S}\left(\mathbf{U}_P^{(k)} \right) \right\} \Delta t. \tag{4a.8}$$

where $\mathbf{U} = [\rho, \ \rho\mathbf{v}, \ \rho E]^T$, and α_i and β_i are coefficients of the scheme with $\sum_{k=1}^{K} \beta_k = 1$. Runge–Kutta methods are usually favored for explicit algorithms for compressible NSE, whereas multistep methods often are used for the semi-implicit algorithms for the incompressible NSE.

4a.3.1 Solution algorithms for the incompressible NSE

The solution of the incompressible NSE is complicated because of the lack of an independent equation for p. The continuity equation does not have a dominant variable – it is a kinematic constraint on $\mathbf{v}$ rather than a dynamic equation. However, by taking the divergence of the momentum equation and using the divergence constraint to simplify the results, we see that it follows that $\nabla^2 p = -\nabla \cdot [\nabla \cdot (\mathbf{v} \otimes \mathbf{v} - \mathbf{S}) + \mathbf{f}]$, which can be solved together with the momentum equation. All available formulations share advantages and disadvantages, and the superiority of each formulation is problem specific. Accuracy and efficiency within each formulation depend on several factors, including the choice of discretization, explicit versus implicit time integration, solver efficiency, implementation details, parallelization, and accuracy. The most widely used formulations are (i) artificial compressibility, (ii) pressure-Poisson formulations, and (iii) projection formulations.

High-resolution methods are most commonly associated with compressible flow solutions. Their introduction to incompressible flow solutions was most directly impacted by their use with a projection method by Bell, Colella, and Glaz (1989), as well as with the artificial compressibility method by Drikakis, Govatsos, and Papantonis (1988). The use of an exact projection method in Bell et al. (1989) parallels the current time, in which these methods are used in incompressible flow calculations as well. The first issue that we emphasize is that incompressible flow is intrinsically multidimensional, which should be directly reflected in the algorithms applied. A multidimensional algorithm thus allows the incompressibility condition to be felt directly. This is difficult to accomplish through the use of directional split algorithms. Operator splitting is often used most effectively for compressible flows, where unsplit methods may show little, or even no, added benefit in many cases.

One can formulate an accurate and efficient algorithm for solving the unsteady incompressible NSE by recognizing that the discretized Poisson equation for the pressure must be derived by a combination of the discretized momentum and continuity equations, Eqs. (4a.7a) and (4a.7b), using appropriate numerical flux functions. To this end we insert numerical flux functions $\mathbf{F}_f^{C,v}$ and $\mathbf{F}_f^{D,v}$ into (4a.7b) and reformulate the resulting equation as an algebraic equation from which $\mathbf{v}_f^{n+m}$ can be obtained by linear interpolation ($\mathbf{v}_P^{n+m} \mapsto \mathbf{v}_f^{n+m}$), which in turn can be combined with (4a.7a) to give the discrete Poisson equation for the pressure,

$$\sum_f \left(\left(a_P^{-1} \right)_f (\nabla p)_f^{n+m} \cdot d\mathbf{A}_f \right) = \sum_f \left(\left(a_P^{-1} \mathbf{H} \right)_f \cdot d\mathbf{A}_f \right) = 0, \qquad (4a.9)$$

where

$$\mathbf{H} = \sum_{i=0}^{m} \left[\sum_{N} \left(a_N^{n+i} \mathbf{v}_N^{n+i} \right) \right] + \sum_{i=0}^{m-1} \left(b_P^{n+i} \mathbf{v}_P^{n+i} \right) - \sum_{i=0}^{m-1} \left[\beta_i (\nabla p)_P^{n+i} \right] + \sum_{i=0}^{m} \left(\beta_i \mathbf{f}_P^{n+i} \right),$$

in which a_N and b_N are matrix coefficients. We can solve these resulting equations, (4a.7b) and (4e.10), by using several types of algorithms – explicit or implicit, as well as simultaneous or segregated.

One such approach is the pressure implicit with splitting of operators (PISO) method (Issa 1986), which is a predictor–corrector algorithm that can be outlined by the following steps. The first is the *momentum-predictor step*. Momentum equation (4a.7b) is solved for $\mathbf{v}_P^*$, using the pressure from the old time step, $p_P^{(n)}$, which does not necessarily satisfy constraint (4a.7a). The second step is the *pressure-corrector step*. Predicted velocity $\mathbf{v}_P^*$ is then used to assemble $\mathbf{H}$ that is needed for pressure equation (4e.10), the solution of which is p_P^*. If the grid is nonorthogonal, iterations over the explicit nonorthogonal source terms in (4e.10) can be repeated until a predefined tolerance is met. The third step is the *momentum-corrector step*. Interpolation $\mathbf{v}_P^{n+m} \mapsto \mathbf{v}_f^{n+m}$ gives a new set of conservative volumetric fluxes consistent with pressure field p_P^*, which is used to do an explicit correction on the velocity resulting in $\mathbf{v}_P^{(n+1)}$, now consistent with $p_P^* \equiv p_P^{(n+1)}$. The corrector steps are repeated until a until a predefined tolerance is met.

4a.3.2 Solution algorithms for the compressible NSE

For the compressible NSE, involving also acoustic and thermal effects, semi-implicit (predictor–corrector) or fully explicit algorithms can be used equally well. For such flows, however, where small time steps are physically required, fully explicit algorithms have some advantages over semi-implicit algorithms in that the semi-implicit algorithms are typically more expensive per time step than are the fully explicit algorithms.

All semi-implicit or predictor–corrector algorithms for compressible flows rely on solving the momentum and energy equations together with a Poisson equation for p, replacing the continuity equation. This Poisson equation for p is more complicated than for the incompressible NSE since it will now also contain a transient term, involving the density. More precisely, combining Eq. (4a.6a) and Eq. (4a.6b) in the same way as for the incompressible NSE results in

$$\sum_{i=0}^{m} \left(\alpha_i \frac{1}{\Delta t} \left(\frac{1}{RT} p \right)_P^{n+i} \right.$$

$$\left. + \frac{\beta_i}{\delta V_P} \sum_f \left[\left(\frac{1}{a_P} \mathbf{H} \right)_f^{n+i} \cdot d\mathbf{A}_f - \left(\frac{1}{a_P} \nabla p \right)_f^{n+i} \cdot d\mathbf{A}_f \right] \right) = 0. \quad (4a.10)$$

A compressible extension to the PISO algorithm is suggested by Issa (1986); it consists of a sequence involving a continuity-predictor step, a momentum-predictor step, an energy-corrector step, a pressure-corrector step, and a momentum-corrector step. Typically, the corrector steps are repeated until a predefined tolerance for the solution is met.

For explicit algorithms, the aforementioned Runge–Kutta methods offer a good alternative. In particular, we favor the multistage method of Jameson et al. (1981), or that of Gottlieb and Shu (1998), the latter being faster, less memory demanding, and somewhat more stable,

$$\mathbf{U}^* = \mathbf{U}^n - \Delta t \left(\frac{1}{\delta V_P} \sum_f \left[\mathbf{F}_f(\mathbf{U}^n) \right] - \mathbf{s}\left(\mathbf{U}_P^n\right) \right),$$

$$\mathbf{U}^{n+1} = \frac{1}{2}(\mathbf{U}^n + \mathbf{U}^*) - \frac{1}{2}\Delta t \left(\frac{1}{\delta V_P} \sum_f \left[\mathbf{F}_f(\mathbf{U}^*) \right] - \mathbf{s}(\mathbf{U}_P^*) \right), \qquad (4a.11)$$

where $\mathbf{U} = [\rho, \ \rho \mathbf{v}, \ \rho E]^T$ is the state vector for Courant numbers smaller than one.

4a.4 Flux reconstruction

In order to close the discretization, we must specify the numerical flux functions that approximate the exact convective, diffusive, and additional fluxes, (4a.3) and (4e.6). These are typically defined at the cell faces, f, whereas the values of the dependent variables are stored at the CV centers, P and N, according to Figure 4a.1. In order to obtain a complete FVM, rules must be prescribed for how these numerical flux functions are to be reconstructed from the stored dependent variable data. Since information propagates at finite speed, it is reasonable to first assume that we can obtain these fluxes based only on the values at P and N, bracketing the cell face, f.

Here, we distinguish between flux reconstruction of the convective and diffusive fluxes since these fluxes describe different physical processes: Diffusion is represented by viscous stress tensor $\mathbf{S} = \frac{1}{3}\mu(\nabla \cdot \mathbf{v})\mathbf{I} + \mu(\nabla \mathbf{v} + \nabla \mathbf{v}^T)$ and heat conduction vector $\mathbf{h} = \kappa \nabla T$ and describes the effects of viscosity and heat conductivity, whereas convection describes the motion of fluid and how compression affects the density. Ideally, the numerical treatment should be regarded as a tool separated from the physics. This viewpoint is theoretically sound and in principle the best choice, since we would like the numerical errors to be easily separable from the physical modeling errors; we can accomplish this by choosing numerical methods with very high order of accuracy and fine grids (e.g., Oran and Boris 2001), but this is often too expensive for practical problems. Alternatively, the flux-reconstruction step for FVM can be seen as one in which physical interpretation and modeling cross paths with the numerical methods; this is the view pioneered by Boris (Boris 1990; Boris et al. 1992), and it provides the basis of the implicit approach to LES.

4a.4.1 Preamble to flux reconstruction of the convective fluxes

The nonlinear tensorial form of convective flux $\mathbf{F}_f^{C,v}$ (and $\mathbf{F}_f^{C,E}$) is complicated, but we can simplify it by observing that, for example, $\mathbf{F}_f^{C,v} = (\mathbf{v} \otimes \mathbf{v})_f d\mathbf{A}_f = F_f^{C,\rho} \mathbf{v}_f$, where $F_f^{C,\rho} = (\mathbf{v} \cdot d\mathbf{A})_f$ is the scalar flux across f. Both terms have to be evaluated before a complete FVM is achieved – a feat that is related to the overall algorithm architecture. For the moment, we consider flux $F_f^{C,\rho}$ to be known, and we focus on the reconstruction of $\mathbf{v}_f$. Any flux function $\mathbf{F}_f^{C,\rho} = (\rho\mathbf{v} \cdot d\mathbf{A})_f = F_f^C \rho_f$, where ρ denotes a generic variable (e.g., $\mathbf{v}$ and E), must satisfy these properties:

- *Conservation*: Fluxes from adjacent control volumes, sharing interface f, must cancel when summed: $\mathbf{F}_{f^-}^{C,\rho}(F_f^C; \rho_P, \rho_N) = -\mathbf{F}_{f^+}^{C,\rho}(F_f^C; \rho_N, \rho_P)$.
- *Consistency*: Consistency is obtained if the numerical flux with identical state arguments reduces to the true flux passing through interface f of that same state.

These properties, although obvious, are insufficient to guarantee convergence to entropy satisfying weak solutions (Barth and Ohlberger 2004), and hence numerical flux restrictions are necessary. This is the subject of Sections 4a.4.2 to 4a.4.5 and leads to high-resolution methods.

To demonstrate the convective flux-reconstruction procedure, and to introduce some basic algorithms, we first consider a reconstruction model based on an assumed piecewise linear variation of ρ between bracketing CV values. This flux reconstruction is a special case of Lagrange's interpolation formula, which for a two-point situation results in the *linear* flux function,

$$\mathbf{F}_f^{C,\rho,\text{linear}} = F_f^C \rho_f^{\text{linear}} \approx F_f^C(\ell_f \rho_P + (1 - \ell_f)\rho_N), \tag{4a.12}$$

where $\ell_f = |\mathbf{x}_f - \mathbf{x}_P|/|\mathbf{x}_N - \mathbf{x}_P|$ is the normalized distance between face f and the neighboring CV, P. We obtain the remainder term associated with using (4a.12) as the difference between the exact and the linearly approximated flux function, $\mathcal{R}_f^{\text{linear}} = \mathbf{F}_f^{C,\rho} - \mathbf{F}_f^{C,\rho,\text{linear}}$, that follows directly from Lagrange's interpolation formula, $\mathcal{R}_f^{\text{linear}} = \alpha F_f^C(\nabla^2\rho)_f \cdot (\mathbf{d} \otimes \mathbf{d})$, where $\mathbf{d} = \mathbf{x}_N - \mathbf{x}_P$ is the vector between neighboring CVs N and P, which for an orthogonal mesh is parallel to $d\mathbf{A}_f$ or $\mathbf{n}_f$ and $\alpha = \frac{1}{2}\ell_f(1 - \ell_f)$. The remainder term, $\mathcal{R}_f^{\text{linear}}$, shows that this linear interpolation is accurate to the second order and dispersive since it is associated with even-order derivatives in ρ. Another well-known flux reconstruction model is based on an upwind-biased piecewise constant variation of ρ between bracketing CV values, resulting in the *upwind-biased* flux function,

$$\mathbf{F}_f^{C,\rho,\text{upwind}} = F_f^C \rho_f^{\text{upwind}} \approx F_f^C(\beta^+ \rho_P + \beta^- \rho_N), \tag{4a.13}$$

where $\beta_f^{\pm} = \frac{1}{2}[\text{sign}(F_f^C) \pm 1]$. We obtain the remainder term associated with using the upwind-biased flux function, (4a.13), as the difference between the exact flux function, $\mathbf{F}_f^{C,\rho} = \mathbf{F}_f^{C,\rho,\text{linear}} + \mathcal{R}_f^{\text{linear}}$, and the upwind-biased flux function, $\mathcal{R}_f^{\text{upwind}} = \mathbf{F}_f^{C,\rho} - \mathbf{F}_f^{C,\rho,\text{upwind}} = (\mathbf{F}_f^{C,\rho,\text{linear}} + \mathcal{R}_f^{\text{linear}}) - \mathbf{F}_f^{C,\rho,\text{upwind}}$, which becomes $\mathcal{R}_f^{\text{upwind}} = F_f^C\{[(1 - \ell_f)\beta_f^- - \ell_f\beta_f^+](\nabla\rho)_f\mathbf{d} + 1/8\alpha(\nabla^2\rho)_f \cdot (\mathbf{d} \otimes \mathbf{d})\}$. This expression indicates that it is accurate to the first order and dissipative since it is dominated by odd-order derivatives in ρ. We can obtain yet another useful flux-reconstruction model by combining (4a.12) and (4a.13), using a local blending parameter, $\Psi = \Psi(\rho, \mathbf{d})$, often referred to as a *flux* or *slope limiter*. In this case we may use flux functions (4a.13) and (4a.14) to express the *hybrid* flux function as

$$\mathbf{F}_f^{C,\rho,\text{hybrid}} = \mathbf{F}_f^{C,\rho,H} - (1 - \Psi_f)\left[\mathbf{F}_f^{C,\rho,H} - \mathbf{F}_f^{C,\rho,L}\right] = F_f^C\left(\rho_f^H - (1 - \Psi_f)\left[\rho_f^H - \rho_f^L\right]\right)$$

$$= F_f^C((\ell_f \rho_P + (1 - \ell_f)\rho_N) - (1 - \Psi_f)[(\ell_f \rho_P + (1 - \ell_f)\rho_N)$$

$$- (\beta^+ \rho_P + \beta^- \rho_N)]), \tag{4a.14}$$

where superscript H refers to a high-order scheme, based on piecewise linear variation of ρ between bracketing CV values as in (4a.13) and L to a low-order scheme, based on the upwind-biased piecewise constant variation of ρ between bracketing CV values as in (4a.14), and Ψ is the blending parameter. We obtain the remainder term associated with using the hybrid flux function, Eq. (4a.14), as the difference between the exact flux function, $\mathbf{F}_f^{C,\rho} = \mathbf{F}_f^{C,\rho,\text{linear}} + \mathcal{R}_f^{\text{linear}}$, and the hybrid flux function, $\mathcal{R}_f^{\text{hybrid}} = (\mathbf{F}_f^{C,\rho,\text{linear}} + \mathcal{R}_f^{\text{linear}}) - \mathbf{F}_f^{C,\rho,\text{hybrid}}$, which becomes

$$\mathcal{R}_f^{\text{hybrid}} = \left(\mathbf{F}_f^{C,\rho,\text{linear}} + \mathcal{R}_f^{\text{linear}}\right) - \left(\mathbf{F}_f^{C,\rho,\text{linear}} - (1 - \Psi_f)\left[\mathbf{F}_f^{C,\rho,\text{linear}} - \mathbf{F}_f^{C,\rho,\text{upwind}}\right]\right)$$

$$= (1 - \Psi_f)F_f^C\left[(1 - \ell_f)\beta_f^+ - \ell_f\beta_f^-\right](\nabla\rho)_f\,\mathbf{d} + \alpha F_f^C(\nabla^2\rho)_f\,(\mathbf{d} \otimes \mathbf{d}) + \cdots . \tag{4a.15}$$

From Eq. (4a.14) and the associated remainder term, (4a.15), we conclude the following: (i) Ψ behaves as a switch between the underlying flux-reconstruction schemes; (ii) the limiting conditions occur when $\Psi = 1$, which implies that the remainder term is $\mathcal{R}_f^{\text{linear}}$, and when $\Psi = 0$, which implies that it becomes $\mathcal{R}_f^{\text{upwind}}$; (iii) Ψ can be a function of the dependent variables, making the switch nonlinear – that is, the entire algorithm becomes nonlinear, even for linear equations, in order to achieve second-order accuracy simultaneously with monotonicity; (iv) depending on the local value of Ψ, the properties of the leading-order truncation error can be adjusted to give the algorithm certain desired features, thus adapting the numerical algorithm based upon the behavior of the local solution; (v) the fundamental aspect of all the methods based on the flux-limiting approach is the goal of producing a nonoscillatory solution. This means that the method does not produce (significant) unphysical oscillations in the numerical solution. This blending parameter is often referred to as the flux (or slope) limiter and is the general nonlinear mechanism that distinguishes modern methods from classical linear schemes. We discuss the development of such limiters later.

4a.4.2 Total-variation diminishing, monotonicity, and flux limiting

Following Harten (1983), we classify as high-resolution methods those with the following properties: (i) they provide at least a second order of accuracy in smooth regions of the flow; (ii) they produce numerical solutions free from spurious oscillations, and (iii) in the case of discontinuities, the number of grid points in the transition zone containing the shock wave is smaller in comparison with that of first-order monotone methods. Such methods are designed from concepts such as total variation and mono-tonicity. Note, however, that in general, these properties only hold for the exact solution of scalar problems. Methods designed according to these guidelines are then extended to systems of equations that do not satisfy properties such as diminishing of the total variation of the solution. Here, we outline the mathematical ingredients for analyzing such schemes. To this end we consider the finite-volume discretization of the (nonlinear) model equation,

$$\partial_t(\rho) + \nabla(\mathbf{F}(\rho, \mathbf{v})) = s(\rho), \quad \text{where } \mathbf{F}(\rho, \mathbf{v}) = \rho\mathbf{v}. \tag{4a.16}$$

Under the assumption of a one-dimensional (1D) domain, covered with a uniform grid, Δx, the application of the finite-volume discretization to (4a.16), using the numerical flux function (4a.14), Euler implicit time integration, and constant velocity $v = v_0$, leads to the 1D explicit FVM,

$$
\begin{aligned}
\rho_P^{n+1} = \rho_P^n - v & \left[-\frac{1}{2}\Psi_{f-}\delta\rho_{f-}^n - \left(1 - \frac{1}{2}\Psi_{f+}\right)\delta\rho_{f+}^n \right. \\
& \left. + \left(1 - \frac{1}{2}\Psi_{f+} + \frac{1}{2}\Psi_{f-}\right)\left(\delta\rho_{f-}^n + \delta\rho_{f+}^n\right) \right] \\
= \rho_P^n - v & \left[\left(1 - \frac{1}{2}\Psi_{f+}\right)\delta\rho_{f-}^n + \left(\frac{1}{2}\Psi_{f-}\right)\delta\rho_{f+}^n \right],
\end{aligned}
\tag{4a.17}
$$

where $v = v_0 \Delta t/\Delta x$ is the Courant number, $\delta\rho_{f-}^n = \rho_P^n - \rho_{N-}^n$ and $\delta\rho_{f+}^n = \rho_{N+}^n - \rho_P^n$ are the local differences, and H denotes the spatial part of the algorithm, which is a special case of the more general algorithm $\rho_P^{n+1} = H(\rho_{P-k}^n, \rho_{P-k+1}^n, \ldots, \rho_{P+k}^n)$. The multidimensional counterpart of (4a.17) is more complicated, but here the 1D simplification is sufficient.

A numerical algorithm is said to be *monotone* if it does not lead to an oscillatory behavior of the numerical solution. The apparently smooth behavior of a numerical solution may result from different conditions, the strongest being monotonicity, while weaker conditions are associated with L_1 contraction, total variation, and monotonic-ity preservation. Mononicity is best explained for the model equation, Eq. (4a.16), discretized by the finite-volume algorithm,

$$\rho_P^{n+1} = H\left(\rho_{P-k}^n, \rho_{P-k+1}^n, \ldots, \rho_{P+k}^n\right). \tag{4a.18}$$

Algorithm (4a.18) is said to be monotone if H is a monotone increasing function of each of its arguments, which results in the following sequence of inequalities:

$$\frac{\partial H}{\partial \rho_Q}\left(\rho_{P-k}^n, \rho_{P-k+1}^n, \ldots, \rho_{P+k}^n\right) \geq 0, \quad \forall \, P - k \leq Q \leq P + k. \tag{4a.19}$$

Since H is defined by the numerical flux functions, the numerical flux of a monotone algorithm is nondecreasing in its first argument and nonincreasing in its last arguments:

$$\frac{\partial F_{f+}}{\partial \rho_{P-k+1}} \geq \quad \text{and} \quad \frac{\partial F_{f-}}{\partial \rho_{P+k}} \leq 0. \tag{4a.20}$$

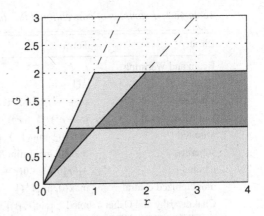

Figure 4a.2. TVD regions for first-order- and second-order-accurate TVD schemes. The light gray and the dark gray regions denote the first-order and the second-order TVD regions, respectively.

Accordingly, for any linear algorithm of the form $\rho_P^{n+1} = \sum_k [\pi_k \rho_{P+k}^n]$, the condition of monotonicity requires all coefficients π_k to be nonnegative. Harten (1983) showed, by using the MEA approach, that monotone schemes for model equation (4a.16) are accurate only to the first order, which represents a severe limitation since first-order accuracy is insufficient for practical purposes – the corresponding schemes being too dissipative. Conditions less severe than monotonicity therefore have to be defined and introduced.

Given model equation (4a.16), Lax (1973) shows that the total variation (TV) of any physically admissible solution, $\mathrm{TV} = \int |\nabla_x \rho| \, dx$, does not increase in time. The TV of a discrete solution to this conservation law is therefore defined by $\mathrm{TV}(\rho) = \sum_P |\rho_{N+} - \rho_P|$. A numerical solution is thus said to have bounded TV if TV is uniformly bounded. Hence, a numerical algorithm is said to be *total-variation diminishing* (TVD) if and only if $\mathrm{TV}(\rho^{n+1}) \leq \mathrm{TV}(\rho^n)$. Following Harten (1983), we see that a numerical algorithm for model equation (4a.16) of the form

$$\rho_P^{n+1} = \rho_P^n - C\delta\rho_{f-}^n + D\delta\rho_{f+}^n \qquad (4a.21)$$

is TVD if and only if $C \geq 0$, $D \geq 0$, and $C + D \leq 1$ for all P.

Considering in particular the finite-volume discretization, Eq. (4a.17), we find that by choosing

$$C = \upsilon \left[\left(1 - \frac{1}{2}\Psi_{f+}\right) + \left(\frac{1}{2}\Psi_{f-}\right) / r \right] \text{ and } D = 0, \qquad (4a.22)$$

where $r = \delta\rho_{f-}^n / \delta\rho_{f+}^n = (\rho_P^n - \rho_{N-}^n)/(\rho_{N+}^n - \rho_P^n)$ is the local ratio of consecutive gradients, we can readily show that the TVD-defining inequalities satisfy

$$0 \leq \left(1 - \frac{1}{2}\Psi_{f+}\right) - \frac{1}{2}\Psi_{f-}/r \leq 1, \qquad \text{for all } P, \qquad (4a.23)$$

from which we conclude, by further assuming that $\Psi = \Psi(r)$, that $|\Psi(r)/r - \Psi(r)| \leq 2$. This is a fairly natural assumption since r is an appropriate way of measuring the smoothness of field ρ. Constraint (4a.24) is graphically illustrated in Figure 4a.2. Note that for any second-order-accurate algorithm, we must have $\Psi(1) = 1$; moreover, Sweby (1984) found that it is best practice to take Ψ to be a convex combination of $\Psi = 1$ and $\Psi = 0$. Other choices give too much compression, and smooth data such as a sine wave

Table 4a.2. *Examples of conventional flux limiters*

Limiter	Form
Beam and Warming	r
Fromm	$\frac{1}{2}(1+r)$
van Leer	$(r+|r|)/(1+|r|)$
van Albada et al.	$(r+r^2)/(1+r^2)$
minmod	$\max[0,\ \min(1,\ r)]$
superbee	$\max\{0, \max[\min(2r,\ 1), \min(r, 2)]\}$
gamma	$\frac{1-k}{k}r[\theta(r) - \theta(r - \frac{k}{1-k})] + \theta(r - \frac{k}{1-k})$
monotonized central	$\max\{0, \min[\frac{1}{2}(1+r), 2, 2r]\}$
Chakravarthy and Osher	$\max[0, \min(r, \beta)]$

tend to turn into a square wave as time evolves. For unstructured grids the definition of r can be generalized to $r = 2(\nabla\rho)_P/(\nabla\rho)_f - 1$, as shown by Darwish (1993).

Algorithms for solving model equation (4a.16) of the particular form of Eq. (4a.18) are said to be *monotonicity preserving* if, given the initial distribution $\rho_P^0 \geq \rho_{P+1}^0$,

$$\rho_P^n \geq \rho_{P+1}^n, \quad \forall P \text{ and time steps } n. \tag{4a.24}$$

Hence, new oscillations cannot arise near an isolated propagating discontinuity for which the initial condition is monotone; that is, monotone profiles are preserved during the time evolution of the discrete solution and overshoots will not be created. Finally, with the aforementioned definitions, we find the following hierarchy between properties:

- All monotone algorithms are TVD.
- All TVD schemes are monotonicity preserving.

4a.4.3 Examples of flux limiters

From Eq. (4a.17) it is evident that $\Psi = 0$ results in a flux-reconstruction algorithm that eventually results in the first-order-accurate upwind scheme whereas $\Psi = 1$ results in a flux-reconstruction algorithm that results in the second-order-accurate Lax–Wendroff scheme (Lax and Wendroff 1964). Other well-known schemes that can be derived from formulation (4a.22) involve the Beam–Warming scheme (Beam and Warming 1976) and the method of Fromm (1968) with $\Psi = r$ and $\Psi = \frac{1}{2}(1+r)$, respectively. Table 4a.2 presents some of the most well-known flux limiters available in the literature, such as the flux limiters introduced by van Leer (1979) and van Albada, van Leer, and van Roberts (1982), the minmod and superbee limiters suggested by Roe (1985), the gamma limiter (Jasak, Weller, and Gosman 1999), the monotonized central-difference limiter of van Leer (1977), and the flux limiter developed by Chakravarthy and Osher (1983). Sweby (1984) studied many of these classical limiters and found that the limits of the TVD region are members of a family of limiters, based on a single parameter, β, in the range $1 < \beta < 2$, such that $\Psi(r) = \max[0, \min(\beta r, 1), \min(\beta, r)]$, where $1 \leq \beta \leq 2$. All these limiters share the symmetry property $\Psi(r)/r = \Psi(1/r)$,

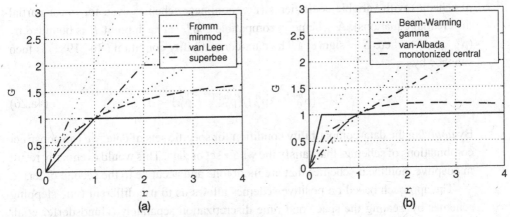

Figure 4a.3. Examples of limiters presented in Table 4a.2.

indicating that forward and backward gradients are treated similarly. Alternately, this symmetry ensures that the limited gradients remain associated with a linear variation of ρ within each cell.

Following Sweby (1984), we can graphically illustrate the aforementioned flux limiters and compare them with the different TVD regions in Figure 4a.3. Note here that the minmod limiter lies along the lower boundary of the second-order TVD region, whereas the superbee limiter lies along the upper boundary. Some of these flux limiters are not smooth at $(r, \Psi) = (1,1)$ since there is a switch in the choice of one-sided approximations used at $r = 1$. This may lead to loss of accuracy, and for full second-order accuracy, Ψ should be smooth near $r = 1$.

4a.4.4 Construction of modern TVD-based flux limiters

The construction of high-order TVD schemes can be achieved by use of the flux-limiter approach (Boris and Book 1973, 1976; Boris, Book, and Hain 1975; Sweby 1985a, 1985b). Typically, the derivation of flux limiters is based on the linear advection equation and explicit discretizations. To this end, we again consider model problem (4a.16). Under the assumption of a 1D domain, covered with a uniform grid, Δx, the application of the finite-volume discretization, using the TVD-based numerical flux function, $F_f^{\text{TVD}} = F_f^L + \Psi_f[F_f^H - F_f^L]$, where $F_f^L = v_x(\alpha_0 \rho_P + \alpha_1 \rho_{N+})$ and $F_f^H = v_x(\beta_0 \rho_P + \beta_1 \rho_{N+})$ are low- and high-order flux functions, respectively, yields

$$\rho_P^{n+1} = \rho_P^N - v[\alpha_0 + \Psi_{f-}(\beta_0 - \alpha_0)]\delta\rho_{f-}^n - v[\alpha_1 + \Psi_{f+}(\beta_1 - \alpha_1)]\delta\rho_{f+}^n$$

$$= \rho_P^N - v C\delta\rho_{f-}^n + v D\delta\rho_{f+}^n. \tag{4a.25}$$

Alternate approaches to limiting advective fluxes – other than the theorem by Harten (1983) – are the data-compatibility condition (Roe 1981), the positive schemes (Hundsdorfer et al. 1995), and the universal limiter (Leonard and Niknafs 1981). Roe's idea was to circumvent Godunov's theorem by constructing adaptive algorithms that would adjust themselves to the local nature of the solution. This leads to the design of schemes with variable coefficients (i.e., nonlinear schemes even for linear partial differential equations). A scheme is compatible if ρ_{N+}^n at each point P is bounded by (ρ_{P-S}^n, ρ_P^n) where $S = \text{sign}(v_x)$. The data-compatibility condition (Roe 1981) is then satisfied by

$$0 < \left(\rho_P^{n+1} - \rho_P^n\right)/\left(\rho_{P-S}^{n+1} - \rho_P^n\right) < 1. \tag{4a.26}$$

By *applying* the data-compatibility condition for specific sets of data, we can construct combinations of schemes that satisfy the whole set of data. This would eventually result in adaptive, nonlinear schemes that are monotone and accurate to the second order.

The approach based on positive schemes allows us to use different time-stepping schemes by treating the space and time discretization separately (Hundsdorfer et al. 1995). Thuburn (1993) has shown that different approaches for constructing limiters can lead to equivalent schemes, at least in the context of the 1D linear advection equation. However, the differences between the aforementioned approaches still remain important, since each of these approaches can be extended and utilized in different ways. For example, the TVD approach can be extended to conservation laws other than the advection equation (e.g., the NSE), and the universal limiter can be implemented to multiple advection on arbitrary meshes (Thuburn 1996).

Next, we present an example for the construction of flux limiters using Godunov's first-order upwind flux and Lax–Wendroff flux, as low- and high-order fluxes (Rogers and Kwak 1990), respectively (see also Drikakis and Rider 2004). The coefficients for these fluxes are

$$\alpha_0 = \frac{1}{2}(1 + S), \quad \alpha_1 = \frac{1}{2}(1 - S), \text{ and } S = \text{sign}(v_x) \tag{4a.27}$$

for the Godunov first-order upwind flux and

$$\beta_0 = \frac{1}{2}(1 + v), \quad \beta_1 = \frac{1}{2}(1 - v) \tag{4a.28}$$

for the Lax–Wendroff flux. For the sake of simplicity, we consider only the case $v_x > 0$, where the coefficients of the Godunov scheme are $\alpha_0 = 1$ and $\alpha_1 = 0$. The coefficients C and D in (4a.21) therefore become $C = 1 + \Psi_{f-}(\beta_0 - 1)$ and $D = -\Psi_{f+}\beta_1$. We can now rewrite (4a.25) such that

$$\rho_P^{n+1} = \rho_P^n - v\hat{C}\delta\rho_{f-}^n, \tag{4a.29}$$

where $\hat{C}$ is a function of the local data such that $\hat{C} = C - D/r$ where $r = \delta\rho_{f-}^n/\delta\rho_{f+}^n$ is the ratio of consecutive gradients. We can obtain the derivation of the limiter by using Harten's (1983) theorem (stating that $C_{N+} \geq 0$, $D_{N+} \geq 0$, and $0 \leq C_{N+} + D_{N+} \leq 1$) such that

$$0 \leq v\left[1 + \Psi_{f-}(\beta_0 - 1) + \Psi_{f+}\beta_1/r\right] \leq 1. \tag{4a.30}$$

Let us now assume that an upper bound Ψ_T and a lower bound Ψ_B to Ψ_f exist, such that $\Psi_B \leq \Psi_{f-} \leq \Psi_T$; then, it is straightforward to show, using (4a.27) and (4a.28), that

$$\Psi_T - 2/(1-v) = \Psi_L \leq \Psi_{f+} \leq \Psi_R = \Psi_B + 2/v, \quad r \geq 0,$$
$$\Psi_B + 2/v = \Psi_R \leq \Psi_{f+} \leq \Psi_L = \Psi_T - 2/(1-v), \quad r < 0. \tag{4a.31}$$

Equation (4a.31) determines the limiter Ψ for different choices of the bottom and top bounds. Note also that (4a.31) is also valid for $v_x < 0$ if v is replaced by $|v|$.

The derivation of flux limiters described in the preceding section is general and applicable to any process governed by advection. However, in the case of compressible flow gradients that arise from shock waves, strong shocks and their interaction with turbulence may pose difficulties to satisfy the TVD properties in the framework of nonlinear equations; note that the derivation was based on a linear advection model. To circumvent this difficulty, we must derive limiters that are derived on the basis of numerical experimentation. These limiters aim at reducing the accuracy of the numerical scheme in the proximity of shocks and in regions of large gradients in general. Here we give an example of limiters derived on the basis of numerical experimentation in conjunction with the MUSCL scheme (van Leer 1979):

$$\rho_{N+}^L = \rho_P + \frac{1}{4}\Psi_P[(1 - k\Psi_P)\nabla + (1 + k\Psi_P)\nabla^2]\rho_P,$$

$$\rho_{N+}^R = \rho_{N+} - \frac{1}{4}\Psi_{N+}[(1 + k\Psi_{N+})\nabla + (1 - k\Psi_{N+})\nabla^2]\rho_{N+}, \tag{4a.32}$$

where parameter k controls different MUSCL realizations – fully upwind for $k = -1$, third order for $k = 1/3$ (strictly third order only for 1D problems), and centered for $k = 1$, and Ψ_f is the Albada limiter (van Albada et al. 1982). In the implementation of high-order interpolation, we need to specify one or two fictitious values (depending on the order of interpolation) at the computational boundaries. Our experience is that the extrapolation of these values from the interior of the domain works satisfactorily. In the case of wall boundaries, these values can still be extrapolated without violating the no-slip and no-penetration condition if the flux component normal to the wall boundary is set equal to zero. Numerical experiments have shown that the MUSCL reconstruction provides accurate results for transonic flows, but it has to be modified for supersonic and hypersonic flows to better satisfy the nonoscillatory TVD property and provision of adequate dissipation. A variant of the MUSCL scheme for calculating the states of the intercell variables in such flows is suggested by Zoltak and Drikakis (1998).

4a.4.5 Flux-corrected transport

Review papers discussing the original as well as the more recent flux-corrected transport (FCT) formulations and applications can be found in Kuzmin, Lohner, and Turek (2005). FCT was originally developed to accurately solve the conservation equations without violating the positivity of mass and energy, particularly in the vicinity of shocks and other discontinuities. For the original FCT, Boris and Book (1973) proposed the use of a three-step scheme involving successive convection, diffusion, and antidiffusion phases, seeking to provide as accurate a solution to the original equation as is consistent with maintaining positivity and local monotonicity everywhere. Zalesak (1979) later showed that any given FCT algorithm could be recast as a hybrid method consisting of three components: (i) a high-order algorithm to which it reduces in smooth parts of the flow, (ii) a low-order algorithm to which it reduces in nonsmooth parts of the flow, and (iii) a flux limiter that calculates the weights assigned to the high- and low-order fluxes in various regions of the flow field. Zalesak (1979) also provided a multidimensional extension of FCT. As subsequently shown here, Zalesak's original FCT limiter preserves positivity but not monotonicity, possibly yielding solutions with numerical ripples; an effective modification of Zalesak's limiter to cure this problem by DeVore (1998) proposes a prelimiting step based on the use of a monotonicity-preserving 1D FCT formulation in each coordinate direction.

With FCT, Boris and Book (1973) resolved the conflicting demands for a positive, monotone numerical solution ρ^L on the one hand and for an accurate, nondiffusive solution ρ^H on the other. Their approach is to limit the antidiffusive fluxes that change ρ^L into ρ^H so that no new extrema are created, nor are any existing extrema enhanced by the application of the corrected fluxes. The FCT prescription for solving equations such as Eq. (4a.16) is as follows.

1. Convect ρ.
2. Numerically diffuse ρ sufficiently to ensure that the result is positive everywhere, if the initial profile ρ^0 is positive.
3. Add sources S to obtain the low-order, provisional solution ρ^L.
4. Compute numerical antidiffusive fluxes that would convert ρ^L to a high-order accurate solution ρ^H.
5. Limit these antidiffusive fluxes so that no new extrema will be created and no existing extrema will be enhanced.
6. Apply the corrected fluxes to ρ^L to obtain the desired solution ρ^n.

Given the provisional profile ρ^L, and the raw antidiffusive fluxes ρ_f^A defined on the cell faces, f, the limiting rule used in Step 5 is

$$\rho_f^C = \sigma_{f+} \max \left(0, \min \left(\left| \rho_{f+}^A \right|, \sigma_{f+} \Delta \rho_{f-}^L, \sigma_{f+} \Delta \rho_{f++}^L \right) \right), \qquad (4a.33)$$

where ρ_f^C is the corrected positivity and monotonicity-preserving antidiffusive flux, $\Delta \rho_f^L = \rho_N^L - \rho_P^L$, and $\sigma_f = \mathrm{sign}(\Delta \rho_f^L)$, with $|\sigma_f| = 1$. This flux limiter has the following consequences.

1. If the ρ^L profile is not monotone between cells N and P, which means that at least one of ρ_P^L and ρ_N^L is a local extremum and $\Delta \rho_f^L$ changes sign, then the flux is canceled.

2. If the ρ^L profile is monotone over that range, then the amplitude of the flux is limited to the smallest of its original magnitude and the two mass changes sufficient to flatten the profile ahead of and behind the interface.

3. If the raw antidiffusive flux is directed down the gradient of ρ^L, so that it smooths rather than steepens the profile, then the sign of the flux is reversed (it is given the sign of $\Delta\rho_f^L$) and the amplitude is limited according to the aforementioned criteria.

In his analysis of flux-limiting formula (4a.33) to derive an extension to two or more spatial dimensions, Zalesak (1979) noted the remarkable fact that each flux is limited independently. The situation is different in two or more dimensions, where multiple fluxes may enter or leave a cell without that cell being a local extremum. As a result, the flux limiter must take into account the consequences of fluxes acting in concert. Zalesak reformulated the 1D flux limiter in a way that preserves the desirable properties of Boris and Book's limiter, but also generalizes immediately to multidimensional problems. His procedure is as follows.

A. Calculate extrema in each cell by $\rho^{min} = \min(\rho_{N-}^L, \rho_P^L, \rho_{N+}^L)$ and $\rho^{max} = \max(\rho_{N-}^L, \rho_P^L, \rho_{N+}^L)$.

B. Reverse the sign of any antidiffusive flux that is directed down the gradient of ρ^L; that is, $\rho_f' = \sigma_f|\rho_f^A|$.

C. Compute the total incoming and outgoing antidiffusive fluxes in each cell, P, according to $\rho_P^{in} = \max(\rho_{f-}', 0) - \min(\rho_{f+}', 0)$ and $\rho_P^{out} = \max(\rho_{f+}', 0) - \min(\rho_{f-}', 0)$, respectively.

D. Determine the fractions of the incoming and outgoing fluxes that can be applied to each cell, P, as $\hat{\rho}_P^{in} = (\rho_P^{max} - \rho_P^L)/\rho_P^{in}$ and $\hat{\rho}_P^{out} = (\rho_P^L - \rho_P^{min})/\rho_P^{out}$, respectively.

E. Limit each antidiffusive flux so that it creates neither an undershoot in the cell it is leaving nor an overshoot in the cell it is entering:

$$\rho_f^C = \sigma_f|\rho_f^A| \begin{cases} \min(\hat{\rho}_P^{out}, \hat{\rho}_P^{in}, 1) & \text{if } \sigma_f|\rho_f^A| > 0, \\ \min(\hat{\rho}_P^{in}, \hat{\rho}_P^{out}, 1) & \text{otherwise.} \end{cases}$$

Zalesak's Step B as given in his 1979 paper is different from the one here, and his is not fully consistent with the original limiter. The discrepancy is that he used the sign of the antidiffusive flux ρ_f^A, rather than the sign of the difference $\Delta\rho_f^L$, for σ in Eq. (4a.33). As he noted, this adjustment is relatively rarely applied in any case – the majority of raw antidiffusive fluxes act to steepen the gradient – but it does serve to suppress the generation of numerical ripples. Zalesak also pointed out and showed through several examples how the limiter can be made more flexible by generalizing the formulae of Step A for the allowed extrema. The flux limiter as reformulated by Zalesak has the following consequences:

• It can be made identical to Boris and Book's limiter in one spatial dimension.
• It generalizes readily to multiple dimensions.
• It enforces positivity if ρ^L is positive and if the allowed extrema ρ^{min} and ρ^{max} are adequately constrained.

- It prevents the creation of new extrema and the enhancement of existing ones.
- However, the creation of new and the enhancement of existing numerical ripples in the solution ρ^n is allowed in multidimensional problems; that is, the limiter is not monotonicity preserving.

An effective modification of Zalesak's limiter that contributes toward preserving monotone profiles introduced by DeVore (1998) consists of applying Boris and Book's limiter along each coordinate direction separately, as a prelimiting step in the FCT procedure (also see the recent discussion by Zalesak 2005). This pre-limiting step replaces and augments the sign-reversal Step B. It is important to note that the remainder of Zalesak's scheme must still be carried out. Antidiffusive fluxes acting together on cells that are not directional extrema of ρ^L could conspire to produce new local peaks in ρ^n.

4a.4.6 Flux reconstruction of the diffusive fluxes

The construction of a numerical function for the diffusive flux defined in Eq. (4a.5) is straightforward, in particular if we assume that $\mathbf{F}_f^{D,v} = \mathbf{F}_f^{D,v}(v; \mathbf{v}_P, \mathbf{v}_N)$, where P and N are neighboring control volumes sharing face f. If the grid is orthogonal, that is, if $\mathbf{d}$ and $d\mathbf{A}_f$ are parallel, the simplest approximation of the inner gradient at f is $(\mathbf{v}_N - \mathbf{v}_P)/|\mathbf{d}_f|$. It therefore follows that

$$\mathbf{F}_f^{D,v,\text{linear}} \approx v(\nabla \mathbf{v})_f d\mathbf{A} = v|d\mathbf{A}_f|(\mathbf{v}_N - \mathbf{v}_P)/|\mathbf{d}|, \qquad (4a.34)$$

which enables us to calculate the face gradient by using just two points around the face. The leading-order truncation error in (4a.34) is $T_f \approx -\frac{1}{24}[(\nabla^3 \mathbf{v})_f d\mathbf{A}_f](\mathbf{d} \otimes \mathbf{d})$. We obtain this expression by using a multidimensional Taylor series expansion around face f, bracketed by P and N. Note also that, for nonorthogonal grids (i.e., $d\mathbf{A}_f \not\parallel \mathbf{d}$), the approximation in (4a.34) and the resulting leading-order truncation error become more complicated because the approximation of $(\nabla \mathbf{v})_f$ has to be divided into orthogonal and nonorthogonal contributions, respectively. Note further that approximation (4a.34) is accurate to the second order and hyperdissipative since it is associated with odd-order derivatives of dependent variable $\mathbf{v}$. For nonorthogonal grids, $(\nabla \mathbf{v})d\mathbf{A}$ is split into two parts: $(\nabla \mathbf{v})d\mathbf{A} = (\nabla \mathbf{v})\mathbf{k}_\perp + (\nabla \mathbf{v})\mathbf{k}_\angle$, where the first part denotes the orthogonal contribution and the second part denotes the nonorthogonal correction. Here the two vectors $\mathbf{k}_\perp$ and $\mathbf{k}_\angle$ have to satisfy the condition $d\mathbf{A} = \mathbf{k}_\perp + \mathbf{k}_\angle$. In addition, $\mathbf{k}_\perp$ is usually chosen to be parallel with $\mathbf{d}$. Several possible decompositions exist, but we mention only three:

The first is minimum correction. Decomposition is performed such as to keep the correction as small as possible; that is, $\mathbf{k}_\perp = [(\mathbf{d} \cdot d\mathbf{A})/|\mathbf{d}|^2]\mathbf{d}$ and $\mathbf{k}_\angle = d\mathbf{A} - \mathbf{k}_\perp$. The two vectors $\mathbf{k}_\perp$ and $\mathbf{k}_\angle$ are orthogonal; both terms are kept as small as possible. It should be noticed that as nonorthogonality increases, the contributions from $\mathbf{v}_P$ and $\mathbf{v}_N$ correspondingly decrease.

The second is orthogonal correction. This approach keeps the contribution from $\mathbf{v}_P$ and $\mathbf{v}_N$ the same as on the orthogonal grid. More precisely, $\mathbf{k}_\perp = (\mathbf{d}/|\mathbf{d}|)|d\mathbf{A}|$ and $\mathbf{k}_\angle = d\mathbf{A} - \mathbf{k}_\perp$. The part dependent on $\mathbf{v}_P$ and $\mathbf{v}_N$ remains the same irrespective of the nonorthogonality; that is, the magnitude of $\mathbf{k}_\perp$ remains the same.

The third is overrelaxed. The importance of the term in $\mathbf{v}_P$ and $\mathbf{v}_N$ increases with the increase of nonorthogonality, and therefore $\mathbf{k}_\perp = [\mathbf{d}/(\mathbf{d} \cdot d\mathbf{A})]|d\mathbf{A}|$ and $\mathbf{k}_\angle = d\mathbf{A} - \mathbf{k}_\perp$.

The difference between the aforementioned algorithms occurs in implicit calculations, when the nonorthogonal correction term is lagged and the term in $\mathbf{v}_P$ and $\mathbf{v}_N$ is treated implicitly. The overrelaxed approach is found to give the smallest number of "nonorthogonal" iterations. Finally, an expression for the diffusive flux can be given by

$$\mathbf{F}_f^{D,v} = \nu|\mathbf{k}_\perp|(\mathbf{v}_N - \mathbf{v}_P)/|\mathbf{d}| + (\nabla \mathbf{v})_f \mathbf{k}_\angle + \frac{1}{24}\nu(\mathbf{d} \otimes \mathbf{d})\nabla^3 \mathbf{v} + \cdots, \qquad (4a.35)$$

where the face interpolate of $(\nabla \mathbf{v})_f$ is calculated according to $(\nabla \mathbf{v})_f = \ell_f(\nabla \mathbf{v})_P + (1 - \ell_f)(\nabla \mathbf{v})_N$.

4a.5 Modified equation analysis

We proceed with our study by using the method of MEA, introduced by Hirt (1968) and Yanenko and Shokin (1969), in the context of the finite-difference method. Briefly stated, given the differential equations of interest and the numerical method to be employed, this method yields the differential equations actually solved numerically. These modified differential equations will thus include the actual differential equations plus additional terms related to the truncation error of the numerical method. More precisely, if the differential equation of interest is $\partial_t(\mathbf{U}) + \nabla \cdot [\mathbf{F}(\mathbf{U})] = \mathbf{s}(\mathbf{U})$, then the modified differential equation, satisfied by the numerical solution, is $\partial_t(\mathbf{U}) + \nabla \cdot [\mathbf{F}(\mathbf{U})] = \mathbf{s}(\mathbf{U}) + \mathcal{T}(\mathbf{U}, \mathbf{d})$, where $\mathcal{T}(\mathbf{U}, \mathbf{d})$ is the truncation error associated with the selected spatial discretization, time integration, and flux-reconstruction process. One key aspect of the MEA compared with other approaches is that the truncation error will include whatever nonlinearity is related to either the differential equation or the numerical method. The objective here is to use MEA to explain the properties of implicit (built-in) SGS models, and their relation to explicit SGS models used in conventional LES and to the leading-order truncation errors of the underlying discretization. The full three-dimensional analysis presented here is used to emphasize the tensorial properties of the implicit SGS models, this being a key property for marginally resolved flows and for flows in complex geometries in which the anisotropy of the smallest resolved scales is important. The earliest version of the present MEA of ILES was presented in Fureby and Grinstein (1998). Further discussion of the MEA relevant to ILES is presented in Chapter 5.

4a.5.1 Incompressible NSEs

The MEA for incompressible NSE (4a.3) is best illustrated, and most easily executed, when the equations are discretized in space only, and when the convective and diffusive flux terms are explicitly expressed in terms of the dependent variables. *It is important to note that different treatments of the convective fluxes employed by different solvers result in different modified equations.* Thus, one should include as many algorithmic details as possible in the analysis if the aim is to analyze the performance of a particular scheme. For the sake of the presentation here, however, we will omit the time-integration details since we are primarily interested in *comparing* explicit and implicit LES, which usually use similar, or even identical, time-integration schemes. We therefore start the MEA with the semidiscretized incompressible NSE written as

$$\partial_t(\mathbf{v}_P) + \frac{1}{\delta V_P} \sum_f [(\mathbf{v} \otimes \mathbf{v})_f \, d\mathbf{A}_f - \nu(\nabla\mathbf{v})_f d\mathbf{A}_f] = -(\nabla p)_P + \mathbf{f}_P. \tag{4a.36}$$

In order to be able to discuss both explicit and implicit LES, we model the convective fluxes in (4a.36), or alternately in (4a.7), using hybrid flux function (4a.14), which is based on a local combination of linear flux function (4a.12) and upwind-biased flux function (4a.13) using the blending parameter or flux limiter $\Psi = \Psi(\mathbf{U}, \mathbf{d})$, for the particular equation $\partial_t(\mathbf{U}) + \nabla \cdot [\mathbf{F}(\mathbf{U})] = \mathbf{s}(\mathbf{U})$, being considered. We model the diffusive fluxes in (4a.36), or alternatively in (4a.7), by using the linear flux function, (4a.34), which is based on central differencing of the inner gradient in the diffusive fluxes. For convenience, the flux functions used here are

$$\mathbf{v}_f^{\text{hybrid}} = \mathbf{v}_f^{\text{linear}} - (1 - \Psi_f)\left[\mathbf{v}_f^{\text{linear}} - \mathbf{v}_f^{\text{upwind}}\right],$$

$$\mathcal{R}_f^{\text{hybrid}} = (1 - \Psi_f)\left[(1 - \ell_f)\beta_f^+ - \ell_f\beta_f^-\right](\nabla\mathbf{v})_f\mathbf{d} + \frac{1}{8}(\nabla^2\mathbf{v})_f(\mathbf{d} \otimes \mathbf{d}), + \cdots, \tag{4a.37a}$$

$$(\nabla\mathbf{v})_f d\mathbf{A}_f \approx |d\mathbf{A}_f|(\mathbf{v}_N - \mathbf{v}_P)/|\mathbf{d}|, \quad \mathcal{T}_f \approx -\frac{1}{24}[(\nabla^3\mathbf{v})_f d\mathbf{A}_f](\mathbf{d} \otimes \mathbf{d}) + \cdots, \tag{4a.37b}$$

where $\mathbf{v}_f^{\text{linear}} \approx \ell_f\mathbf{v}_P + (1 - \ell_f)\mathbf{v}_N$ and $\mathbf{v}_f^{\text{upwind}} \approx \beta_f^+\mathbf{v}_P + \beta_f^-\mathbf{v}_N$. For the hybrid approximation to the convective term $\mathbf{v}_f \otimes \mathbf{v}_f$ in (4a.36) we have

$$\mathbf{v}_f^{\text{hybrid}} \otimes \mathbf{v}_f^{\text{hybrid}} = \left(\mathbf{v}_f - \mathcal{R}_f^{\text{hybrid}}\right) \otimes \left(\mathbf{v}_f - \mathcal{R}_f^{\text{hybrid}}\right)$$

$$= \mathbf{v}_f \otimes \mathbf{v}_f - \left[\mathbf{C}_f(\nabla\mathbf{v})_f^T + (\nabla\mathbf{v})_f\mathbf{C}_f^T\right] + \chi_f^2((\nabla\mathbf{v})_f\mathbf{d}) \otimes ((\nabla\mathbf{v})_f\mathbf{d})$$

$$+ \frac{1}{64}((\nabla^2\mathbf{v})_f(\mathbf{d} \otimes \mathbf{d})) \otimes ((\nabla^2\mathbf{v})_f(\mathbf{d} \otimes \mathbf{d})) - \frac{1}{8}\big[\mathbf{v}_f \otimes ((\nabla^2\mathbf{v})_f(\mathbf{d} \otimes \mathbf{d}))$$

$$+ ((\nabla^2\mathbf{v})_f(\mathbf{d} \otimes \mathbf{d})) \otimes \mathbf{v}_f\big] + \frac{1}{8}\chi_f\big[((\nabla^2\mathbf{v})_f(\mathbf{d} \otimes \mathbf{d})) \otimes ((\nabla\mathbf{v})_f\mathbf{d})$$

$$+ ((\nabla\mathbf{v})_f\mathbf{d}) \otimes ((\nabla^2\mathbf{v})_f(\mathbf{d} \otimes \mathbf{d}))\big] + \cdots, \tag{4a.38}$$

where $\chi_f = (1 - \Psi_f)[(1 - \ell_f)\beta_f^+ - \ell_f\beta_f^-]$ and $\mathbf{C}_f = \chi_f(\mathbf{v}_f \otimes \mathbf{d})$. Inserting (4a.38) and (4a.37b) into (4a.36) results in the semidiscretized modified incompressible NSE:

$$\partial_t(\mathbf{v}_P) + \frac{1}{\delta V_P}\sum_f \left[(\mathbf{v}_f \otimes \mathbf{v}_f)dA_f - \nu\left((\nabla\mathbf{v})_f dA_f\right)\right]$$

$$+ \frac{1}{\delta V_P}\sum_f \left[\left(-\left[\mathbf{C}_f(\nabla\mathbf{v})_f^T + (\nabla\mathbf{v})_f\mathbf{C}_f^T\right] + \chi_f^2\left((\nabla\mathbf{v})_f\mathbf{d}\right) \otimes \left((\nabla\mathbf{v})_f\mathbf{d}\right)\right)dA_f\right]$$

$$- \frac{1}{\delta V_P}\sum_f \left[\left(\frac{1}{8}\left[\mathbf{v}_f \otimes \left((\nabla^2\mathbf{v})_f(\mathbf{d} \otimes \mathbf{d})\right) + \left((\nabla^2\mathbf{v})_f(\mathbf{d} \otimes \mathbf{d})\right) \otimes \mathbf{v}_f\right]\right)dA_f\right]$$

$$+ \frac{1}{\delta V_P}\sum_f \left[\left(\frac{1}{64}\left((\nabla^2\mathbf{v})_f(\mathbf{d} \otimes \mathbf{d})\right) \otimes \left((\nabla^2\mathbf{v})_f(\mathbf{d} \otimes \mathbf{d})\right)\right)dA_f\right]$$

$$+ \frac{1}{\delta V_P}\sum_f \left[\left(\frac{1}{8}\chi_f\left[\left((\nabla^2\mathbf{v})_f(\mathbf{d} \otimes \mathbf{d})\right) \otimes \left((\nabla\mathbf{v})_f\mathbf{d}\right)\right.\right.\right.$$

$$\left.\left.\left. + \left((\nabla\mathbf{v})_f\mathbf{d}\right) \otimes \left((\nabla^2\mathbf{v})_f(\mathbf{d} \otimes \mathbf{d})\right)\right]\right)dA_f\right]$$

$$- \frac{1}{\delta V_P}\sum_f \left[\frac{1}{24}\nu[(\nabla^3\mathbf{v})_f dA_f](\mathbf{d} \otimes \mathbf{d})\right] + \cdots = -(\nabla p)_P + \mathbf{f}_P. \tag{4a.39}$$

We now convert the semidiscrete modified equations (4a.39) to the corresponding continuous modified equations by using the Gauss theorem in combination with the localization theorem (Kellog 1929),

$$\partial_t(\mathbf{v}) + \nabla \cdot (\mathbf{v} \otimes \mathbf{v})$$

$$= \nabla \cdot (\nu\nabla\mathbf{v}) - \nabla p + \mathbf{f} + \nabla \cdot \left(\mathbf{C}(\nabla\mathbf{v})^T + (\nabla\mathbf{v})\mathbf{C}^T - \chi^2\left((\nabla\mathbf{v})\mathbf{d}\right) \otimes \left((\nabla\mathbf{v})\mathbf{d}\right)\right)$$

$$+ \nabla \cdot \left(\frac{1}{8}\left[\mathbf{v} \otimes \left((\nabla^2\mathbf{v})(\mathbf{d} \otimes \mathbf{d})\right) + \left((\nabla^2\mathbf{v})(\mathbf{d} \otimes \mathbf{d})\right) \otimes \mathbf{v}\right]\right)$$

$$- \nabla \cdot \left(\frac{1}{8}\chi\left[\left((\nabla^2\mathbf{v})(\mathbf{d} \otimes \mathbf{d})\right) \otimes \left((\nabla\mathbf{v})\mathbf{d}\right) + \left((\nabla\mathbf{v})\mathbf{d}\right) \otimes \left((\nabla^2\mathbf{v})(\mathbf{d} \otimes \mathbf{d})\right)\right]\right)$$

$$- \nabla \cdot \left(\frac{1}{64}\left((\nabla^2\mathbf{v})(\mathbf{d} \otimes \mathbf{d})\right) \otimes \left((\nabla^2\mathbf{v})(\mathbf{d} \otimes \mathbf{d})\right)\right)$$

$$+ \nabla \cdot \left(\frac{1}{24}\nu(\nabla^3\mathbf{v})(\mathbf{d} \otimes \mathbf{d})\right) + \cdots, \tag{4a.40}$$

where we note that gridding features directly affect the various terms in (4a.40) through vector field $\mathbf{d}$ and tensor field $\mathbf{d} \otimes \mathbf{d}$. In particular, anisotropies introduced by eventual nonuniform (e.g., adaptive) gridding will directly reflect as contributions to implicit SGS tensor $\mathbf{T}$ through grid tensor $\mathbf{d} \otimes \mathbf{d}$; clearly, this is an additional way in which *good or bad* SGS physics can be implicitly designed. We note that, for a uniform Cartesian grid, with mesh spacing Δx, Δy, and Δz, grid tensor $\mathbf{d} \otimes \mathbf{d}$ degenerates into a diagonal tensor with components Δx^2, Δy^2, and Δz^2.

In the limit of perfectly smooth conditions $\Psi = 1$, $\chi = 0$, and $\mathbf{C} = \mathbf{0}$, and we recover the modified equations for the high-order scheme,

$$\partial_t(\mathbf{v}) + \nabla \cdot (\mathbf{v} \otimes \mathbf{v}) = \nabla \cdot (\nu \nabla \mathbf{v}) - \nabla p + \mathbf{f}$$

$$+ \nabla \cdot \left(\frac{1}{8} \left[\mathbf{v} \otimes ((\nabla^2 \mathbf{v})(\mathbf{d} \otimes \mathbf{d})) + ((\nabla^2 \mathbf{v})(\mathbf{d} \otimes \mathbf{d})) \otimes \mathbf{v} \right] \right)$$

$$- \nabla \cdot \left(\frac{1}{64} ((\nabla^2 \mathbf{v})(\mathbf{d} \otimes \mathbf{d})) \otimes ((\nabla^2 \mathbf{v})(\mathbf{d} \otimes \mathbf{d})) \right)$$

$$+ \nabla \cdot \left(\frac{1}{24} \nu (\nabla^3 \mathbf{v})(\mathbf{d} \otimes \mathbf{d}) \right) + \cdots , \tag{4a.41}$$

commonly used in explicit LES. From the leading-order terms in (4a.40) we can identify the built-in SGS model appearing in ILES when we are using algorithms based on the hybrid flux function,

$$\mathbf{T} = \mathbf{C}(\nabla \mathbf{v})^T + (\nabla \mathbf{v})\mathbf{C}^T - \chi^2((\nabla \mathbf{v})\mathbf{d}) \otimes ((\nabla \mathbf{v})\mathbf{d})$$

$$+ \frac{1}{8} [\mathbf{v} \otimes ((\nabla^2 \mathbf{v})(\mathbf{d} \otimes \mathbf{d})) + ((\nabla^2 \mathbf{v})(\mathbf{d} \otimes \mathbf{d})) \otimes \mathbf{v}], \tag{4a.42}$$

where we note that the last term on the right-hand side of (4a.42), present in both (4a.40) and (4a.41), is independent of the limiter and is due only to the second-order finite-volume discretization.

Other than discussing the perfectly smooth solution case ($\Psi = 1$) just shown, we have made no specific assumptions on Ψ, or equivalently on χ and $\mathbf{C}$. This is a particularly sensitive aspect of the analysis, given that (i) the MEA relies on the existence and use of Taylor series expansions (for which solution smoothness is *in principle* a requirement), and (ii) the local (grid-dependent) flow-adaptive features of Ψ (and their contribution to the series expansion) are intimately connected not only with its specific nature but also with its three-dimensional implementation in the algorithm under consideration. We discuss in detail such properties of the implicit (or built-in) SGS model, Eq. (4a.42), implied by specific assumptions in Section 4a.5.3. However, it is readily apparent that $\mathbf{T}$ has similarities with the commonly used SGS viscosity models of explicit LES, but with grid properties entering in a tensorial form rather than in the common scalar form.

4a.5.2 Compressible Navier–Stokes equations

The analysis of the compressible NSE, Eq. (4a.2), follows the outlined procedure, but it is technically more involved because of the larger number of terms as compared with the incompressible NSE. The results are, however, similar in structure. The semidiscretized compressible NSE is

$$\partial_t(\rho_P) + \frac{1}{\delta V_P} \sum_f \left[(\rho \mathbf{v})_f \cdot d\mathbf{A}_f \right] = 0,$$

$$\partial_t(\rho_P \mathbf{v}_P) + \frac{1}{\delta V_P} \sum_f \left[(\rho \mathbf{v} \otimes \mathbf{v})_f d\mathbf{A}_f - (\mu \nabla \mathbf{v})_f d\mathbf{A}_f + \left(\frac{1}{3}\mu(\nabla \cdot \mathbf{v}) \right)_f d\mathbf{A}_f \right]$$
$$= -(\nabla p)_P + (\rho \mathbf{f})_P,$$

$$\partial_t(\rho_P E_P) + \frac{1}{\delta V_P} \sum_f \left[(\rho \mathbf{v} E)_f \cdot d\mathbf{A}_f - \left(\frac{\kappa}{C_V} \nabla E \right)_f d\mathbf{A}_f - (p\mathbf{v})_f d\mathbf{A}_f \right.$$
$$\left. + \left(\frac{1}{3}\mu(\nabla \cdot \mathbf{v})\mathbf{I} - \mu(\nabla \mathbf{v} + \nabla \mathbf{v}^T) \right)_f d\mathbf{A}_f \right] = (\rho \sigma)_P. \qquad (4a.43)$$

In order to be able to discuss both explicit and implicit LES, we model the convective fluxes in (4a.43), or alternatively in (4a.2), using hybrid flux function (4a.14), which is based on a local combination of linear flux function (4a.12) and the upwind-biased flux function, (4a.11), using blending parameter or flux limiter Ψ. We model the diffusive fluxes in (4a.43), or alternately in (4a.2), by using linear flux function (4a.34), which is based on central differencing of the inner gradient in the diffusive fluxes. Performing the same analysis as described in Section 4a.5.1, we find that the modified equation for the compressible NSE becomes

$$\partial_t(\rho) + \nabla \cdot (\rho \mathbf{v})$$
$$= \nabla \cdot \left(\rho \chi (\nabla \mathbf{v})\mathbf{d} + \frac{1}{8}\rho(\nabla^2 \mathbf{v})(\mathbf{d} \otimes \mathbf{d}) \right) + \cdots,$$

$$\partial_t(\rho \mathbf{v}) + \nabla \cdot (\rho \mathbf{v} \otimes \mathbf{v})$$
$$= -\nabla p + \nabla \cdot \left(\frac{1}{3}\mu(\nabla \cdot \mathbf{v})\mathbf{I} - \mu(\nabla \mathbf{v} + \nabla \mathbf{v}^T) \right) + \rho \mathbf{f}$$
$$+ \nabla \cdot (\rho[\mathbf{C}(\nabla \mathbf{v})^T + (\nabla \mathbf{v})\mathbf{C}^T + \chi^2(\nabla \mathbf{v})\mathbf{d} \otimes (\nabla \mathbf{v})\mathbf{d}])$$
$$+ \nabla \cdot \left(\frac{1}{8}\rho[\mathbf{v} \otimes ((\nabla^2 \mathbf{v})(\mathbf{d} \otimes \mathbf{d})) + ((\nabla^2 \mathbf{v})(\mathbf{d} \otimes \mathbf{d})) \otimes \mathbf{v}] \right)$$
$$- \nabla \cdot \left(\frac{1}{8}\rho \chi[((\nabla^2 \mathbf{v})(\mathbf{d} \otimes \mathbf{d})) \otimes ((\nabla \mathbf{v})\mathbf{d}) + ((\nabla \mathbf{v})\mathbf{d}) \otimes ((\nabla^2 \mathbf{v})(\mathbf{d} \otimes \mathbf{d}))] \right)$$
$$- \nabla \cdot \left(\frac{1}{64}\rho((\nabla^2 \mathbf{v})(\mathbf{d} \otimes \mathbf{d})) \otimes ((\nabla^2 \mathbf{v})(\mathbf{d} \otimes \mathbf{d})) \right)$$
$$+ \nabla \cdot \left(\frac{1}{24}\mu(\nabla^3 \mathbf{v})(\mathbf{d} \otimes \mathbf{d}) \right) + \cdots,$$

$$\partial_t(\rho E) + \nabla \cdot (\rho \mathbf{v} E)$$
$$= \nabla \cdot \left(-p\mathbf{v} + \frac{\kappa}{C_V}(\nabla E)\mathbf{v} + \frac{1}{3}\mu(\nabla \cdot \mathbf{v})\mathbf{v} - \mu(\nabla \mathbf{v} + \nabla \mathbf{v}^T)\mathbf{v} \right) + \rho \sigma$$
$$+ \nabla \cdot (\rho[\mathbf{C}\nabla E + E\chi(\nabla \mathbf{v})\mathbf{d} + \chi^2(\nabla E \cdot \mathbf{d})(\nabla \mathbf{v})\mathbf{d}])$$
$$+ \nabla \cdot \left(\frac{1}{8}\rho[(\nabla^2 E)(\mathbf{d} \otimes \mathbf{d})\mathbf{v} + (\nabla^2 \mathbf{v})(\mathbf{d} \otimes \mathbf{d})E] \right)$$
$$+ \nabla \cdot \left(\frac{1}{8}\rho \chi[(\nabla^2 E)(\mathbf{d} \otimes \mathbf{d})(\nabla \mathbf{v})\mathbf{d} + (\nabla^2 \mathbf{v})(\mathbf{d} \otimes \mathbf{d})(\nabla E)\mathbf{d}] \right)$$
$$- \nabla \cdot \left(\frac{1}{64}\rho[(\nabla^2 \mathbf{v})(\mathbf{d} \otimes \mathbf{d})(\nabla^2 E)(\mathbf{d} \otimes \mathbf{d})] \right) + \nabla \cdot \left(\frac{1}{24}\kappa(\nabla^3 E)(\mathbf{d} \otimes \mathbf{d}) \right) + \cdots.$$
$$(4a.44)$$

If $\Psi = 1$ (and $\chi = 0$, $\mathbf{C}_f = 0$), we again recover the modified equations of the high-order linear scheme. From the leading-order terms in (4a.44) we can identify the built-in SGS model in the ILES based on the hybrid flux function.

We note that implicit SGS models are present not only on the right-hand side of the momentum and energy equations, but also in the continuity equation, as a consequence of the flux reconstruction of the velocity field. This is contrast with classical approaches to LES of compressible flow (e.g., Chapter 3), where use of mass weighted filtered velocities prevents the appearance of SGS terms in the continuity equation. The term on the right-hand side of the continuity equation in (4a.44) is particularly interesting from two perspectives. First, it has the potential of providing a useful model for the effects of SGS mass density fluctuations – the role of which becomes increasingly more important for high-Mach-number flow regimes (e.g., Van der Bos and Geurts 2006). Moreover (on a still controversial note at present), recent arguments on the kinematics of *volume transport* (Brenner 2005) have led to the recognition that (finite scale) mass diffusion effects, which should be clearly absent in the strictly incompressible limit, could, however, be expected in the continuity equation in compressible regimes; in this context, the implicitly provided source term in the mass equation can be viewed as providing some necessary SGS modeling for simulations involving high (but finite) Reynolds numbers – based on mass diffusion, much like when we formally base ILES on the Euler equations. Further research is required to assess these possibilities.

Otherwise, focusing on the momenta and energy equations, we see that the leading terms in the implicit counterparts of the (conventional LES) SGS flux vector and stress tensors are given as

$$\mathbf{T} = \rho[\mathbf{C}(\nabla\mathbf{v})^T + (\nabla\mathbf{v})\mathbf{C}^T + \chi^2(\nabla\mathbf{v})\mathbf{d} \otimes (\nabla\mathbf{v})\mathbf{d}]$$

$$\times \frac{1}{8}\rho[\mathbf{v} \otimes ((\nabla^2\mathbf{v})(\mathbf{d} \otimes \mathbf{d})) + ((\nabla^2\mathbf{v})(\mathbf{d} \otimes \mathbf{d})) \otimes \mathbf{v}],$$

$$\mathbf{t} = \rho[\mathbf{C}\nabla E + E\chi(\nabla\mathbf{v})\mathbf{d} + \chi^2(\nabla E \cdot \mathbf{d})(\nabla\mathbf{v})\mathbf{d}] \qquad (4a.45)$$

$$\times \frac{1}{8}\rho[(\nabla^2 E)(\mathbf{d} \otimes \mathbf{d})\mathbf{v} + (\nabla^2\mathbf{v})(\mathbf{d} \otimes \mathbf{d})E],$$

which have the same mathematical structure, and physical interpretation, as the implicit SGS model in (4a.42) for the incompressible ILES model. In particular, we note, as before, that the last terms in the expressions of $\mathbf{t}$ and $\mathbf{T}$ in (4a.45) depend only on the second-order finite-volume discretization (and not on the limiter).

4a.5.3 Detailed properties of the built-in SGS models of ILES

Next we discuss some key properties of the built-in (or implicit) SGS model in ILES. For this purpose we focus, for simplicity, on the incompressible flow model shown in Eq. (4a.3). The built-in SGS tensor, (4a.42), can be split into three components: $\mathbf{T}^1 = \mathbf{C}(\nabla\mathbf{v})^T + (\nabla\mathbf{v})\mathbf{C}^T$, $\mathbf{T}^2 = \chi^2[(\nabla\mathbf{v})\mathbf{d} \otimes (\nabla\mathbf{v})\mathbf{d}]$, and $\mathbf{T}^3 = \frac{1}{8}\{\mathbf{v} \otimes [(\nabla^2\mathbf{v})(\mathbf{d} \otimes \mathbf{d})] + [(\nabla^2\mathbf{v})(\mathbf{d} \otimes \mathbf{d})] \otimes \mathbf{v}\}$. The first component is symmetric and linear in the velocity gradient tensor, $\nabla\mathbf{v}$, and can therefore be viewed as the implicit counterpart to a generalized conventional SGS viscosity model used in explicit LES (Sagaut, 2005). The second component is also symmetric but quadratic in $\nabla\mathbf{v}$ and can be interpreted as an implicit counterpart to the Clark model (Clark, Ferziger, and Reynolds 1979), which has been generalized by Aldama (1993). These two terms ($\mathbf{T}^1 + \mathbf{T}^2$) provide an implicit analogue of a mixed (SGS viscosity + scale-similar) model. The third component ($\mathbf{T}^3$) is symmetric and, as we already noted, it is associated with the high-order part of the numerical scheme and thus would also be present in a comparable explicit LES framework (competing with the explicit SGS model). The decomposition in $\mathbf{T}^1$, $\mathbf{T}^2$, and $\mathbf{T}^3$ is also attractive, considering the decomposition into rapid and slow parts (Shao, Sarkar, and Pantano 1999). In ILES, the rapid part that cannot be captured by isotropic models relates to $\mathbf{T}^2$, while the slow part relates to $\mathbf{T}^1$. Borue and Orszag (1998) presented evidence that a $\mathbf{T}^2$-type term improves the correlations between the exact and modeled SGS stress tensor.

4a.5.3.1 Implications of a specific model for the limiter

Up to this point and in previously published MEA studies for this class of schemes (e.g., Fureby and Grinstein 1998, 2002), we have made no explicit use in the analysis of any specifics on the nature of flux limiter Ψ. As we noted, the impact of the local features of Ψ on the MEA are intimately connected with their nature and actual implementation in the algorithm under consideration; in particular, this information must be specified to determine the actual contribution of the limiter to the Taylor series expansions used in the analysis.

In order to assess the possible implications of including such detailed limiter information, let us assume a simple model for the limiter at the smooth end of its range; that is, assume $\Psi \approx 1 - \theta|\mathbf{d}|$, where $\theta > 0$ has dimensions of reciprocal length. Such a model is relevant, in particular, when using directional split integrations, since it can be realized in one dimension. Specifically, for the van Leer limiter (cf. Table 4a.2), we have $\Psi = \Psi(r) = 2r/(1 + r)$, with $r = (\rho_i - \rho_{i-1})/(\rho_{i+1} - \rho_i) > 0$, and then, assuming smooth conditions in the neighborhood of $x = x_i$, we see $r \approx 1 - h\partial_{xx}\rho/\partial_x\rho$ and $\Psi \approx 1 - \frac{3}{2}h\partial_{xx}\rho/\partial_x\rho$, where $h = x_{i+1} - x_i = x_i - x_{i-1}$ (L. Margolin, private communication, December 2005), and $\theta = \frac{3}{2}\partial_{xx}\rho/\partial_x\rho$ is associated with large-scale variations of ρ. It is interesting to note here that when ρ is the streamwise velocity, and x denotes a cross-stream variable, the reciprocal of θ has the form of a mixing length used in conjunction with a classical similarity hypothesis (e.g., Schlichting 1979). We now consider the inclusion of this simple model ($\Psi \approx 1 - \theta|\mathbf{d}|$) and focus on the lowest-order implied approximations, that is, $\chi \approx \theta|\mathbf{d}|[(1 - \ell)\beta^+ - \ell\beta^-]$ and $\mathbf{C} \approx \theta|\mathbf{d}|[(1 - \ell]\beta^+ - \ell\beta^-(\mathbf{v} \otimes \mathbf{d})$, which in turn lead to the following estimates of the implicit SGS tensor:

$$\mathbf{T}^1 \approx [\theta|\mathbf{d}|((1-\ell)\beta^+ - \ell\beta^-)(\mathbf{v}\otimes\mathbf{d})](\nabla\mathbf{v})^T + (\nabla\mathbf{v})[\theta|\mathbf{d}|((1-\ell)\beta^+ - \ell\beta^-)(\mathbf{v}\otimes\mathbf{d})]^T,$$

$$\mathbf{T}^2 \approx [\theta|\mathbf{d}|((1-\ell)\beta^+ - \ell\beta^-)]^2((\nabla\mathbf{v})\mathbf{d} \otimes (\nabla\mathbf{v})\mathbf{d}),$$

$$\mathbf{T}^3 = \frac{1}{8}[\mathbf{v} \otimes ((\nabla^2\mathbf{v})(\mathbf{d} \otimes \mathbf{d})) + ((\nabla^2\mathbf{v})(\mathbf{d} \otimes \mathbf{d})) \otimes \mathbf{v}]. \tag{4a.46}$$

From (4a.46) we find that in regimes of smooth flow, the leading $O(|\mathbf{d}|^2)$ contribution to the implicit SGS model T comes from $\mathbf{T}^1 + \mathbf{T}^3$ (dissipative and dispersive terms, proportional to $|\nabla \mathbf{v}|$ and $|\nabla^2 \mathbf{v}|$, respectively), whereas $\mathbf{T}^2$ is $O(|\mathbf{d}|^4)$ and proportional to $|\nabla \mathbf{v}|^2$. The latter (higher-order) $\mathbf{T}^2$ term is a dispersive implicit analogue of a (rapid) scale-similar component, which (as noted) is known to improve on the anisotropic predictions of SGS models.

In closing, we note the limitations of this analysis. First, FCT (and other popular) limiters cannot be expressed as simple functions $\Psi(r)$ even in one dimension, and the specifics of full (unsplit) multidimensional implementations will have a role that also must be addressed. Moreover, associated with our smooth flow assumption $\Psi \approx 1 - \theta|\mathbf{d}|$, the MEA results presented here do not provide insights regarding behaviors on the other end of the limiter range, that is, in the nonsmooth flow regime largely impacted by the low-order part of the hybrid scheme – where the leading-order terms in the Taylor series expansions are likely not very meaningful by themselves. Alternative procedures for processing information in a series expansion, such as those based on using Padé approximants (e.g., Baker 1975), typically have much larger convergence domains and could provide greater applicability of MEA. The appropriateness of such techniques has yet to be evaluated in the MEA context, and their use might allow for better understanding on how nonoscillatory finite-volume methods work when the limiters are active. This is clearly an area where the analysis has to be extended and improved.

4a.6 The Taylor–Green Vortex problem

The Taylor–Green Vortex (TGV) system is a well-defined flow that has been used as a prototype for vortex stretching, instability, and production of small-scale eddies to examine the dynamics of transition to turbulence based on DNS (Brachet et al. 1983). As such, it can also be effectively used (as done here) as a convenient case to test the ability of explicit and implicit SGS modeling to allow the simulation of the basic empirical laws of turbulence (e.g., Frisch 1995), namely the existence of an inertial subrange on the energy spectra for sufficiently high Reynolds number and the finite (viscosity-independent) energy-dissipation-limit law; see also Chapter 6.

The TGV case is also used here to revisit the ability of ILES based on locally monotonic, nonoscillatory finite-volume methods to reproduce established features of decaying turbulence. Successful such demonstrations using the piecewise parabolic method (Chapter 7, this volume) and FCT (Fureby and Grinstein 2002) have been reported. Positive evaluations of ILES in the turbulence-decaying case using a certain class of nonmonotonic nonoscillatory finite-volume methods are discussed in Chapters 5 and 14. In contrast, poor ILES performances noted elsewhere in this context by Garnier et al. (1999), using other popular shock-capturing schemes – such as MUSCL (van Leer 1979) with a minmod limiter – indicate that inherently more diffusive features of certain nonoscillatory finite-volume algorithms are not suitable for ILES. A further discussion characterizing desirable numerics for ILES is presented in Chapter 5.

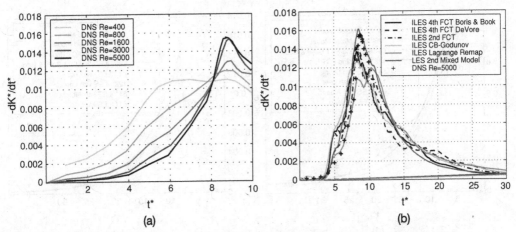

Figure 4a.4. Temporal evolution of the kinetic energy dissipation rate $-dK/dt$ from (a) the DNS of Brachet et al. (1983) and Brachet (1991), and (b) the explicit and implicit 128^3 LES calculations.

It is important to note here that the conventional wisdom (classical LES) is that numerical diffusion effects at the convection stage are undesirable and should be avoided. Specifically, the kinetic energy can be damped only by resolved (large-scale) viscous effects, or by those introduced through the explicit SGS models. Kinetic energy should otherwise be conserved. In this context, integral measures in the TGV case have been previously used as a reference to assess (the presumably unwanted) effects of numerical dissipation (Benhamadouche and Laurence 2002; Mossi 1999; Shu et al. 2005). This is in contrast with the ILES perspective here and elsewhere in this volume, where we demonstrate how convective numerical diffusion effects of certain algorithms can be effectively used *by themselves* to emulate the dominant SGS physics in the high-Reynolds-number applications.

The TGV configuration considered here involves triply-periodic boundary conditions enforced on a cubical domain with a box side length of 2π cm, using 64^3 or 128^3 evenly spaced computational cells. The flow is initialized with the solenoidal velocity components, $u = u_o \sin(x)\cos(y)\cos(z)$, $v = -u_o \cos(x)\sin(y)\cos(z)$ and $w = 0$, with pressure given by a solution of the Poisson equation for the aforementioned given velocity field, that is, $p = p_o + [\rho(u_o)^2/8][1 + \cos(2z)][\cos(2x) + \cos(2y)]$, where we further select $p_o = 1.0$ bar, mass density $\rho = 1.178$ kg/m^3, $u_o = 10,000$ cm/s (corresponding to a Mach number of M = 0.28), and an ideal gas equation of state for air.

Using truncated series analysis techniques, Morf, Orszag, and Frisch (1980) identified an inviscid instability for the TGV system, with estimated onset at a nondimensional time $t^* = ku_o t \approx 5.2$ (wave number k is unity here). These results were later questioned in Brachet (1991), where it was pointed out that the accuracy in the analytic continuation procedure used by Morf et al. (1980) deteriorates too quickly to lead to a definite conclusion regarding their early prediction. Further estimates of t^* based on the DNS of Brachet et al. (1983) and Brachet (1991) reported a fairly consistent dissipation peak at $t^* \approx 9$, for Re = 800, 1600, 3000, and 5000, where Re was based on the integral scale. The almost indistinguishable results for Re = 3000 and 5000, such as in Figure 4a.4(a), suggested that they may be close to the viscosity-independent limit.

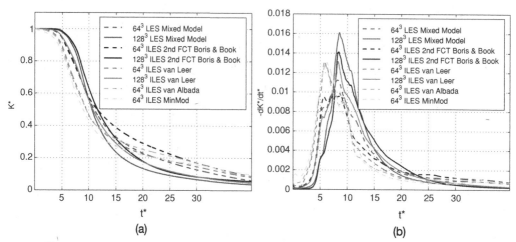

Figure 4a.5. Temporal evolution of the turbulent kinetic energy, K, and its dissipation rate $-dK/dt$ for a range of schemes and for 64^3 and 128^3 resolutions.

Whether or not this finite-time singularity exists for the purely inviscid case remains unsettled and controversial (e.g., Majda and Bertozzi 2002, Shu et al. 2005).

ILES was tested on this TGV case by the use of various algorithms, including FCT, characteristics-based (CB) – Godunov, and Lagrangian remap (LR) methods, discussed herein and in Chapter 4c, as well as conventional LES (Drikakis et al. 2007) . The FCT schemes considered involved the standard fourth-order FCT algorithm (Boris and Book 1973), the fourth-order 3D monotone limiter FCT (DeVore 1998), and a second-order FCT using a hybridization of first-order upwind and second-order central differences. The CB schemes (Bagabir and Drikakis 2004; Zoltak and Drikakis 1998) used are of the Godunov type (Drikakis and Rider 2004; Eberle 1987), and of the hybrid TVD type (Bagabir and Drikakis 2004; Zoltak and Drikakis 1998). The fluxes are discretized at the cell faces by using the values of the conservative variables along the characteristics. Third-order variants of the fluxes can be obtained through flux limiting, based on the squares of second-order pressure or energy derivatives. Examples of flux limiting in connection with the CB scheme can be found in Eberle (1987) and in Drikakis and Rider (2004). The LR method (Youngs 1991) uses a nondissipative finite-difference method plus quadratic artificial viscosity in the Lagrange phase, and a third-order van Leer monotonic advection method (van Leer 1977) in the remap phase.

The evolution in time of the kinetic energy dissipation $-dK/dt$, where $K = \frac{1}{2}\langle \mathbf{v}^2 \rangle$ and $\langle \cdot \rangle$ denotes mean (volumetric average), is demonstrated in Figure 4a.4(b) for the 128^3 resolution, for FCT, CB–Godunov, LR, and the conventional LES, using the mixed model of Bardina, Ferziger, and Reynolds (1980). The fastest K decay at the dissipation peak (and peak mean enstrophy – Drikakis et al. 2006) corresponds to the onset of the inviscid TGV instability at $t^* \approx 9$. Figure 4a.5 also shows results from the 64^3 grid simulations, and further comparisons of the kinetic energy and its dissipation using gamma (Jasak et al. 1999), minmod (Roe 1985), and (Albada et al. 1982) flux limiting (also described herein), as well as the conventional LES mixed-model approach.

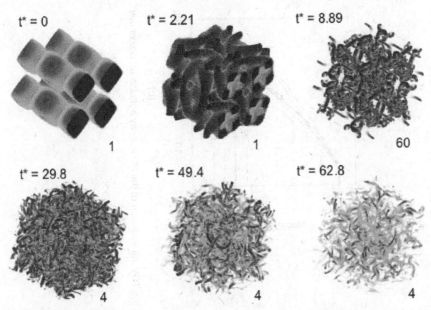

Figure 4a.6. Flow visualizations of the TGV flow, using volume renderings of λ_2. The results shown here (representative of all methods discussed) are from the fourth-order 3D monotone FCT on the 128^3 grid. (See color inset, Plate 1).

The observed qualitative agreement among ILES, mixed-model LES, and previous high-Reynolds-number DNS is quite good at predicting both the time and height of the dissipation peak for the high-Reynolds-number limit; agreement is particularly good when we are comparing the (present) 128^3 results with the highest-Reynolds-number DNS cases. We can use an effective Reynolds number increasing with grid resolution to consistently explain the lower characteristic t^* at the dissipation peaks predicted by the 64^3 simulations, as in the lower-Reynolds-number DNS results in Figure 4a.4(a).

Some obvious differences between the present ILES and the conventional LES compared with DNS, such as the additional structure near $t^* \approx 5$–6 in Figure 4a.4(b), not present in the DNS results as well as other observed finer structure in the details of the results are tied with specifics of the various limiting algorithms or their implementation. A more systematic analysis of these features will be reported separately. For example, the double-peaked structure of the dissipation near $t^* \approx 9$ predicted by the LR method is likely due to the dispersive properties of this scheme compared with the less dispersive FCT and CB–Godunov schemes. In contrast, the LR method is also the least dissipative of the compared methods, especially at the early stages of the flow development, exhibiting a relatively slower decay rate of K (in Figure 4a.7, discussed later).

The fluid dynamics underlying the dissipation results shown herein is analyzed in what follows. Figure 4a.6 compares instantaneous flow visualizations ranging from the initial TGV at $t^* = 0$, the transition to increasingly smaller-scale (but organized) vortices (top row), to the fully developed (disorganized) decaying worm-vortex-dominated flow regime (bottom row), as characteristic of developed turbulence. The particular

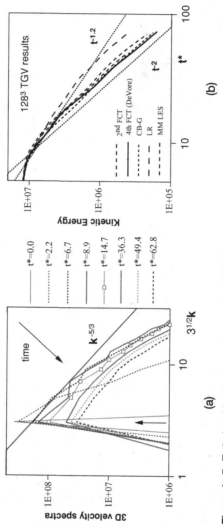

Figure 4a.7. Turbulent kinetic energy decay in the TGV: (a) evolution of the 3D velocity spectra; (b) power law evolution of the volumetric mean.

simulation was the one run with the fourth-order three-dimensional monotone FCT on the 128^3 grid. The flow visualizations are based on (ray-tracing) volume renderings of λ_2 – the second-largest eigenvalue of the velocity gradient tensor (Jeong and Hussain 1995), where hue and opacity maps were chosen the same for all times, except for peak magnitude values (normalized by value at $t^* = 0$), indicated at the lower right of each frame. As the size of the smallest-scale structures approaches the cutoff resolution, kinetic energy is removed at the grid level through numerical dissipation.

Corresponding 3D velocity spectra are shown in Figure 4a.7. The peak in the velocity spectra around the $k = \sqrt{3}$ wave-number shell reflects the imprint leftover by the chosen initial (TGV) conditions ($t^* = 0$). Higher wave-number modes are populated in time ($t^* = 2.2$), by the virtually inviscid cascading process. As the size of the smallest-scale structures approaches the cutoff resolution, kinetic energy is removed at the grid level through numerical dissipation. Figure 4a.7 shows that the spectra consistently emulate a $(-5/3)$ power law inertial subrange and self-similar decay for the times $t^* > 6$. As the Kolmogorov spectra become more established in time ($t^* > 15$ in Figure 4a.7), it is associated at the smallest resolved scale with the more disorganized worm-vortex-dominated flow regime (bottom row of Figure 4a.6) – as characteristic of developed turbulence (e.g., Jimenez et al. 1993).

The depicted self-similar decay in Figure 4a.7 suggests that the removal of kinetic energy by numerical dissipation may occur at a physically suitable rate. The decay rates are examined with more detail on the basis of selected representative 128^3 data in Figure 4a.7, where, for reference, we have indicated slopes corresponding to power laws with exponents -1.2 and -2 through the mean value of K at the observed dissipation peaks (Figure 4a.4). Despite the unavoidable degree of subjectivity introduced by choice of origin times when such power law fits are attempted, all compared methods show decay rates consistent with each other and with the -1.2 law for times immediately after that of the dissipation peak at $t^* \approx 9$, and with the -2 exponent for much later times. The power law with the -1.2 exponent is the one generally accepted as characteristic of decaying turbulence, whereas the *later* -2 exponent can be understood in terms of the expected saturation of the energy-containing length scales, reflecting that eddies larger than the simulation box side length cannot exist (Skrbek and Stalp 2000). The differences between the LR results and the other schemes employed here can be attributed to the Lagrangian step, which tends to be less dissipative and more dispersive than the FCT and Godunov-type schemes.

4a.7 Concluding remarks

In this chapter we have outlined the basic principles for flux-limiting-based numerical methods particularly suited for ILES, with particular emphasis on unstructured finite-volume-based LES codes. We discussed the general principles for finite-volume discretization of both the incompressible and the compressible NSE, and time integration by means of both explicit and implicit methods. The concept of flux limiting is introduced as a systematic way of constructing high-resolution schemes with particular properties that have the ability to adapt to the local features of solution. We used MEA to show that the leading-order truncation error terms introduced by such methods provide implicit SGS models similar in form to those of conventional mixed SGS models.

We tested the behavior of ILES based on different high-resolution monotonic schemes in the simulation of the TGV. The results show that numerical schemes such as those tested here can provide a fairly robust LES computational framework to capture the physics of flow transition and turbulence decay without resorting to an explicit SGS model and using relatively coarse grids. The results also show that the kinetic energy-dissipation specifics depend somewhat on the detailed aspects of the SGS model provided implicitly by the various high-resolution algorithms (or explicitly by the conventional LES methods), and by their actual implementations.

ILES performance enhancements are possible through improved design of the SGS physics-capturing capabilities. Thus, further investigations seeking a better understanding of the specific dissipation and dispersion properties of the different high-resolution schemes – and suitable testing and validation frameworks to establish a physical basis for the various possible choices – are clearly warranted in this context.

REFERENCES

Aldama A. A., 1993 Leonard and cross-term approximations in the anisotropically filtered equations of motion, in *Large Eddy Simulation of Complex Engineering and Geophysical Flows*, ed. B. Galperin & S. A. Orszag. New York: Cambridge University Press.

Bagabir, A., & Drikakis, D. 2004 Numerical experiments using high-resolution schemes for unsteady, inviscid, compressible flows. *Comput Meth. Appl. Mech. Eng* **193**, 4675–4705.

Baker, G. A. 1975 *Essentials of Padé Approximants.* New York: Academic Press.

Bardina, J., Ferziger, J. H., & Reynolds, W. C. 1980 Improved subgrid scale models for large eddy simulations. Paper AIAA 80-1357.

Barth, T., & Ohlberger, M. 2004 Finite volume methods: Foundation and analysis, in *Encyclopedia of Computational Mechanics*, Ed. E. Stein, R. de Borst, & T. J. R. Hughes. New York: Wiley.

Beam, R. M., & Warming, R. F. 1976 An implicit finite-difference algorithm for hyperbolic system in conservation law form. *J. Comput. Phys.* **22**, 87–110.

Bell, J. B., Colella, P., & Glaz, H. M. 1989 A second-order projection method of the incompressible Navier–Stokes equations. *J. Comput. Phys.* **85**, 257–283.

Benhamadouche, S., & Laurence, D. 2002 Global kinetic energy conservation with unstructured meshes. *Int. J. Numer. Meth. Fluids* **40**, 561–571.

Boris J. P. 1990 On large eddy simulation using subgrid turbulence models, in *Whither Turbulence? Turbulence at the Crossroads*, ed. J. L. Lumley. New York: Springer, 344.

Boris, J. P., & Book, D. L. 1973 Flux-corrected transport I. SHASTA, a fluid transport algorithm that works. *J. Comput. Phys.* **11**, 38–69.

Boris, J. P., & Book, D. L. 1976 Flux-corrected transport III. Minimal-error FCT algorithms. *J. Comput. Phys.* **20**, 397–431.

Boris, J. P., Book, D. L., & Hain, K. 1975 Flux-corrected transport II: Generalizations of the method. *J. Comput. Phys.* **18**, 248–283.

Boris, J. P., Grinstein, F. F., Oran, E. S., & Kolbe, R. L. 1992 New insights into large eddy simulation. *Fluid Dyn. Res.* **10**, 199–229.

Borue, V., & Orszag S. A. 1998 Local energy flux and subgrid-scale statistics in three dimensional turbulence. *J. Fluid Mech.* **366**, 1–31.

Brachet, M. E. 1991 Direct numerical simulation of three-dimensional turbulence in the Taylor–Green vortex, *Fluid Dyn. Res.* **8**, 1–8.

Brachet, M. E., Meiron, D. I., Orszag, S. A., Nickel, B. G., Morg, R. H., & Frisch, U. J. 1983 Small-scale structure of the Taylor–Green vortex. *J. Fluid Mech.* **130**, 411–452.

Brenner, H. 2005 Kinematics of volume transport, *Physica A* **349**, 11–59.

Chakravarthy, S. R., & Osher, S. 1983 High resolution applications of the Osher upwind scheme for the Euler equations. Paper AIAA 83-1943.

Clark, R. A., Ferziger, J. H., & Reynolds, W. C. 1979 Evaluation of sub-grid scale models using an accurately simulated turbulent flow. *J. Fluid Mech.* **118**, 1–16.

Darwish, M. S. 1993 A new high-resolution scheme based on the normalized variable formulation. *Num. Heat Transfer B* **24**, 353–371.

DeVore, C. R. 1998 *An Improved Limiter for Multidimensional Flux-Corrected Transport.* Washington, DC: NRL Report NRL-MR-6440-98-8330.

Drikakis, D., Fureby, C., Grinstein, F. F., & Youngs, D. 2007 Simulation of transition and turbulence decay in the Taylor–Green vortex. *J Turb.*, **8**, 020.

Drikakis, D., Govatsos, P. A., & Papantonis, D. E. 1994 A characteristic-based method for incompressible flows. *Int. J. Numer. Meth. Fluids* **19**, 667–685.

Drikakis, D., & Rider, W. 2004 *High-Resolution Methods for Incompressible and Low-Speed Flows.* New York: Springer-Verlag.

Eberle, A. 1987 Characteristic flux averaging approach to the solution of Euler's equations. Report 1987-04, *VKI Lecture Series, Computational Fluid Dynamics.*

Frisch, U. 1995 *Turbulence.* New York: Cambridge University Press, 57.

Fromm, J. E. 1968 A method for reducing dispersion in convective difference schemes. *J. Comput. Phys.* **3**, 176–189.

Fureby C., & Grinstein, F. F. 1998 Monotonically integrated large eddy simulation of free shear flows, Paper AIAA 98-0537.

Fureby C., & Grinstein, F. F. 2002 Large eddy simulation of high Reynolds-number free and wall bounded flows. *J. Comp. Phys.* **181**, 68–97.

Garnier, E., Mossi, M., Sagaut, P., Comte, P., & Deville, M. 1999 On the use of shock-capturing schemes for large eddy simulation, *J. Comput. Phys.* **153**, 273–311.

Gear, C. W. 1971 *Numerical Initial Value Problems in Ordinary Differential Equations.* Englewood Cliffs, NJ: Prentice-Hall.

Godunov, S. K. 1959 A finite difference method for the numerical computation of discontinuous solutions of the equations of fluid dynamics. *Mat. Sb.* **47**, 271–306.

Gottlieb, S., & Shu, C. W. 1998 Total variation diminishing Runge–Kutta schemes. *Math. Comp.* **67**, 73–85.

Gurtin, M. E. 1981 *An Introduction to Continuum Mechanics.* Orlando: Academic Press.

Harten, A. 1983 High resolution schemes for hyperbolic conservation laws. *J. Comput. Phys.* **49**, 357–393.

Hartwich, P.-M., Hsu, C.-H., & Liu, C. H. 1988 Vectorizable implicit algorithms for the flux difference split three-dimensional Navier–Stokes equations. *J. Fluids Eng.* **110**, 297–305.

Hirt, C. W. 1968 Heuristic stability theory for finite difference equations. *J. Comput. Phys.* **2**, 339–355.

Hundsdorfer, W., Koren, B., van Loon, M., & Verwer, J. G. 1995 A positive finite difference advection scheme. *Appl. Math. Comput.* **117**, 35–46.

Issa, R. I. 1986 Solution of the implicitly discretised fluid flow equations by operator splitting. *J. Comput. Phys.* **62**, 40–65.

Jameson, A., Schmidt, W., & Turkel, E. 1981 Numerical simulation of the Euler equations by finite volume methods using Runge–Kutta time stepping schemes. Paper AIAA 81-1259.

Jasak, H., Weller, H. G., & Gosman, A. D. 1999 High resolution NVD differencing scheme for arbitrarily unstructured meshes. *Int. J. Numer. Meth. Fluids* **31**, 431–437.

Jeong, J., & Hussain, F. 1995 On the identification of a vortex. *J. Fluid Mech.* **285**, 69–94.

Jimenez, J., Wray, A., Saffman, P., & Rogallo, R. 1993 The structure of intense vorticity in isotropic turbulence. *J. Fluid Mech.* **255**, 65–90.

Kellogg, O. D. 1929 *Foundations of Potential Theory*. New York: Springer-Verlag (reprinted 1959, Dover).

Kuzmin, D., Lohner, R., & Turek, S. 2005 *High-Resolution Schemes for Convection Dominated Flows: 30 Years of FCT*. New York: Springer-Verlag.

Kwak, D., Chang, J. L. C., Shanks, S. P., & Chakravarthy, S. R. 1986 A three-dimensional incompressible Navier–Stokes flow solver using primitive variables. *AIAA J.* **24**, 390–396.

Lambert, J. D. 1973 *Computational Methods in Ordinary Differential Equations*. New York: Wiley.

Lax, P. D. 1973 *Hyperbolic Systems of Conservation Laws*. Philadelphia: SIAM.

Lax, P. D., & Wendroff, B. 1964 Difference schemes for hyperbolic equations with high order of accuracy. *Commun. Pure Appl. Math.* **17**, 381.

Leonard, B. P., & Niknafs, H. S. 1991 Sharp monotonic resolution of discontinuities without clipping of narrow extrema. *Comput. Fluids* **19**, 141–154.

LeVeque, R. J. 1992 *Numerical Methods for Conservation Laws*. Berlin: Birkhäuser Verlag 2nd ed.

Majda, A. J., & Bertozzi, A. L. 2002 *Vorticity and Incompressible Flow*. Cambridge Texts in Applied Mathematics, Vol. 27. New York: Cambridge University Press.

Morf, R. H., Orszag, S. A., & Frisch, U. 1980 Spontaneous singularity in three-dimensional, inviscid, incompressible flow. *Phys. Rev.* **44**, 572–575.

Mossi, M. 1999 *Simulation of Benchmark and Industrial Unsteady Compressible Turbulent Fluid Flows*. PhD. thesis, École Polytechnique Fédérale de Lausanne.

Oran, E. S., & Boris, J. P. 2001 *Numerical Simulation of Reactive Flow*. New York: Cambridge University Press, 2nd ed.

Roe, P. L. 1981 Numerical algorithms for the linear wave equation. Royal Aircraft Establishment, UK: Technical Report 81047.

Roe, P. L. 1985 Some contributions to the modeling of discontinuous flows. *Lectures Appl. Math.* **22**, 163.

Rogers, S. E., & Kwak, D. 1990 Upwind differencing scheme for the time-accurate incompressible Navier–Stokes equations. *AIAA J.* **28**, 253–262.

Sagaut, P. 2005 *Large Eddy Simulation for Incompressible Flows*. New York: Springer, 3rd ed.

Schlichting, H. 1979 *Boundary Layer Theory*. New York: McGraw-Hill, 585.

Skrbek, L., & Stalp, S. R. 2000 On the decay of homogeneous isotropic turbulence. *Phys. Fluids* **12**, 1997–2019.

Shao L., Sarkar, S., & Pantano, C. 1999 On the relationship between the mean flow and subgrid stresses in large eddy simulation of turbulent shear flows. *Phys. Fluids*, **11**, 1229–1248.

Shu, C-W., Don, W-S., Gottlieb, D., Schilling, O., & Jameson, L. 2005 Numerical convergence study of nearly incompressible, inviscid Taylor–Green vortex flow. *J. Sci. Comput.* **24**, 1–27.

Sweby, P. K. 1984 High resolution schemes using flux limiters for hyperbolic conservation laws. *SIAM J. Numer. Anal.* **21**, 995–1011.

Sweby, P. K. 1985a Flux limiters, in *Numerical Methods for the Euler Equations of Fluid Dynamics*, ed. F. Angrand, A. Dervieux, J. A. Desideri, & R. Glowinski Philadelphia, SIAM, 48–65.

Sweby, P. K. 1985b High resolution TVD schemes using flux limiters, in *Lectures in Applied Mathematics*, *22*, ed. B. Engquist, La Jolla, 289–309.

Thuburn, J. 1993 TVD schemes, positive schemes, and the universal limiter. *Mon. Weather Rev* **125**, 1990–1993.

Thuburn, J. 1996 Multidimensional flux-limited advection schemes. *J. Comput. Phys.* **123**, 74–83.

van Albada, G. D., van Leer, B., & Roberts, W. W. 1982 A comparative study of computational methods in cosmic gas dynamics. *Astron. Astrophys.* **108**, 76–84.

Van der Bos, F., & Geurts, B. J. 2006 Computational turbulent stress closure for large eddy simulation of compressible flow. *J. Turb.*, **7**, no. 9. 1–16.

Van der Houwen, P. J. 1977 *Construction of Integration Formulas for Initial Value Problems*. New York: North-Holland.

van Leer, B. 1977 Towards the ultimate conservative difference scheme IV. A new approach to numerical convection. *J. Comput. Phys.* **23**, 276–299.

van Leer, B. 1979 Towards the ultimate conservative difference scheme V. A second order sequel to Godunov's method. *J. Comput. Phys.* **32**, 101–136.

Yanenko, N. N., & Shokin, Y. I. 1969 First differential approximation method and approximate viscosity of difference schemes. *Phys. Fluids* **12**, 28–34.

Youngs, D. L. 1991 Three-dimensional numerical simulation of turbulent mixing by Rayleigh–Taylor instability. *Phys. Fluids* **A3**, 1312–1320.

Zalesak, S. T. 1979 Fully multidimensional flux-corrected transport algorithms for fluids. *J. Comput. Phys.* **31**, 335–362.

Zalesak, S. T. 2005 The design of flux-corrected transport (FCT) algorithms for structured grids, in *High-Resolution Schemes for Convection Dominated Flows: 30 Years of FCT*, ed. D. Kuzmin, R. Lohner, & S. Turek. New York: Springer-Verlag, 30–78.

Zoltak, J., & Drikakis, D. 1998 Hybrid upwind methods for the simulation of unsteady shock-wave diffraction over a cylinder. *Comput. Meth. Appl. Mech. Eng.* **162**, 165–185.

4b The PPM Compressible Gas Dynamics Scheme

Paul R. Woodward

4b.1 Introduction

The development of the Piecewise Parabolic Method (PPM) gas dynamics scheme grew out of earlier work in the mid-1970s with Bram van Leer on the MUSCL scheme (van Leer 1977, 1979; van Leer and Woodward 1980). The work of Godunov (1959) inspired essential aspects of MUSCL. Features that I introduced into the MUSCL code – directional operator splitting and a Lagrangian step followed by a remap – were inspired by my work with the BBC code (Sutcliffe 1973) at Livermore. These features of BBC were derived in turn from the work in the 1960s by DeBar on the KRAKEN code (DeBar 1974). At Livermore, in the late 1970s, the use of moments of the internal cell distributions, which makes MUSCL resemble finite element schemes, was abandoned for reasons of compatibility with Livermore production codes. In order to recapture the accuracy of MUSCL while using only cell averages as the fundamental data, I developed PPM in collaboration with Phil Colella (Woodward and Colella 1981, 1984; Colella and Woodward 1984; Woodward 1986). Over the past 20 years, PPM has evolved considerably in order to address shortcomings of the scheme as it was laid out in 1984 and 1986. A complete description of the scheme has not appeared in print, but descriptions on the LCSE Web site associated with the PPMLib subroutine library exist (Edgar, Woodward, and Anderson 1998). PPM has also been extended to magnetohydrodynamics in collaboration with Wenlong Dai (Dai and Woodward 1994a,b,c, 1995a, 1998a,b); applied extensively to the study of jets and supersonic shear layers in collaboration with Karl-Heinz Winkler, Steve Hodson, Norm Zabusky, Jeffrey Pedelty, and Gene Bassett (Woodward 1985, 1986; Pedelty and Woodward 1991; Bassett and Woodward 1995a,b); used to simulate flow around stationary and moving objects in collaboration with B. Kevin Edgar (Edgar and Woodward 1993); applied extensively to convection and turbulence problems in collaboration with David Porter, Annick Pouquet, Igor Sytine, and the Livermore ASCI turbulence team (see references in Chapters 7 and 15 of this volume); and, most recently, extended to multifluid gas dynamics problems. Implicit versions of the method have also been worked on in collaboration with Bruce Fryxell and, later, Wenlong Dai (Fryxell et al. 1986; Dai and Woodward 1995b, 1996a,b, 1998c, 2000). A cell-based adaptive mesh refinement (AMR) version of PPM has been under development in collaboration with Dennis Dinge (see Dinge and Woodward 2003). A complete description of PPM algorithm at the time of this writing can also be found on the LCSE Web site (Woodward 2005). Versions of PPM gas dynamics scheme have become incorporated into three community codes aimed at astrophysics applications: FLASH (Fryxell et al. 2000, Calder et al. 2002), ENZO (Bryan et al. 1995, Kritsuk et al. 2006), and VH-1 (Blondin and Lufkin 1993, Mahinthakumar et al. 2004).

Because the several improvements and modifications to the 1984 or 1986 versions of PPM in the literature have not been published, here I lay out the present scheme for simple, single-fluid, ideal gas dynamics in some detail for those aspects that most directly affect simulation of turbulent flows. I do not describe the aspects of the scheme that allow it to accurately capture and follow shocks, because of the lack of space. For these aspects of PPM algorithm, which are rather complex, the reader is referred to the Web documents (Woodward 2005), because this treatment has changed since the scheme was last described completely in the literature.

4b.2 Design constraints

Before launching into a systematic description of the PPM algorithm, I find it worth-while to first explain the goals and constraints that have influenced its design. These are as follows:

- directional operator splitting;
- robustness for problems involving very strong shocks;
- contact discontinuity steepening;
- fundamental data in the form of cell averages only;
- minimal dissipation;
- numerical errors nevertheless dominated by dissipation, as opposed to dispersion;
- preservation of signals, if possible, even if their shapes are modified, as long as they travel at roughly the right speeds; and
- minimal degradation of accuracy as the Courant number decreases toward zero.

The first of these design constraints was to guarantee very high processing speeds on all CPU designs. It has the side benefit of allowing the PPM scheme to employ complicated techniques to achieve high accuracy that would be impractical in a truly multidimensional implementation, especially a three-dimensional (3D) one. Accuracy is still limited in principle by the use of one-dimensional (1D) sweeps, but, in practical problems, errors produced by the use of 1D sweeps have rarely, if ever, appeared to be important. The demand that the scheme be robust in the presence of very strong shocks has produced a general orientation that dissipation, in the right places, times, and amounts is good rather than bad. Early experience with Glimm's random-choice scheme for multidimensional compressible gas dynamics in the 1970s illustrated the error of letting the formal dissipation of the scheme vanish in smooth flow when the simulated flow contains shocks. In low-speed turbulent flows, the need for dissipation in the numerical scheme is less obvious but no less real. Nevertheless, a design goal has always been to minimize the dissipation consistent with an accurate solution. The treatment of contact discontinuities involves a steepening algorithm that has an antidiffusion effect. This contact discontinuity steepening is included in PPM in order to allow multifluid problems to be treated simply through the addition of passively advected constitutive properties, such as the constants of an analytic representation of the equation of state. This steepening of contact discontinuities and the action of monotonicity constraints together serve to preserve signals that can be moved across the grid accurately, even though this sometimes requires a falsification of the signal shape. Such preserved but falsified signals travel at the fluid velocity, since very-short-wavelength sound waves, with the exception of shocks, are difficult to propagate at the right speeds. In PPM we take the view that signals that cannot be propagated at the right speeds are best dissipated. This view has led to an intolerance in PPM for sound signals with very short wavelengths, such as, for example, four cell widths. We achieve the final design goal on the aforementioned list, that the accuracy of the scheme should be maintained in the limit of very many very small time steps, by increasing the formal order of the interpolation discussed in the following paragraphs as the edge of a grid cell is approached.

The attitude that has been taken in PPM design toward numerical dissipation is the most relevant of the aforementioned goals and constraints for the use of PPM in simulating turbulent flows. Therefore, in this chapter I describe the two portions of PPM algorithm that most directly address numerical dissipation – the interpolation scheme used in PPM and the manner in which it is applied to constructing a picture of the structure of the flow within a grid cell. The remainder of the scheme that will not be described here – the approximate Riemann solver used in PPM, the construction of fluxes of conserved quantities at grid cell interfaces in Eulerian coordinates, the application of the conservation laws, and the application of a smart diffusion at shocks, and only at shocks – has much less impact on PPM's numerical treatment of turbulence. As far as these other aspects of the scheme are concerned, it should suffice to point out that shocks are thin and obey the conservation laws that determine both their jumps and their propagation speeds and also to point out that these conservation laws assure that the numerical dissipation of turbulent kinetic energy by the PPM scheme results in the appropriate amount of heat being deposited locally, where the dissipation occurs.

4b.3 PPM interpolation

The heart of many numerical schemes is its approach to interpolation. PPM utilizes the cell averages of various quantities in a fairly broad stencil surrounding a cell in three dimensions in order to construct a picture of the internal structure of the cell. Because the equations of hydrodynamics are coupled, PPM interpolation of the relevant quantities is also coupled. However, to see how the interpolation process works, it is easiest to consider it first for the simple case of a single variable, which we represent by the symbol a. We represent the cell averages of a by $\langle a \rangle_i$. We write the value of a at the left- and right-hand interfaces of the cells as $a_{L,i}$ and $a_{R,i}$. We determine the coefficients of a parabola $a(\tilde{x}) = a_0 + a_1\tilde{x} + a_2\tilde{x}^2$ representing the distribution of the variable a within the cell in terms of a cell-centered local coordinate $\tilde{x} = (x - x_M)/\Delta x$. Here x_M is the cell center and Δx is the cell width. For simplicity, we first assume a uniform grid. (In modern AMR codes, interpolation on nonuniform grids is unnecessary.) Our first task is to determine whether or not the function $a(x)$ is smooth in the region near our grid cell. To do this, we will essentially compare the first and third derivatives of the function in our cell. This process can be formulated in several different ways, all of which boil down to essentially the same criterion. We begin by determining the unique parabola that has the prescribed cell averages in our cell of interest and its two nearest neighbors.

If we write

$$\Delta a_{L,i} = \langle a \rangle_i - \langle a \rangle_{i-1},$$

then we have

$$a_1 = (\Delta a_L + \Delta a_R)/2; \quad a_2 = (\Delta a_R - \Delta a_L)/2$$

Here we have dropped the subscript i by writing Δa_R for $\Delta a_{L,i+1}$. Where possible, we try to eliminate confusing subscripts by such devices. However, all our subscript indices will refer to cells, and none to interfaces (no half-indices will be used). If no subscript index is given, then the subscript i for our cell of interest will always be implied. In PPM, interpolated variables are discontinuous at interfaces, which makes the use of index subscripts such as $i + 1/2$ ambiguous. Thus, $a_{R,i}$, or simply a_R, is not equivalent to $a_{L,i+1}$.

We consider a function to be smooth near our grid cell if the aforementioned unique parabola, when extrapolated to the next nearest neighbor cells, gives a reasonable approximation to the behavior of the function there. This will not be true, of course, if the third derivative of the function is sufficiently large. We will consider the function to be smooth in our cell of interest if the parabola in its neighbor on the left, when extrapolated into our cell's neighbor on the right gives an accurate estimate of that cell's average, and if also that cell's parabola gives a good estimate of the average in our cell's neighbor on the left. Writing the average of the extrapolated parabola in the neighbor cell as $\langle a_i \rangle_{i\pm1}$ and in the next nearest neighbor cell as $\langle a_i \rangle_{i\pm2}$, we find that

$$\langle a_{i-1} \rangle_{i+1} - \langle a \rangle_{i+1} = 2(a_{2,i-1} - a_2); \quad \langle a \rangle_{i-1} - \langle a_{i+1} \rangle_{i-1} = 2(a_2 - a_{2,i+1}).$$

Here we have used the fact that, by construction of these parabolae, $\langle a_i \rangle_{i\pm1} = \langle a \rangle_{i\pm1}$. We denote the fractional error in this extrapolation by f_{err}, given by

$$f_{err} = \max\{|(a_{2,i+1} - a_2)/[\Delta a_R + \text{sign}(\alpha, \Delta a_R)]|,$$

$$|(a_2 - a_{2,i-1})/[\Delta a_L + \text{sign}(\alpha, \Delta a_L)]|\}.$$

Here sign is the FORTRAN sign transfer function, which applies the sign of its second argument to the absolute value of its first argument. We also denote a trivial value of the quantity a by α. Thus the term with the sign function is simply used to protect the divide operation in a FORTRAN program. Of course, we ignore this error estimate when the slope of the function is very small, since in that case it is of no consequence whether we get a "good" extrapolation or not. To get some perspective on this measure, f_{err}, of the smoothness of the function a on our grid, it is instructive to evaluate f_{err} for sine waves of wavelengths $n \Delta x$. For the case where $a(x) = \sin[2\pi x/(n \Delta x)]$, we have $\langle a \rangle = s\, a_0$, where a_0 is the value at the center of the cell and $s = (2/\Delta x) \sin(\Delta x/2)$. The higher-order coefficients of the interpolation parabola are given by $a_1 = s\, \sin(\Delta x) \cos[2\pi x_M/(n \Delta x)]$, where x_M is the coordinate of the cell center, and by $a_2 = s\, (\cos(\Delta x) - 1)\, a_0$. We therefore find that for values of n ranging from 4 to 15, our fractional error, equal simply to $1 - \cos(\Delta x)$, assumes the values: 1.00, 0.69, 0.50, 0.38, 0.29, 0.23, 0.19, 0.16, 0.13, 0.11, 0.10, and 0.09. If we assert that a sine wave over 14 cells is smooth and one over 10 cells is not, then we may conclude that it is reasonable to assert that a function is smooth if $f_{err} \leq 0.10$ and it is not smooth if $f_{err} \geq 0.20$. In between these two values, we may treat the function as partially smooth, by taking a linear combination of a smooth and an unsmooth interpolation parabola. The choice of which functions we will treat as smooth and which not comes from experience. Using only cell averages, we need five cell values to determine, for example, whether one is really in a shock or not or, as we have just seen, whether the function is locally smooth or not. Under such conditions, it is simply unreasonable to believe that we can properly treat a sine wave with a wavelength of only six cells. There are those in the community who believe this can be done, but PPM assumes that this is impossible without further independent information provided upon which the numerical scheme can operate. Experience also shows that numerical noise, which can originate from a number of sources, tends to show up principally at wavelengths between four and eight cells. Signals that appear at these wavelengths are thus quite possibly noise, and therefore deserve to be treated as if they might actually be noise. PPM takes the view that a little noise is not bad, but too much is intolerable. This view is motivated by much experience indicating that nearly total elimination of noise is usually accompanied by partial collateral elimination of the signal.

In PPM, we interpolate several different functions. Some of them, like the Riemann invariants, are defined only as differences (the Riemann invariants are defined by inexact differentials). Other functions, like the transverse velocities, are defined by cell averages. Some may have sharp transitions, associated with contact discontinuities, that deserve special treatment. Here I describe the most general algorithm, for the special case of a uniform grid. This is the algorithm that PPM uses to interpolate a subgrid structure for the Riemann invariant associated with the entropy, for which the differential form is given by

$$dA = d\rho - dp/c^2.$$

From this most complicated interpolation algorithm, all other, simpler forms used in PPM can be derived as special cases. I therefore describe the interpolation for the entropy Riemann invariant in detail, and later briefly note in which respects the other interpolations are degenerate cases of this one.

We begin the interpolation of dA by defining a variable, A, that we derive from it by determining the arbitrary constant of integration such that the cell average of A vanishes. Thus each cell has its own local scaling for A in which the cell average vanishes but the derivatives are correct. This local scaling is much like the local, cell-centered coordinate, $\tilde{x}$, that we use to define the interpolation parabola in a cell. There is a discontinuous jump in the integration constant when we go from one cell to its neighbor. This jump is

$$\Delta A_L = \Delta \rho_L - \Delta p_L / c_L^2 = (\langle \rho \rangle - \langle \rho \rangle_{i-1}) - (\langle p \rangle - \langle p \rangle_{i-1}) / c_L^2.$$

For the gamma-law equation of state, we estimate c_L^2 by means of

$$c_L^2 = \frac{1}{2} (\langle c \rangle_{i-1}^2 + \langle c \rangle^2) = \frac{1}{2} \left[\frac{\gamma \langle p \rangle_{i-1}}{\langle \rho \rangle_{i-1}} + \frac{\gamma \langle p \rangle}{\langle \rho \rangle} \right].$$

Here we see a common practice used in PPM, namely the evaluation of an estimate of the cell average of a function of primitive fluid state variables as the function of the cell averages. Experience shows that attempts to do better than this are usually unrewarded by compensating increases in simulation accuracy. In the specific case of the kinetic energy, PPM does do a better job, but the effect on the flow accuracy is only marginal in this case, and the cost is not insignificant.

Using the aforementioned definitions, we evaluate a fractional error estimate as follows:

$$f_{err} = \frac{1}{2} \frac{|\Delta A_{L,i+2} - 2\Delta A_R + \Delta A_L| + |\Delta A_R - 2\Delta A_L + \Delta A_{L,i-1}|}{|\Delta A_L| + |\Delta A_R| + d\alpha}.$$

Here $d\alpha$ represents a trivial Riemann invariant difference that is used to protect the divide in the FORTRAN program. Note that this error estimate is very similar to but not exactly the same as the one discussed earlier. Using this error estimate, we construct a measure, Ω, varying between 0 and 1, of the roughness of the function near this grid cell as follows:

$$\Omega = \min \{1, \max \{0, 10 (f_{err} - 0.1)\}\}.$$

We now construct estimates A_{s1} and A_{m1} of the first-order parabola coefficient, which gives the change of the variable across the cell. The first, A_{s1}, applies to the parabola that has the three prescribed average values in this and the nearest neighbor cells. The second, A_{m1}, applies to this parabola after the application of a monotonicity constraint. We will use these quantities in constructing further quantities later. We have already seen that $A_{s1} = (\Delta A_L + \Delta A_R)/2$. Defining a sign variable, s, of absolute value unity to have the sign of A_{s1}, we may then compute A_{m1} in the following sequence of steps:

$$\tilde{A}_{m1} = 2 \min \{s\Delta A_L, s\Delta A_R\}, \quad \tilde{\tilde{A}}_{m1} = s \max\{0, \min\{s A_{s1}, \tilde{A}_{m1}\}\},$$

$$A_{m1} = \Omega A_{s1} + (1 - \Omega) \tilde{\tilde{A}}_{m1}.$$

(The superiority of FORTRAN over mathematical notation for the description of numerical algorithms is becoming apparent here.) We now define two estimates, A_{sL} and A_{mL}, of the value at the left-hand cell interface. The first of these, A_{sL}, corresponds to an unconstrained interpolation polynomial, while A_{mL} corresponds to one that is constrained:

$$A_{sL} = \frac{1}{6}(A_{s1,i-1} - A_{s1}) - \frac{1}{2}\Delta A_L, \quad A_{mL} = \frac{1}{6}(A_{m1,i-1} - A_{m1}) - \frac{1}{2}\Delta A_L.$$

Note that these interface value estimates involve the cell averages in two cells on either side of the interface. The unconstrained estimate is the value of the unique cubic curve that has the prescribed average values in these four cells. The constrained value is guaranteed to lie within the range defined by the cell averages adjacent to the interface. Of course, both these values refer to the choice of integration constant that gives a vanishing cell average to the right of this interface (in the cell of interest, i). We desire a linear combination of these two interface value estimates, which we will regard as a provisional estimate $A_{pL} = (1 - \Omega)A_{sL} + \Omega A_{mL}$. Rather than build this value A_{pL} by blending the unconstrained and constrained estimates, we find it more useful to build blended parabola coefficients, $A_{p1} = (1 - \Omega)A_{s1} + \Omega A_{m1}$, and to form A_{pL} from these. This procedure allows us to make use of the coefficients A_{p1} later in our contact discontinuity detection and steepening algorithm. We now proceed to modify our provisional interface values A_{pL} in cells where the function is not smooth. First, we set our interpolation function to a constant in cells where extrema occur, unless of course the function is smooth there. Thus, if $A_{pL} A_{pR} \geq 0$, we set new provisional values, indicated by a subscript q, as follows: $A_{qL} = (1 - \Omega) A_{pL}$, $A_{qR} = (1 - \Omega) A_{pR}$. Otherwise we simply set $A_{qL} = A_{pL}$ and $A_{qL} = A_{pL}$. Now, in cells where the function is not smooth, we constrain the internal structure to be monotone. We must make sure in this process not to revise the cell structure further in the cells containing extrema. We therefore construct limiting values, A_{lL} and A_{lR}, which we must be careful to set to zero in cells containing extrema. In the remaining cells, these values are $A_{lL} = -2A_{pR}$ and $A_{lR} = -2A_{pL}$. This reflects the fact that a parabola having zero slope at one side of the cell will assume at the opposite interface the negative of twice this value, as long as the cell average is zero. If these limiting values are exceeded in cells where the function is not smooth, we must reset them to these limits. Therefore we construct new, but still provisional, values A_{rL} and A_{rR} (in FORTRAN, of course, no new names are required), which reflect this additional constraint:

if $(A_{qR} - A_{qL})(A_{qL} - A_{lL}) < 0$;   then $A_{rL} = (1 - \Omega)A_{qL} + \Omega A_{lL}$,

otherwise $A_{rL} = A_{qL}$;

if $(A_{qR} - A_{qL})(A_{lR} - A_{qR}) < 0$;   then $A_{rR} = (1 - \Omega)A_{qR} + \Omega A_{lR}$,

otherwise $A_{rR} = A_{qR}$.

From these provisional interface values, we construct provisional interpolation parabola coefficients:

$$A_{r1} = A_{rR} - A_{rL}, \quad A_{r2} = 3(A_{rL} + A_{rR}), \quad A_{r0} = -A_{r2}/12.$$

For interpolation of sound-wave Riemann invariant differences, for which we do no contact discontinuity steepening, these values are our final results. However, for the entropy Riemann invariant differences, we continue as subsequently described. We also note that for interpolation of a variable, such as the transverse velocity component, which is not expressed as a differential form, we may first construct differences of the cell averages and then proceed as outlined herein.

We wish to detect cells that are inside sharp jumps in the entropy Riemann invariant that are associated with contact discontinuities. In such cells, it is inappropriate to try to fit smooth curves to determine the subgrid structure, since it is actually discontinuous (PPM is solving the Euler equations). For the purpose of computation, we will nevertheless require that the distribution inside the grid cell be a parabola; however, we may use the knowledge that the actual structure is discontinuous to construct an appropriate choice for this parabola. The idea here is quite simple. We build a test that detects cells that are within contact discontinuity structures. Then we obtain estimates of the interface values for such cells by extrapolating to the cell interfaces presumably smoother structures from outside the cell and hence, we hope, from outside the discontinuity. Together with the prescribed cell averages, these more appropriate cell interface values allow us to build an improved, steeper parabola to describe the distribution within the cell. Obviously, the function in such cells is not smooth, so we apply monotonicity constraints to the new edge values and also to this improved parabola. This procedure works very well in practice. A similar procedure can be adapted for use with other numerical schemes that employ linear interpolation functions for the cell structures, but it is less effective. Apparently, the curvature provided by the use of parabolae allows the dimensionless constants that will be introduced in the following paragraphs to make the steepening process effective for contact discontinuities without requiring that the steepening process be applied for marginal cases, which can occasionally turn out not to actually be contact discontinuities. That is, using parabolae we are able to apply the steepening process only in unequivocal circumstances and thus to avoid steepening structures by mistake that should not actually be steepened.

We begin, as with our test for function smoothness, with a measurement of the size of the third derivative of the function relative to its first derivative. Reusing the symbol s for a different sign variable, we build a steepness measure, S, in a sequence of steps as follows:

$$S_1 = (A_{sR} - A_{sL} - A_{s1})/(A_{s1} + 10 s \, d\alpha).$$

Here s has absolute value unity and has the same sign as A_{s1} in order to protect the divide without altering the sign of the overall expression. As before, $d\alpha$ is a trivial Riemann invariant difference. We must take care to realize that the numerator in this expression is not necessarily zero, since the edge values that appear there come from cubic curves rather than the parabolae that define the A_{s1} values. We must also take care to realize that because each cell has its own value of the integration constant for this Riemann invariant's differential form, we must set $A_{sR} = A_{sL, i+1} + \Delta A_R$.

With those cautions, we see that this expression for S_1 is very similar to our measure of the lack of function smoothness, except that no absolute value signs appear in it. The first expression in the numerator for the variable difference across the cell involves cubic interpolation, which accounts for the third derivative of the function. From this we subtract A_{s1}, which removes the part that is due to the first two function derivatives. To obtain S_1, we divide by an estimate of the first derivative. The lack of absolute value operations in this formula for S_1 is caused by our desire to detect sharp jumps in the function, and to reject sudden flat spots. Flat spots will correspond to negative values of S_1. We now scale S_1 and limit it to the range from 0 to 1:

$$S_2 = \max \{0, \ \min \{1, 20 \ (S_1 - 0.05)\}\}.$$

To guard against applying our contact discontinuity steepening anywhere but at true sharp jumps in the entropy, we reset S_2 to zero where the second derivative of the function does not change sign and also where the amplitude of the jump is not sufficient to warrant this special treatment (we do not want to preserve, or worse, to amplify small numerical glitches). Thus:

$$\text{if} \quad A_{s2, i-1} \, A_{s2, i+1} \geq 0, \quad \text{then} \quad S_3 = 0; \quad \text{otherwise} \quad S_3 = S_2.$$

Here we use the earlier definition of the coefficients of the parabola that has the prescribed three cell averages, writing $A_{s2} = (\Delta A_R - \Delta A_L)/2$. To eliminate trivial jumps from consideration, we form S_4:

$$\text{if} \quad |\Delta A_L| + |\Delta A_R| < d\alpha, \quad \text{then} \quad S_4 = 0; \quad \text{otherwise} \quad S_4 = S_3.$$

Before proceeding with the steepening algorithm, we make one further test to make certain that steepening is appropriate. This test makes sense for contact discontinuities, but for other variables, such as a fractional volume of a second fluid, it might not. Therefore this last test is optional, but we do use it for the entropy. The calling program to the interpolation routine provides two variable differences along with the differences ΔA_L. If the jump is one of the type we seek to detect, then the differences ΔB will be very small compared with the differences ΔD. Both these sets of differences are cell centered and have the same units. We enforce this demand by constructing our final steepness measure, S, as follows:

$$\text{if} \quad \Delta D = 0 \quad \text{or} \quad \Delta B/\Delta D \geq 0.1, \quad \text{then} \quad S = 0; \quad \text{otherwise} \quad S = S_4.$$

In previous versions of PPM, the ΔB were pressure differences and the ΔD were estimates of $c^2 \Delta \rho$, while the density was the quantity being interpolated with potential contact discontinuity steepening. We still perform this additional test, although it is somewhat redundant when interpolating the entropy Riemann invariant, for which the differential form is just $\Delta A = \Delta \rho - \Delta p/c^2$.

With our contact discontinuity detection process complete, and all cells inside such structures marked by S, varying from 0 to 1, we are at last ready to begin the steepening operation. Much earlier, we computed constrained variable differences A_{m1} across the

cells. In principle, these variable differences could be only partially constrained, but
it is safe to assume that if our cell is in a contact discontinuity, the parameter Ω that
measured the function roughness assumes the value unity in this and the neighboring
cells. We define cell interface values, A_{cL} and A_{cR}, appropriate for a contact discontinuity
structure by extrapolating the constrained slopes of the function in neighboring cells
to the interfaces of our cell of interest: $A_{cL} = -\Delta A_L + A_{m1, i-1}/2$ and $A_{cR} = \Delta A_R - A_{m1, i+1}/2$. We then construct our nearly final estimates for the interface values as

$$A_{tL} = (1 - S)A_{rL} + SA_{cL}, \quad A_{tR} = (1 - S)A_{rR} + SA_{cR}.$$

We must once again apply the constraints on the implied parabola and then compute
the coefficients of that parabola, just as before (however, hardly any cells are steepened,
and hence this extra work occurs in a scalar loop for only these cells, which incurs
hardly any additional computational cost). These operations will not be repeated here.
They are just like those we performed to arrive at first A_{qL} and then A_{rL} beginning
with A_{pL}.

It is worthwhile to note the ways in which the aforementioned interpolation algorithm addresses the previously listed design constraints for PPM. Some of the points
are obvious, but others are less so. In particular, it is not obvious that dissipation errors will dominate dispersion errors, although this is in fact the case. The reason for
this is the action of the monotonicity constraints on all short wavelength disturbances,
which always fail our test for function smoothness. The monotonicity constraints are
nonlinear, in that they alter the shape of a sine wave by introducing higher frequency
components through effects such as the clipping of extrema. Their overall effect is
strongly dissipative for very short wavelength signals, which of course are the ones
for which the numerical scheme would otherwise introduce significant dispersion errors. This dominance of dissipation over dispersion error as a design goal may seem
incompatible with the goal of minimal dissipation. Our task is to deliver as little dissipation as possible while still having dissipation overwhelm dispersion. PPM's use of
parabolae rather than the more commonly used linear interpolation functions is meant
to address this design goal. The use of cubic interpolation functions to help define these
parabolae through interpolation of cell interface values reduces dissipation errors still
further, since not all parabolae provide equally good fits. However, what is even more
important is that this use of cubic interpolation in PPM preserves the accuracy of the
scheme in the limit of vanishing Courant number. Since, as is common practice, we
will determine a single time step value for the entire grid based upon Courant number
limitations for the single most demanding cell, the bulk of the cells we update will have
very small Courant numbers. When AMR is added to the scheme, the Courant numbers
used for the bulk of the cells are likely to become even smaller. Therefore it is very
important for the accuracy of the calculation to hold up in this limit. The use of cubic
interpolation of cell interface values is our way of addressing this issue in PPM. Over
many years of applying PPM to a wide variety of flow problems, very little Courant
number sensitivity has ever appeared.

The monotonicity constraints and contact discontinuity detection and steepening algorithms incorporated in PPM have important consequences for the propagation of barely resolved or effectively unresolved signals. Our design goal is to preserve these signals, as long as we can propagate them at roughly the proper speeds, rather than to destroy them through numerical diffusion processes. The monotonicity constraints and, to a far greater degree, the contact discontinuity steepening in PPM have this effect. They act upon passively advected signals, such as entropy variations, that can easily be propagated at the correct speeds. They apply an artificial compression along with their other effects, such as maintaining positivity where appropriate. This tends to maintain signal amplitude while altering signal shape when the signal is not adequately resolved. As the description of the contact discontinuity detection algorithm shows, we take great care to apply the steepening method only where appropriate, since some small signals, better described as numerical glitches of various types and causes, deserve to be dissipated.

4b.4 Using the interpolation operators to build a subgrid-scale model for a cell

The aforementioned interpolation process is quite elaborate. Nevertheless, it is not the entire story of PPM interpolation. As I remarked earlier, the equations of gas dynamics are coupled, and therefore PPM interpolations of the fluid state variables are also coupled. The goal is to come up with a complete and consistent picture of the internal structure of a grid cell – that is, to come up with a subgrid-scale model. This is not a turbulence model, but it *is* a subgrid-scale model. This model must satisfy, for consistency, certain primary constraints. First, the integrated mass, momenta, and total energy must be consistent with the prescribed cell averages of these quantities. This reflects the need for the numerical scheme to be in strict conservation form in order to correctly capture and propagate shocks. PPM regards the fundamental cell averages on which it operates to be the cell averages of density and pressure (both volume weighted) and those of the three velocity components (all mass weighted). From these, PPM defines the mass, momenta, and total energy of a cell to be

$$\Delta m = \langle \rho \rangle \, \Delta x, \, \langle u_x \rangle \, \Delta m, \langle u_y \rangle \, \Delta m, \langle u_z \rangle \, \Delta m,$$

$$\text{and } \frac{\langle p \rangle}{\gamma - 1} \Delta x + \frac{1}{2} \left[\langle u_x \rangle^2 + \langle u_y \rangle^2 + \langle u_z \rangle^2 \right] \Delta m.$$

Here we do not bother to include the cell widths in the transverse directions, Δy and Δz, because they cancel out of all our equations for the x-pass. The last expression, for the cell's total energy, is misleading. At the beginning of the grid cell update for a 1D pass, we compute the total energy of the cell in just this way. However, once we have done this, we make a more accurate estimate of the cell's kinetic energy and then revise its thermal energy, keeping this total constant. We do this by using nearest neighbor cells to compute cell-centered slope estimates for all three velocity components. We then apply the standard monotonicity constraints to these slopes. All these computations are similar to the following steps, for the slope of u_y in the z direction:

$$\Delta_z u_y = \frac{1}{2} \left(\langle u_y \rangle_{k+1} - \langle u_y \rangle_{k-1} \right),$$

$$(\Delta_z u_y)_{max} = 2 \min \left\{ s \left(\langle u_y \rangle_{k+1} - \langle u_y \rangle \right), \, s \left(\langle u_y \rangle - \langle u_y \rangle_{k-1} \right) \right\},$$

$$\Delta_{mz} u_y = s \max \left\{ 0, \min \left\{ s \Delta_z u_y, (\Delta_z u_y)_{max} \right\} \right\}.$$

Here s has absolute value unity and the same sign as $\Delta_z u_y$. We now estimate the cell's average kinetic energy to be given by

$$2 \langle E_{kin} \rangle = \langle u_x \rangle^2 + \frac{1}{12} \left[(\Delta_{mx} u_x)^2 + (\Delta_{my} u_x)^2 + (\Delta_{mz} u_x)^2 \right]$$

$$+ \langle u_y \rangle^2 + \frac{1}{12} \left[(\Delta_{mx} u_y)^2 + (\Delta_{my} u_y)^2 + (\Delta_{mz} u_y)^2 \right]$$

$$+ \langle u_z \rangle^2 + \frac{1}{12} \left[(\Delta_{mx} u_z)^2 + (\Delta_{my} u_z)^2 + (\Delta_{mz} u_z)^2 \right].$$

This process captures the first correction to the average kinetic energy that arises from the internal structure of the velocity within the grid cell. The method would still be second-order accurate without this correction, but this correction proves to be of some marginal value (relative to its implementation cost) in flows involving strong shear layers. Terms of this sort make up part of some turbulence closure models, although we might argue that this is inappropriate, since they have nothing specifically to do with turbulence. In PPM we must compute estimates for cell averages of several other quantities, and including terms like these for all those computations would roughly double the cost of the scheme. This has been tried, and little noticeable benefit is delivered in practical problems. Therefore only these velocity terms are included in PPM. We can see why these terms are important as follows. The velocity slopes can be very large in a cell, because shear layers that are stretching as a result of the local flow or their own instability naturally become ever thinner. The contribution to the kinetic energy from the velocity slope terms can therefore be significant, and can thus have a significant effect upon the pressure (since the total energy is prescribed), and hence upon the dynamics. Nevertheless, this effect upon the dynamics is strongly localized, so that omitting the terms, as was done in PPM for many years, only tends to increase by roughly 50% the numerical friction in thin, strong (roughly sonic or supersonic) shear layers.

In order to propagate signals, we need to interpolate their subgrid structures. We therefore apply the previously described interpolation algorithm to the five Riemann invariants of the Euler equations for 3D compressible flow. For the two transverse components of the velocity, u_y and u_z, which are advected passively with the fluid velocity in the x-pass, we form differences and then apply the interpolation scheme described herein. We do no contact discontinuity detection or steepening for these variables, even though they may in fact jump at such discontinuities. We have already discussed in detail the treatment for the entropy Riemann invariant. The sound-wave

Riemann invariants, which we denote by $R_\pm$, are defined by the following differential forms and their corresponding numerical approximations:

$$dR_\pm = du_x \pm \frac{dp}{C},$$

$$\Delta R_{\pm L} = \Delta u_{xL} \pm \frac{\Delta p_L}{C_L} = \langle u_x \rangle - \langle u_x \rangle_{i-1} \pm 2 \left(\langle p \rangle - \langle p \rangle_{i-1} \right) / \left(\langle C \rangle + \langle C \rangle_{i-1} \right).$$

Here we introduce the Lagrangian sound speed $C = \rho\, c$, and we approximate its cell average as $\langle C \rangle = \langle \rho \rangle \langle c \rangle = \sqrt{\gamma \langle p \rangle \langle \rho \rangle}$. The pressure, of course, is p, and only the x-component of velocity appears because we are describing the x-pass of the directionally split PPM algorithm. It is disturbances in R_+ that propagate to the right at speed $s_+ = u_x + c$ and disturbances in R_- that propagate to the left at speed $s_- = u_x - c$. We interpolate parabolae to describe the structure of these sound wave signals in the grid cells, using $\Delta R_{\pm L}$ in place of ΔA_L in the algorithm previously described, and of course performing no contact discontinuity detection or steepening. Performing interpolations for the Riemann invariants attempts to uncouple the gas dynamic equations as much as possible. However, we will not work in terms of these variables directly. Instead, we will immediately go about constructing interpolation parabolae for the primary fluid state variables – the density, pressure, and three components of velocity – in a manner that attempts to achieve as much consistency as possible with the interpolated structures of the Riemann invariant signals. To perform our hydrodynamical cell update, we will require internal cell structures for all these variables as well as the total energy. Any choice of structures for five variables will imply structures for all the others, but those implied structures may not satisfy reasonable constraints, such as monotonicity or positivity when appropriate. PPM attempts to achieve consistency between interpolated and implied structures by first interpolating constrained parabolae for the Riemann invariant signals, constructing the implied parabolae for the primary state variables, and then constraining those implied parabolae where appropriate. In a sense this is an attempt to "have it both ways," which is of course impossible. However, experience has shown that the extra labor involved in this process delivers additional accuracy and robustness of a value worthy of its computational and programming cost.

From the interpolated parabolae for the sound-wave Riemann invariants, we may compute the cell interface values of the pressure and x-velocity as follows:

$$p_L = \langle p \rangle + C_L (R_{+L} - R_{-L})/2, \quad p_R = \langle p \rangle + C_R (R_{+R} - R_{-R})/2,$$

$$u_{xL} = \langle u_x \rangle + (R_{+L} + R_{-L})/2, \quad u_{xR} = \langle u_x \rangle + (R_{+R} + R_{-R})/2.$$

We also compute a measure, Ω, of the lack of smoothness of the associated functions as the maximum of the values found for the two sound wave Riemann invariants separately during the process of their interpolation: $\Omega = \max\{\Omega_{R_+}, \Omega_{R_-}\}$. We will use this measure Ω to control the application of constraints to our interpolation parabolae for p and u_x. We first apply monotonicity constraints, controlled by Ω, to the cell interface pressures and x-velocities already obtained. When $\Omega = 1$, we demand that the interface values lie within the ranges defined by the averages in adjacent cells. We then, again when

$\Omega = 1$, demand that the parabolae defined by these constrained interface values and the cell averages are monotone. These constraints are essentially the same as those described earlier for the interpolation of the entropy Riemann invariant, and hence they are not stated in detail here. Once the interpolation parabola for the pressure has been so defined, we may use the interpolated interface values for the entropy to determine interface values for the density. Using the maximum of the Ω measures for the pressure and for the entropy, we may then constrain the interface densities and ultimately the parabolae that they and the cell averages define.

At the end of this lengthy process, we have interpolation parabolae defined for the density, pressure, and all three velocity components. The formulae presented assume uniform cell sizes, although more general formulae are easily derived and were presented in the description of PPM in Colella and Woodward (1984). We can think of these interpolations as occurring in a cell number variable, so that they are valid, although potentially somewhat less accurate, even if the cell size is smoothly varying. Nevertheless, we must take care in interpreting these parabolae. Some of the variables described by them, such as the velocity components, must be considered to vary according to mass fraction across each cell, since the (mass-weighted) cell average is associated directly with the conserved cell momentum. For the velocities, therefore, we take the interpolation variable $\tilde{x}$ within the cell to have its origin at the center of mass and to describe fractions of the cell mass rather than of the cell volume. For the pressure or the density, whose volume integrals are directly associated with the mass and internal energy, we must interpret $\tilde{x}$ as having its origin at the center of the cell in a volume coordinate and to describe volume fractions within the cell. Our use of the uniform grid formulae therefore reflects an assumption that all interpolated quantities are smoothly varying in a cell fraction variable, be it a volume or a mass fraction. We know this assumption to be false at contact discontinuities and slip surfaces (for some of the variables) and at shocks, but we have augmented our interpolation procedure to deal with these discontinuities. The interpolation is therefore valid, as long as we interpret it properly.

The process just described has enabled us to construct a subgrid-scale model for each cell. Since each fluid state variable varies as a simple parabola inside a cell, this subgrid-scale model cannot possibly describe subgrid-scale turbulent motions. To perform that function, the model would have to be augmented by the addition of one or more new state variables, such as a subgrid-scale turbulent kinetic energy variable, which could also be given an interpolation parabola within the cell. Nevertheless, our demand that the structure of a velocity component within a cell be no more complicated than a parabola, and our constraints on that parabola provide a powerful mechanism to dissipate kinetic energy of small-scale motions unresolvable or marginally resolvable on our grid into heat. This is one of the important functions of turbulence closure models, especially when incorporated into difference schemes that lack any other method to accomplish this very necessary dissipation. In this respect, turbulence models can be used to stabilize otherwise nonlinearly unstable numerical schemes, a role that has nothing to do with turbulence itself and that confuses the purpose and function of a turbulence model. As we will see later in Chapters 7 and 15, with PPM we may compute turbulent flow directly from the Euler equations that govern it, and we accomplish the dissipation of small-scale motions through the truncation errors of our numerical scheme rather than from differencing an explicit viscous diffusion term. We take the view that the details of this dissipation are unimportant as long as kinetic energy turning up on the smallest possible scales in our computation is dissipated into heat with no unphysical side effects. We presume that in a far more expensive and careful simulation of all relevant physics, this same kinetic energy would be dissipated in any event, and the same amount of heat generated (since total energy is conserved), but the (hopefully unimportant) details on tiny length and time scales would differ. The assumption here is that the kinetic energy that appears on scales where the numerical dissipation comes into play has arrived there due to a physical process, such as a turbulent cascade, that is correctly simulated in our calculation based on the Euler equations because it has nothing to do with the molecular viscosity, which has been omitted in deriving the Euler equations from the more general Navier–Stokes equations.

4b.5 Summary

This chapter has given a complete description of the PPM interpolation scheme. This is the part of PPM scheme that is most important in determining the fashion in which PPM simulates low-speed turbulent flows. Because of the complexity of PPM and the limitations of space, I have chosen a complete description of part of this scheme over a partial description of all of it. I have presented no simple examples of the accuracy the scheme delivers in practice, although such examples can be found in the references that follow. The reader can, however, go to the LCSE Web site at www.lcse.umn.edu, download a Windows application that runs PPM in simple situations, consult the user guide, and run test problems of various classic case studies. At this same Web site, a complete description of PPM scheme can be found (Woodward 2005). PPM achieves its robustness and accuracy through complexity, and this has its cost. From the description here, that cost might not be clear. On a difficult flow problem involving many strong shocks, on the average each cell update for a 1D pass involves 921 flops and 249 vectorizable logical operations (which do not count as flops). This work is performed at a speed of 1255 Mflop/s on a 1.7 GHz Intel Pentium-M CPU working from its cache memory in a laptop machine. This laptop performance degrades to 954 Mflop/s for a full 128^3 3D grid brick update, so that full 128^3 test runs can be performed overnight at a hotel. These speeds are for 32-bit arithmetic, which is all that PPM ever requires. On a 3.2 GHz Intel Pentium-4 CPU, these 32-bit rates become 1607 Mflop/s and 1273 Mflop/s, respectively, while the corresponding rates for 64-bit arithmetic, which is unnecessary but quoted for comparison purposes with other applications, are 1132 Mflop/s and 938 Mflop/s, respectively. PPM performance has scaled essentially linearly on every multiprocessing machine on which the code has ever been implemented, including ASCI machines at DoE labs with over 6000 CPUs (Woodward and Anderson 1999, Woodward and Porter 2005, Woodward et al. 2004).

Acknowledgment

This work has been performed with support over many years from the MICS program of the Department of Energy's Office of Science. The present grant from this program is DE-FG02-03ER25569.

REFERENCES

Bassett, G. M., & Woodward, P. R. 1995a Numerical simulation of nonlinear kink instabilities of supersonic shear layers. *J. Fluid Mech.* **284**, 323–340.

Bassett, G. M., & Woodward, P. R. 1995b Simulation of the instability of Mach 2 and Mach 4 gaseous jets in 2 and 3 dimensions. *Astrophys. J.* **441**, 582–602.

Blondin, J. M., & Lufkin, E. A. 1993 The piecewise-parabolic method in curvilinear co-ordinates. *Astrophys. J. Suppl.* **88**, 589–594. VH-1 code available at http://wonka.physics.ncsu.edu/pub/VH-1. Enhanced Virginia Hydrodynamics #1 benchmark code, EVH1, available at http://www4.ncsu.edu/~gmkumar/perc/evh1/evh1.htm.

Bryan, G. L., Norman, M. L., Stone, J. M., Cen, R., & Ostriker, J. P. 1995 A piecewise-parabolic method for cosmological hydrodynamics. *Comput. Phys. Commun.* **89**, 149–168; ENZO code available at http://cosmos.ucsd.edu/enzo.

Calder, A. C., Fryxell, B., Plewa, T., Rosner, R., Dursi, L. J., Weirs, V. G., Dupont, T., Robey, H. F., Kane, J. O., Remington, B. A., Drake, R. P., Dimonte, G., Zingale, M., Timmes, F. X., Olson, K., Ricker, P., MacNeice, P., & Tufo, H. M. 2002 On validating an astrophysical simulation code *Astrophys. J. Suppl.* **143**, 201–229.

Colella, P., & Woodward, P. R. 1984 The piecewise-parabolic method (PPM) for gas dynamical simulations. *J. Comput. Phys.* **54**, 174–201.

Dai, W., & Woodward, P. R. 1994 An approximate Riemann solver for ideal magnetohydrodynamics. *J. Comput. Phys.* **111**, 354–372.

Dai, W., & Woodward, P. R. 1994 Extension of the piecewise-parabolic method to multidimensional ideal magnetohydrodynamics. *J. Comput. Phys.* **115**, 485–514.

Dai, W., & Woodward, P. R. 1994 Interactions between magnetohydrodynamical shocks and denser clouds. *Astrophys. J.* **436**, 776–783.

Dai, W., & Woodward, P. R. 1995 A simple Riemann solver and high-order Godunov schemes for hyperbolic systems of conservation laws. *J. Comput. Phys.* **121**, 51.

Dai, W., & Woodward, P. R. 1995 A High-Order Iterative Implicit-Explicit Hybrid Scheme for 2-D Hydrodynamics, in *Numerical Methods for Fluid Dynamics V*, pp. 377–383, eds. K. W. Morton and M. J. Baines, Oxford Science Publications.

Dai, W., & Woodward, P. R. 1996 Iterative implementation of an implicit-explicit hybrid scheme for hydrodynamics. *J. Comput. Phys.* **124**, 217–229.

Dai, W., & Woodward, P. R. 1996 A second-order iterative implicit-explicit hybrid scheme for hyperbolic systems of conservation laws. *J. Comput. Phys.* **128**, 181–196.

Dai, W., & Woodward, P. R. 1998, A simple finite difference scheme for multidimensional magneto-hydrodynamical equations. *J. Comput. Phys.* **142**, 331–369.

Dai, W., & Woodward, P. R. 1998 On the divergence-free condition and conservation laws in numerical simulations for supersonic magnetohydrodynamical flows. *Astrophys. J.*, **493**.

Dai, W., & Woodward, P. R. 1998 Numerical simulations for radiation hydrodynamics. I. Diffusion limit. *J. Comput. Phys.* **141**, 182–207.

Dai, W., & Woodward, P. R. 2000 Numerical simulations for radiation Hydrodynamics II. Transport limit. *J. Comput. Phys.* **157**, 199–233.

DeBar, R. 1974 *Fundamentals of the KRAKEN Code.* Lawrence Livermore National Laboratory Report UCIR-760.

Dinge, D., & Woodward, P. R. 2003 A parallel cell-by-cell AMR method for the PPM hydrodynamics code. *Int. J. Modern Phys. C* **14**, 1–24.

Edgar, B. K., & Woodward, P. R. 1993 Diffraction of a shock wave by a wedge: Comparison of PPM simulations with experiment. *AIAA Paper 91-0696,* and *Video J. Eng. Res.* **3**, 25–33.

Edgar, B. K., Woodward, P. R., & Anderson, S. E. 1998. PPM code library, with documentation and examples, available at www.lcse.umn.edu/PPMlib.

Fryxell, B., Olson, K., Ricker, P., Timmes, F. X., Zingale, M., Lamb, D. Q., MacNeice, P., Rosner, R., Truran, J. W., & Tufo, H. 2000 FLASH: An adaptive mesh hydrodynamics code for modeling astrophysical thermonuclear flashes. *Astrophys. J. suppl.* **131**, 273–334.

Fryxell, B., Woodward, P. R., Colella, P., & Winkler, K.-H. 1986 An implicit-explicit extension of the PPM scheme for lagrangian hydrodynamics in one dimension. *J. Comput. Phys.* **63**, 283.

Godunov, S. K. 1959 A finite-difference method for the numerical computation and discontinuous solutions of the equations of fluid dynamics. *Mat. Sb.* **47**, 271.

Kritsuk, A. G., Norman, M. L., & Padoan, P. 2006 Adaptive mesh refinement for supersonic molecular cloud turbulence. *Astrophys. J.* **638**, L25–L28.

Mahinthakumar, G., Sayeed, M., Blondin, J., Worley, P., Hix, W. R., & Mezzacappa, A., 2004 performance evaluation and modeling of a parallel astrophysics application, in Proc. 2004 High Performance Computing Symposium; available at www.scs.org.getDoc.cfm?id=1681.

Pedelty, J. A., & Woodward, P. R. 1991 Numerical simulations of the nonlinear kink modes in linearly stable supersonic slip surfaces. *J. Fluid Mech.*, **225**, 101–120.

Sutcliffe, W. G. 1973 *BBC Hydrodynamics Lawrence.* Livermore National Laboratory Report UCID-17013.

van Leer, B. 1977 Towards the ultimate conservative difference scheme IV. A new approach to numerical convection. *J. Comput. Phys.*, **23**, 276–299.

van Leer, B. 1979 Towards the ultimate conservative difference scheme V. A second-order sequel to Godunov's method. *J. Comput. Phys.* **32**, 101.

van Leer, B., & Woodward, P. R. 1980 The MUSCL code for compressible flow: Philosophy and results. *Computational Methods in Nonlinear Mechanics: TICOM Second Intnatl. Conf.*, Austin, Texas, March, 1979, edited by J. T. Oden (Amsterdam: North-Holland), 234.

Woodward, P. R. 1985 Simulation of the Kelvin-Helmholtz instability of a supersonic slip surface with the piecewise-parabolic method (PPM) in *Numerical Methods for the Euler Equations of Fluid Dynamics*, eds. Angrand, Dervieux, Desideri, and Glowinski, SIAM.

Woodward, P. R. 1986 Numerical methods for astrophysicists, in *Astrophysical Radiation Hydrodynamics*, eds. K.-H. Winkler & M. L. Norman. Dordrecht: Reidel, 245–326.

Woodward, P. R. 2005 A Complete Description of the PPM Compressible Gas Dynamics Scheme. LCSE internal report available from the main LCSE page at www.lcse.umn.edu.

Woodward, P. R., & Anderson, S. E. 1999 Portable Petaflop/s Programming: Applying Distributed Computing Methodology to the Grid Within a Single Machine Room, in Proc. of the 8th IEEE International Conference on High Performance Distributed Computing, Redondo Beach, Calif., Aug., 1999; available at www.lcse.umn.edu/HPDC8.

Woodward, P. R., Anderson, S. E., Porter, D. H., & Iyer, A. 2004 Cluster computing in the SHMOD framework on the NSF TeraGrid, LCSE internal report, April, 2004, available on the Web at www.lcse.umn.edu/turb2048.

Woodward, P. R., & Colella, P. 1981 High-resolution difference schemes for compressible gas dynamics. *Lecture Notes Phys.* **141**, 434.

Woodward, P. R., & Colella, P. 1984 The numerical simulation of two-dimensional fluid flow with strong shocks. *J. Comput. Phys.* **54**, 115–173.

Woodward, P. R., & Porter, D. H. 2005 PPM Code Kernel Performance, LCSE internal report available at www.lcse.umn.edu.

4c The Lagrangian Remap Method

David L. Youngs

4c.1 Overview of the numerical method

This section gives a summary of the Lagrangian remap method used in the TUR-MOIL3D hydrocode for simulation of turbulent mixing that is due to Rayleigh–Taylor and Richtmyer–Meshkov instabilities. The Lagrangian remap method (as a way of constructing an essentially Eulerian hydrocode) originated from the work at Los Alamos National Laboratory in the 1960s on the PIC and FLIC techniques (Gentry, Martin, and Daly 1966), and it was further developed at Lawrence Livermore National Laboratory in the 1970s (DeBar 1974; Sutcliffe 1973). The arbitrary Lagrangian–Eulerian, or ALE, technique (Hirt, Amsden, and Cook 1974) extended the idea to enable simulations that were intermediate between pure Eulerian and pure Lagrangian. The Lagrangian remap method of van Leer (1979) used a second-order Godunov method in the Lagrange phase and a monotonic advection method (van Leer 1977) in the remap phase. The Lagrangian remap method has subsequently been widely used, and a review of more recent research is given by Benson (1992).

In my earlier Lagrangian remap method (Youngs 1982), a simple finite-difference formulation is used for the Lagrangian phase. In the remap phase, an accurate interface reconstruction method is introduced and the monotonic advection method of van Leer (1977) is used. This relatively simple two-dimensional technique has since been very successfully used for a wide range of complex problems. TURMOIL3D (see Youngs 1991) uses a simplified three-dimensional (3D) version of the method, and this provides an effective way of calculating many relatively complex turbulent flows.

I use an explicit compressible finite-difference method. The technique described here is mainly Eulerian. However, I note one key extension later in this section. A staggered grid is used, with density (ρ), specific internal energy (e), pressure (p) defined at cell centers, and velocity components (u, v, w) defined at cell corners. My earlier interface-tracking technique (Youngs 1982) is not used; I assume that the two fluids (1 and 2) are miscible and that mixing occurs at a subgrid scale. Instead of interface tracking, I treat the mass fraction of Fluid 1, m_1, as a continuous variable. The numerical method solves the Euler equations plus an equation for m_1:

$$\frac{\partial(\rho m_1)}{\partial t} + \text{div}\,(\rho m_1 \underline{u}) = 0.$$

Each fluid has a perfect gas equation of state, and the mixture in each cell is assumed to be in pressure and temperature equilibrium. The pressure is given by

$$p = (\gamma - 1)\rho e, \quad \text{where} \quad \gamma = \frac{m_1 \gamma_1 c_{v1} + m_2 \gamma_2 c_{v2}}{m_1 c_{v1} + m_2 c_{v2}} \quad \text{and} \quad m_2 = 1 - m_1,$$

and $\gamma_1, \gamma_2, c_{v1}$, and c_{v2} are the adiabatic constants and specific heats for each fluid.

I divide the calculation for each time step into two parts, a Lagrangian phase and a remap (or advection) phase. The Lagrangian phase calculates the changes in velocity and internal energy that are due to the pressure terms, and the remap phase calculates the transport of mass, internal energy, and momentum across cell boundaries. One calculates X, Y, and Z advection in separate steps by using the monotonic advection method of van Leer (1977). This minimizes numerical diffusion while preventing spurious overshoots and undershoots. Further details of the advection technique are given in the next section.

The Lagrangian phase consists of three steps. The time-integration scheme is as follows. The first step is the calculation of the half-time-step pressures:

$$\rho^n \frac{e^{n+1/2} - e^n}{\frac{1}{2}\Delta t} = -(p^n + q^n)\,\text{div}\,\underline{u}^n,$$

$$\rho^{n+1/2} = \rho^n \Big/ \left(1 + \frac{1}{2}\Delta t \quad \text{div} \quad \underline{u}^n\right), \quad p^{n+1/2} = P\left(\rho^{n+1/2}, e^{n+1/2}\right).$$

The second step is the velocity update:

$$\rho \frac{\tilde{u}^n - u^n}{\Delta t} = -\nabla\left(p^{n+1/2} + q^n\right).$$

The third step is the density and internal energy update:

$$\tilde{\rho}^n = \rho^n \Big/ \left\{1 + \Delta t\,\text{div}\,\frac{1}{2}(\tilde{\underline{u}}^n + \underline{u}^n)\right\},$$

$$\rho \frac{\tilde{e}^n - e^n}{\Delta t} = -\left(p^{n+1/2} + q^n\right)\text{div}\,\frac{1}{2}(\tilde{\underline{u}}^n + \underline{u}^n).$$

Here $\tilde{\rho}^n$, $\tilde{e}^n$, and $\tilde{u}^n$ denote the fluid variables at the end of the Lagrangian phase.

For each of these steps, one calculates all three directions simultaneously. For low-Mach-number flows, one may divide the Lagrangian phase into several subtime steps. The overall time step is then less constrained by the sound speed, and computational efficiency is increased. The explicit numerical technique is ideally suited to parallel computing, and incompressible turbulent mixing can be efficiently modeled by low-Mach-number compressible simulations.

Some Lagrangian remap methods (e.g., van Leer 1979) use a Godunov technique in the Lagrange phase, whereas TURMOIL3D uses a finite-difference formulation. Hence some form of artificial viscosity (q) is needed for flows involving shocks. Two types of quadratic artificial viscosity are used.

A simple divergence form is used for weakly compressible flows:

$$q = c\,\rho\,(\min(\Delta x, \Delta y, \Delta z))^2 (\operatorname{div} \underline{u})^2 \quad \text{for div } \underline{u} < 0.$$

This form tends to inhibit compression in converging flows. Hence, for highly compressible flows, a more complex form (Wilkins 1980) is used:

$$q = c\,\rho\,\Delta s^2 S^2 \quad \text{for div } \underline{u} < 0 \quad \text{and } S < 0.$$

Here, $S = n_i \varepsilon_{ij} n_j$ is the strain rate in direction n_i of the pressure gradient and Δs is the mesh width in this direction. For both forms, c is constant and usually set to 2.0.

In order to analyze the numerical dissipation present in the scheme, it is useful to quantify that which is due to the artificial viscous pressure:

$$D_q = -\int q \operatorname{div} \underline{u}\, dV dt.$$

TURMOIL3D is essentially an Eulerian hydrocode. However, a simple extension may be used to give a semi-Lagrangian calculation (a special case of the ALE technique). In this case, the x-direction mesh moves with mean x velocity

$$\bar{\underline{u}} = \int \underline{u}\, dy\, dz \Big/ \iint dy\, dz.$$

Some "smoothing" is required so that the meshes in the x direction do not collapse to zero. The semi-Lagrangian facility involves a very simple change to the numerical technique – remap to a displaced mesh instead of the initial mesh. It is particularly useful for shock-tube-mixing experiments. The region of 3D meshing follows the turbulent mixing zone as it moves along the shock tube. This reduces the number of meshes required and also minimizes numerical errors caused by advecting turbulence through the mesh.

4c.2 Monotonicity properties

The essence of the remap method is that the "real changes" to the flow are calculated in the Lagrange phase. All the remap phase does is to transfer the problem defined on the mesh displaced by the Lagrangian motion back to the original mesh. Hence the required monotonicity properties are that values of all fluid variables at the end of the time step should not overshoot or undershoot the values at the end of the Lagrange phase ($\tilde{\rho}^n, \tilde{e}^n, \tilde{u}^n$). A version of the van Leer advection method is used to achieve this. Splitting the remap phase into X, Y, and Z steps makes exact monotonicity simple to achieve. In general, there are two main cases to consider – quantities defined per unit volume (e.g., density) and quantities defined per unit mass (e.g., mass fraction or velocity, i.e., momentum per unit mass). Let us consider the former first.

The volume of a cell at the end of the Lagrange phase is

$$\tilde{V} = V\left[1 + \Delta t \operatorname{div}\frac{1}{2}(\underline{\tilde{u}}^n + \underline{u}^n)\right] = V + \delta V_x^+ - \delta V_x^- + \delta V_y^+ - \delta V_y^- + \delta V_z^+ - \delta V_z^-,$$

where the δVs denote the cell-face volume fluxes. We remap this is to the initial volume, V, in three one-dimensional steps:

$$\tilde{V} \rightarrow V + \delta V_y^+ - \delta V_y^- + \delta V_z^+ - \delta V_z^- \rightarrow V + \delta V_z^+ - \delta V_z^- \rightarrow V.$$

The remapping order, XYZ (as shown here) or ZYX, is reversed every time step. For simplicity, let us consider only the X step, which suffices for the Y and Z directions (which are dropped):

$$V_{i+\frac{1}{2}}^1 = V_{i+\frac{1}{2}}^0 + \delta V_i - \delta V_{i+1}.$$

Here $V_{i+\frac{1}{2}}^0$ and $V_{i+\frac{1}{2}}^1$ denote cell volumes before and after the remap. If, for example, the X remapping is done first, then

$$V_{i+\frac{1}{2}}^0 = \tilde{V}_{i+\frac{1}{2}}, \quad \delta V_i = \delta V_x^-, \quad \delta V_{i+1} = \delta V_x^+ \quad \text{and} \quad \phi_{i+\frac{1}{2}}^0 = \tilde{\phi}_{i+\frac{1}{2}}^n.$$

The remap for a quantity ϕ, defined per unit volume is then given by

$$\phi_{i+\frac{1}{2}}^1 V_{i+\frac{1}{2}}^1 = \phi_{i+\frac{1}{2}}^0 V_{i+\frac{1}{2}}^0 + \phi_i^x \delta V_i - \phi_{i+1}^x \delta V_{i+1} \quad \text{where} \quad \phi_i^x = \phi_b^0 + \frac{1}{2}(1 - \varepsilon)\Delta\phi^{\text{VL}}.$$

$$\tag{4c.1}$$

Here a, b, and c denote $i - \frac{3}{2}$, $i - \frac{1}{2}$, and $i + \frac{1}{2}$ if $\delta V_i > 0$, and they denote $i + \frac{3}{2}$, $i + \frac{1}{2}$, and $i - \frac{1}{2}$ if $\delta V_i \leq 0$. The upwind cell is then denoted by b, and $\Delta\phi^{\text{VL}}$ denotes the van Leer limited variation of ϕ^0 across the upwind cell. Note that $\varepsilon = |\delta V_i|/V_b$ is the fraction of the upwind cell that is remapped. The form used for $\Delta\phi^{\text{VL}}$ is

$$\Delta\phi^{\text{VL}} = S \min\left\{|\Delta\phi|, 2\left|\phi_b^0 - \phi_a^0\right|/\varepsilon, 2\left|\phi_c^0 - \phi_b^0\right|/(1 - \varepsilon)\right\}$$

$$\text{where } S = \operatorname{sign}(\Delta\phi) \quad \text{if} \quad \left(\phi_b^0 - \phi_a^0\right)\left(\phi_c^0 - \phi_b^0\right) > 0, \quad S = 0 \text{ otherwise};$$

$$\Delta\phi = \Delta x_b \left\{ p\frac{\phi_b^0 - \phi_a^0}{\frac{1}{2}(\Delta x_a + \Delta x_b)} + (1 - p)\frac{\phi_c^0 - \phi_b^0}{\frac{1}{2}(\Delta x_b + \Delta x_c)} \right\} \text{ with}$$

$$p = \frac{\Delta x_c + \varepsilon\Delta x_b}{\Delta x_a + \Delta x_b + \Delta x_c}.$$

$$\tag{4c.2}$$

The third-order approximation used here for $\Delta\phi$ is preferred to a second-order approximation because it gives reduced numerical dispersion (Youngs 1982). It is based on the use of a quadratic distribution that reproduces the cell-averaged values, ϕ_a^0, ϕ_b^0, and ϕ_c^0. This is related to but not identical to the piecewise parabolic method advection method of Colella and Woodward (1984); see also Chapter 4b. Monotonicity now follows, provided that $\varepsilon < 1$:

$$\min\left(\phi_b^0, \phi_c^0\right) \leq \phi_{i+\frac{1}{2}}^1 \leq \max\left(\phi_b^0, \phi_c^0\right).$$

$$\tag{4c.3}$$

A condition corresponding to (4c.3) applies to each of the three one-dimensional remap steps; this ensures that, for the remap phase as a whole, the values of ϕ^{n+1} lie between the maximum and minimum values of a range of neighboring values of post-Lagrangian phase values, $\tilde{\phi}^n$.

For a cell-centered quantity, ψ, defined per unit mass, the procedure is as follows. First we remap the density field by using (4c.1) with $\phi =$ density, giving

$$M^1_{i+\frac{1}{2}} = M^0_{i+\frac{1}{2}} + \delta M_i - \delta M_{i+1}, \text{ where } M^1_{i+\frac{1}{2}} = \rho^1_{i+\frac{1}{2}} V^1_{i+\frac{1}{2}}, \ldots. \qquad (4c.4)$$

Then, for remapping ψ, we use

$$\psi^1_{i+\frac{1}{2}} M^1_{i+\frac{1}{2}} = \psi^0_{i+\frac{1}{2}} M^0_{i+\frac{1}{2}} + \psi^x_i \delta M_i - \psi^x_{i+1} \delta M_{i+1}, \qquad (4c.5)$$

where we calculate ψ^x_i by using Eq. (4c.2) with $\varepsilon = |\delta V_i| / V_b$ replaced by $\varepsilon = |\delta M_i| / M_b$. The monotonicity properties (4c.3) then apply to variable ψ. The mass fractions, m_1, are treated in this manner. For internal energy, we may use two approaches. Equation (4c.5) may be used with $\psi = e$, the specific internal energy, or Eq. (4c.1) may be used with $\phi = E = \rho e$, the internal energy per unit volume. The latter is preferred if the two fluids have the same value of γ, as E is then continuous across a fluid interface. Fortuitously, for shock-tube experiments using air and SF_6 (see Chapter 13), which have very different values of γ, the specific internal energy, e, has only a small discontinuity across an interface in temperature equilibrium. The use of Eq. (4c.5) then gives satisfactory results. If more complex equations of state were needed, it would be necessary to revert to the treatment used earlier (Youngs 1982), that is, separate internal equations for each fluid. However, this additional complexity has not been necessary for the TURMOIL3D applications considered so far.

Velocity components are defined at cell corners. Hence for momentum advection, we use corner-based versions of Eqs. (4c.4) and (4c.5):

$$\hat{M}^1_i = \hat{M}^0_i + \delta\hat{M}_{i-\frac{1}{2}} - \delta\hat{M}_{i+\frac{1}{2}}, \qquad (4c.6)$$

$$\psi^1_i \hat{M}^1_i = \psi^0_i \hat{M}^0_i + \psi^x_{i-\frac{1}{2}} \delta\hat{M}_{i-\frac{1}{2}} - \psi^x_{i+\frac{1}{2}} \delta\hat{M}_{i+\frac{1}{2}}, \qquad (4c.7)$$

where ψ stands for one of the velocity components. In a 3D situation, we obtain Eq. (4c.6) by averaging Eq. (4c.4) for the eight surrounding cells.

4c.3 Dissipation

The Lagrangian phase conserves the sum of kinetic and internal energies exactly (unless gravity is present). Moreover, it is nondissipative in the absence of shocks. However, the remap phase conserves the mass of each fluid, internal energy, and momentum but dissipates some kinetic energy. DeBar (1974) showed that it is possible to explicitly calculate the kinetic energy dissipation for each cell. There are nine contributions corresponding to the X, Y, Z remap of u, v, and w momentum. For the X remap of

u momentum, it is simply a matter of calculation u_i^1 from Eq. (4c.7), rearranging the sum

$$\sum_i \frac{1}{2} \left[\hat{M}_i^0 \left(u_i^0 \right)^2 - \hat{M}_i^1 \left(u_i^1 \right)^2 \right] \rightarrow \sum_i \Delta K_{i+\frac{1}{2}}^{xu} = \sum_i \Delta_{i+\frac{1}{2}}^{xu} \delta \hat{M}_{i+\frac{1}{2}}$$

where

$$\Delta_{i+\frac{1}{2}}^{xu} = \left(u_{i+1}^0 - u_i^0 \right) \left\{ \frac{1}{2} \left(u_{i+1}^0 + u_i^0 \right) - u_{i+\frac{1}{2}}^x \right\} - \frac{1}{2} \gamma_i \left(u_{i+\frac{1}{2}}^x - u_i^0 \right) + \frac{1}{2} \gamma_{i+1} \left(u_{i+1}^0 - u_{i+\frac{1}{2}}^x \right)$$

and

$$\gamma_i = \left(u_i^0 - u_{i-\frac{1}{2}}^x \right) \frac{\delta \hat{M}_{i-\frac{1}{2}}}{\hat{M}_i} + \left(u_{i+\frac{1}{2}}^x - u_i^0 \right) \frac{\delta \hat{M}_{i+\frac{1}{2}}}{\hat{M}_i}.$$

We use the nine similar ΔK contributions to calculate the kinetic energy dissipation associated with a particular cell and then add them to the internal energy. This gives total energy conservation and is in fact necessary to calculate shock propagation correctly. The so-called DeBar dissipation is also significant in turbulent flow, and I argue here that, in monotone integrated large eddy simulations, it corresponds to the dissipation provided by the subgrid-scale model used in conventional large eddy simulations.

For $\Delta t \rightarrow 0$,

$$\Delta K_{i+\frac{1}{2}}^{xu} = \left(u_{i+1}^0 - u_i^0 \right) \left\{ \frac{1}{2} \left(u_{i+1}^0 + u_i^0 \right) - u_{i+\frac{1}{2}}^x \right\} \delta \hat{M}_{i+\frac{1}{2}}.$$

Hence dissipation is zero for small time steps and central differencing. Van Leer limiting gives upwind differencing if there are steep gradients. Then

$$\Delta K_{i+\frac{1}{2}}^{xu} = \frac{1}{2} \left(u_{i+1}^0 - u_i^0 \right)^2 \left| \delta \hat{M}_{i+\frac{1}{2}} \right| > 0.$$

For diagnostic purposes, we sum the DeBar dissipation over space and time to give a total of D_{DB}.

4c.4 Inclusion of subgrid-scale models

For simulation of Rayleigh–Taylor and Richtmyer–Meshkov instabilities, monotonicity is considered essential for the initially discontinuous scalar fields, density, and mass fraction. However, one can argue that monotonicity is not essential for the velocity components. Hence an alternative approach is available for comparison purposes. Van Leer limiting is switched off and a subgrid-scale viscosity introduced. Moreover, a second-order approximation may be used in Eq. (4c.2), $\Delta \phi = \phi_c^0 - \phi_b^0$. This gives central differencing and hence minimal DeBar dissipation if small time steps are used. The effects of viscosity are calculated in a separate step after the end of the remap phase. An explicit finite-difference approximation is used and the time step is subdivided if required for numerical stability.

A simple Smagorinsky viscosity is currently used:

$$\nu_{SGS} = \left[C_s \min(\Delta x, \Delta y, \Delta z) \right]^2 \sqrt{2\varepsilon'_{ij}\varepsilon'_{ij}}, \quad \text{where} \quad \varepsilon'_{ij} = \varepsilon_{ij} - \delta_{ij}\,\varepsilon_{kk}$$

is the strain rate deviator. There is now explicit dissipation in the simulation, and the total dissipation of kinetic energy is given by

$$D = D_q + D_{DB} + D_v, \quad \text{where} \quad D_v = \int \rho\, \nu_{SGS}\, \varepsilon_{ij}\varepsilon_{ij}\, dV dt. \tag{4c.8}$$

The total kinetic energy dissipation rate is defined to be $\varepsilon = \dot{D}$.

REFERENCES

Benson, D. J. 1992 Computational methods in Lagrangian and Eulerian hydrocodes. *Comput. Meth. Appl. Mech. Eng.* **99**, 235–394.

Colella, P., & Woodward, P. R. 1984 The piecewise parabolic method (PPM) for gas-dynamical simulations. *J. Comput. Phys.* **54**, 174–201.

Debar, R. B. 1974 *A Method in Two-D Eulerian Hydrodynamics*. Lawrence Livermore National Laboratory: LLNL Report UCID-19683.

Gentry, R. J., Martin, R. E., & Daly, B. J. 1966 An Eulerian differencing method or unsteady compressible flow problems. *J. Comput. Phys.* **1**, 87–118.

Hirt, C. W., Amsden, A. A., & Cook, J. L. 1974 An arbitrary Lagrangian–Eulerian computing method for all flow speeds. *J. Comput. Phys.* **14**, 227–253.

Sutcliffe, W. G. 1973 *BBC Hydrodynamics*. Lawrence Livermore National Laboratory: LLNL Report UCID-17013.

Van Leer, B. 1979 Towards the ultimate conservative difference scheme. V. A second-order sequel to Godunov's method. *J. Comput. Phys.* **32**, 101–136.

Van Leer, B. 1977 Towards the ultimate conservative difference scheme. IV. A new pproach to numerical convection. *J. Comput. Phys.* **23**, 276–299.

Wilkins, M. L. 1980 Use of artificial viscosity in multidimensional fluid dynamic alculations. *J. Comput. Phys.* **36**, 281–303.

Youngs, D. L. 1982 Time-dependent multi-material flow with large fluid distortion, in *Numerical Methods for Fluid Dynamics*, ed. K. W. Morton & M. J. Baines, 273–85. New York: Academic Press.

Youngs, D. L. 1991 Three-dimensional numerical simulation of turbulent mixing by Rayleigh–Taylor instability. *Phys. Fluids A* **3**, 1312–1320.

4d MPDATA

Piotr K. Smolarkiewicz and Len G. Margolin

4d.1 Introduction

MPDATA (multidimensional positive definite advection transport algorithm; Smolarkiewicz 1983, 1984) is a family of finite-difference approximations to the advective terms in the conservative (flux) formulation of fluid equations. In general, MPDATA is akin to the dissipative Lax–Wendroff schemes (also known as Taylor–Galerkin in the finite-element literature), which can be derived from the first-order-accurate upwind algorithm (alias donor cell) by subtracting from the right-hand side a space-centered representation of the first-order error. In MPDATA, however, the compensation is based on a nonlinear estimate of the truncation error. The compensation is achieved by the iterative application of upwind differencing, where in the second and following iterations the leading truncation-error terms (of the upwind scheme) are cast into the form of advective fluxes, defined as products of the current solution iterate and a suitably defined velocity field. The resulting algorithm is accurate to the second order, conservative, fully multidimensional (i.e., free of the splitting errors), and computationally efficient; yet it maintains the signature properties of upwind differencing such as the strict preservation of sign of the transported field and relatively small phase error.

The theoretical foundation of MPDATA – the modified equation approach (see Chapter 5) – facilitates generalizations of the scheme to transport problems beyond elementary advection. In the early 1980s, the algorithm was conceived as an inexpensive alternative to flux-limited schemes for evaluating the advection of nonnegative thermodynamic variables (such as water substance) in atmospheric cloud models. Since then, a variety of options have been documented that extend MPDATA to full monotonicity preservation, to third-order accuracy, and to variable sign fields (such as momentum). MPDATA was generalized to be a complete fluid solver in the early 1990s (Smolarkiewicz 1991). In analyzing the truncation error of approximations to the momentum equation, one finds error terms that depend on the interaction of advection with the forcing terms, including the pressure gradient. Many implementations of nonoscillatory algorithms treat advection separately from the forcings, leaving this error uncompensated, thereby reducing the order of accuracy of the solution and potentially leading to oscillations and even instability (Smolarkiewicz and Margolin 1993). In MPDATA, this error is compensated by effectively integrating the forcing terms along a flow trajectory rather than at a grid point. A comprehensive review of MPDATA, including both the underlying concepts and the details of implementation, can be found in Smolarkiewicz and Margolin (1998). The overview of MPDATA as a general, nonoscillatory approach for simulating geophysical flows – namely high-Reynolds-number and low-Mach-number flows – from microscales to planetary scales, has been presented in Smolarkiewicz and Prusa (2002).

MPDATA is implemented in the three-dimensional (3D) program EULAG for simulating rotating, stratified flows in complex, time-dependent curvilinear geometries (Prusa and Smolarkiewicz 2003; Smolarkiewicz and Prusa 2002, 2005; Wedi and Smolarkiewicz 2004). The name EULAG alludes to the capability to solve the fluid equations in either an Eulerian (flux form; Smolarkiewicz and Margolin 1993) or a Lagrangian (advective form; Smolarkiewicz and Pudykiewicz 1992) framework. EULAG can be used for incompressible or anelastic fluids; in either case, an elliptic equation for pressure is solved using a preconditioned nonsymmetric Krylov solver. EULAG is fully parallelized by use of message passing and runs efficiently on a variety of platforms.

MPDATA has proven successful in simulations of geophysical flows using single block, structured, topologically rectangular meshes (hereafter Cartesian meshes), while employing diffeomorphic mappings to accommodate time-dependent curvilinear domains. Its potential for unstructured-grid approximations has been realized only recently. Bacon et al. (2000) implemented the basic MPDATA advection algorithm in the multiscale environmental model OMEGA for operational forecast of weather and pollutant dispersion. Independently, Margolin and Shashkov (2003) drew inspiration from the MPDATA approach to develop a second-order, sign-preserving conservative interpolation for remapping two-dimensional (2D) arbitrary Lagrangian–Eulerian (ALE) grids. In MPDATA, as in any Taylor-series-based integration method for differential equations, the choice of data structure has a pronounced impact on the details of the algorithm and its computational implications (Mavriplis 1992). Aiming at a broad range of applications involving complex geometries and inhomogeneous flows, Smolarkiewicz and Szmelter (2005a, 2005b) developed a general, compact edge-based unstructured-mesh formulation of MPDATA and extended it for aerodynamics flows Szmelter and Smolarkiewicz (2006). Their development differs from the approach adopted in Bacon et al. (2000), where the focus on meteorological applications dictated an unstructured-mesh discretization only in the horizontal, with cell-centered and face-centered control-volume staggering of scalar and vector dependent variables, respectively.

Unlike most nonoscillatory methods, MPDATA is based directly on the *convexity* of upwind advection,* rather than on the idea of flux limiting. In practical terms, the

---

* Numerical solutions remain bounded by the surrounding local values of the preceding time step, given solenoidal advecting flow and an adequately limited time step; for arbitrary flows, the weaker condition of sign preservation can be ensured (cf. Smolarkiewicz and Szmelter 2005b).

algorithm consists of a series of donor-cell steps; the first step provides a first-order-accurate solution while subsequent steps compensate leading-order truncation errors, derived analytically from a modified equation analysis of the upwind scheme. The iterative application of upwinding in MPDATA greatly simplifies the task of designing higher-order schemes without the necessity of knowing details of the resulting truncation error. Since the errors are cast in the form of the advective fluxes, the convexity of upwinding ensures their compensation while preserving the sign of the transported field. In consequence, the linear computational stability of the first donor-cell step implies the nonlinear stability of the entire MPDATA advection, a property essential for simulating turbulent flows.

4d.2 Basic scheme

Introductions to MPDATA typically start (cf. Smolarkiewicz and Margolin 1998) with a one-dimensional flux-form advection equation,

$$\frac{\partial \Psi}{\partial t} = -\frac{\partial}{\partial x}(u\Psi), \tag{4d.1}$$

where u is the flow velocity and Ψ a scalar field that is assumed to be nonnegative. Then, a donor-cell (or upstream) approximation to (4d.1) can be written in flux form as

$$\Psi_i^{n+1} = \Psi_i^n - \left[F\left(\Psi_i^n, \Psi_{i+1}^n, U_{i+1/2}\right) - F\left(\Psi_{i-1}^n, \Psi_i^n, U_{i-1/2}\right) \right], \tag{4d.2}$$

where the flux function F is defined in terms of the local Courant number U by

$$F(\Psi_L, \Psi_R, U) \equiv [U]^+ \Psi_L + [U]^- \Psi_R,$$

$$U \equiv \frac{u\delta t}{\delta x}; \quad [U]^+ \equiv 0.5(U + |U|); \quad [U]^- \equiv 0.5(U - |U|). \tag{4d.3}$$

The integer and half-integer indices correspond to the cell centers and cell walls, respectively. Here δt is the computational time step, δx is the length of a cell, and $[U]^+$ and $[U]^-$ are the nonnegative and nonpositive parts of the Courant number, respectively. A simple truncation error analysis[†] reveals that the first-order-accurate scheme in (4d.2) approximates, to the second order in δx and δt, the diffusion-advection equation

$$\frac{\partial \Psi}{\partial t} = -\frac{\partial}{\partial x}(u\Psi) - \frac{\partial}{\partial x}\left(-K\frac{\partial \Psi}{\partial x}\right), \tag{4d.4}$$

where

$$K = \frac{(\delta x)^2}{2\delta t}(|U| - U^2). \tag{4d.5}$$

The key idea of MPDATA is to rewrite the Fickian flux in the error term on the right-hand side of (4d.4) as a convective flux $-K(\partial \Psi/\partial x) \equiv u_d \Psi$, thereby defining

---

[†] All dependent variables are expanded in a Taylor series about a common point, say (x_i, t^n), and all temporal derivatives are expressed in terms of the spatial derivatives using the governing equation.

the diffusive pseudo-velocity,

$$u_d \equiv -\frac{(\delta x)^2}{2\delta t}(|U| - U^2)\frac{1}{\Psi}\frac{\partial \Psi}{\partial x}. \tag{4d.6}$$

In order to subsequently compensate the first-order error of (4d.2), one again uses the donor-cell scheme but with the *antidiffusive* velocity, $\widetilde{u} = -u_d$, in lieu of u and with the value of Ψ^{n+1} already updated in (4d.2) in lieu of Ψ^n. Choosing a suitable approximation of the ratio $(1/\Psi)(\partial \Psi/\partial x)$ in (4d.6), for example,

$$\frac{1}{\Psi}\frac{\partial \Psi}{\partial x} \approx \frac{2}{\delta x}\frac{\Psi_{i+1}^{n+1} - \Psi_i^{n+1}}{\Psi_{i+1}^{n+1} + \Psi_i^{n+1} + \varepsilon}, \tag{4d.7}$$

ensures the stability of the corrective iteration for arbitrary small $\varepsilon > 0$ and $|U| \le 1$.

Extending MPDATA to multiple dimensions is straightforward, except that one needs to account for the cross-derivative terms that appear in the truncation error of the donor-cell scheme. For example, in two dimensions the elementary advection equation, (4d.1), becomes

$$\frac{\partial \Psi}{\partial t} = -\frac{\partial}{\partial x}(u\Psi) - \frac{\partial}{\partial y}(v\Psi) \equiv -\nabla \cdot (\mathbf{v}\Psi), \tag{4d.8}$$

and its donor-cell approximation

$$\Psi_{i,j}^{n+1} = \Psi_{i,j}^n - \left[F\left(\Psi_{i,j}^n, \Psi_{i+1,j}^n, U_{i+1/2,j}\right) - F\left(\Psi_{i-1,j}^n, \Psi_{i,j}^n, U_{i-1/2,j}\right)\right]$$
$$- \left[F\left(\Psi_{i,j}^n, \Psi_{i,j+1}^n, V_{i,j+1/2}\right) - F\left(\Psi_{i,j-1}^n, \Psi_{i,j}^n, V_{i,j-1/2}\right)\right], \tag{4d.9}$$

where now U and V are the dimensionless Courant numbers

$$U \equiv \frac{u\delta t}{\delta x} \quad \text{and} \quad V \equiv \frac{v\delta t}{\delta y} \tag{4d.10}$$

and the flux function F has been defined in (4d.3). Here, the flow velocity is assumed constant for simplicity of presentation. The truncation-error analysis reveals that (4d.9) approximates to the second order the advection-diffusion equation,

$$\frac{\partial \Psi}{\partial t} = -\frac{\partial}{\partial x}(u\Psi) - \frac{\partial}{\partial y}(v\Psi)$$
$$+ \frac{(\delta x)^2}{2\delta t}(|U| - U^2)\frac{\partial^2 \Psi}{\partial x^2} + \frac{(\delta y)^2}{2\delta t}(|V| - V^2)\frac{\partial^2 \Psi}{\partial y^2} - \frac{UV\delta x\delta y}{\delta t}\frac{\partial^2 \Psi}{\partial x\partial y}. \tag{4d.11}$$

As in one dimension, the diffusive fluxes are cast in convective form, thereby defining antidiffusive pseudo-velocities in x and y directions; for example,

$$\widetilde{u} = \frac{(\delta x)^2}{2\delta t}(|U| - U^2)\frac{1}{\Psi}\frac{\partial \Psi}{\partial x} - \frac{UV\delta x\delta y}{2\delta t}\frac{1}{\Psi}\frac{\partial \Psi}{\partial y},$$

$$\widetilde{v} = \frac{(\delta y)^2}{2\delta t}(|V| - V^2)\frac{1}{\Psi}\frac{\partial \Psi}{\partial y} - \frac{UV\delta x\delta y}{2\delta t}\frac{1}{\Psi}\frac{\partial \Psi}{\partial x}. \tag{4d.12}$$

By selecting uniformly bounded approximations to the ratios $(1/\Psi)\nabla\Psi$, one ensures the stability of the corrective iteration.

4d.3 Accuracy, stability, and benchmark results

The basic MPDATA schemes described in the preceding section are constructed from the classical upwind scheme, which is consistent, conditionally stable, and first-order accurate. These properties, together with the algorithm's design, predetermine the consistency, stability, and accuracy of MPDATA (Smolarkiewicz 1984). In particular, since the antidiffusive velocities (4d.12) tend to zero as spatial and temporal increments decrease, the consistency of MPDATA is implied by that of the upwind scheme. Similarly, since the corrective upwind iteration compensates the first-order leading error of the preceding upwind step, with accuracy to the first order at least, the uncompensated portion of the upwind error remains at second order. The latter suffices to increase the formal accuracy order of upwinding on a Cartesian mesh – second- and third-order asymptotic convergence rates of various MPDATA options have been documented for Cartesian meshes (cf. Margolin and Smolarkiewicz 1998; Smolarkiewicz and Margolin 1998). For arbitrary meshes, however, this is not necessarily the case, since the formal accuracy of the centered scheme – a target of the MPDATA derivation; compare Eq. (12) in Smolarkiewicz and Szmelter (2005b) – is not mesh independent (Roe 1987). Notwithstanding, Bacon et al. (2000) demonstrated second-order convergence of their cell-centered unstructured MPDATA using a standard "rotating-cone" benchmark while invoking intensive local mesh refinement; Smolarkiewicz and Szmelter (2005b) have shown second-order accuracy for their edge-based MPDATA formulation on both quality and skewed unstructured grids.

The stability of MPDATA also follows from that of the upwind scheme, but it deserves comment, since there are a few subtleties involved. In one spatial dimension, (4d.6) together with the stability and positivity of the original upwind scheme imply

$$|\widetilde{U}_{i+1/2}| \leq |U_{i+1/2}| - (U_{i+1/2})^2 \leq |U_{i+1/2}|; \widetilde{U} \equiv \frac{\widetilde{u}\delta t}{\delta x}, \qquad (4d.13)$$

so that the stability of the upwind approximation ensures the stability of MPDATA. A similar result occurs in multidimensional problems, but the presence of cross-derivatives makes formal proof difficult. In Smolarkiewicz (1984), it has been proven that the stability of the upwind scheme implies that of MPDATA, but with the caveat that the time step used is smaller than that allowed for the upwind alone by the factor $2^{-1/2}$ and 2^{-1}, in two and three spatial dimensions, respectively. This particular time-step requirement follows from assuming a worst-case scenario in which the velocity components flip sign across the cell. Since the latter is a rare event in CFD applications, the heuristic limit recommended for all structured-mesh MPDATA extensions has been the same as that which is valid for the upwind scheme. A similar result has been argued and verified for the unstructured-mesh formulation (Smolarkiewicz and Szmelter 2005b).

We illustrate the performance of basic MPDATA using a standard solid-body rotation test (Smolarkiewicz 1983, 1984; Smolarkiewicz and Margolin 1998). A cone of base radius 15 and height 4, centered initially at (75, 50), is rotating counterclockwise around the center of a $[0, 100] \times [0, 100]$ domain with angular velocity $\omega = 0.1$. Figure 4d.1 displays the isolines of the exact result and of two MPDATA solutions after

Table 4d.1. *Error norms for solid-body rotation test*

Scheme	Formulation (grid)	Max	Min	L_2
MPDATA	Cartesian	2.18	0.	0.47×10^{-3}
MPDATA	unstructured (squares)	2.18	0.	0.47×10^{-3}
MPDATA	unstructured (triangles)	2.19	0.	0.47×10^{-3}
Upwind	Cartesian	0.27	0.	1.21×10^{-3}
Upwind	unstructured (squares)	0.28	0.	1.04×10^{-3}
Upwind	unstructured (triangles)	0.25	0.	1.06×10^{-3}
Leapfrog	Cartesian	3.16	-0.62	0.62×10^{-3}
Leapfrog	unstructured (squares)	3.11	-0.67	0.64×10^{-3}
Leapfrog	unstructured (triangles)	3.11	-0.69	0.65×10^{-3}

Note: Norms quantify Cartesian and unstructured-grid formulations of MPDATA; the classical upwind and centered-in-time-and-space leapfrog schemes are included for reference.

six rotations. These results were generated with the edge-based unstructured-grid formulation of MPDATA (Smolarkiewicz and Szmelter 2005a, 2005b). The solution in the central plate uses 10^4 squared cells, whereas the solution in the right plate uses a triangular grid with a similar number and distribution of points. For reference, all parameters of the test and of the display are identical to those in Figure 4d.1 in Smolarkiewicz and Margolin (1998), which showed similar results generated with a Cartesian-mesh formulation of MPDATA. The solution using square cells is indistinguishable from the corresponding result in their Figure 4d.1. The accuracy of the displayed results is quantified in Table 4d.1, where the corresponding values for the classical upwind and centered-in-time-and-space leapfrog schemes are included for the sake of reference.

4d.4 Extensions

The basic MPDATA scheme can be supplemented by numerous extensions that either enhance the accuracy and generality of the MPDATA advection, or expand its capabilities beyond advective transport to alternate partial differential equations (PDEs) and to complete flow solvers. A comprehensive review of existing MPDATA options can be found in Smolarkiewicz and Margolin (1998). Here, we summarize a few

Figure 4d.1. Isolines of cone advected through six rotations around the center of the lower frame (only a quarter of the domain is shown). The contour interval is 0.25, and the zero contour line is not shown. Left plate, the analytic solution; center plate, the unstructured-grid MPDATA on a regular square mesh; right plate, as in the center but for a triangular mesh.

selected extensions that are particularly important for many applications and are routinely employed in flow simulations discussed in other chapters.

4d.4.1 Generalized transport equation

It is instructive to present selected MPDATA options in terms of a generalized transport equation,

$$\frac{\partial G\Psi}{\partial t} + \nabla \cdot (\mathbf{v}\Psi) = GR, \tag{4d.14}$$

where $G = G(\mathbf{x}, t)$, $\mathbf{v} = \mathbf{v}(\mathbf{x}, t)$, and $R = R(\mathbf{x}, t)$ are assumed to be known functions of the coordinates. In fluid dynamics applications, G may play the role of the Jacobian of the coordinate transformation from a Cartesian $(\mathbf{x}_C, t_C)$ to a time-dependent curvilinear framework $(\mathbf{x}, t)$ or, alternately, it may be a product of the Jacobian and the fluid density. The latter distinction depends on the elastic versus inelastic character of the governing fluid equations[‡] and determines the interpretation of Ψ. In elastic systems, Ψ represents a fluid property per unit of volume – that is, a density type of variable – whereas in inelastic systems, Ψ represents a fluid property per unit of mass – that is, a mixing ratio type of variable. This freedom of interpretation benefits the efficacy of the Taylor-series-based flow solvers such as MPDATA. In particular, in inelastic systems it suffices to cancel truncation errors depending on the flow divergence, regardless of the complexity of the accompanying mass continuity equation (Smolarkiewicz and Prusa 2002). Depending upon the definition of G, $\mathbf{v} \equiv G\dot{\mathbf{x}}$ should be viewed as either a generalized "advective" velocity or a momentum vector; R combines all forcings, sources, or both. In general, both $\mathbf{v}$ and R will be functionals of the dependent variables.

In order to design a fully second-order MPDATA flow solver, it is important to derive leading truncation-error terms for (4d.14) rather than to hastily combine the basic scheme with, for example, a time-centered representation of the right-hand side forcing. Such derivations (Smolarkiewicz and Margolin 1998; Smolarkiewicz and Prusa 2002) reveal terms that depend on the interaction of advection with the forcing, on the coordinate transformation as well as on the time dependence of the flow. Since these terms are rooted in the uncentered time discretization, they are independent of spatial discretization, whereupon all related developments (cf. Smolarkiewicz and Margolin 1998 and Smolarkiewicz and Prusa 2002 for discussions) are common to structured- and unstructured-mesh frameworks. For example, discretizing (4d.14) in time as

$$\frac{G\Psi^{n+1} - G\Psi^n}{\delta t} + \nabla \cdot \left(\mathbf{v}^{n+1/2}\Psi^n\right) = GR^{n+1/2} \tag{4d.15}$$

[‡] The terms *elastic* and *inelastic* distinguish between a compressible versus an incompressible character of the PDE used to describe the flow. For example, consider the shallow water equations – a long-wave approximation for incompressible stratified flows – that are mathematically akin to the compressible-flow Euler equations.

leads to the modified equation

$$\frac{\partial G\Psi}{\partial t} + \nabla \cdot (\mathbf{v}\Psi) = GR - \nabla \cdot \left[\frac{1}{2}\delta t \frac{1}{G}\mathbf{v}(\mathbf{v} \cdot \nabla\Psi) + \frac{1}{2}\delta t \frac{1}{G}\mathbf{v}\Psi(\nabla \cdot \mathbf{v}) \right]$$

$$+ \nabla \cdot \left(\frac{1}{2}\delta t \mathbf{v} R \right) + \mathcal{O}(\delta t^2), \tag{4d.16}$$

where all $\mathcal{O}(\delta t)$ errors that are due to the uncentered time differencing in (4d.15) are now expressed as spatial derivatives. Choosing the time levels of both the advective velocity and the forcing term as $n + 1/2$ in (4d.15) is important, as it eliminates $\mathcal{O}(\delta t)$ truncation errors proportional to their temporal derivatives (Smolarkiewicz and Margolin 1998); any $\mathcal{O}(\delta t^2)$ approximations to $\mathbf{v}^{n+1/2}$ and $R^{n+1/2}$ will suffice for compensating these error terms. The term on the right-hand side of (4d.16) proportional to the convective flux of the forcing, R, couples advection and forcing. Many implementations of nonoscillatory algorithms treat advection separately from the forcings, naively adopting the experience with centered-in-time-and-space methods. However, leaving this error uncompensated not only reduces the order of accuracy of the solution but also amplifies oscillations and can even lead to instability (Smolarkiewicz and Margolin 1993). In MPDATA, this piece of the error is typically compensated (Smolarkiewicz and Margolin 1998) by effectively integrating the forcing terms along a flow trajectory rather than at a grid point (Smolarkiewicz and Margolin 1993), thereby paraphrasing the Strang splitting (Strang 1968). Ultimately, this leads to compact-form algorithms for integrating systems of generalized transport equations (4d.14). For example, integrating the anelastic system of equations of motion cast in time-dependent curvilinear coordinates – an effective strategy for modeling a broad class of geophysical or astrophysical flows (cf. chapter 14) – can be pursued with an MPDATA advection scheme applied judiciously as follows:

$$\Psi_i^{n+1} = \frac{G_i^n}{G_i^{n+1}} \mathcal{M}_i \left(\Psi^n + 0.5\delta t R^n, \ \mathbf{v}^{n+1/2}, \ G^n \right) + 0.5\delta t R_i^{n+1}. \tag{4d.17}$$

Here, $\mathcal{M}_i$ symbolizes the output from a homogeneous advection MPDATA module, applied to an auxiliary input variable $\Psi^n + 0.5\delta t R^n$ defined at each data point i. The only two extensions of the basic MPDATA required for (4d.17) are (i) the G^{-1} factors in the antidiffusive velocity terms indicated in (4d.16); and (ii) a generalized form of the antidiffusive velocity in (4d.12) allowing the advection of fields of variable sign such as momenta.

4d.4.2 Transporting fields of variable sign

So far, we have assumed that transported field Ψ is exclusively either nonnegative or nonpositive. This assumption is important for the stability, accuracy, and, generally speaking, for the design of MPDATA. However, it enters MPDATA schemes explicitly only in the pseudo-velocity formulae, in the $\sim \Delta\Psi / \sum \Psi$ terms of the discrete approximations to the components of $\sim \Psi^{-1}\nabla\Psi$ ratios – compare (4d.6) and (4d.12).

These terms are bounded when Ψ is of a constant sign, but can lead to arbitrarily large pseudo-velocities and unstable schemes when Ψ changes sign. MPDATA can be extended to the transport of variable-sign fields in several ways. Here we outline two that have proven useful in applications.

The simplest and most common way is to replace all Ψs in (4d.6) and (4d.12) with $|\Psi|$s, exploiting the relationship[§]

$$\frac{1}{\Psi}\frac{\partial \Psi}{\partial x} \equiv \frac{1}{2\mu}\frac{1}{(\Psi^2)^\mu}\frac{\partial (\Psi^2)^\mu}{\partial x}\Big|_{\mu=1/2} = \frac{1}{|\Psi|}\frac{\partial |\Psi|}{\partial x}. \qquad (4d.18)$$

The results are, practically speaking, insensitive to the value of μ; however, $\mu = 1/2$ is the optimal choice as it only requires replacing Ψ with $|\Psi|$ in the pseudo-velocity formulae derived for the constant-sign fields and is, furthermore, computationally the most efficient.

An alternate approach exploits the mass continuity equation (Section 4 in Smolarkiewicz and Clark 1986). Multiplying (4d.14) – with $\Psi \equiv \chi$ being the fluid density (elastic systems), or with $\Psi \equiv \chi \equiv 1$ and a steady reference density included in G (inelastic systems) – by an arbitrary constant c and adding the resulting equation to (4d.14) leads to

$$\frac{\partial G(\Psi + c\chi)}{\partial t} + \nabla \cdot [\mathbf{v}(\Psi + c\chi)] = GR, \qquad (4d.19)$$

which illustrates yet another degree of freedom in MPDATA. The arbitrary constant c can be chosen to ensure positivity of Ψ^n, while making MPDATA susceptible to asymptotic linear analysis as $c \nearrow \infty$ (Smolarkiewicz and Clark 1986). Furthermore, MPDATA itself can be linearized around an arbitrary large constant, leading straightforwardly to a two-pass scheme that differs technically from the basic algorithm in only two details: at the second iteration, the donor-cell flux function in (4d.3) takes the value unity in its two Ψ arguments, and the pseudo-velocities in (4d.6) and (4d.12) replace each Ψ with unity in all "$\sum \Psi$" denominators. This asymptotic form of MPDATA – often referred to as the "infinite gauge" – is a realization of the classical Lax–Wendroff algorithm (cf. Section 4 in Smolarkiewicz and Clark 1986). Combined with the nonoscillatory option, it makes a viable scheme for transporting momenta in fluid models.

4d.4.3 Nonoscillatory option

The basic MPDATA scheme described in Section 4d.2 preserves the sign but not the monotonicity of the transported variables (Smolarkiewicz 1984; Smolarkiewicz and Clark 1986; Smolarkiewicz and Grabowski 1990), and, in general, the solutions are not free of spurious extrema (see Figure 4d.2). This is because the antidiffusive velocity is

---

[§] For a discussion of some formal issues at $\Psi \to 0$, see Section 3.2 in Smolarkiewicz and Clark (1986).

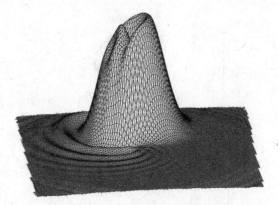

Figure 4d.2. Unstructured-grid MPDATA solution for the solid-body rotation benchmark from Section 4d.3 but with the cone replaced by the cylinder placed on a large constant background.

not necessarily solenoidal, even for a solenoidal physical flow. In many studies of natural flows, the preservation of sign is adequate. However, when required, one can make MPDATA fully monotone (Smolarkiewicz and Grabowski 1990) by adapting the flux-corrected transport (FCT) formalism (Zalesak 1979) to limit the pseudo-velocities (cf. Figure 4d.3). It can be argued (Smolarkiewicz and Grabowski 1990) that MPDATA is particularly well suited for this adaptation for several reasons. First, the initial MPDATA iteration is the upwind scheme – a low-order monotone scheme commonly used as the reference in the FCT design. Second, ensuring monotonicity of the subsequent iterations provides a higher-order-accurate reference solution for the next iteration with the effect of improving the overall accuracy of the resulting FCT scheme. Third, since all MPDATA iterations have similar low phase errors characteristic of the upwind scheme (Smolarkiewicz and Clark 1986), the FCT procedure mixes solutions with consistent phase errors. This benefits the overall accuracy of the resulting FCT scheme (see Figure 4d.5 in Smolarkiewicz and Grabowski 1990, and the accompanying discussion).

The FCT extension for Cartesian-mesh MPDATA has been presented in Smolarkiewicz and Grabowski (1990) together with an algebraic theory of FCT limiting; the algebraic formalism has proven useful for synchronous FCT where physical bounds imposed on functions of transported fields alter the standard limiters of the individual fields (Grabowski and Smolarkiewicz 1990; Schär and Smolarkiewicz 1996). Recently, a technical summary of relevant formulae and details of implementation as well as

Figure 4d.3. As in Figure 4d.2, but for the nonoscillatory-option MPDATA solution.

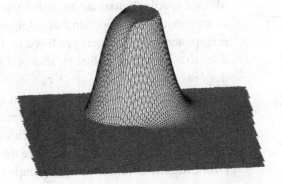

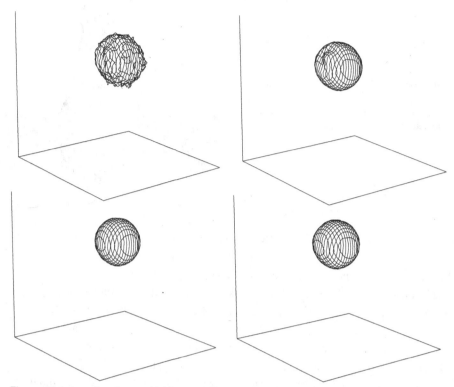

Figure 4d.4. Leapfrog-trapezoidal (upper left), leapfrog-trapezoidal-FCT (upper right), basic MPDATA (lower left), and infinite-gauge-FCT MPDATA (lower right) solutions after one revolution of the sphere.

3D benchmark calculations were presented in Smolarkiewicz and Szmelter (2005b), in the context of the edge-based unstructured-grid algorithm; Figure 4d.4 shows four of their solutions. The test case adopted is an extension of the rotating-cylinder problem from Figures 4d.2 and 4d.3 to three spatial dimensions: A sphere with radius 15 and constant density 4, placed initially at $\mathbf{x}_o = (25, \ 75, \ 75)$, rotates with angular velocity $\omega = 0.1 \times 3^{-1/2}(1, \ 1, \ 1)$ around a diagonal of the cuboidal domain $[0, \ 100] \times [0, \ 100] \times [0, \ 100]$. Two solutions in the upper left and right plates are, respectively, for the leapfrog-trapezoidal scheme and its FCT version (Zalesak 1979), while the two solutions in the lower plates are for the basic and infinite-gauge-FCT MPDATA. All solutions use an unstructured grid with background spacing 2, a roughly twice coarser resolution than that employed in the 2D examples discussed earlier. The relative accuracy measures are listed in Table 4d.2. All solutions are evaluated after $T = 10 \times 2\pi \approx 556\delta t$, that is, after one revolution of the sphere around the domain diagonal.

Table 4d.2 lists three standard norms of the solution error, where both the L_2 and L_1 norms are normalized to reflect the rms and absolute value of the truncation error per cell and the unit of time. Inasmuch as L_2 better reflects the overall accuracy, the remaining measures tend to emphasize the amplitude and phase errors (viz. behavioral errors; Roache 1972). The leapfrog-trapezoidal scheme evinces well-known dispersive

Table 4d.2. *Relative accuracy measures for the solutions displayed in Figure 4d.4*

Scheme	Max	Min	L_∞	L_2	L_1
Leapfrog-Trp.	7.1	−2.5	4.4	5.3×10^{-3}	2.5×10^{-3}
Leapfrog-Trp. FCT	4.4	0.0	3.0	3.5×10^{-3}	5.4×10^{-4}
MPDATA basic	4.8	0.0	2.9	4.1×10^{-3}	7.8×10^{-4}
MPDATA$_\infty$ FCT	4.4	0.0	2.8	3.2×10^{-3}	4.1×10^{-4}

Note: The first column identifies the scheme used, and the following five columns provide norms of the solutions' departures from the analytic result.

oscillations characteristic of all higher-order-accurate linear schemes. The remaining three schemes are nonlinear and suppress the oscillations. While the basic MPDATA is only sign preserving, that is, nonoscillatory near "zeros," the two FCT schemes are formally monotone. The overshoots with respect to the initial amplitude 4, a cumulative effect $\sim \Psi \nabla \cdot \mathbf{v}$ due to the residual flow divergence,[‖] are consistent with the FCT formalism that defines monotonicity relatively to a smooth low-order solution (here upwind) and vanish for strictly incompressible flows.

4d.5 Concluding remarks

In general, nonlinear algorithms are roughly three to four times as expensive computationally as linear schemes. Although the nonlinear schemes are more accurate overall (cf. Tables 4d.1 and 4d.2), preserving the monotonicity or sign of transported variables appears cost ineffective for pure advection. However, in numerical simulation of fluids, the monotone or sign-preserving advection may not be optional, but rather a necessary prerequisite of solution realizability (e.g., for reactive or multiphase flows). While arguments can be made that the basic MPDATA scheme is of comparable accuracy to the classical leapfrog-trapezoidal FCT scheme, the monotone infinite-gauge MPDATA is the most accurate in all norms (Table 4d.2). The judgment as to which scheme should be the method of choice in not straightforward, as in practice there are factors other than the standard norms and the computational efficiency of the sole advection.

Over the past two decades, MPDATA has been frequently compared with other transport schemes, primarily in the context of passive scalar advection. The assessments of MPDATA's relative strengths and weaknesses reported in the literature depend very much on the schemes included in comparisons, choice of test problems, MPDATA's options, and details of implementation. The most common complaints are that the basic MPDATA is too diffusive, while enhanced MPDATAs are too expensive. The most often acknowledged virtues are MPDATA's multidimensionality, robustness, and its underlying conceptual simplicity. These advantages carry over to complete fluid models, where the relative efficiency of advection becomes less important with the

---

‖ Since the actual rotational flow is prescribed analytically, the maximal Courant-number divergence is 1.6×10^{-3}, only 3 orders of magnitude smaller than the Courant number itself.

increasing complexity of other model physics. In particular, the accuracy measures of representative schemes listed in Table 4d.2 stimulate a question: which scheme is optimal? In our experience, there is no straightforward answer. Consider, for instance, that for implicit flow solvers the efficacy of a fluid model depends critically on the efficiency of the associated elliptic solver. The latter, in turn, depends on the spectral composition of the right-hand side, depending, in turn, on the spectral properties of the advective transport (Smolarkiewicz, Grubišić, and Margolin 1997). Our years of experience with MPDATA indicate that several combinations of options may have merit for an accurate simulation of differing complex high-Reynolds-number flows (cf. Smolarkiewicz, Margolin, and Wyszogrodzki 2001).

REFERENCES

Bacon, D. P., & coauthors. 2000 A dynamically adapting weather and dispersion model: The operational environment model with grid adaptivity (OMEGA). *Mon. Weather Rev.* **128**, 2044–2076.

Grabowski, W. W., & Smolarkiewicz, P. K. 1990 Monotone finite-difference approximations to the advection-condensation problem. *Mon. Weather Rev.* **118**, 2082–2097.

Mavriplis, D. J. 1992 Three dimensional unstructured multigrid for the Euler equations. *AIAA J.* **30**, 1753–1761.

Margolin, L. G., & Shashkov, M. 2003 Second-order sign-preserving conservative interpolation (remapping) on general grids. *J. Comput. Phys.* **184**, 266–298.

Margolin, L. G., & Smolarkiewicz, P. K. 1998 Antidiffusive velocities for multipass donor cell advection. *SIAM J. Sci. Comput.* **20**, 907–929.

Prusa, J. M., & Smolarkiewicz, P. K. 2003 An all-scale anelastic model for geophysical flows: Dynamic grid deformation. *J. Comput. Phys.* **190**, 601–622.

Roache, P. J. 1972 *Computational Fluid Dynamics.* Albuquerque, NM: Hermosa Publishers.

Roe, P. L. 1987 Error estimates for cell-vertex solutions of the compressible Euler equations. Institute for Computer Application in Science and Engineering: ICASE Report 87-6.

Schär, C., & Smolarkiewicz, P. K. 1996 A synchronous and iterative flux-correction formalism for coupled transport equations. *J. Comput. Phys.* **128**, 101–120.

Smolarkiewicz, P. K. 1983 A simple positive definite advection scheme with small implicit diffusion. *Mon. Weather Rev.* **111**, 479–486.

Smolarkiewicz, P. K. 1984 A fully multidimensional positive definite advection transport algorithm with small implicit diffusion. *J. Comput. Phys.* **54**, 325–362.

Smolarkiewicz, P. K. 1991 On forward-in-time differencing for fluids. *Mon. Weather Rev.* **119**, 2505–2510.

Smolarkiewicz, P. K., & Clark, T. L. 1986 The multidimensional positive definite advection transport algorithm: Further development and applications. *J. Comput. Phys.* **67**, 396–438.

Smolarkiewicz, P. K., & Grabowski, W. W. 1990 The multidimensional positive definite advection transport algorithm: Nonoscillatory option. *J. Comput. Phys.* **86**, 355–375.

Smolarkiewicz, P. K., Grubišić, V., & Margolin, L. G. 1997 On forward-in-time differencing for fluids: Stopping criteria for iterative solutions of anelastic pressure equations. *Mon. Weather Rev.* **125**, 647–654.

Smolarkiewicz, P. K., & Margolin, L. G. 1993 On forward-in-time differencing for fluids: Extension to a curvilinear framework. *Mon. Weather Rev.* **121**, 1847–1859.

Smolarkiewicz, P. K., & Margolin, L. G. 1998 MPDATA: A finite-difference solver for geophysical flows. *J. Comput. Phys.* **140**, 459–480.

Smolarkiewicz, P. K., Margolin, L. G., & Wyszogrodzki, A. A. 2001 A class of nonhydrostatic global models. *J. Atmos. Sci.* **58**, 349–364.

Smolarkiewicz, P. K., & Prusa, J. M. 2002 Forward-in-time differencing for fluids: Simulation of geophysical turbulence, in *Turbulent Flow Computation*, eds. D. Drikakis & B. J. Guertz. Boston: Kluwer Academic, 279–312.

Smolarkiewicz P. K., & Prusa, J. M. 2005 Towards mesh adaptivity for geophysical turbulence. *Int. J. Numer. Meth. Fluids* **47**, 1369–1374.

Smolarkiewicz, P. K., & Pudykiewicz, J. A. 1992 A class of semi-Lagrangian approximations for fluids. *J. Atmos. Sci.* **49**, 2082–2096.

Smolarkiewicz P. K., & Szmelter, J. 2005a Multidimensional positive definite advection transport algorithm (MPDATA): An edge-based unstructured-data formulation. *Int. J. Numer. Meth. Fluids* **47**, 1293–1299.

Smolarkiewicz P. K., & Szmelter, J. 2005b MPDATA: An edge-based unstructured-grid formulation. *J. Comput. Phys.* **206**, 624–649.

Strang, G. 1968 On the construction and comparison of difference schemes. *SIAM J. Numer. Anal.* **5**, 506–517.

Szmelter, J., & Smolarkiewicz, P. K. 2006 MPDATA error estimator for mesh adaptivity, *Int. J. Numer. Meth. Fluids*, **50**, 1269–1293.

Wedi, N. P., & Smolarkiewicz, P. K. 2004 Extending Gal–Chen and Somerville terrain-following coordinate transformation on time dependent curvilinear boundaries. *J. Comput. Phys.* **193**, 1–20.

Zalesak, S. T. 1979 Fully multidimensional flux-corrected transport algorithms for fluids. *J. Comput. Phys.* **31**, 335–362.

4e Vorticity Confinement

John Steinhoff, Nicholas Lynn, and Lesong Wang

4e.1 Introduction

We describe the vorticity confinement (VC) method for efficiently treating thin vortical structures in high-Reynolds-number incompressible flows. This forms the basis of a VC-based implicit large eddy simulation (ILES) method for turbulent flow simulation. Earlier applications of VC have included capturing attached and separating boundary layers and convecting vortex sheets and filaments. These have served as a substitute for much more expensive Reynolds-averaged Navier–Stokes (RANS) schemes, and they are described in the literature (Dietz 2004a). The present study focuses on turbulent flow simulations, where VC is used in the modeling and computation of the thin vortex filaments and sheets, which we assume to be the smallest resolved scales.

The principal objective of VC is to capture the *essential* features of these small-scale vortical structures. By essential features, we mean vortex filaments that may convect over long distances with no significant spreading and can change topology and merge or reconnect and can absorb large-scale energy by stretching; we also mean vortex sheets that can become unstable and break up into filaments. Thus, VC can be thought of as "physical structure preserving." This is effected with a very efficient difference method *directly* on a (fixed Eulerian) computational grid. We argue that this is more effective than first formulating a model partial differential equation (PDE) for these small scales and then making a discrete approximation, since it allows thin vortical regions to be implicitly modeled, spread over as few as one to three grid cells, and convected over arbitrarily long distances with no effects from numerical error. The irrotational part of the flow, as well as any larger-scale vortical structures, can then be solved with sufficient resolution with conventional, discretized PDEs (Euler equations). For these regions, the method can reduce to conventional computational fluid dynamics (CFD).

Besides allowing simple coarse grids to be used, the method can be used effectively with low-order time and space discretization of the basic equations, since it eliminates artificial spreading caused by numerical diffusion for the small vortical scales. This greatly simplifies boundary conditions and reduces computational time further.

Our purpose in this chapter is to explain the basic rationale behind using VC, to describe the main ideas behind VC in more detail than has been available in previous papers, and finally to explore the possibilities that it offers toward more efficient LES computations of turbulent flows (also see Chapter 12).

Basic rationale Our basic overall objectives for turbulent flow simulations are described in what follows. The goal of any large eddy simulation (LES) method is, of course, to solve for the larger scales of turbulent flow and obtain a solution of a filtered, or convolved, flow field with the small scales filtered out. There are two basic concepts in an LES method that we assume to be important and that serve as the rationale for using VC.

First, for small-scale vortices, VC allows us to reduce numerical dissipation so that it minimizes the effect on the numerical solution, down to scales of a small number (about two) of grid cells sizes (h), so that other nearby scales, as small as approximately 3–4 h, are not significantly contaminated. In this way, VC allows us to maximize h for a given resolution, or effective filter width. This is important because every factor of 2 increase in h leads to a factor of 16 decrease in computing time. For general small vortical scales, it is not possible to avoid numerical error over long convection times, even if expensive high-order conventional CFD schemes are used. However, if the small scales in the field consist of vortex sheets and filaments, as is often assumed, this is easily accomplished – with VC.

Second, even in the limit of vanishing viscosity, just eliminating numerical errors, such as diffusion, and accurately solving the Euler equations for the filtered field is not sufficient. A filtered field computed in this way would not experience instabilities as strong as the unfiltered field would. This reduction of instabilities can have strong consequences, even in two-dimensional (2D) flows. Thus, these instabilities must be restored by a destabilizing term, which can be accomplished with VC. The basic idea will be (qualitatively) discussed below for Kelvin–Helmholtz (KH) flows, as an example. This issue essentially involves approximating backscatter.

Kelvin–Helmholtz instability Consider a flat 2D contact discontinuity (vortex sheet) along the x axis, with velocity $u = +1$ above and -1 below, and $v = 0$. The kinetic energy density is

$$e = \vec{q}^2/2 = u^2/2 = 1/2$$

both above and below the sheet. Now consider a filtered field (Figure 4e.1); e will have the form in Figure 4e.2 with the dashed and the solid curves representing the unfiltered and the filtered fields, respectively. The area between the solid and dashed curves represents the subgrid scale, or "latent" energy that is not present in the filtered field. This energy should be considered since it can be added to the filtered field as it evolves to energize the KH instabilities that have been damped by the filtering.

The unfiltered sheet will, of course, be unstable to an imposed perturbation, which will initially grow exponentially and then tend to saturate (by the sheet rolling up).

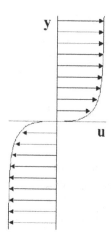

Figure 4e.1. Velocity profile: filtered shear layer.

The physical realization of this is the set of small, rolling-up vortex sheets sketched in Figure 4e.3. The LES goal is, of course, to evolve the filtered velocity field to represent this phenomenon. Thus, an initially thick vortex layer (representing the filtered initial sheet) should be unstable and evolve to a set of vorticity concentrations, which represent the filtered spirals. These should also initially grow exponentially and then saturate, as sketched by the vorticity contours in Figure 4e.4. Computations of a similar case are described in Section 4e.3.1.3.

However, even if the filtered field evolves exactly according to the Euler equations, with no numerical error, it will have to evolve over a longer time period than the unfiltered field before exhibiting such instabilities. Also in 2D inviscid, incompressible flow (obeying Euler equations alone), it is well known that vorticity is only convected without changing magnitude. Thus regions of increased magnitude cannot develop. We conclude that the filtered field should evolve according to the Euler equations, but with a negative dissipation added that will also cause it to be unstable initially, but that will

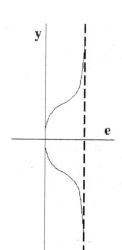

Figure 4e.2. Energy profile: filtered and unfiltered shear layer.

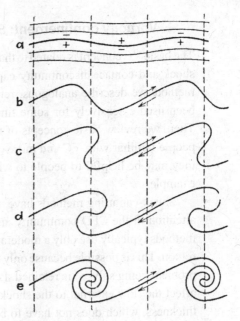

Figure 4e.3. KH instability from Prandtl and Tietjens (1934).

eventually saturate and not diverge. This extra term should involve an acceleration of the flow or energizing of the developing vorticity concentrations, leading eventually to a set of separated vortices. The total energy will then have increased over that of the initial thick, flat sheet. A possible approximation, or basis for this model, would be that the final energy gain of the filtered field, which is determined by the initial field and the final vortices, should represent the subgrid or "latent" energy present initially. Currently, we use this idea only as a qualitative guide, but we are beginning to develop more quantitative methods.

The main point is that the VC method represents a very simple way to add the required energy and, at the same time, eliminate the spreading effects of numerical dissipation.

Figure 4e.4. Vorticity contours: filtered shear layer.

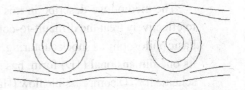

4e.2 Vorticity confinement: Basic concepts

The basic VC concept is related to that of similar methods also involving thin structures – shock and contact-discontinuity capturing. Accordingly, before we describe the VC method, we describe analogous, relevant features of these methods, since they have been used extensively for some time and are very familiar to the CFD community. Then, we review basic concepts of the new method (VC). These points are known to people familiar with VC and more conventional discontinuity-capturing methods, but they may be helpful to people to whom it is new. We will use shock capturing as an example.

Shock-capturing methods have, of course, received an extremely large amount of attention in the CFD community and have proven to be extremely important. These methods typically use only a moderately sized inviscid computational grid in the shock region. This is possible because only the *essential* physics of the shock (as far as the flow problem being solved) is retained. By "essential physics" we mean those features that affect the flow external to the shock interior. These features include computed shock thickness, which does not have to be as small as the physical thickness but, like the physical thickness, must be small compared with the main length scales of the problem. They also include the requirement that conservation laws, integrated through the shock, are preserved. In this way, for many problems that do not depend on the details of the shock internal structure, accurate flow solutions have been obtained with specially developed numerical shock-capturing algorithms. In these methods the detailed, accurate solution of PDEs (e.g., Navier–Stokes equations) for the internal shock structure has been avoided. This has been important since it avoids the requirements of a very fine computational grid within the structure, and very time-consuming viscous computations there. These ideas, which go back to Von Neumann and Richtmyer (1950), Lax (1957), and others, involve the concept of "weak solutions" of PDEs where, in the inviscid limit, discontinuous features can be treated. After discretization these ideas should allow shocks to be approximated over as few as one to three grid cells.

The question naturally arises as to whether similar efficient "capturing" treatments of thin vortical features are also possible, which would result in similar benefits. One difference, however, is that with shocks, unlike vortical regions, characteristics slope inward toward the shock, which naturally tends to become steeper during a computation. As a result, modeling shocks is simpler than modeling thin vortical structures and other contact-like discontinuities, which naturally tend to spread because of numerical discretization errors and the need for stabilizing numerical diffusion. This results in the requirement that a "steepening" or "confinement" term be added to prevent artificial spreading.

The method we describe in this chapter – VC – has been specifically formulated to effectively treat the difficult-to-compute concentrated vortical regions with the same basic philosophy as shock capturing. Although developed independently, the method, in its one-dimensional (1D) form, has some relation to Harten's "artificial compression" scheme for 1D compressible flow (Harten 1978). However, the VC formulation is much simpler (at least for incompressible flows). Also, an important feature for capturing thin vortical regions is that VC is intrinsically multidimensional and rotationally invariant (Harten 1978; van Leer 1974). A number of recent papers (Steinhoff, Senge, and Wenren 1990; Steinhoff et al. 1992, 1997, 2003a; Steinhoff and Underhill 1994; Steinhoff, Wenren, and Wang 1999; Wang, Steinhoff, and Wenren 1995; Wenren et al. 2003), mentioned in what follows, describe the use of VC for incompressible flow. Further extensions to compressible flow have recently been developed (Costes and Kowani 2003; Dadone, Hu, and Grossman 2001; Dietz 2004a; Hu, Grossman, and Steinhoff 2002) but will not be described here, since we want to concentrate on turbulent flow with as few extra complications as possible. As with shock capturing, it is understood that the details of the internal structure of thin vortical regions will not be accurately treated, unless special models are developed for them. We assume here that these details are not important, other than possibly subgrid-scale energy of thin sheets, and that simple capturing alone is sufficient.

As background, we first mention some previous (non-LES) examples for which VC alone is effective. These include convecting vortex rings, which can be convected with no spreading yet can merge with no requirement for special logic (Steinhoff and Underhill 1994). They also include thin shed wingtip vortices, which can be computed over arbitrarily long distances (Haas 2003; Wenren, Steinhoff, and Robins 1995) and exhibit Crow instability, including merging (Wenren, Steinhoff, and Robins 1995). Computations of both of these phenomena show close agreement with experiment. For trailing vortex convection over very long distances (many kilometers), the method can serve as a zeroth-order approach in this case, since turbulence eventually induces a very slow spreading. This effect can then be simply modeled within the VC framework, again without using very fine grids or high-order methods. Examples also include a very simple and inexpensive RANS substitute for attached and separating boundary layers. This is described below in Section 4e.3.1.4 and elsewhere (Dietz 2004a; Dietz et al. 2001; Fan et al. 2002; Wenren et al. 2001).

An important point that should be emphasized is that, at high Reynolds numbers, most vortical regions will be turbulent. Hence, *any* computational method must involve, explicitly or implicitly, a numerical model for the small-scale structure, since it is not feasible to directly solve the Navier–Stokes equations for this structure. The structure obtained with VC when a small-scale vortical region or "eddy" is captured can, thus, be thought of as just such a model, but one that is very efficient to compute. Further, this model is intrinsically discrete, defined over only a few grid cells, and is not meant to be an accurate solution of a model PDE. The rationale for taking this approach is that it is difficult to resolve PDEs for a thin vortical structure over long distances, even with higher-order methods, if it is spread over only a small number (two to four) of grid cells. This is due to the well-known fact that the accuracy, or *order*, of a method is only an *asymptotic* estimate of the behavior of the error, valid for large N, the number of grid cells across the vortical region; $N = 2$–3 is not sufficient to apply such an estimate. Since VC is meant to *capture* the feature, and not accurately solve a model PDE, it gets around this problem. Further, many features of the flow external to the core or interior of a vortical region are not sensitive to the details of the internal structure. For example, in 2D cases, vortices tend to evolve to an axially symmetric state (Melander, McWilliams, and Zabusky 1987). Then, the only requirements for accurately determining the induced flow external to the initial core are that the total circulation is conserved, that the vortex centroid have the correct location, and that the core does not spread as a result of numerical effects.

As stated, VC even involves a negative, though nondiverging, *total* diffusion at certain length scales. Similar observations about the need for such a negative total dissipation, or eddy viscosity, in LES for certain regions of the flow have been made by Ferziger (1996). This author explained that, in those regions, popular models for the eddy viscosity cannot be used because conventional numerical methods for solving the flow equations would diverge there. As a result, the eddy viscosity is often arbitrarily set to zero in those regions. One of the main advantages of VC is that it allows an overall negative eddy viscosity to be implemented – which ultimately saturates, rather than diverges. This approach should be much simpler and more direct than other approaches used to treat negative diffusion, or, effectively, energize the small-scale vortices, such as "stochastic forcing."

There are currently other efforts to capture small scales directly on the grid for use in LES turbulence simulations. These often involve combinations of 1D operators, as in original discontinuity-capturing schemes (Boris and Book 1973; Harten 1978; van Leer 1974), and they are the particular subject of Chapters 4a–4d and elsewhere in this volume. As opposed to VC, the emphasis in these other ILES approaches is typically not on vorticity. Furthermore, their main goal is to cancel numerical diffusion as much as possible, whereas for VC it is to specifically treat thin vortical structures with a controlled "model" structure.

4e.2.1 Illustrative one-dimensional example

As explained in the last section, the basic concept of VC is that thin vortices are captured, like shocks, over only a few grid cells. This means, of course, that the discrete equations do not represent an accurate solution of a simple PDE for the internal structure, although they can be considered as approximations to singular (weak) solutions. The goal in these cases (for small scales) is to accurately treat only certain integral quantities, and to be physical structure preserving (i.e., ensure that the vortex remains thin).

There is a simple example of this concept in one dimension involving the convection of a passive scalar, or "pulse," that is concentrated in a small region. A goal, then, could be to preserve the total amplitude of this scalar, to have the centroid move at the correct convection speed, and to ensure that it remains compact – essentially spread over a small number of cells. Additional moments of this pulse representing additional structure could, if desired, also be transported by using additional fields, but no attempt is made at convecting an accurate, pointwise representation of the internal structure.

It will be seen that the solution of the "confined" equations is, effectively, a nonlinear *solitary wave* that "lives" on the lattice indefinitely. The same applies to the VC solutions of convecting vortices, although, of course, they cannot be analyzed in as much detail as this 1D example.

We then consider a scalar (ϕ) advecting at a speed c that, in the continuum case, would satisfy the PDE

$$\partial_t \phi = -\partial_x c\phi. \tag{4e.1}$$

Symbolically, we then define a discretized equation,

$$\phi_j^{n+1} = \phi_j^n - \Delta t C\phi_j^n + E_j^n, \tag{4e.2}$$

where C is a conventional conservative discrete convection operator, j denotes the spatial grid index, n is the time step ($t = n\Delta t$), and E_j^n is a nonlinear term designed to keep the pulse compact.

There are requirements for E_j^n if the total pulse amplitude and speed are to be conserved, independent of the pulse amplitude:

1. E_j^n must be (at least) a first difference to conserve the total amplitude (which is assumed to vanish rapidly away from the centroid).
2. E_j^n must be (at least) a second derivative so that, each time step, the centroid position is not changed by E_j^n and, hence, the speed of the centroid is correct, as given by the original PDE. Then,

$$E_j^n = \delta_j^2 F_j^n, \tag{4e.3}$$

   where δ_j^2 is a discrete second-difference operator and F_j^n is a function of ϕ and its differences, which vanishes in the far field.
3. F_j^n must be homogeneous of degree 1 in ϕ, as the other terms in the equation, so that there is no dependence on the amplitude, or scale of ϕ.
4. $\delta_j^2 F_j^n$ must represent a negative diffusion if the pulse is spread over too large an area, so that it contracts and relaxes to a fixed shape.
5. $\delta_j^2 F_j^n$ must become a positive diffusion if the pulse is too thin, for the same reasons.
6. Property 4 requires that F_j^n be nonlinear. If it were linear, the negative diffusion would result in divergence: Any modes that initially increase in amplitude would continue to increase and eventually diverge since they would be uncoupled and would evolve independently.

A simple formulation that satisfies these requirements is

$$C\phi_j^n = \frac{\delta_j^c c_j \phi_j^n}{h}, \tag{4e.4}$$

$$F_j^n = \mu\phi_j^n - \varepsilon\Phi_j^n, \tag{4e.5}$$

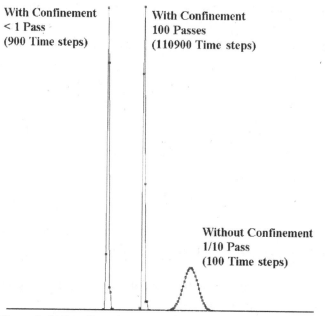

Figure 4e.5. One-dimensional scalar convection with and without confinement.

where δ_j^c is a central difference, h is the grid cell size, μ and ε are constants, and Φ_j^n is a harmonic mean:

$$\Phi_j^n = \left[\frac{\sum\limits_{j-1}^{j+1} \left(\phi_l^n \right)^{-1}}{3} \right]^{-1}. \tag{4e.6}$$

There are many other formulations for F that can also be used. The first term in F_j^n acts to stabilize the central difference operator, C, and also satisfies Conditions 1–3 and 5, and the second term satisfies Conditions 1–4 and 6.

The resulting difference equation can be written as

$$\phi_j^{n+1} = \phi_j^n - \frac{1}{2} \left(v_{j+1}\phi_{j+1}^n - v_{j-1}\phi_{j-1}^n \right) + \delta_j^2 \left(\mu\phi_j^n - \varepsilon\Phi_j^n \right), \tag{4e.7}$$

where δ_j^2 is a second-difference operator and

$$v_j = c_j \Delta t / h. \tag{4e.8}$$

Results of a computation (on a periodic – 256-cell grid) with no confinement, using $\mu = 0.2$ and $\varepsilon = 0$ after 100 time steps (1/10th pass through the grid) are shown in Figure 4e.5. (Of course, higher-order conventional CFD methods could result in less diffusion than shown here. However, compared to confinement, these would all require more grid cells within the pulse and, eventually, spread it over even more cells as a result of accumulated numerical error.) Results are also shown in Figure 4e.5 with confinement for 1 and 100 passes through the grid for $\mu = 0.2$ and $\varepsilon = 0.5$.

For all of these cases, $v = \sqrt{2}/6$, chosen to be irrational to prevent coincidental error cancellation.

The exact "sampled" pulse height at any given time step depends on the centroid position within a grid cell, almost as if an approximately fixed pulse shape were "moved" through the grid and the values sampled at each time step. The pulse, however, remains confined indefinitely, except for a very small effect (after $\sim 10^6$ time steps) that is due to the finite precision of the computer. This implies that there is an approximate, smooth solution to Eq. (4e.7), $\phi(x, t)$, in terms of a similarity variable, $z = x - ct$. For a range of initial conditions that have the form of a thin pulse, ϕ will relax to this particular "solitary wave" pulse solution. This has been shown numerically. However, stability of the solitary wave and the set of initial conditions that are attracted to it have not been mathematically derived.

It has been shown (Steinhoff et al. 2003b) that, for a range of values of μ, ε, and v, the sign of ϕ cannot change. Thus for an initial nonnegative pulse, the maximum cannot exceed the sum, which is conserved. This prevents the solution from diverging.

In the small Δt limit, for constant c, ϕ satisfies

$$\delta_j^2 \left(\mu \phi_j^n - \varepsilon \Phi_j^n \right) = 0. \tag{4e.9}$$

A solution is then

$$\phi_j^n = A \operatorname{sech} \left[\gamma \left(x - x_0 - ct \right) \right], \tag{4e.10}$$

where A and x_0 depend on initial amplitude and centroid, and

$$\cosh (\gamma) = \left\{ \left[(3\varepsilon/\mu) - 1 \right] /2 \right\}. \tag{4e.11}$$

We will show in Section 4e.3.2 that Eqs. (4e.6) and (4e.7), when summed over $j' < j$, represent a propagating step function–like discontinuity, rather than a pulse. The summed form then exhibits some similarity to schemes used by van Leer (1974), Harten (1978), and others to capture 1D contact discontinuities and shocks in compressible flow.

4e.3 Vorticity confinement: Methodology

Two formulations of VC have been developed that have similar properties: The first, VC1, involves first derivatives of velocity (Steinhoff, Mersch, and Wenren 1992; Steinhoff, Senge, and Wenren 1990), while the second, VC2, involves second derivatives (Steinhoff et al. 2003b). Starting from an initial condition more spread than the final structure, VC1 acts essentially as an inward convection and relaxes to the final structure more quickly than VC2, which initially acts as a negative, second-order diffusion. The two versions are described here.

The basic principle in VC, as in the 1D convecting scalar example already described, is that there is a solution with a stable structure that can be propagated indefinitely. Although the VC equations can be written as a discretization of a PDE (described later), the resulting *solution* at the small scales (within the structure) is not meant to be an accurate, or even approximate, solution of the original PDE. This is because VC is meant to capture, or model, the small-scale features over only a couple of grid cells, so that the discretization "error" is O(1) there. As explained in Section 4e.1, the features essential for the problem, however, are still preserved. Thus, the captured feature is actually a nonlinear *solitary wave* that "lives" on the grid lattice. (There is currently a large amount of work being done on intrinsically discrete, or *difference*, as opposed to *finite difference,* equations; examples can be found in Bohr et al. 1998.) In smooth regions (or for large scales), in contrast, VC can be made to automatically revert to conventional CFD where the PDEs are then accurately and efficiently approximated.

For thin vortical regions in incompressible flow, we use essentially the same approach as in the aforementioned 1D example: "confinement" terms are added to the conventional, discretized momentum equation. Although we use a primitive variable, and not a vorticity, formulation, we will see that if we look at the resulting vorticity transport equation (for the VC2 version, to be described later), it has the identical confinement terms as a multidimensional extension of the 1D scalar transport equation described in Section 4e.2. For general unsteady incompressible flows, the governing equations with VC are discretizations of the continuity and momentum equations, with added terms:

$$\nabla \cdot \vec{q} = 0, \tag{4e.12}$$

$$\partial_t \vec{q} = -\vec{\nabla} \cdot (\vec{q}\vec{q}) + \left[\mu \nabla^2 \vec{q} - \varepsilon \vec{s} \right], \tag{4e.13}$$

where $\vec{q}$ is the velocity vector, p is the pressure, ρ is the density, and μ is a diffusion coefficient that includes numerical effects that are due, for example, to discretization of the first right-hand-side (convection) term. (We assume that the Reynolds number is large and that physical diffusion is much smaller than the added terms.) For the last term, $\varepsilon \vec{s}$, ε is a numerical coefficient that, together with μ, controls the size and time scales of the convecting vortical regions or vortical boundary layers and $\vec{s}$ is defined later. For this reason, we refer to the two terms in the brackets as the confinement terms. Vector $\vec{s}$ is different for the two VC formulations and is defined below.

Equation (4e.13) involves constant μ and ε, which is sufficient for many problems. If these are not constant, such as, for example, when the grid spacing is not constant or models are used for them, then these quantities can be taken inside the differential operators in the corresponding terms, to maintain explicit momentum conservation (in the VC2 formulation).

As in the 1D example, the *pair* of confinement terms, which represent spreading, or positive diffusion and contraction, or negative diffusion, together create the confined structures. Stable solutions result when the two terms are approximately balanced. In this way, corrections are made each time step to compensate for any perturbations to the vortical structure caused by convection in a nonconstant external velocity, a discretization error in the convection operator, or the pressure correction. The parameters μ and

ε then essentially determine the thickness of the resulting vortical structure and the relaxation rate to that state. It should be emphasized that stable, equilibrium structures result for a wide range of values of these parameters.

In general, for boundary layers and isolated, convecting vortex filaments, computed flow fields *external* to the vortical regions are not sensitive to the internal structures, and hence to parameters ε and μ, over a wide range of values. For example, a general thin, concentrated vortex will physically tend to evolve to an axisymmetric configuration (Melander, McWilliams, and Zabusky 1987). Further, even a rapidly rotating nonsymmetric configuration will be approximately axisymmetric when averaged over a short time (Misra and Pullin 1997). Then, it is well known that the flow outside an axisymmetric 2D vortex core is independent of the vortical distribution, and hence will not depend on ε and μ as long as the core is thin (and the filament curvature is large, so that the flow is approximately 2D in a plane normal to the filament). Therefore, the issues involved in setting these parameters will be similar to those involved in setting numerical parameters in other standard CFD schemes, such as artificial dissipation in many conventional shock-capturing schemes, which, as explained, are closely analogous. Further, for turbulent wake flows, preliminary studies – described in Chapter 12 – suggest that ε can be used to parameterize finite-Reynolds-number effects, since it controls the intensity of the smallest resolved vortical scales (this is the subject of current research; Fan and Steinhoff 2004). This parameter can also be used to adjust the confinement level according to the amount of available subgrid energy (this is also a subject of current research).

An important feature of the VC method is that, for incompressible flow, the confinement terms are nonzero only in the vortical regions, since both the diffusion term and the contraction term vanish outside those regions. Thus, even if there is a second-order isotropic numerical diffusion associated with the convection operator, and the diffusion operators are only second order, outside the vortical regions the resulting accuracy of these terms can be third or fourth order, since this diffusion is just the negative curl of the vorticity.

A final point concerns the total change induced by the VC correction in mass, vorticity, and momentum, integrated over a cross section of a convecting vortex. A pressure–projection method (Kim and Moin 1985) is used to solve Eqs. (4e.12) and (4e.13), so that mass is automatically conserved. Vorticity is explicitly conserved because of the vanishing of the correction outside the vortical regions. Finally (in the VC2 formulation), momentum is also exactly conserved (Steinhoff et al. 2003b) because the VC terms added to the momentum equations have a spatial derivative operator in front. In the 1D example just described, this ensures that the pulse centroid convects with a weighted average of the imposed velocity. This should also be true for a confined vortex convecting in an imposed external velocity field. Then, the vortex centroid will move with a weighted average of the velocity of the "background" flow, with no effect from self-induced flow (at least in two dimensions). This has been demonstrated numerically (Fan et al. 2002; Steinhoff, Fan, and Wang 2003c; see Section 4e.3.1.2). This is not exactly satisfied in the VC1 formulation, but errors that are due to the lack of momentum conservation have been shown numerically to be small in most cases (see Section 4e.3.1.1).

Many basic numerical methods could be used for space and time discretization. We use a simple first-order Euler integration in time and second order in space with, as stated, a pressure–projection method to enforce mass conservation. In conventional CFD schemes, higher-order methods often must be used, usually to reduce numerical diffusion and hence attempt to reduce spreading of thin vortical regions. Vorticity confinement eliminates this problem for many cases and avoids the boundary condition complexity and computational cost of the higher-order methods. (It should be mentioned, however, that the second confinement, or contraction, term involves a larger difference stencil than the other terms.)

Another numerical issue involves the regularity of the grid. It is important to realize that, since a convecting vortex or separated boundary layer is captured directly on the grid, over a few grid cells, large grid aspect ratios or rapidly varying cell sizes should not be used. If these are avoided, VC will result in a dynamics that is close to rotationally invariant. These issues also occur, of course, in shock capturing. Some modifications can be made, however, to accommodate nonuniform grids if the aspect ratio is not too large (Morvant 2004; Morvant et al. 2003; Steinhoff and Raviprakash 1995).

4e.3.1 Basic formulation

As explained, the two different formulations, VC1 and VC2, have somewhat different dynamics, since they differ in the order of the derivative in the contraction term. The one developed initially (VC1) has been described in a number of publications, and we present only a few details here.

4e.3.1.1 VC1 formulation

This formulation involves an expression for the contraction term, $\vec{s}$, that does not explicitly conserve momentum:

$$\vec{s} = \hat{n} \times \vec{\omega}. \tag{4e.14}$$

For convecting vortices,

$$\hat{n} = \vec{\nabla}\eta \big/ |\vec{\nabla}\eta|, \tag{4e.15}$$

where

$$\eta = |\vec{\omega}|.$$

For boundary layers, $\hat{n}$ is a unit vector parallel to the local normal. This term essentially convects vorticity within a thin vortical region either along its own gradient or along the local normal, from the edge, or region of lower magnitude, toward the center, or region of larger magnitude. As the structure contracts and the gradient increases, the expansion term, which is a linear diffusion, increases until a balance is reached. (This is a well-known property of convection–diffusion phenomena.) Because of the rapid rotation of convecting concentrated vortices, any non-conservative momentum errors are almost completely canceled, and the method has proved to be sufficiently accurate for many problems.

A technicality in applying this method is often overlooked by people using it, and this has been described elsewhere (Steinhoff, Mersch, and Wenren 1992). Since vorticity is convected along $\hat{n}$, upwind (in $\hat{n}$) values of ω should be used in the contraction term to avoid creating "downwind" values of vorticity with an opposite sign. This is easily accomplished with weighting factors at each node that depend on $\hat{n}$ and unit vectors to neighboring grid nodes.

Most of the VC results presented in the literature use the VC1 formulation. However, they do not involve very slow background flow and do not involve the momentum conservation issue discussed here. An important point, however, is that exact momentum conservation, in some cases, may not be as important as other features (such as, in our case, ensuring that a convecting vortex remain thin) and should not be regarded as an absolute requirement (see, e.g., the basic CFD textbook – Tannehill, Anderson, and Pletcher 1997, p. 60).

4e.3.1.2 VC2 formulation

Only for very accurate long-term trajectory determination of vortices convecting in a slow background velocity field has the momentum-conserving VC2 formulation been found to be necessary (for incompressible flow). This ensures that the contribution of the self-induced velocity to the vortex motion is completely canceled.

The VC2 formulation involves

$$\vec{s} = \vec{\nabla} \times \vec{w}^n. \tag{4e.16}$$

We can also combine the dissipation and the confinement into a single term:

$$\mu \nabla^2 \vec{q} - \varepsilon \vec{s} = \nabla \times (\mu \vec{\omega}^n - \varepsilon \vec{w}^n), \tag{4e.17}$$

where

$$\vec{\omega}^n = \vec{\nabla} \times \vec{q}^n \tag{4e.18}$$

and

$$\vec{w}^n = \frac{\vec{\omega}^n}{\tilde{\omega}^n} \left[\frac{\sum_l \left(\tilde{\omega}_l^n \right)^{-1}}{N} \right]^{-1}, \tag{4e.19}$$

$$\tilde{\omega}_l^n = \left| \vec{\omega}_l^n \right| + \delta. \tag{4e.20}$$

Equation (4e.17) has some numerical advantages over the form just before it, since the same difference operator acts on $\vec{\omega}$ and $\vec{w}$. Also, the second confinement term (4e.18) is the sum over the stencil that consists of the central node (where $\vec{w}$ is computed) and its neighboring $(N-1)$ nodes, and δ is a small positive constant $(\sim 10^{-8})$ to prevent problems caused by finite precision.

When two approximately oppositely directed vortices are close to each other, there can be grid cells in between in which $\vec{w}$ is not well defined, which may cause oscillations. To prevent this, if the scalar product of any of the other vorticity vectors in the stencil with the central node is negative, $\vec{w}$ is set to zero.

To see the action of VC2, we take the curl of Eq. (4e.17) We then get a transport equation for vorticity. For example, in two dimensions,

$$\partial_t \omega = -\nabla \cdot (\omega \vec{q}) + \nabla^2 \left[\mu \omega - \varepsilon \Phi(\omega)\right]. \tag{4e.21}$$

This equation, including the confinement term, is exactly a multidimensional, rotationally invariant generalization of the 1D scalar advection equation shown to be effective in Section 4e.2. Of course, the solution will still reflect the fourfold symmetry of the grid. This effect, however, vanishes rapidly away from a vortical region. Further, the rotating flow around a vortex core actually allows a simpler discretization of Eq. (4e.21), compared with an axisymmetric convecting passive scalar distribution. This is explained in Steinhoff et al. (2003b).

Equation (4e.19) is a harmonic mean. It is chosen to weigh the small values in the stencil more heavily. As is well known, this term vanishes when any of the values of its argument vanish, preventing creation of values of opposite sign (for a range of parameters). With the use of VC, the total vorticity in a region surrounding a vortex is conserved, since it is a local term. This means that the vorticity cannot diverge as a result of this term, since the maximum absolute value cannot be greater than the absolute value of sum when all values have the same sign. (This is also a property of the 1D scalar advection example.)

There are a large number of alternative forms that would work as well as the harmonic mean. We believe that the term should have a smooth algebraic form, however, to give smooth results. This should be more appropriate for multidimensional applications than forms involving logic functions, such as "minmod," which give good results in 1D applications. As already discussed, a harmonic mean term was used by van Leer (1974) as a limiter, but mostly in 1D compressible flow, and, to our knowledge, not as a function of vorticity. (The VC1 and VC2 methods were developed independently, as multidimensional, rotationally invariant confinement techniques specifically for thin vortical regions.)

4e.3.1.3 Examples of VC results for convecting vortices

Vorticity contours from computations on a 128×128 uniform Cartesian grid for two vortices of the same strength rotating around each other in 2D are presented in Figure 4e.6 for a sequence of "snapshots" (Steinhoff et al. 2003a). In Figure 4e.6(a), we used no confinement. The large dissipation of the low-order convection numerical method is apparent. In Figures 4e.6(b) and 4e.6(c), we used VC2. The thin white lines across

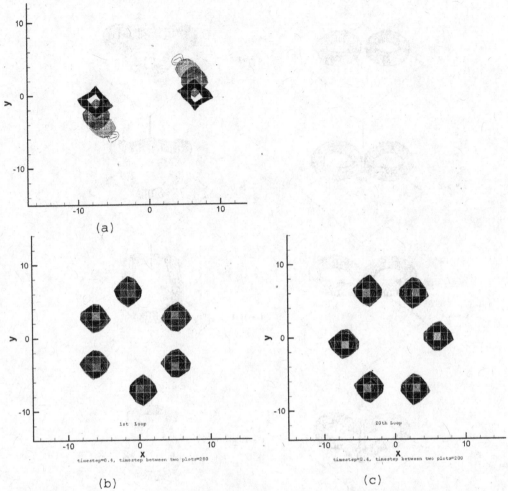

Figure 4e.6. Vorticity contour plots of self-induced flow by two vortices (14 cells apart) with the same sign: (a) without VC; (b) with VC, first loop; (c) with VC, after 20 loops.

the cores represent grid lines. The ability of VC2 to maintain very compact vortices, even after 20 full orbits around each other, is apparent. VC1 shows essentially the same behavior but exhibits some "drifting" over long periods, as expected. It should be emphasized that we obtained these results without resorting to high-order methods (which would have been futile on the coarse grid).

As another example, computations on a $48 \times 48 \times 48$ uniform Cartesian grid for two interacting, initially coplanar 3D vortex rings, using VC1 (Steinhoff and Underhill 1994) are discussed next. A vorticity isosurface is shown in Figure 4e.7, for a sequence of times. The cores can be seen to be confined to about two grid cells. In this case, the vortices merge and relink, with no requirement for special logic. The vorticity isosurfaces from an experiment of the same flow (Oshima and Izutsu 1988), shown in Figure 4e.8, can be seen to compare very favorably. As before, we used only low-order numerical methods. The use of VC2 results in slightly larger (but still constant) core sizes, and essentially the same behavior.

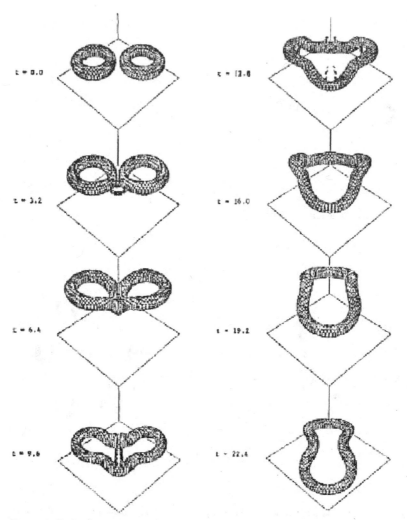

Figure 4e.7. Coplanar vortex rings: computation.

One other illustrative example involves the KH instability described in Section 4e.1.1. We computed a simple flow with, initially, a 2D stadium shaped, thin vortex sheet with zero velocity inside. This would be expected to roll up, in the zero thickness limit, into spirals. We computed the evolution of the filtered field, using VC1 on a 128 × 128 grid. The initial vorticity contours are shown in Figure 4e.9. As expected, the flow becomes unstable, but soon saturates into concentrated vortices, as shown in Figure 4e.10.

4e.3.1.4 Boundary layer models

This section does not directly involve LES. However, the turbulent wakes described in the results in Chapter 12 originate from boundary layers (BLs) separating from blunt bodies. For this reason, we describe the general use of VC for creating implicit models for these BLs. Here, VC provides a very simple, low-cost substitute for RANS models.

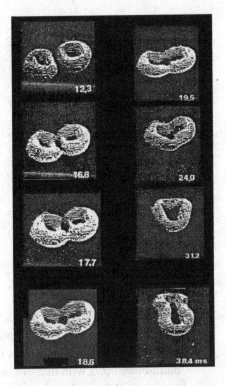

Figure 4e.8. Coplanar vortex rings: experiment (Oshima).

However, it should have comparable accuracy for these applications, since the BL treatment is consistent with the VC treatment of the vortices in the wake. We describe two approaches: the use with immersed BLs, and the use with surface-conforming grids. Both enforce the necessary no-flow-through conditions. They also enforce no-slip conditions. This latter feature ensures that the resulting BL has the correct total vorticity, which is just the difference between the velocity at the outer edge of the BL and the inner velocity, which is zero. This, in turn, ensures that a separating BL has the correct total vorticity. Since VC also ensures that a separating BL will subsequently remain thin (but can still roll up or lead to large-scale separation), this treatment should be accurate, since when the separating BL is thin, the details of the internal structure should not have an important influence. Both of these approaches involve coarse, inviscid-size grids.

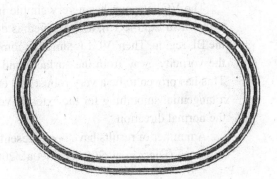

Figure 4e.9. Initial vorticity contours of a 2D computation.

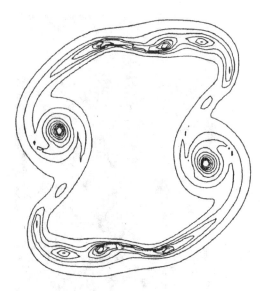

Figure 4e.10. Vorticity contours showing instability and formation of concentrated vortices.

These models do not involve determining a detailed time-averaged velocity profile, as in RANS schemes, which would require a very fine, body-fitted grid: Instead, they model the profile over only a few coarse grid cells. Thus, they are meant to be useful for blunt-body flows with massive separation, where the internal structure of this profile, as well as skin friction, are of secondary importance compared with the location and strength of the separating BL.

With a conventional CFD solution without VC, even for attached flow, the BL vorticity would quickly convect and diffuse away from the surface regions as a result of the large numerical errors at the boundary resulting from the coarse and possibly nonconforming grid, destroying the accuracy of the outer solution. However, the use of VC confines vorticity to one to three grid cells along the surface, when it is attached. Just outside this layer, the velocity is smooth and close to tangent to the adjacent surface (ARO 2000). The important feature here is that an attached BL, even with constant confinement parameters, maintains a constant thickness. This is close to the very slow thickness growth ($x^{1/7}$) of physical turbulent BLs in constant pressure gradients (on flat plates). This simple BL can still separate, however, especially at edges and in regions of strong adverse pressure gradient.

The VC1 version has a very simple interpretation for attached BLs: In this case, vector $\hat{n}$ in Eq. (4e.15), as explained, is defined to be locally normal to the surface in the BL region. Then, VC1 is simply a combination of positive diffusion (which spreads the vorticity away from the surface) and convection of vorticity toward the surface. This has proven to be a very robust and efficient way of modeling the BL, combining a tangential smoothing for the external velocity and a compression of the vorticity in the normal direction.

A number of results have been presented that demonstrate the effectiveness of this approach (Dietz et al. 2004; Fan et al. 2002; Fan and Steinhoff 2004; Steinhoff et al.

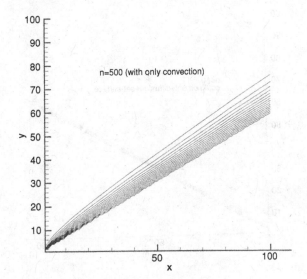

Figure 4e.11. Immersed surface without VC.

2003b). In cases where the BL separates, it can be seen to remain thin because of VC. Other immersed surface methods apparently result in numerical diffusion and numerical thickening of the separating BLs (Mittal and Iaccarino 2005).

4e.3.1.4.1 Immersed BL model

To enforce no-slip boundary conditions on immersed surfaces, first the surface is represented implicitly by a smooth "level set" function, F, defined at each grid point. This is just the (signed) distance from each grid point to the nearest point on the surface of an object – positive outside, negative inside. Then, at each time step during the solution, velocities in the interior are simply set to zero. In a computation using VC, this results in a thin vortical region along the surface, which is smooth in the tangential direction, with no "staircase" effects.

The important point is that no special logic is required in the "cut" cells, unlike many conventional schemes: Only the same VC equations are applied, as in the rest of the grid, but with a different form for $\hat{n}$. In addition, unlike many conventional immersed surface schemes, which are inviscid because of cell-size constraints, there is effectively a no-slip boundary condition, which results in a BL with well-defined total vorticity and that, because of VC, remains thin, even after separation.

The method is especially effective for complex configurations with separation from sharp corners. In addition, even with constant coefficients, it can approximately treat separation from smooth surfaces, as shown in Chapter 12.

Results (tangential velocity contours) are presented in Figures 4e.11 and 4e.12, for a computation on a uniform Cartesian with 128×128 cells, for flow over an immersed, oblique flat plate in zero pressure gradient. The resulting large, diffusive numerical errors can be seen in Figure 4e.11, for the case with no VC. For this flow, the velocity is

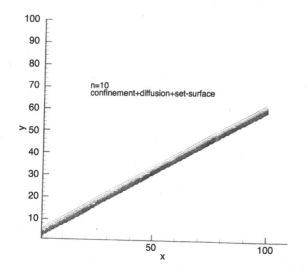

Figure 4e.12. Immersed surface with VC.

simply set to zero at nodes below the surface. (The small wiggles are from the spline fit in the contour plotter.) It can be seen that these are eliminated, to plottable accuracy, for the case with VC1, plotted in Figure 4e.12. Here, as in section 4e.3.1.3, only low-order numerical methods were used with constant confinement coefficients.

4e.3.1.4.2 Conforming grid BL model

We are beginning to develop detailed models for turbulent boundary layers, within the VC framework. This involves modeling the evolution of the confinement parameters so that, for example, separation is accurately predicted, even on smooth surfaces in time-dependent adverse pressure gradients. Results of some new, exploratory studies for these cases are also described in Chapter 12. There is still research to be done on these models, but the capabilities of the basic approach appear to be very promising.

An important point is that this VC-based method is fundamentally different from conventional RANS schemes, which typically use an eddy viscosity (EV) type of term and discretize a (modified) Navier–Stokes type of equation on a very fine grid, in order to model the time-averaged velocity. A very important feature of VC here is that it greatly expands our modeling capability, compared with EV–type schemes: Typically, these latter schemes can only accommodate positive values of EV. If the EV is negative over significant regions of space and time, they tend to diverge as a result of numerical instability (Ferziger 1996). This means that the modeled BL can only directly be made to expand, or diffuse, and not to contract. (Of course, slower expansion rates can be obtained and smaller BL thickness, but a finer grid is then required and a smaller value of the EV.) VC, in contrast, can directly model contraction, and, unlike a conventional scheme with a negative EV, VC will not diverge. This is very useful, for example, in turbulent BL separation from a smooth surface at moderate values of the Reynolds

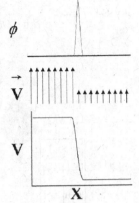

Figure 4e.13. Comparison between 1D discontinuity-capturing schemes and pulse advection.

number: physically, the separating layer then tends to transition and quickly reattach. A contraction term such as VC easily models this effect (Dietz 2004b).

Another point is that, for conforming grids with variable cell sizes, a scaling factor must be applied to μ and ε that depends on this size (Steinhoff and Raviprakash 1995). This is not a large correction, since inviscid-size grids that do not have large aspect ratios are used.

A final point involves the use of VC in retarding separation in adverse pressure gradient regions. For example, as is well known, a turbulent BL tends to separate later (in an adverse pressure gradient) than a laminar one. One can easily use VC to simulate this by increasing the confinement strength, again without very fine grids (Fan et al. 2002).

4e.3.2 Comparison of the VC2 formulation with direction-split discontinuity-steepening schemes

In this section, we describe advantages of VC over direction-split discontinuity-steepening schemes. We first reformulate the 1D scalar pulse equation of Section 4e.2 as a steepening method for velocity contact discontinuities. The result has some similarity to forms that have been developed over a number of years to keep gradients steep and overcome the smoothing that results from the convection terms. As already explained, these schemes have typically involved 1D compressible flows and have a number of essential differences, compared with VC.

If we consider the integral of the 1D pulse of Section 4e.2 (and change the sign), we have a propagating step function that remains steep for arbitrarily long times (see Figure 4e.13). Making the substitution,

$$\phi_j^n \equiv \delta_j V_j^n = V_j^n - V_{j-1}^n, \tag{4e.22}$$

we see that Eq. (4e.6) becomes

$$\Phi_j^n = \Phi_j^n \left(\{ \delta_j V_j^n \} \right). \tag{4e.23}$$

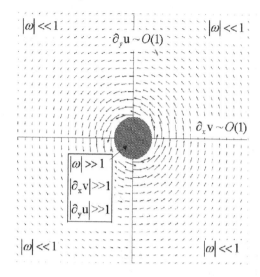

Figure 4e.14. Velocity vector field around a vortex displaying the basis of choosing a rotationally invariant steepener.

Partially summing Eq. (4e.7) over j, we then have (for constant c)

$$V_j^{n+1} = V_j^n - \frac{v}{2}\left(V_{j+1}^n - V_{j-1}^n\right) + \mu\delta_j^2 V_j^n - \varepsilon\delta_j\,\Phi_j^n. \tag{4e.24}$$

In this form, confinement is, effectively, a 1D steepening scheme. However, real flows never have a single region in which they have a steep gradient, and exactly constant properties everywhere else. In general, there are $O(1)$ (smooth) gradients away from the discontinuity region. In one dimension, these smooth gradients are also acted on by the steepener, causing errors, unless special logic is used to cut off the steepener.

Using such 1D schemes along each coordinate axis to keep a convecting vortex core compact by maintaining the steep gradients there would cause the same problem, since the velocities vary inversely with the radius away from the core (see Figure 4e.14). However, we do not do this! The important point is that we should not use the exact 1D confinement terms, but should only keep their basic mathematical structure – that they are functions of first derivatives of velocities. In developing a formulation for multiple dimensions, we should then use only rotationally invariant quantities. The only quantities, for example in 2D incompressible flow, that are first derivatives of a velocity are

$$\omega = \vec{\nabla}\times\vec{q}\,|_k \tag{4e.25}$$

and

$$D = \vec{\nabla}\cdot\vec{q}, \tag{4e.26}$$

where k denotes the out-of-plane direction. But $D = 0$ for incompressible flow, so we have only one choice:

$$\Phi = \Phi\left(\omega\right). \tag{4e.27}$$

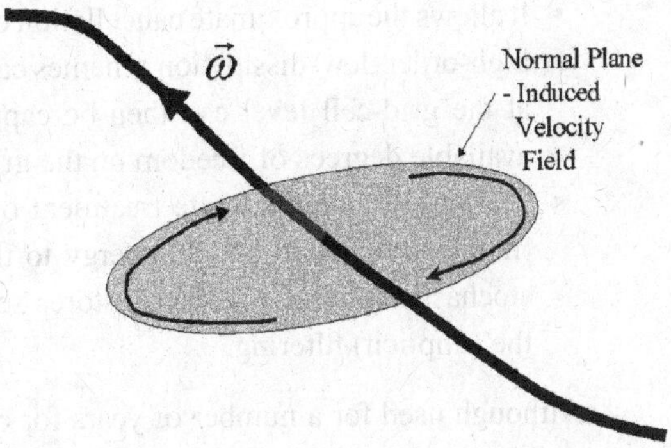

Figure 4e.15. Confinement operator in three dimensions.

This eliminates any problems with gradients away from the core because, just outside the core, $\omega \to 0$, even though both $\partial_x v$ and $\delta_y u$ are O(1) there. In addition, this choice results in a much simpler formulation than using separate operators along each axis. Finally, for a vortex filament in three dimensions, we consider it sufficient to confine in a 2D plane normal to the vortex, as depicted in Figure 4e.15. In this way we arrive at the momentum-conserving VC formulation.

4e.4 Conclusions

We describe a computational method that has been designed to capture thin vortical regions in high-Reynolds-number incompressible flows. The principal objective of the method – VC – is to capture the *essential* features of these small-scale vortical structures and model them with a very efficient difference method *directly* on an Eulerian computational grid. Effectively, the small vortical scales are treated as *nonlinear solitary waves* that live on the lattice indefinitely. The method allows isolated, convecting structures to be modeled over as few as two grid cells with no numerical spreading as they convect over arbitrarily long distances, with no special logic required for merging or reconnection. It also serves as a very efficient substitute for RANS models of attached and separating BLs and vortex sheets and filaments. Further, the method easily allows boundaries with no-slip conditions to be treated as immersed surfaces in uniform, nonconforming grids, with no requirements for complex logic involving cut cells.

In this chapter, we give a description of the basic VC method. This description is more comprehensive than that which has been previously available. There are close analogies between VC and well-known shock- and contact-discontinuity-capturing methodologies. These are discussed to explain the basic ideas behind VC, since it is somewhat different than conventional CFD methods. Some of the possibilities that VC offers toward very efficient computation of turbulent flows in LES approximations are also explored. These stem from the ability of VC to act as a negative dissipation at scales just above a grid cell, but that saturates and does not lead to divergence. This feature does the following:

- It allows the approximate cancellation of numerical diffusion, so that more complex, high-order–low-dissipation schemes can be avoided. Small-scale vortical structures at the grid-cell level can then be captured, resulting in very efficient use of the available degrees of freedom on the grid.
- It allows the approximate treatment of backscatter. This involves the addition of (modeled) subgrid kinetic energy to the flow in a natural way, without requiring stochastic forcing, and that restores some of the instabilities that are removed by the (implicit) filtering.

Although used for a number of years for complex, attached and separating flows, and trailing vortices, VC's use as an LES method is relatively recent. In Chapter 12, results of initial applications of VC-based LES will be presented.

Acknowledgments

Funding from a number of sources is gratefully acknowledged: primarily, the U.S. Army Research Office and the Army Aeroflightdynamics Directorate, which have supported the development of vorticity confinement from its beginning. Additional help has been provided by the Institute für Luft und Raumfahrt at the Technical University of Aachen, and the University of Tennessee Space Institute. Numerous discussions with Frank Caradonna at Moffett Field and with Stanley Osher and Barry Merriman at UCLA are also acknowledged.

REFERENCES

ARO 2000 *Proceedings of the 4th ARO Workshop on Vorticity Confinement*. Tullahoma, TN: University of Tennessee Space Institute (UTSI preprint).

Bohr, T., Jensen, M., Paladin, G., & Vulpiani, A. 1998 *Dynamical Systems Approach to Turbulence*. New York: Cambridge University Press.

Boris, J., & Book, D. 1973 Flux-corrected transport: I. SHASTA. A fluid transport algorithm that works. *J. Comput. Phys.* **11**, 38–69.

Costes, M., & Kowani, G. 2003 An automatic anti-diffusion method for vortical flows based on vorticity confinement. *Aerospace Sci. Technol.* **7**, 11–21.

Dadone, A., Hu, G., & Grossman, B. 2001 Towards a better understanding of vorticity confinement methods in compressible flow. Paper AIAA-2001-2639.

Dietz, W. 2004a Application of vorticity confinement to compressible flow. Paper AIAA-2004-0718.

Dietz, W. 2004b *Analysis, Design, and Test of Low Reynolds Number Rotors and Propellers*, SBIR Final Report, Topic no. A03-077, proposal no. A2-1452.

Dietz, W., Fan, M., Steinhoff, J., & Wenren, Y. 2001 Application of vorticity confinement to the prediction of the flow over complex bodies. Paper AIAA-2001-2642.

Dietz, W., Wang, L., Wenren, Y., Caradonna, F., & Steinhoff, J. 2004 The development of a CFD-based model of dynamic stall. Presented at the AHS 60th Annual Forum, June 7–10, Baltimore, MD.

Fan, M., & Steinhoff, J. 2004 Computation of blunt body wake flow by vorticity confinement. Paper AIAA-2004-0592.

Fan, M., Wenren, Y., Dietz, W., Xiao, M., & Steinhoff, J. 2002 Computing blunt body flows on coarse grids using vorticity confinement. *J. Fluids Eng.* **124**, 876–885.

Ferziger, J. 1996 Large eddy simulation, in *Simulation and Modeling of Turbulent Flows*, ed. T. B. Gatski, M.Y. Hussaini, & J. L. Lumley. New York: Oxford University Press.

Haas, S. 2003 *Computation of Trailing Vortex Flow over Long Distances Using Vorticity Confinement*. Master's thesis, Institut fur Luft- und Raumfahrt, RWTH Aachen, Germany.

Harten, A. 1978 The artifical compression method for computation of shocks and contact discontinuities III. Self-adjusting hybrid schemes. *Math. Comput.* **32**, 363–389.

Hu, G., Grossman, B., & Steinhoff, J. 2002 A numerical method for vortex confinement in compressible flow. *AIAA J.* **40**, 1945–1953.

Kim, J., & Moin, P. 1985 Application of a fractional-step method to incompressible Navier–Stokes equations. *J. Comput. Phys.* **59**, 308–323.

Lax, P. D. 1957 Hyperbolic systems of conservation laws II. *Comm. Pure Appl. Math* **10**, 537–566.

Melander, M. V., McWilliams, J. C., & Zabusky, N. J. 1987 Axisymmetrization and vorticity-gradient intensification of an isolated two-dimensional vortex through filamentation. *J. Fluid Mech.* **178**, 137–159.

Misra, A., & Pullin, D. 1997 A vortex-based stress model for large-eddy simulation. *Phys. Fluids* **9**, 2443–2454.

Mittal, R., & Iaccarino, G. 2005 Immersed boundary methods. *Annu. Rev. Fluid Mech.* **37**, 239–261.

Morvant, R. 2004 *The Investigation of Blade-Vortex Interaction Noise Using Computational Fluid Dynamics*. PhD dissertation, University of Glasgow, United Kingdom.

Morvant, R., Badcock, K., Barakos, G., & Richards, B. 2003 Aerofoil–vortex interaction simulation using the compressible vorticity confinement method, presented at the 29th European Rotorcraft Forum, September 16–18, Friedrichshafen, Germany, paper 3.

Oshima Y., & Izutsu N. 1988 Cross-linking of two vortex rings. *Phys. Fluids* **31**, 2401–2403.

Prandtl, L., & Tietjens, O. G. 1934 *Fundamentals of Hydro- and Aeromechanics*. New York: Dover.

Steinhoff, J., Fan, M., Wang, L., & Dietz, W. 2003a Convection of concentrated vortices and passive scalars as solitary waves. *J. Sci. Comput.* **19**, 457–478.

Steinhoff, J., Dietz, W., Haas, S., Xiao, M., Lynn, N., & Fan, M. 2003b Simulating small scale features in fluid dynamics and acoustics as nonlinear solitary waves. Paper AIAA-2003-0078.

Steinhoff, J., Fan, M., & Wang, L. 2003c Vorticity confinement – recent results: Turbulent wake simulations and a new, conservative formulation, in *Numerical Simulations of Incompressible Flows*, ed. M. M. Hafez. Singapore: World Scientific.

Steinhoff, J., Mersch, T., Underhill, D., Wenren, Y., & Wang, C. 1992 Computational vorticity confinement: A non-diffusive Eulerian method for vortex dominated flows. Tullahoma, TN: University of Tennessee Space Institute (UTSI preprint).

Steinhoff, J., Mersch, T., & Wenren, Y. 1992 Computational vorticity confinement: Two dimensional incompressible flow, in *Developments in Theoretical and Applied Mechanics*, Proceedings of the Sixteenth Southeastern Conference on Theoretical and Applied Mechanics, Manchester, TN: Beaver Press, III.II.73–III.II.82.

Steinhoff, J., Puskas, E., Babu, S., Wenren, Y., & Underhill, D. 1997 Computation of thin features over long distances using solitary waves, in *AIAA Proceedings of the 13th Computational Fluid Dynamics Conference*. Washington, DC: AIAA, 743–759.

Steinhoff, J., & Raviprakash, G. 1995 Navier–Stokes computation of blade–vortex interaction using vorticity confinement. Paper AIAA-95-0161.

Steinhoff, J., Senge, H., & Wenren, Y. 1990 Computational vortex capturing. Tullahoma, TN: University of Tennessee Space Institute (UTSI preprint).

Steinhoff, J., & Underhill, D. 1994 Modification of the Euler equations for "vorticity confinement" – application to the computation of interacting vortex rings. *Phys. Fluids* **6**, 2738–2743.

Steinhoff, J., Wenren, Y., & Wang, L. 1999 Efficient computation of separating high Reynolds number incompressible flows using vorticity confinement. Paper AIAA-99-3316-CP.

Tannehill, J., Anderson, D., & Pletcher, R. 1997 *Computational Fluid Mechanics and Heat Transfer*. Washington, DC: Taylor & Francis.

van Leer, B. 1974 Towards the ultimate conservative difference scheme. II. Monotonicity and conservation combined in a second-order scheme. *J. Comput. Phys.* **14**, 361–370.

Von Neumann, J., & Richtmyer, R. D. 1950 A method for the numerical calculation of hydrodynamic shocks. *J. Appl. Phys.* **21**, 232–237.

Wang, C., Steinhoff, J., & Wenren, Y. 1995 Numerical vorticity confinement for vortex–solid body interaction problems. *AIAA J.* **33**, 1447–1453.

Wenren, Y., Fan, M., Dietz, W., Hu, G., Braun, C., Steinhoff, J., & Grossman, B. 2001 Efficient Eulerian computation of realistic rotorcraft flows using vorticity confinement: A survey of recent results. Paper AIAA-2001-0996.

Wenren, Y., Fan, M., Wang, L., & Steinhoff, J. 2003 Application of vorticity confinement to the prediction of flow over complex bodies. *AIAA J.* **41**, 809–816.

Wenren, Y., Steinhoff, J., & Robins, R. 1995 Computation of aircraft trailing vortices. City: SBIR Final Report, NAS 1-20358.

5 Numerical Regularization: The Numerical Analysis of Implicit Subgrid Models

Len G. Margolin and William J. Rider

5.1 Introduction

In this chapter we extend our study of the underlying justification of implicit large eddy simulation (ILES) to the numerical point of view. In Chapter 2 we proposed that the finite-volume equations, found by integrating the governing partial differential equations (PDEs) over a finite region of space and time, were more appropriate models for describing the behavior of discrete parcels of fluid, including computational cells in numerical simulation. However, effective simulation of turbulent flows must consider not only issues of accuracy but also those of computational stability. Here we introduce and apply the machinery of modified equation analysis (MEA) to identify the properties of discrete algorithms and to compare different algorithms. We then apply MEA to several of the nonoscillatory finite-volume (NFV) methods described in Chapter 4, with the goal of identifying those elements essential to successful ILES. In the process we make connections to the some of the explicit subgrid models discussed in Chapter 3, thus demonstrating that many subgrid models implicit within NFV methods are closely related to existing explicit models. MEA is also applied with the methods description in Chapter 4a.

We consider the answer to this question: What are the essential ingredients of a numerical scheme that make it a viable basis for ILES? Many of our conclusions are based on MEA, a technique that processes discrete equations to produce a PDE that closely represents the behavior of a numerical algorithm (see Hirt 1968; Fureby and Grinstein 2002; Margolin and Rider 2002; Grinstein and Fureby 2002; Margolin and Rider 2005). Unlike many techniques used for numerical analysis, MEA can treat the nonlinear character of an algorithm. This is crucial because turbulence and its modeling are both nonlinear at their core. MEA enables the comparison of a numerical algorithm with physical theory as well as with conventional LES approaches. We describe MEA in some detail, including techniques for treating the nonlinear functions upon which many nonoscillatory schemes are based. We also demonstrate the ability of the direct numerical simulation of the modified equation to reproduce the behavior of the discrete algorithm.

The main results of this chapter draw together threads of the previous four chapters. Our principal result in Chapter 2 is an augmented finite-volume equation that was derived to represent the effects of finite scales of measurement when the Navier–Stokes equations form the underlying continuum description. In that chapter, we also exhibited the modified equation of a particular NFV method, MPDATA (see Chapter 4d). On the basis of the qualitative similarities between the physical equation and the numerical algorithm, we concluded that the success of MPDATA as an ILES algorithm follows because MPDATA is solving the finite-volume equations. In this chapter, we make these conclusions more quantitative and generalize them for additional NFV schemes.

In Chapter 3, we described the approach to conventional LES modeling. One of the important strategies of LES is that *all* the physics of the unresolved scales is incorporated into the explicit subgrid-scale model, implying that the numerical model itself should be as free of truncation error as possible. We noted that dissipative models, such as that of Smagorinsky, reproduce the dissipative and diffusive effects of turbulence, while the self-similar models are more theoretically designed to capture the inertial dynamics of energy transfer. However, we also noted that the self-similar models by themselves are not sufficiently dissipative, and that better results are obtained with hybrid models that combine the two approaches. By analogy, we show here that it is possible to separate the implicit subgrid-scale models of ILES (i.e., the truncation terms that appear from the MEA) into two pieces, one of which corresponds to a self-similar model (in Meneveau and Katz 2000; Sagaut 2003) and the other to a dissipative model (Smagorinsky 1963).

When high-frequency errors grow to a level that threatens computational stability in NFV simulations, adaptive dissipation intervenes, invoking strongly dissipative mechanisms to stabilize the calculation. That is, an ILES simulation is in many ways equivalent to a explicit hybrid LES approach.

Our analysis focuses on the transport (advective or hyperbolic) terms. We begin with detailed analyses of several NFV schemes in one dimension with general flux functions, isolating those elements common to successful ILES algorithms and identifying the differences that characterize those schemes that do not provide good ILES. We then proceed to fully multidimensional analyses for both the incompressible and compressible Navier–Stokes equations.

5.2 Modified equation analysis

MEA was first developed to assess the stability of numerical algorithms (Hirt 1968). Basically, the analysis consists of Taylor series expansion about a relevant mesh spacing, $\Delta_x \to 0$, applied to the discrete terms of the algorithm as if the equation (and its solution) were continuous. The main assumption is the use of Taylor series analysis itself, which implies restrictions on the smoothness of the function. In particular, the Taylor series analysis produces an infinite series of terms, but we form the modified equation by truncating this series and keeping only the lowest-order terms, which are assumed to dominate the numerical effects. A discussion of the uses of MEA and caveats of its use can be found in Knoll et al. (2003).

The modified equation for a *consistent* algorithm consists of the modeled PDE plus additional terms, each proportional to a power of the computational time step, Δt, or of the computational cell size, Δx, or possibly both. The smallest power among these terms determines the order of accuracy of the scheme. In general, if the PDE has the form

$$\frac{\partial \mathbf{U}}{\partial t} + \nabla \cdot \mathbf{F}(\mathbf{U}) = 0, \tag{5.1}$$

then the modified equation of a consistent algorithm will have the form

$$\frac{\partial \mathbf{U}}{\partial t} + \nabla \cdot \mathbf{F}(\mathbf{U}) = \mathbf{T}(\mathbf{U}, \Delta x, \Delta t), \tag{5.2}$$

where $\mathbf{T}$ is the truncation term. More specifically, when finite-volume differencing* is employed, the modified equation has the form

$$\frac{\partial \mathbf{U}}{\partial t} + \nabla \cdot \mathbf{F}(\mathbf{U}) = \nabla \cdot \boldsymbol{\tau}(\mathbf{U}, \Delta x, \Delta t). \tag{5.3}$$

Thus, all algorithms based on finite-volume differencing have a truncation term in the form of the divergence of a subgrid-scale model. The conditions under which this form is useful for ILES is the main subject of this chapter.

We next turn our attention to issues more directly concerned with NFV methods. The class of high-resolution Godunov methods (see Chapter 4a) employs nonlinear limiter functions. These limiters must be expressed in terms of differentiable continuous functions to enable the MEA. A particular example that we consider in some detail is the minmod function, which is a popular element of many limiters and is used in many methods such as FCT, TVD, MUSCL, and UNO. The minmod function has two arguments, and it returns the smaller value in terms of its absolute value if the terms have the same sign. If the signs of the two terms differ, the value zero is returned. The definition of minmod uses the elementary functions min, max, sgn, and abs. All of these functions can be defined in terms of differentiable continuous functions almost everywhere (except at $x = 0$, where the Taylor series are not impacted in a substantial way because the expansions are around $\Delta x = 0$),

$$|x| = \sqrt{x^2}; \quad \text{sign}(x) = \frac{x}{|x|},$$

$$\min(a, b) = \frac{a+b}{2} - \frac{|a-b|}{2}; \quad \max(a, b) = \frac{a+b}{2} + \frac{|a-b|}{2},$$

which can readily be expanded in Taylor series. Once we make these definitions, the analysis of a broad spectrum of high-resolution methods proceeds without difficulty (Rider and Margolin 2003). We can conduct this analysis effectively by using a symbolic algebra package such as Mathematica.

In the next section, we show detailed examples of the truncation terms that result from several high-resolution Godunov methods. We close this section by considering

---

* Finite-volume differencing is also known as *conservative form* and as *flux form differencing*. We use these terms interchangeably.

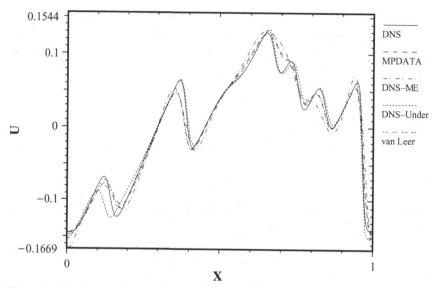

Figure 5.1. Here we demonstrate the ability of both ILES and its modified equation to effectively simulate Burgers' equation with small viscosity. The solid line is the DNS on a fine mesh. The other two lines are the MPDATA method and the DNS of its modified equation as defined on the coarse MPDATA mesh (reprinted from Margolin and Rider 2002). Included in this figure are an underresolved DNS calculation and the Van Leer scheme that is analyzed in the next section.

the validity of MEA applied to NFV methods. A detailed study of a flow simulation governed by Burgers' equation was described in Rider and Margolin (2002). The study compared an MPDATA simulation with a well-resolved calculation using the MPDATA modified equation. The actual problem followed the evolution of an inviscid flow from initial conditions, including wave steeping and multiple interactions between shocks and rarefactions, and it was designed to be an idealized example of high-Reynolds-number turbulent flow. The differences between the two results are only evident when we are examining the flow at the smallest scales in the regions of local maxima. Further, both results compared well with a highly resolved direct numerical simulation (DNS) of the viscous Burgers' equation. These results are shown in Figure 5.1.

The previous example illustrates the ability of the modified equation to approximate the numerical algorithm; however, in this chapter we are more concerned with the quantification of error. We present the following example to address this point. We compare the relative error of two algorithms for the linear advection of a Gaussian pulse. The model PDE is

$$\partial_t U + \partial_x U = 0. \tag{5.4}$$

We can devise a simple second-order numerical method by defining cell-edge values as

$$U_{j+1/2}^{n+1/2} = U_j + \frac{1}{2} S_j (1 - \nu),$$

```
abs[x_] := Sqrt[x^2]
U[m_] := u[x + mh]
Sl[m_] := U[m] - U[m - 1]
Sr[m_] := U[m + 1] - U[m]
Sv[m_] := (abs[Sr[m]]Sl[m] + abs[Sl[m]]Sr[m])/(abs[Sl[m]] + abs[Sr[m]])
Uev[m_] := U[m] + 1/2 Sv[m]
duev = (Uef[0] - Uef[-1])/h;
Series[duev, {h, 0, 4}];
Collect[Simplify[Expand[Normal[%]]], h];
Collect[Expand[Integrate[%, x]], h]
```

Figure 5.2. The Mathematica commands used to produce the MEA for van Leer's scheme.

where $S_j \approx \partial_x U \Delta x$ and ν is the Courant number (here $\Delta t / \Delta x$). We then update cells to the next time level by using

$$U_j^{n+1} = U_j^n - \nu\left(U_{j+1/2}^{n+1/2} - U_{j-1/2}^{n+1/2}\right). \tag{5.5}$$

The first algorithm is the Fromm scheme, which is a linear method:

$$S^{\mathrm{Fr}} \approx \frac{U_{j+1} - U_{j-1}}{2}. \tag{5.6}$$

Fromm's method produces a spatial error with the following form:

$$\frac{\partial}{\partial x}\left[-\frac{\Delta x^2}{12}\frac{\partial^2 U}{\partial x^2} + \frac{\Delta x^3}{8}\frac{\partial^3 U}{\partial x^3} - \frac{13\Delta x^4}{240}\frac{\partial^4 U}{\partial x^4}\right]. \tag{5.7}$$

The second algorithm is the van Leer scheme, which is a nonlinear method based on the use of a harmonic mean to approximate the spatial differences:

$$S^{\mathrm{VL}} \approx \frac{|U_j - U_{j-1}|(U_{j+1} - U_j) + |U_{j+1} - U_j|(U_j - U_{j-1})}{|U_j - U_{j-1}| + |U_{j+1} - U_j|}. \tag{5.8}$$

Van Leer's scheme produces a modified-equation-based form:

$$\frac{\partial}{\partial x}\left\{-\frac{\Delta x^2}{12}\frac{\partial^2 U}{\partial x^2} + \frac{\Delta x^3}{8}\left[\frac{\partial^3 U}{\partial x^3} - \left(\frac{\partial^2 U}{\partial x^2}\right)^2\left(\frac{\partial U}{\partial x}\right)^{-1}\right]\frac{1}{8}\frac{\partial^2 U}{\partial x^2}\frac{\partial^3 U}{\partial x^3}\left(\frac{\partial U}{\partial x}\right)^{-1}\right.$$
$$\left. -\frac{1}{16}\left(\frac{\partial^2 U}{\partial x^2}\right)^3\left(\frac{\partial U}{\partial x}\right)^{-2} - \frac{13\Delta x^4}{240}\frac{\partial^4 U}{\partial x^4}\right\}. \tag{5.9}$$

Figure 5.2 shows the Mathematica commands that will produce this result for van Leer's scheme.

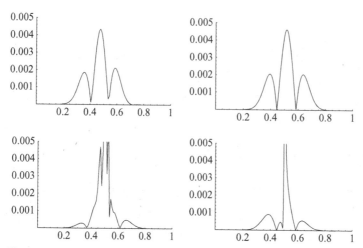

Figure 5.3. The plots show a comparison of an actual calculation using two different methods and the predicted structure of the numerical error from the MEA. The top two figures are for Fromm's method and the bottom two are for van Leer's method. The first and third figures are from the MEA analysis and the second and fourth are from computations.

A detailed MEA yields the form for the leading-order difference between the methods as

$$U^{\text{VL}} - U^{\text{Fr}} \approx \frac{\Delta x^3 (\partial_{xx} U)^2}{8 \partial_x U} + \text{HOT}, \tag{5.10}$$

where "HOT" denotes higher-order term. Note that on the right-hand side of this equation, we have not distinguished between the two solutions. To evaluate this term, we compute the advection of a Gaussian function, $\exp[-100(x - 1/2)]$, with both methods. We have also used this form to evaluate the modified equations for these methods in (5.10). The comparison between the numerically generated data and the MEA result is shown in Figure 5.3. As these figures demonstrate, the MEA provides all of the qualitative features of the numerical approximations and provides a quantitative evaluation of the flow in smooth regions away from critical points in the functions being approximated. Both of these features are important to keep in mind when we are considering the results of the analyses that follow.

5.3 MEA of a high-resolution method

We begin our analysis of NFV schemes by using a Godunov-type framework in one spatial dimension. The generic PDE model is given in (5.1). We note that the results that follow in the rest of this chapter are all written in terms of a general flux function, $F(U)$. For compressible gas dynamics,[†] where U is the column vector, $(\rho, \rho u, \rho E)^T$,

---

[†] For the compressible Euler equations, the variables are defined as follows: ρ is the density, u is the velocity (ρu is the momentum density), E is the total energy, and p is the pressure.

flux function $F(U)$ would be $(\rho u, \rho u^2 + p, \rho u E + pu)^T$. However, for the Burgers equation, or incompressible Navier–Stokes, U is the velocity and $F(U) = \frac{1}{2} U^2$. In a Godunov framework, given a cell-centered value of dependent variable U_j, we construct the second-order mean-preserving interpolant

$$U(x) = U_j + S_j \frac{x - \langle x \rangle_j}{\Delta x}. \tag{5.11}$$

There are many ways to choose slope S_j; we subsequently describe several of these. It is instructive to recognize that the spatial reconstruction using slopes is effectively a deconvolution method and, as we show, produces similar subgrid effects (see Chapter 6 for a more detailed discussion of deconvolution).

The solution is advanced in time by a Riemann solver (Hirsch 1999; Drikakis and Rider 2004), in which the interface flux is estimated as

$$F_{j+1/2} = \frac{1}{2} \left[F \left(U_{j+1/2,j} \right) + F \left(U_{j+1/2,j+1} \right) \right]$$

$$- \frac{1}{2} |\partial_U F| \left[U_{j+1/2,j+1} - U_{j+1/2,j} \right]. \tag{5.12}$$

The Riemann solution is a method for producing an upwind difference for nonlinear systems of equations. Here, the notation $U_{j+1/2,j} \equiv U_j + S_j(x_{j+1/2} - x_j)$; in words, it is the value of the interpolant of cell j at its right-hand edge $j + 1/2$. The right-hand side consists of two terms – a centered term that describes the inertial transport, and a dissipative term that regularizes the flow. As noted, since our scheme is written in finite-volume form, it is possible to write the truncation error as the divergence of a subgrid stress, τ; further, the separation of the flux form into two pieces allows us to define two contributions to τ. To be explicit, the MEA of the centered term yields $[\frac{1}{2} U^2 - \tau_{\text{hyp}}(U)]$ while that of the dissipative term yields $[\tau_{\text{diss}}(U)]$. In terms of LES, this constitutes a mixed model. The two subgrid-stress terms describe different aspects of the model, and they will facilitate our categorization of what constitutes effective ILES methods.

We now proceed to consider the results of different choices of the interpolant slope, S_j. When the interpolant used is piecewise constant, $S_j = 0$, the method is first order and the dissipative error $\nabla \cdot \tau_{\text{diss}}$ dominates inertial error $\nabla \cdot \tau_{\text{hyp}}$. In particular, $\nabla \cdot \tau_{\text{diss}} \sim \Delta x$ whereas $\nabla \cdot \tau_{\text{hyp}} \sim \Delta x^2$. Thus, in general, the dissipation dominates the inertial dynamics, which explains the failure of first-order methods to adequately simulate turbulent flows.

In general, one can make Godunov methods accurate to the second order by including nonzero estimates of the slope in (5.11). A linear estimate of the slope will produce an oscillatory solution (see Godunov's theorem in Chapter 2) in which energy is not dissipated at a physically sufficient rate at the small scales, but builds up and gives an incorrect solution. Nonlinear Godunov methods result from the nonlinear (i.e., solution-dependent) limiting of the linear estimate of the slope. Perhaps the simplest nonlinear limiter is based on the minmod function described in the previous section. A limited value of the slope based on minmod, but written in terms of more elementary functions, is

$$S_j = (\text{sgn } \Delta_- U) \max \{0, \min [|\Delta_- U|, (\text{sgn } \Delta_- U)\Delta_+ U]\}, \tag{5.13}$$

where $\Delta_- U = U_j - U_{j-1}$ and $\Delta_+ U = U_{j+1} - U_j$.

To form the modified equation, we take all variables in discrete form, expand them in Taylor series about the cell center, x_j, and collect terms in powers of Δx. We define the velocity at the cell boundary (expanded to either edge) as

$$U_{j \pm 1/2, j} = U_j \pm S_j/2.$$

We form upwind approximations to the Riemann solution with the cell boundary values of the adjacent cell, and we use these values to form a finite-volume approximation to the derivative in a zone, $\partial \mathbf{F}(\mathbf{U})/\partial x$, which includes both the analytic flux derivative and the truncation term:

$$\frac{F_{j+1/2} - F_{j-1/2}}{\Delta x} = \frac{1}{2}\frac{\partial U^2}{\partial x} + \frac{\partial \tau(U)}{\partial x}. \tag{5.14}$$

Finally, we expand this expression and simplify by collecting terms with common powers of Δx. For the minmod slope of (5.13), the leading-order terms are

$$\frac{\partial \tau(U)}{\partial x} = \Delta x^2 \left\{ \frac{1}{24}\partial_{UU}F\partial_x U\partial_{xx}U \right\} + \frac{1}{24}\partial_{UUU}F(\partial_x U)^3$$

$$+ \frac{1}{12}\partial_U F(\partial_{xxx}U) + \frac{1}{4}\frac{|\partial_U F(\partial_x U\partial_{xx}U)|}{\partial_x U\partial_{xx}U}\partial_{xxx}U + \text{HOT}, \tag{5.15}$$

where HOT indicates terms of higher order in Δx. We can then integrate this expression in space to obtain an expression for $\tau(U)$. For the minmod slope of (5.13), we finally derive

$$\tau_{\text{hyp}} = -\Delta x^2 \left[\frac{1}{24}\partial_{UU}F(\partial_x U)^2 + \frac{1}{12}\partial_U F\partial_{xx}U \right] + \text{HOT}, \tag{5.16}$$

and

$$\tau_{\text{diss}} = \frac{\Delta x^2}{4}\frac{|\partial_U F(\partial_x U\partial_{xx}U)|}{\partial_x U} + \text{HOT}. \tag{5.17}$$

This method has the correct inertial range term in τ_{hyp}, in the form identical to the scale-similarity term of Chapter 2 and also as found in Clark's model (discussed in Meneveau and Katz 2000; Sagaut 2003), which is an approximation to the scale-similarity model. However, here the nonlinear dissipation directly competes with other terms in the leading-order truncation error in its scaling with Δx. The resulting inertial range dissipation does not match the expectations from Kolmogorov's theory. Further, its nondimensional coefficient is too large, resulting in a strongly nonlinear dissipation at all scales. In summary, this model is too diffusive, which is confirmed by experience and as detailed later in this chapter by our validation of ILES with decaying turbulence.

The minmod models described herein are characterized by a limiter that is always "on," although its effect on the larger scales of the flow is small. One should not conclude that all minmod-based limiters are inappropriate for ILES. It is possible to construct more effective limiters based on minmod that have higher-order dissipation. The underlying idea for this construction is to design the limiter to be less restrictive – that is, to allow the underlying high-order (unlimited) method to apply over a larger region of the flow. It is possible to design another class of limiters whose action is restricted to the smallest scales. This is analogous to the explicit multiscale models of Hughes and collaborators (Hughes et al. 2001).

Mathematically, we can accomplish this strategy by using the double minmod where the arguments that are used to bound the high-order flux are twice larger than the local gradients. We can easily write the form of this limiter as two invocations of a median function, where this is defined (Huynh 1995) as

$$\text{median}\,(a, b, c) = a + \text{minmod}\,(b - a, c - a).$$

The limited slope is then

$$S_{\text{mono}} = 2\,\text{median}\,(0, \Delta_- U, \Delta_+ U),$$
$$S_{\text{high order}} = \frac{\Delta_- U + \Delta_+ U}{2} = \frac{U_{j+1} - U_{j-1}}{2},$$
$$S_j = \text{median}\,(0, S_{\text{mono}}, S_{\text{high order}}). \tag{5.18}$$

This gives a large region where the high-order flux can be used and leaves the effective scale-self-similar model as the dominant implicit model. The nonlinear dissipation associated with the limiter is one order higher in Δx and consequently more scale selective. The effective subgrid model for this limiter takes the form

$$\tau_{\text{hyp}} = -\frac{\Delta x^2}{24} \partial_{UU} F (\partial_x U)^2 + \frac{\Delta x^2}{12} \partial_U F \partial_{xx} U + \text{HOT}, \tag{5.19}$$

and

$$\tau_{\text{diss}} = \frac{\Delta x^3}{8} |\partial_U F| \partial_{xxx} U + \text{HOT}. \tag{5.20}$$

Another example is the median limiter of the UNO scheme, which is constructed to use the full unlimited method where the flow is resolved (Huynh 1995). This method uses the smoothest second-order slope locally, defined mathematically by a series of invocations of the median limiter,

$$q_- = \text{median}\,(p_-, p_o, \Delta_-), \quad q_+ = \text{median}\,(p_o, p_+, \Delta_+),$$

$$S_j = \text{minmod}\,(q_-, q_+),$$

where $p_- = (3U_j - 4U_{j-1} + U_{j-2})/2$, $p_o = (U_{j+1} - U_{j-1})/2$, and $p_+ = (-3U_j + 4U_{j+1} - U_{j+2})/2$. This is an example of a Godunov method that has good ILES properties without necessarily imposing monotonicity – the minmod final step can be replaced with a weaker limiter that does not drop to first order at extrema in the solution. The overall effect is to keep the strongly dissipative effects localized to the highest wave numbers. This provides a dissipation that is very scale selective, a variational multiscale method in an implicit rather than explicit manner. This method has the same τ_{hyp} as the double-minmod method, but its dissipative effective subgrid model differs:

$$\tau_{\text{diss}} = \frac{\Delta x^3}{8} |\partial_U F| \frac{|\partial_x U \partial_{xxx} U|}{\partial_x U}. \tag{5.21}$$

This is similar to the form for the implicit model found if the van Leer limited slope S^{VL} is used, whose dissipative subgrid stress is

$$\tau_{\text{diss}} = \frac{\Delta x^3}{8} |\partial_U F| \frac{(\partial_{xx} U)^2}{\partial_x U}. \tag{5.22}$$

The lack of the scale-self-similar model in a ILES can explain another failure by a notable high-resolution method. The fifth-order WENO method produced quite poor results for decaying isotropic turbulence in a paper by Garnier et al. (1999). An analysis of this method shows that the high-order differencing does not produce the scale-self-similar model despite the use of the conservation form. The leading-order error is dissipative,

$$\tau(U) = \frac{\Delta x^5}{10} |\partial_U F| \frac{(\partial_{xxx} U)^2}{\partial_x U} + \frac{\Delta x^5}{60} |\partial_U F| \partial_{xxxx} U + \text{HOT}, \tag{5.23}$$

a combination of "linear" hyperviscosity and a nonlinear viscosity associated with the effective limiting strategy used in the stencil weighting that determines the finite differences. This leads to nonlinear dissipation on all spatial scales and produces the wrong scaling for a turbulent flow.

The "problem" with WENO can be traced to its goal of producing formal high-order differencing in a finite-volume method. The high-order interpolation is applied to the fluxes rather than to the reconstruction of variables, and the fluxes produce fifth-order nonlinear finite-volume fluxes. If instead the WENO procedure were applied to the reconstruction of variables that are then used to define fluxes, the second-order finite-volume terms reappear in the expansion. Under these conditions the leading-order subgrid model would match the double-minmod scheme and only contain τ_{hyp}. The next order in the expansion would be nonpositively dissipative as well at Δx^4 with absolutely dissipative terms entering at Δx^5 in the form of Eq. (5.23). We expect that this method would be more effective as an ILES method than the flux-based WENO method.

Our last example is the MPDATA scheme, which does not fit into the Godunov framework but is amenable to a similar analysis. MPDATA is not a monotonicity-preserving scheme in its basic form; rather it is a positive-definite, or sign-preserving, scheme (see Chapter 4d). Nevertheless, it is an effective ILES method. The flux has the form

$$F_{j+1/2}^{\text{MPDATA}} = F_{j+1/2}^{\text{1st}} + \Psi \left(F_{j+1/2}^{\text{2nd}} - F_{j+1/2}^{\text{1st}} \right), \tag{5.24}$$

where

$$\Psi = \frac{|U_{j+1} - U_j|}{|U_{j+1}| + |U_j|},$$

and it is accurate to the second order everywhere (Margolin and Rider 2002). A MEA of the leading-order error shows that the nonlinear dissipation

$$\tau_{\text{diss}}(U) = \frac{\Delta x^2}{4} |\partial_x U| \partial_x U + \text{HOT} \tag{5.25}$$

enters at the same order as the scale-similar term:

$$\tau_{\text{hyp}}(U) = \Delta x^2 \left[\frac{1}{12} \partial_{UU} F (\partial_x U)^2 + \frac{1}{3} \partial_U F \partial_{xx} U \right] + \text{HOT}. \tag{5.26}$$

Thus it competes with the scale-self-similar term, as did the basic minmod scheme, but its dissipation is less dominant, allowing the scale-similarity term to operate effectively.

In recent years, deconvolution has become an increasingly popular method for modeling turbulent flows in LES. The deconvolution approach to LES and its connection to high-resolution methods is explored more fully in Chapter 6 of this book. It is instructive to examine the portions of high-resolution methods that act as effective deconvolution operators. High-resolution methods effectively produce a deconvolution of the cell-averaged data to a spatial distribution throughout the computational cell. An essential aspect of deconvolution is the realization that deconvolution is underdetermined and thus is not unique. Therefore, a number of different interpolations are equally valid (mathematically) for a given flowfield. The high-resolution Godunov methods blend accuracy with conditions that determine whether a cell reconstruction produces a physically realizable representation of a flow. The determination of the physical realizability of the reconstruction constitutes an important aspect of the regularization of the flow. The regularization is then completed through the Riemann solver (upwind differencing). In standard deconvolution for LES, the necessary regularization is applied is a separate step from the deconvolution.

Godunov-type high-resolution methods produce a deconvolution through a reconstruction of variables in a zone by means of piecewise interpolation. Most commonly the interpolant is piecewise linear or parabolic. The linear or parabolic profile is chosen by use of local data and its value limited by a nonlinear stability principle such as monotonicity. Usually the interpolation leaves the system unclosed because the interpolant will (or can) be multivalued at cell interfaces. Then some sort of Riemann solution, or other nonlinear dissipation mechanism, must be used to resolve the multivalued nature of the reconstruction in space. Together these two elements can contribute the majority of the implicit subgrid modeling to a Godunov-type method.

5.4 Energy analysis and the relation of LES and ILES

In this section, we draw some qualitative conclusions about the relative advantages of conventional LES and ILES. In the previous section where we analyzed the momentum equation, it became clear that the modified equations of NFV methods have truncation terms that closely resemble explicit subgrid-scale models. In particular, we showed that both explicit, self-similar models and successful ILES models contain the scale-similar term, also derived in Chapter 2, which ensures the correct finite-scale inertial dynamics. Here, we explore the energy equation, and the connection of physical dissipation to computational stability. We focus on the particular choice of velocity, $U = u$, and the flux vector, $F = \frac{1}{2}U^2$. To derive a generic equation for the time evolution of kinetic, energy E, we multiply the modified equation of a finite-volume algorithm (5.3) by U, to derive

$$
\begin{aligned}
\frac{\partial E}{\partial t} &= -\left[\frac{1}{3}u^3\right]_x + u\tau_x \\
&= -\left[\frac{1}{3}u^3 - u\tau\right]_x - u_x\tau .
\end{aligned}
\tag{5.27}
$$

To test global stability, we now integrate the time derivative of energy over the problem domain. In the absence of boundary sources, the terms inside the total derivative do not contribute and we have

$$
\int_{\text{domain}} \frac{\partial E}{\partial t}\, dx' = -\int_{\text{domain}} u\tau_{x'}\, dx'.
\tag{5.28}
$$

If the right-hand side of (5.28) is negative, we say that the equation is nonlinearly stable. In general, it is not known whether the stability of the modified equation is sufficient to guarantee the stability of the discrete algorithm. Recall that the modified equation is a truncation of the infinite series in which only lowest-order terms are kept. This is sufficient to study questions of accuracy. However, instability may result from higher-order terms. In contrast, NFV schemes are, by construction, nonlinearly stable.

In Section 5.3, we wrote the implicit subgrid-scale stress as $\tau = \tau_{\text{hyp}} + \tau_{\text{diss}}$. For successful ILES, we require that $(\tau_{\text{hyp}})_x$ contains a term $\sim \Delta x^2 u_x\, u_{xx}$. Based purely on dimensional analysis, τ_{hyp} must contain terms like $\Delta x^2(u\, u_{xx})$ or $\Delta x^2(u_x)^2$. Furthermore, we note that

$$
u\, u_{xx} = (u\, u_x)_x - (u_x)^2 ,
\tag{5.29}
$$

so that, from the point of view of energy analysis, there is only one term to consider:

$$
\int_{\text{domain}} u_{x'}\tau_{\text{hyp}}\, dx' = \int_{\text{domain}} (u_{x'})^3 dx'.
\tag{5.30}
$$

Finally, we note that, theoretically, the right-hand side of (5.30) is negative for the Navier–Stokes equations, a rigorous consequence of Kolmogov's "4/5" law (clearly discussed in Frisch 1995). We conclude then that the stability of the modified equation *with respect to this term* depends on the physical correctness of the solution itself. In Chapter 16, we show that an MPDATA simulation of decaying turbulence does in fact reproduce the 4/5 law. As pointed out earlier, the actual result of a MEA of a NFV algorithm contains higher-order terms, say $\sim \Delta x^3$, that may affect the stability of the equation. From dimensional analysis, such third-order terms will lead to a term $\Delta x^3 (u_x)^2 u_{xx}$ integrated over the domain in the energy analysis.

Because NFV schemes are nonlinearly stable by construction, we anticipate that the modified equation also contains an implicit subgrid-scale stress, τ_{diss}, that is absolutely dissipative – that is, whose dissipative character does not depend on an accurate solution. More specifically, in terms of (5.28), an absolutely dissipative subgrid-scale stress will give a negative contribution to the integral *in every cell*. In the previous section, we showed the example τ_{diss} of both second order and third order in Δx. Using similar arguments of dimensional analysis and integration by parts as already given, we see that these terms must have the form $\tau_{\text{diss}} \sim \Delta x^2 |u_x|^3$ or $\tau_{\text{diss}} \sim \Delta x^3 (u_x)^2 |u_{xx}|$.

Let us now consider the first case, of second-order dissipation, in more detail. Local stability where the second-order dissipation controls the third-order term would require

$$c_2 |u_x|^3 \Delta x^2 \geq c_3 (u_x)^2 u_{xx} \Delta x^3, \tag{5.31}$$

where c_2 and c_3 are nondimensional positive coefficients. Let us consider a sequence of three computational cells, with consecutive velocities u_L, u_C, and u_R. Estimating

$$u_x \approx \frac{u_R - u_L}{2\Delta x}, \quad u_{xx} \approx \frac{u_R - 2u_C + u_L}{\Delta x^2}$$

and inserting these into the inequality (5.31), we derive

$$\alpha |u_r - u_L| \geq (u_R - 2u_C + u_L), \tag{5.32}$$

where $\alpha = \frac{c_2}{2 c_3} > 0$. Defining the ratio

$$R = \frac{u_R - u_C}{u_C - u_L},$$

we have

$$R(\alpha - 1) \geq -(1 + \alpha). \tag{5.33}$$

Thus, on one hand, when R is positive, so that the velocity field is monotonically increasing or decreasing, we have the constraint $\alpha \geq 1$. On the other hand, when R is negative, we have a local maximum or minimum; in this case, limiters like those based on minmod become zero. The previous analysis illustrates the close connection between the mathematical condition of ensuring nonoscillatory solutions and the physical constraint of the second law of thermodynamics. It also shows that the preservation of monotonicity is equivalent to controlling the smoothness of the solution. From the point of view of stability, we see that monotone schemes enforce a sufficient, but not a necessary condition. In other words, the ILES technique guarantees a stable calculation for every problem; however, it may be overly dissipative for a particular problem, especially if one has particular knowledge to incorporate. For example, one may want to introduce other physical length scales into the model.

To summarize, ILES may be more robust and more generally applicable to a wide range of flows, whereas the ability to vary an explicit model may make LES more flexible. The analysis of Section 5.3 is only one dimensional. When multidimensionality is considered, new considerations arise. For example, the interpretation of the velocity as a volume integral implies that the finite-volume equations, and hence the implicit subgrid scale, will depend on the details of the integration volume shape and orientation. Even the analytic example in Chapter 2, in which the integration volume is a simple rectangle, shows that the scale-similarity term is not tensor invariant (in the sense that it depends on the geometry of the computational cell). The situation becomes much more complicated for irregular grids, such as the terrain-following grids used in geophysics (see Chapter 14) or the body-fitted grids used in aerospace simulations (see Chapter 10). In these cases, the variability of the cell size and shape is automatically taken into account by ILES. The connection between explicit LES and ILES exposes other possibilities.

As another example, one might consider using the implicit subgrid model, as identified by a MEA of an ILES method, as an explicit subgrid-scale model. The MEA might be complex, especially on an irregular mesh, but the analysis could be facilitated by automatic (symbol) algorithms. In such a strategy, one might be concerned by the effects of the finite truncation of the modified equation. Conversely, one could conceive of constructing a NFV method whose implicit subgrid-scale model corresponds to a chosen explicit subgrid-scale model. Although this appears to be more complex, some effort has been made in this direction. This process, termed *reverse engineering*, was explored in one dimension in Margolin and Rider (2004).

5.5 Validation of the analysis

To validate the analysis of ILES using MEA presented herein, we simulate an experiment by Kang et al. (2003) and make a comparison with their high-quality data. The decay of isotropic turbulence is a fundamental benchmark flow for idealized turbulence. The classical experiment for decaying turbulence, the Comte–Bellot–Corrsin test, is a commonly used test of turbulence models. In 2003, a group at John Hopkins University published the results of a similar experiment, but at a much larger Reynolds number (Kang et al. 2003). In the more recent John Hopkins experiment, a wind tunnel was utilized with an active grid to produce a decaying and nearly isotropic turbulent velocity field with an initial Taylor microscale Reynolds number in excess of 700. Along the length of the wind tunnel, four stations with X-wire probes produced measurements of the flowfield. These stations were placed at $x/M = 20, 30, 40$, and 48, where x is the downstream position and M is the spacing of the grid. The computational results

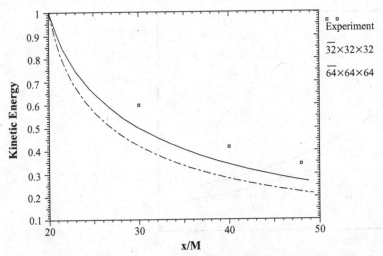

Figure 5.4. The plot shows a comparison of kinetic energy decay between the experiment and simulation, using the minmod method with both 32^3 and 64^3 meshes. The kinetic energy levels and times have been normalized to refer to the same dimensionless time.

are compared with the measurements at these stations by use of Taylor's hypothesis to normalize the computations to the experimental conditions.

We will use several of the published measurements to compare the simulations with the experiment. We will compare results at both resolutions by using four experimentally measured metrics: total flow kinetic energy (normalized), the flow three-dimensional energy spectrum, and the longitudinal and transverse PDFs of the velocity increments. These are derived from the experimental data filtered at different length scales. Our initial condition is taken from the conditions given in Kang et al. (2003), and we measure time by using the flow velocity through relation $x = Ut$, where U is the downstream velocity in the experiment. The data plotted as the experimental PDFs of the velocity increments can be found online at the Web site for the Johns Hopkins turbulence research group. All of the simulations shown here employed grids of 32^3 and 64^3. We use four different high-resolution methods to compute the flow: the minmod total-variation diminishing limiter (Chapter 4a); the van Leer piecewise linear method (PLM, Chapter 4a); xPPM, which is a version of the piecewise parabolic method (PPM), extended to provide higher-order accuracy when the standard method would degenerate to first order (in Rider, Greenough and Kamm 2005, 2006) (see also Chapter 4b); and the MPDATA method with three iterations (Chapter 4c, IORD = 3).

Figures 5.4–5.7 show the decay of kinetic energy computed with these methods compared with the experiment. The PLM and xPPM methods produce a good comparison with the experimentally observed energy decay at either the coarse or the fine grid. Both MPDATA and especially the minmod method are too dissipative on the grids utilized here. In both cases these methods diffuse too much kinetic energy.

The comparison of the simulated power spectra is shown in Figures 5.8–5.11. This measure complements the kinetic energy comparison with the absolute magnitude of the

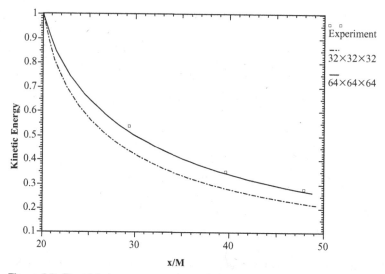

Figure 5.5. The plot shows a comparison of kinetic energy decay between the experiment and simulation, using the PLM method with both 32^3 and 64^3 meshes. The kinetic energy levels and times have been normalized to refer to the same dimensionless time.

spectrum at low wave numbers paralleling the total energy content computed with each method. At high wave numbers, the results with each method are somewhat different, with MPDATA showing the shallowest decay. The minmod method begins to decay strongly at a relatively low wave number, as might be predicted from the numerical analysis of the method. The minmod method's spectrum on the coarse grid differs

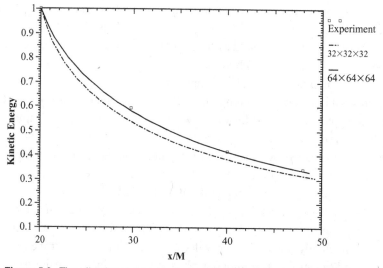

Figure 5.6. The plot shows a comparison of kinetic energy decay between the experiment and simulation, using the xPPM method with both 32^3 and 64^3 meshes. The kinetic energy levels and times have been normalized to refer to the same dimensionless time.

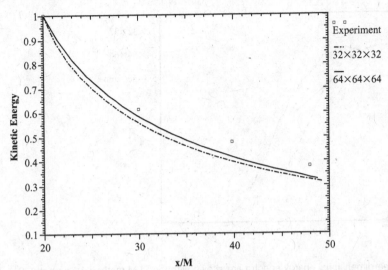

Figure 5.7. The plot shows a comparison of kinetic energy decay between the experiment and simulation, using the MPDATA method with both 32^3 and 64^3 meshes. The kinetic energy levels and times have been normalized to refer to the same dimensionless time.

almost entirely from the fine-grid spectrum. In contrast, MPDATA's spectra are close to the same on both grids. The variation of the spectrum with mesh resolution is intermediate with the PLM and xPPM methods.

Next, we move to a comparison of the PDFs of the velocity increments using our different ILES methods and the experimentally measured PDFs. In Kang et al. (2003). LES comparisons of velocity increments have shown good agreement with the small

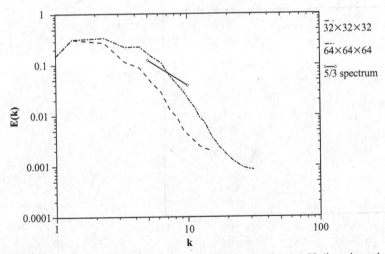

Figure 5.8. The three-dimensional energy spectra are shown with the minmod method, using both 32^3 and 64^3 meshes at $x/M = 48$.

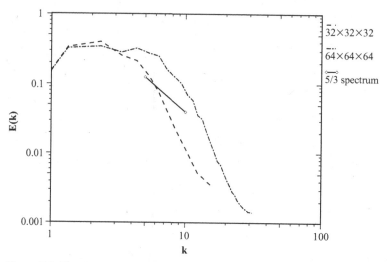

Figure 5.9. The three-dimensional energy spectra are shown with the PLM method, using both 32^3 and 64^3 meshes at $x/M = 48$.

increments while not capturing the tails of the distribution. This indicates that the simulations do not show as much intermittency as the data. For our simulations, we compute the velocity increments by using backward differences. We then construct the PDFs from these differences. The longitudinal velocity increments are $u_i(x_i + \Delta x_i) - u_i(x_i)$, and the transverse velocity increments are $u_i(x_j + \Delta x_j) - u_i(x_j)$, for $i \neq j$. With ILES simulations we find that the intermittency is quite well reproduced in both the longitudinal and transverse velocity increments. We display the comparison between the experiment and simulations in Figures 5.12–5.15. It is important to note that the

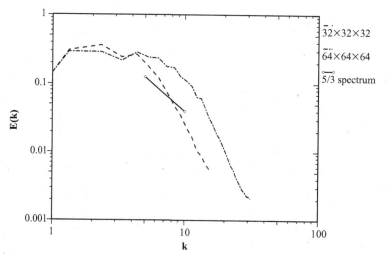

Figure 5.10. The three-dimensional energy spectra are shown with the xPPM method, using both 32^3 and 64^3 meshes at $x/M = 48$.

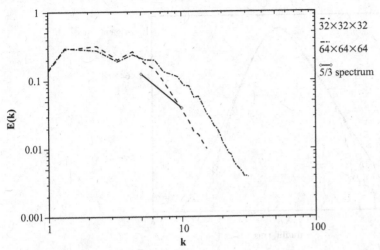

Figure 5.11. The three-dimensional energy spectra are shown with the MPDATA method, using both 32^3 and 64^3 meshes at $x/M = 48$.

experimental data with which we compare have been filtered with a narrower-width filter than that shown in earlier comparisons. These results seem to indicate that ILES methods are effective in simulating the mean behavior of classical turbulence and have a superior capability in capturing the turbulence's intermittency.

Despite this generally good behavior, there are differences among the four methods. The minmod method computes the least intermittent flow, as might be expected to be consistent with other measures. Interestingly, the MPDATA method compares favorably with the two other methods in the tails of the PDFs and if anything provides more intermittency. This is despite its otherwise more dissipative nature compared with PLM or xPPM. One might conclude that the reason for the good behavior of MPDATA is the lack of a monotonicity-based limiter that allows better preservation of high-wave-number flow structures. The overall assessment from this validation provides good evidence that ILES is useful for computing turbulent flows, but we must take some care in choosing the method. While the minmod results are passable, the other methods all provide a better comparison with the data. This is especially true for either xPPM or MPDATA, both of which demonstrate that moving to less restrictive limiters that do not enforce local monotonicity is important for many applications. This must be balanced with the level of robustness that comes with more dissipative methods.

5.6 Analysis of multidimensional equations

It is straightforward to extend the one-dimensional analysis we presented earlier to fully three-dimensional algorithms. In this section, we show the general results of assuming the use of second-order, but otherwise unspecified flux limiters for both incompressible and compressible flows in three dimensions. First, represent the three-dimensional incompressible Euler equations by using a general limiter that provides a nonlinear

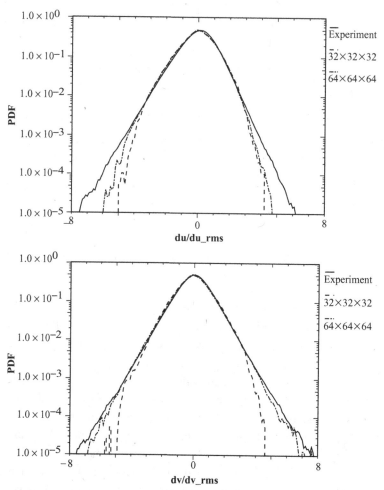

Figure 5.12. The PDFs computed using the minmod method on both 32^3 and 64^3 meshes and experimental velocity increments are compared in this figure. The experimentally measured PDFs are quite well reproduced by the ILES methods. This includes the magnitude of the velocity increments in the tails of the PDF.

combination of a first-order and second-order method. We follow this with a presentation of the modified equation for the fully compressible version of the same method.

We derive the modified equation by using symbolic algebra software, specifically the same Mathematica package used in our one-dimensional analyses. We use tensor notation with implied summation over repeated indices. For the case of incompressible flow, we then specialize our results to the two-dimensional case on a rectangular mesh of cells Δx by Δy. This will allow detailed comparisons of our results with those in Chapter 2 for the finite-scale equations and for the modified equation of the MPDATA algorithm.

The incompressible Euler equations can be written compactly as

$$\frac{\partial u_m}{\partial t} + \frac{\partial u_n u_m}{\partial x_m} + \frac{\partial p}{\partial x_m} = 0, \tag{5.34}$$

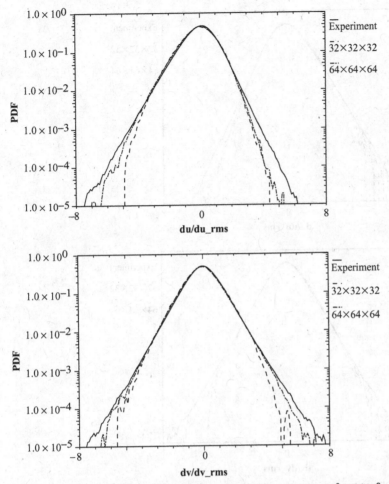

Figure 5.13. The PDFs computed using the PLM method on both 32^3 and 64^3 meshes and experimental velocity increments are compared in this figure. The experimentally measured PDFs are quite well reproduced by the ILES methods. This includes the magnitude of the velocity increments in the tails of the PDF.

where indices $m, n = 1, 2, 3$ indicate summation over those indices. The constraint of incompressibility is expressed by

$$\frac{\partial u_n}{\partial x_n} = 0. \tag{5.35}$$

The compressible equations can also be written abstractly. Our notation uses the vector flux function of a vector of conserved quantities:

$$\frac{\partial \mathbf{u}}{\partial t} + \frac{\partial \mathbf{f}(\mathbf{u})}{\partial x_n} = 0. \tag{5.36}$$

This represents a system of conservation laws.

In the analysis that follows, we define the pressure field as a potential field that acts to ensure that the velocity field is divergence free (i.e., solenoidal). Our algorithm

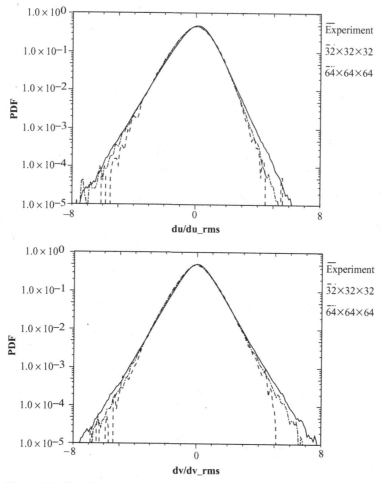

Figure 5.14. The PDFs computed using the xPPM method on both 32^3 and 64^3 meshes and experimental velocity increments are compared in this figure. The experimentally measured PDFs are quite well reproduced by the ILES methods. This includes the magnitude of the velocity increments in the tails of the PDF.

consists of the combination of a first-order- and a second-order-accurate method, with relative weights of the two schemes determined by a flux-limiter function. We construct a second-order-accurate method by defining a set of edge variables as the averages of the cell-centered values across a cell edge:

$$u_{m,i+1/2,j,k} = \frac{1}{2}\left(u_{m,i,j,k} + u_{m,i+1,j,k}\right),$$

$$u_{m,i,j+1/2,k} = \frac{1}{2}\left(u_{m,i,j,k} + u_{m,i,j+1,k}\right),$$

$$u_{m,i,j,k+1/2} = \frac{1}{2}\left(u_{m,i,j,k} + u_{m,i,j,k+1}\right). \tag{5.37}$$

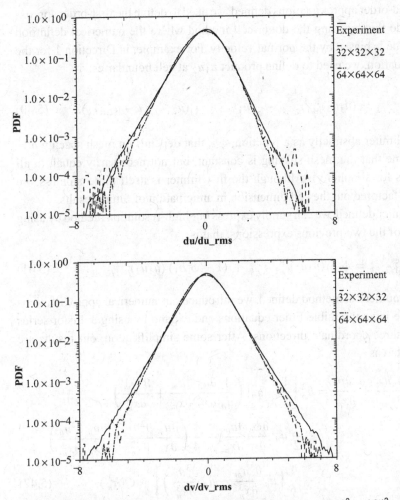

Figure 5.15. The PDFs computed using the MPDATA method on both 32^3 and 64^3 meshes and experimental velocity increments are compared in this figure. The experimentally measured PDFs are quite well reproduced by the ILES methods. This includes the magnitude of the velocity increments in the tails of the PDF.

Here i, j, k are the grid indices associated with the cell centers. The numerical approximation for the divergence is

$$\frac{u_{1,i+1/2,j,k} - u_{1,i-1/2,j,k}}{\Delta x_1} + \frac{u_{2,i,j+1/2,k} - u_{2,i,j-1/2,k}}{\Delta x_2}$$

$$+ \frac{u_{3,i,j,k+1/2} - u_{3,i,j,k-1/2}}{\Delta x_3} = 0. \tag{5.38}$$

The second-order expression for the nonlinear product that appears in the advective terms is simply constructed as the product of the edge values, for example,

$$(u_1 u_2)_{i+1/2,j,k}^{\text{2nd}} = u_{1,i+1/2,j,k} u_{2,i+1/2,j,k}. \tag{5.39}$$

With our second-order approximations defined, we need to define the first-order approximations. We do this by using the donor-cell method where the numerical defintiion of the edge value is biased by the normal velocity. For example, in Direction 1 for the two-motion equation, we need to define product $u_1 u_2$ at cell boundaries:

$$(u_1 u_2)^{1st}_{i+1/2,j,k} = (u_1 u_2)^{2nd}_{i+1/2,j,k} - \frac{1}{2}|u_{1,i+1/2,j,k}|(u_{2,i+1,j,k} - u_{2,i,j,k}). \tag{5.40}$$

We define our limiter abstractly as a function, ϕh, that depends on mesh spacing $h = \Delta x_m$ (we assume that the mesh spacing is constant, but not necessarily equal, in all three directions for simplicity). Although the flux limiter is itself dimensionless, we have explicitly factored out the cell dimension in anticipation of future results.

With the limiter defined, we can simply express our high-resolution method by using a combination of the two previous expressions, that is,

$$(u_1 u_2)^{hires}_{i+1/2,j,k} = \phi_1 h_1 (u_1 u_2)^{1st}_{i+1/2,j,k} + (1 - \phi_1 h_1)(u_1 u_2)^{2nd}_{i+1/2,j,k}. \tag{5.41}$$

With our basic numerical method defined, we introduce our numerical approximations into the discrete incompressible Euler equations and expand by using a Taylor series in each of the three coordinate directions. After some simplification, our results are compactly written as

$$\begin{aligned}
\frac{\partial u_m}{\partial t} + \frac{\partial u_n u_m}{\partial x_n} + \frac{\partial p}{\partial x_m} = h_n^2 \Bigg[&\frac{1}{2}\phi_n |u_n| \left(\frac{1}{u_n}\frac{\partial u_n}{\partial x_n}\frac{\partial u_m}{\partial x_n} + \frac{\partial^2 u_m}{\partial x_n^2} \right) \\
&+ \frac{1}{2}|u_n| \frac{\partial \phi_n}{\partial x}\frac{\partial u_m}{\partial x_n} - \frac{1}{4}\left(\frac{\partial u_m}{\partial x_n}\frac{\partial^2 u_n}{\partial x_n^2} + \frac{\partial u_n}{\partial x_n}\frac{\partial^2 u_n}{\partial x_n^2} \right) \\
&- \frac{1}{6}\left(u_n \frac{\partial^3 u_m}{\partial x_n^3} + u_m \frac{\partial^3 u_n}{\partial x_n^3} \right) \Bigg] + \mathcal{O}(h_n^3),
\end{aligned} \tag{5.42}$$

where again the pressure gradient is simply the gradient of the potential function that enforces the discrete divergence condition, (5.38). This form is similar to the one-dimensional forms in the presence of the self-similar terms and in the leading-order impact of the limiter on the dissipation.

We can apply the same basic differencing for compressible flows. The key differences are the lack of the divergence-free constraint and the application of the differencing to a vector function with an abstractly defined flux function. We define the high-order flux (second order) by using the face-averaged value as before, but now is we write

$$\mathbf{f}^{2nd}_{i+half,j,k} = (\mathbf{u}_{i+1/2,j,k}). \tag{5.43}$$

We define the first-order flux with the help of the flux Jacobian, $\mathbf{f}' = \partial \mathbf{f}/\partial \mathbf{u}$, as

$$\mathbf{f}^{1st}_{i+half,j,k} = \mathbf{f}^{2nd}_{i+half,j,k} - \frac{1}{2}|\mathbf{f}'|(\mathbf{u}_{i+1,j,k} - \mathbf{u}_{i,j,k}). \tag{5.44}$$

Substituting these expressions combined by use of our abstactly defined limiter that blends the first- and second-order fluxes and expanding the expression in a three-dimensional Taylor series produces

$$\frac{\partial \mathbf{u}}{\partial t} + \frac{\partial \mathbf{f}(u)}{\partial x_n} = h_n^2 \frac{\partial}{\partial x_n} \left[\frac{1}{2} \phi_n |\mathbf{f}'| \cdot \frac{\partial \mathbf{u}}{\partial x_n} - \frac{1}{6} \mathbf{f}' \frac{\partial^2 \mathbf{u}}{\partial x_n^2} - \frac{1}{24} \mathbf{f}'' \left(\frac{\partial \mathbf{u}}{\partial x_n} \right)^2 \right] + \mathcal{O}\left(h_n^3 \right). \quad (5.45)$$

This form is also similar to the one-dimensional compressible analysis having the self-similar terms proportional to the convexity of the flux function ($\mathbf{f}''$). The limiter enters in to the analysis as expected at second order.

For purposes of comparison with the results of Chapter 2, we evaluate (5.42) for the particular case of a uniform rectangular mesh in two spatial dimensions. Then each cell has dimensions Δx by Δy. Returning to the notation of Chapter 2, where the velocity vector has components (u, v) in the (x, y) directions, respectively, we see that (5.42) becomes

$$\frac{\partial u}{\partial t} + \frac{\partial uu}{\partial x} + \frac{\partial uv}{\partial y} + \frac{\partial p}{\partial x} = \frac{\Delta x^2}{2} \left[\frac{\partial}{\partial x} (\phi^x |u| u_x) \right] + \frac{\Delta y^2}{2} \left[\frac{\partial}{\partial y} (\phi^y |v| u_y) \right]$$
$$- \frac{\Delta x^2}{6} \left[\frac{\partial}{\partial x} \left(uu_{xx} + (u_x)^2 \right) \right]$$
$$- \frac{\Delta y^2}{12} \left[\frac{\partial}{\partial y} \left(uv_{yy} + vu_{yy} + 2u_y v_y \right) \right] \quad (5.46)$$

and

$$\frac{\partial v}{\partial t} + \frac{\partial uv}{\partial x} + \frac{\partial vv}{\partial y} + \frac{\partial p}{\partial y} = \frac{\Delta y^2}{2} \left[\frac{\partial}{\partial y} (\phi^y |v| v_y) \right]$$
$$= \frac{\Delta x^2}{2} \left[\frac{\partial}{\partial x} (\phi^x |u| v_x) \right] - \frac{\Delta y^2}{6} \left[\frac{\partial}{\partial y} (vv_{yy} + (v_y)^2) \right]$$
$$- \frac{\Delta x^2}{12} \left[\frac{\partial}{\partial x} (vu_{xx} + uv_{xx} + 2u_x v_x) \right]. \quad (5.47)$$

Note that the truncation terms are written in divergence form, with implicit subgrid-stress tensor $\mathcal{T}$ taking the form

$$\mathcal{T}_{xx} = \left[-1/2\phi^x |u| u_x + \frac{1}{6} uu_{xx} + \frac{1}{6} (u_x)^2 \right] \Delta x^2, \quad (5.48)$$

$$\mathcal{T}_{xy} = \left[-1/2\phi^y |v| u_y + \frac{1}{12} uv_{yy} + \frac{1}{12} u_{yy} v + \frac{1}{6} u_y v_y \right] \Delta y^2, \quad (5.49)$$

$$\mathcal{T}_{yx} = \left[-1/2\phi^x |u| v_x + \frac{1}{12} uv_{xx} + \frac{1}{12} u_{xx} v + \frac{1}{6} u_x v_x \right] \Delta x^2, \quad (5.50)$$

$$\mathcal{T}_{yy} = \left[-1/2\phi^y |v| v_y + \frac{1}{6} vv_{yy} + \frac{1}{6} (v_y)^2 \right] \Delta y^2. \quad (5.51)$$

A direct comparison with the implicit subgrid-stress tensor of MPDATA in Chapter 2.4.1 shows that the high-resolution algorithm contains the same terms originating from the high-order method, the same asymmetry, and the same dependence on Δx and Δy in each component. The dimensionless coefficients are different, though of similar order of magnitude.

5.7 Summary

In this chapter, we have used MEA to derive a theoretical connection between explicit LES models and the implicit modeling provided by nonoscillatory methods. In particular, we have developed a structural explanation of why some numerical methods work well as implicit subgrid models whereas others are inadequate. The MEA identifies the key property that determines the ability of a numerical algorithm to regularize a turbulent flow, in addition to providing computational stability. That key property is the leading order of the dissipative truncation terms. When deriving the global evolution of energy for a numerical method, the scaling of this term with zone size must reproduce the observed scaling of a turbulent fluid in order to be effective as an implicit model. Not all nonoscillatory methods provide the proper scaling. Many second-order methods can provide this scaling if the dissipative terms are of third order or higher. Positive-definite schemes are a distinguishable counterexample to this generalization. In contrast to the dissipative properties, the dynamical properties must be of second order in the mesh spacing; this result arises naturally from conservative second-order-accurate approximations to the inertial term.

In closing, it is important to consider the advice given by Occam: "It is vain to do with more what can be done with less." It may be perceived as arguable as to whether explicit or implicit turbulence modeling is simpler. The weight of experience and familiarity favor the explicit models. We offer two reasons to counterbalance this conclusion. One is practical and one is more philosophical.

From a practical point of view, high-resolution NFV methods are already accepted as an accurate and efficient tool for simulating laminar flows. Thus their application to turbulent flows means that only one code is necessary, and further that the user need not determine a priori whether a particular flow is turbulent. In addition it is extremely difficult, if not impossible, to separate modeling errors from numerical errors in underresolved flows. As we have demonstrated in this chapter, the subgrid models are of the same order in their dependence on mesh quantities as the numerical truncation terms. Nevertheless, in classic LES, the development of subgrid models is approached independent of the fluid solver. From a more philosophical viewpoint, we recognize that the PDEs themselves, like the numerical programs, are just models of physical reality. However, reality is understood through experiments, and experiments are necessarily carried out at discrete scales determined by the measuring device. Techniques of measurement themselves depend on the finite length and time scales of observation, with their measurements being the integrated value of quantities over these intervals. Thus one might conclude that it is the discrete codes that better model experiments.

A first principle understanding of turbulence has proven to be a nearly intractable problem. Thus the success of implicit turbulence modeling as a practical predictive tool in engineering and geophysical flows cannot be ignored, despite its apparent lack of congruence with standard practice. As the results of this chapter demonstrate, this lack of congruence is nonexistent, only hidden below the surface. The deeper issue to explore is the underlying reasons for this success and to understand what limitations may exist and what improvements may be possible.

REFERENCES

Drikakis, D., & Rider, W. 2004 *High-Resolution Methods for Incompressible and Low-Speed Flows*. New York: Springer-Verlag.

Frisch, U. 1995 *Turbulence: The Legacy of A. N. Kolmogorov*. New York: Cambridge University Press.

Fureby, C., & Grinstein, F. F. 2002 Large eddy simulation of high reynolds number free and wall bounded flows. *J. Comp. Phys.* **181**, pp. 68–97.

Garnier, E., Mossi, M., Sagaut, P., Comte, P., & Deville, M. 1999 On the use of shock-capturing scheme for large-eddy simulation. *J. Comput. Phys.* **153**, 273.

Grinstein, F. F., & Fureby, C. 2002 Recent progress on MILES for high Re flows. *J. Fluids Eng.* **124**, 848.

Hirsch, C. 1999 *Numerical Computation of Internal and External Flows*. New York: Wiley.

Hirt, C. W. 1968 Heuristic stability theory for finite difference equations. *J. Comput. Phys.* **2**, 339–355.

Hughes, T. J. R., Mazzei, L., Oberai, A. A., & Wray, A. 2001. The multiscale formulation of large eddy simulation: Decay of homogeneous isotropic turbulence. *Phys. Fluids* **13**, 505–512.

Huynh, H. T. 1995 Accurate upwind methods for the Euler equations. *SIAM J. Numer. Anal.* **32**, 1565–1619.

Kang, H.S., Chester, S., & Meneveau, C. 2003 Decaying turbulence in an active-grid-generated flow and comparisons with large-eddy simulation. *J. Fluid Mech.* **480**, 129–160.

Knoll, D., Mousseau, V., Margolin, L., & Chacon, L. 2003 On balanced approximations for time integration of multiple time scale systems. *J. Comput. Phys.* **185**, 583–611.

Margolin, L. G., & Rider, W. J. 2002 A rationale for implicit turbulence modeling. *Int. J. Numer. Meth. Fluids* **39**, 821–841.

Margolin, L. G., & Rider, W. J. 2005 The design and construction of implicit subgrid scale models. *Int. J. Numer. Meth. Fluids* **47**, 1173–1179.

Meneveau, C., & Katz, J. 2000 Scale-invariance and turblence models for large-eddy simulation. *Annu. Rev. Fluid Mech.* **32**, 1–32.

Rider, W., & Margolin, L. 2003 From numerical analysis to implicit subgrid turbulence modeling. Paper AIAA-2003-4101.

Rider, W. J., Greenough, J., & Kamm, J. R. 2005 Combining high-order accuracy with non-oscillatory methods through monotonicity preservation. *Int. J. Numer. Meth. Fluids* **47**, 1253–1259.

Rider, W. J., Greenough, J. & Kamm, J. R. 2006 Extrema, accuracy and monotonicity preserving methods through adaptive nonlinear hybridizations. *J. Comp. Phy.*, revised.

Sagaut, P. 2005 *Large eddy simulation for Incompressible Flows*, 3rd Edition. New York: Springer-Verlag.

Smagorinsky, J. 1963 General circulation experiments with the primitive equations. I. The basic experiment. *Mon. Weather Rev.* **101**, 99–164.

6 Approximate Deconvolution

Nikolaus A. Adams, S. Hickel, and J. A. Domaradzki

6.1 Introduction

In this chapter we make a connection between the *filtering approach* (Leonard 1974) and the *averaged-equation approach* (Schumann 1975) to large eddy simulation (LES). With the averaged-equation approach, the discrete system for evolving a grid-function approximation of the continuous solution is considered directly as a truncated representation of the continuous system. With the filtering approach, a continuous filtered system is considered as an approximation; the numerical error in solving this continuous system is considered to be negligibly small. The filtering approach provides an analytic framework for deriving LES equations and commonly is employed as a basis for the development of functional and structural models (Sagaut 2005) and Chapter 3 of this book. In practice, models derived on the basis of the filtering approach were plagued by the problem that the numerical error in most cases was nonnegligible. The effect of discretizing the filtered continuous equations on the subgrid-scale (SGS) force was analyzed in detail for the first time by Ghosal (1996). It was revealed that, over a large wave-number range, the truncation error of commonly employed nonspectral discretizations can be as large as the SGS stress, if not larger.

During the attempt of improving eddy-viscosity-based models, it was revealed that the correlation of predicted SGS stresses with the exact SGS stresses is much less than unity. This fact is reviewed by Meneveau and Katz (2000) on the basis of experimental data. A much larger correlation is achieved by the scale-similarity model (Bardina, Ferziger, and Reynolds 1983), which does, however, underpredict SGS dissipation. Shah and Ferziger (1995) found that the scale-similarity model lends itself to a straightforward extension by improving the simple estimate, $u_\alpha \sim \bar{u}_\alpha$, with an approximate inversion of the applied top-hat filter. This idea was later generalized by Geurts (1997). Stolz and Adams (1999) devised a systematic method (the approximate deconvolution model, or ADM) for obtaining a regularized inverse-filter operation suitable for any graded filter, that is, finite-order filter (see Section 6.2). One obtains the approximate inverse only by applying additional filtering operations to the filtered solution so that the approximate deconvolution operator is linear. By using the ADM, one can obtain an approximation of the unfiltered solution of any desired order. However, since the filter inversion is an ill-conditioned operation, high-order approximants are undesirable. For the ADM, a chosen finite deconvolution order serves as a regularization parameter.

We now develop the framework for deconvolution methods in SGS modeling. For brevity of notation, we perform most of the analyses in the following sections for the one-dimensional case and a generic scalar conservation law. In Section 6.5.1, we outline the extension to the Navier–Stokes equations. For a given generic nonlinear transport equation

$$\frac{\partial u}{\partial t} + \frac{\partial F(u)}{\partial x} = 0, \ 0 \le x \le L, \tag{6.1}$$

we obtain a filtered transport equation,

$$\frac{\partial \overline{u}}{\partial t} + \frac{\partial F(\overline{u})}{\partial x} = \mathcal{G}, \tag{6.2}$$

by convolution with a homogeneous filter, G,

$$\overline{u}(x) = \int_{-\infty}^{+\infty} G(x - x')u(x')dx' = G * u, \tag{6.3}$$

where

$$\mathcal{G} = \frac{\partial F(\overline{u})}{\partial x} - G * \frac{\partial F(u)}{\partial x} \tag{6.4}$$

is an error term that is due to filtering. Equation (6.2) is the modified differential equation for $\overline{u}$, the solution of which would be identical to the filtered solution of (6.1) if $\mathcal{G}$ could be computed exactly.

Numerical discretizations of (6.2) carry wave numbers k up to the *Nyquist wave number*, or *numerical cutoff wave number*, $k_h = \pi/h$, where $h = L/N$ is the uniform grid spacing, and N is the number of intervals into which the domain $[0, L]$ is partitioned. Wave numbers k with $0 \le |k| \le k_h$ are the *represented wave numbers*. Contributions to solution u in this wave number range constitute the *represented scales*. A numerical discretization inevitably introduces another filter, the grid filter, which we consider here to be a low-pass projection filter (spectral cutoff),

$$u_N = P_N[u] = \sum_{k=-k_N}^{k_N} \hat{u}(k)e^{ikx}, \tag{6.5}$$

where $\hat{u}(k)$ denotes the Fourier transform of $u(x)$. Operator $G * P_N$ we denote as G_N; that is, $\overline{u}_N = G_N * u$. Now we can write the discretized equation, (6.2), as

$$\frac{\partial \overline{u}_N}{\partial t} + D * \frac{\partial F(\overline{u}_N)}{\partial x} = \mathcal{G}_1 + \mathcal{G}_2 + \mathcal{G}_3. \tag{6.6}$$

Note that common linear discretization schemes of the differentiation operator appear as convolution operators in the modified differential equation for $\overline{u}_N$. Their Fourier transform, $\hat{D}(k) = \tilde{k}/k$, can be expressed by modified wave number $\tilde{k}$ (Lele 1992;

Vichnevetsky and Bowles 1982). Error terms $\mathcal{G}_1$, $\mathcal{G}_2$, and $\mathcal{G}_3$ are given by

$$\mathcal{G}_1 = D * \frac{\partial F(\bar{u}_N)}{\partial x} - G_N * D * \frac{\partial F(u_N)}{\partial x}, \tag{6.7}$$

which represents the contribution to the SGS terms by filtering of the discrete solution;

$$\mathcal{G}_2 = G_N * D * \frac{\partial F(u_N)}{\partial x} - G_N * D * \frac{\partial F(u)}{\partial x}, \tag{6.8}$$

which singles out the effect of truncating the solution to u_N; and

$$\mathcal{G}_3 = G_N * D * \frac{\partial F(u)}{\partial x} - G_N * \frac{\partial F(u)}{\partial x}, \tag{6.9}$$

which is the truncation error of the discretization scheme. Sum $\mathcal{G}_1 + \mathcal{G}_2 + \mathcal{G}_3$ corresponds to the divergence of the SGS-stress tensor for the Navier–Stokes equations. In deconvolution-based SGS modeling, either generalized scale-similarity approaches can be employed or the approximation of the unfiltered solution is inserted directly as an argument for the flux function (see Section 6.4). If the deconvolution operation is exact on u_N, the first error term vanishes; $\mathcal{G}_1 = 0$. For finite-volume schemes and nonlinear flux functions, F, the discretization cannot be written as linear operator D. In this case it should be interpreted symbolically, also implying reconstruction or deconvolution of the unfiltered solution and the approximation of the physical flux function by a numerical flux function.

6.2 Filter-kernel definitions

If no explicit filtering is performed, the chosen discretization scheme implies a certain filter kernel that can be derived from operator D in Eq. (6.6). The particular form of this filter kernel is obvious only for spectral schemes (spectral cutoff filter), Eq. (6.5), or for finite-volume schemes (top-hat filter). For a theoretical analysis, one often uses analytic kernels with well-known properties such as the Gauss function. In real space and one dimension, the Gauss filter is often defined as

$$G(x - x') = \sqrt{\frac{8}{\pi} \frac{1}{\Delta}} e^{-8 \frac{(x - x')^2}{\Delta^2}}, \tag{6.10}$$

where Δ is the filter width that can be related to the mesh width. The Fourier transform of (6.10) is given by

$$\hat{G}(k) = e^{-\left(\frac{\Delta k}{4\sqrt{2}}\right)^2}. \tag{6.11}$$

A filter is of order m if the first nonvanishing derivative of its transfer function is of order m at $k = 0$. According to this definition, the Gauss filter has order 2 whereas a spectral-cutoff filter in Fourier dual space has order ∞.

For finite-order filter kernels, such as the Gauss filter, the definition of filter width Δ is somewhat arbitrary. A possible choice is, for example,

$$\hat{G}(k_C) = \frac{1}{2}, \tag{6.12}$$

where cutoff wave number k_C for filters other than the Gauss filter can be related to the filter width by

$$k_C = \frac{2\pi}{\Delta}. \tag{6.13}$$

A numerical evaluation of the filter operation with these analytically defined filters requires the application of a quadrature rule or a Fourier transform of the solution, so that, using the convolution theorem, we can replace Eq. (6.3) by

$$\hat{\bar{u}}(k) = \hat{G}(k)\hat{u}(k). \tag{6.14}$$

Standard numerical quadrature rules, such as the trapezoidal rule, can alter the transfer function of the discrete filter operator considerably from its continuous transfer function. If a good representation of a given analytical filter is desired, one should tune a quadrature rule so that discrete and continuous filter transfer functions match closely. Note that, with standard quadrature rules, accuracy cannot be increased arbitrarily since no intermediate breakpoints are available aside of the given mesh points. The absolute error often increases if one chooses a higher-order quadrature rule. A good choice is to start from the general compact implicit discrete filter definitions of Lele (1992), which include any compact explicit discrete filter definitions. With these formulations, one can obtain additional free parameters by which the filter transfer function can be tuned to match the continuous filter transfer function accurately. These discrete filters are formulated as follows. Let $\mathbf{f}$ be an $N + 1$ vector containing the values of the grid function, $f_i = f(x_i)$, that one obtains by sampling a continuous function $f(x)$ at a set of equally spaced nodes $x_j = x_0 + jh$, $0 \le j \le N$. Let $\bar{\mathbf{f}}$ denote the vector of filtered values that one obtains by applying discrete filter $\mathbf{G}_N$ to $\mathbf{f}$, in matrix-vector notation $\bar{\mathbf{f}} = \mathbf{G}_N\mathbf{f}$. Considering the special case in which $\mathbf{G}_N = \mathbf{M}_l^{-1}\mathbf{M}_r$, and $\mathbf{M}_l$, $\mathbf{M}_r$ are tridiagonal matrices (circulant for periodic domains), we see that a one-parameter family of second-order filters is given by

$$a\bar{f}_{j-1} + \bar{f}_j + a\bar{f}_{j+1} = bf_j + \frac{c}{2}(f_{j-1} + f_{j+1}), \tag{6.15}$$

where $b = (1/2 + a)$ and $c = b/2$ (Lele 1992). For a finite domain, various treatments are possible at boundary points $j = 0$ and $j = N$. Often it is sufficient to impose no filtering at boundary points, which is consistent with a filter width that shrinks to zero at the boundary. The kernel of a Gauss filter with filter width $\Delta = 4h$ is well approximated by (6.15) in spectral space if one chooses $a = -0.2$.

6.3 Averaged equation and filtering approach

A finite-volume discretization of Eq. (6.1) corresponds to a convolution with the top-hat filter,

$$G_N(x - x_j; h) = \begin{cases} 1/h, & |x - x_j| \leq h/2 \\ 0, & \text{otherwise} \end{cases}, \tag{6.16}$$

where h equals Δ. On grid $x_N = \{x_j\}$, an application of the filter operation (6.16) to a function $u(x)$ returns the filtered solution in terms of a grid function u_j at x_j:

$$\bar{u}_j = G_N * u = \frac{1}{h} \int\limits_{x_{j-1/2}}^{x_{j+1/2}} u(x') dx'. \tag{6.17}$$

For the top-hat filter with a filter width equal to the mesh width, the filter operation results in the finite-volume discretization,

$$G_N * \left. \frac{\partial F}{\partial x} \right|_{x_j} = \frac{F_{j+1/2} - F_{j-1/2}}{h}, \tag{6.18}$$

which is the flux difference across a computational mesh cell, as is well known for finite-volume schemes. The cell-face values of the flux function, $F_{j\pm1/2} = F[u(x_{j\pm1/2})]$, require an approximation $\tilde{u}_N(x_{j\pm1/2})$ for $u(x_{j\pm1/2})$ from $\bar{u}_N$. In the terminology of numerical analysis, this step is called *reconstruction*. Note that the problem of reconstruction is identical to that of deconvolution for LES. The essential difference is that reconstruction is considered in the limit where h is small with respect to relevant flow scales (the only exception are shocks and interfaces), whereas in LES h is at best of the order of scales in the inertial subrange of the turbulence cascade. Therefore, with numerical schemes, a high-approximation order p for the deconvolution is desired, $\|\tilde{u}_N - u\| = \mathcal{O}(h^p)$, which ensures convergence to the strong solution, if it is sufficiently smooth, with $h \to 0$. Convergence to the correct weak solution requires additional constraints, which is the subject of standard textbooks on numerical analysis (see, e.g., LeVeque 1992). For SGS modeling, the analysis has to consider finite h, which prevents the use of linearizations.

After application of the top-hat filter, we obtain a finite-volume approximation of Eq. (6.1) by

$$\frac{\partial \bar{u}_N}{\partial t} + G_N * \frac{\partial F(\tilde{u}_N)}{\partial x} = 0, \tag{6.19}$$

where $\tilde{u}_N \doteq u_N$ results from an approximate inversion of the filtering, $\bar{u}_N = G_N * u$. At this point we introduce the additional freedom not to employ the physical flux function, $F(u)$, but a consistent *numerical flux function*, $\tilde{F}_N(\ldots, v_j, v_{j+1}, \ldots)$, which can be a functional of a grid function, v_N, spanning over a stencil of defined extent. In the following we use grid functions and their continuous extensions by Whittaker's cardinal functions synonymously (refer to Vichnevetsky and Bowles 1982). The consistence condition on the numerical flux function is $\tilde{F}_N(\ldots, u, u, \ldots) = F(u)$. With the numerical flux function inserted, the finite-volume approximation becomes

$$\frac{\partial \bar{u}_N}{\partial t} + G_N * \frac{\partial \tilde{F}(\tilde{u}_N)}{\partial x} = 0. \tag{6.20}$$

Following our terminology, we see that Eq. (6.20) is the *averaged equation* corresponding to Eq. (6.1). The *filtered equation* is

$$\frac{\partial \bar{u}_N}{\partial t} + G_N * \frac{\partial F(u_N)}{\partial x} = \mathcal{G}_2 + \mathcal{G}_3 = \mathcal{G}_{SGS}. \tag{6.21}$$

We can rewrite Eq. (6.20) trivially as

$$\frac{\partial \bar{u}_N}{\partial t} + G_N * \frac{\partial F(u_N)}{\partial x} = G_N * \frac{\partial F(u_N)}{\partial x} - G_N * \frac{\partial \tilde{F}(\tilde{u}_N)}{\partial x} = \mathcal{G}_N. \tag{6.22}$$

At this point we have connected filtered and averaged equations. If $\mathcal{G}_N \approx \mathcal{G}_{SGS}$, then the averaged equation contains an SGS model implicitly. The shape of this implicit SGS model obviously depends on the deconvolution operation, $\tilde{u}_N(\bar{u}_N) \approx u_N$, and on the numerical flux function, $\tilde{F} \approx F$. The objective of the following section is to elaborate on this fact and expose how deconvolution and numerical flux function can be tuned for implicit SGS modeling. Such an approach is similar to the "reverse engineering" of Margolin and Rider (2002, 2005; also see Chapters 2 and 5), where the modified equation of a nonoscillatory finite-volume scheme is matched to an evolution equation with a specified SGS-model term.

6.4 Subgrid-scale approximation

Although the inverse-filtering operation is ill posed, one can obtain approximation $\tilde{u}_N$ of u on grid x_N by regularized deconvolution; see Domaradzki and Adams (2002). The simplest approach is to set $\tilde{u}_N = \bar{u}_N$. Inserted into the SGS-stress tensor, this results in the scale-similarity model. Another often-used procedure is to invoke a singular-value decomposition, as discussed by Adams and Stolz (2002); that is, remove the null space of filter G_N from u before the approximate inversion of $\bar{u}_N = G_N * u = G * P_N * u = G * u_N$. Given a filter kernel with positive transfer function, that is, Fourier transform $\hat{G}(k)$, this amounts to simple inversion in Fourier dual space on wave number range $|k| < k_h = \pi/h$. In the following, we argue that other regularized filter inversions lend themselves to implicit SGS modeling. One such regularization method is the van Cittert approach, as introduced by Stolz and Adams (1999), in which a linear approximate deconvolution operator is defined from

$$\tilde{u}_N = \tilde{G}^{-1} * \bar{u}_N = \left[\sum_{v=0}^{M} (I - G)^v \right] * \bar{u}_N. \tag{6.23}$$

The series expansion for the inverse operator on the right-hand side is truncated at M, which regularizes the inverse. M is a problem-dependent regularization parameter

(Domaradzki and Adams 2002; Stolz and Adams 1999). An explicit SGS model based on this linear approximate deconvolution operation was proposed by Stolz and Adams (1999). The approach has since been employed for the modeling of turbulent and scalar mixing, such as that by Stolz, Adams, and Kleiser (2001a) and by Mellado Sarkar, and Patano (2003). The deconvolved solution, $\tilde{u}_N$, can be either inserted as an approximation for u into SGS stress (here one dimensional) $\tau_N = P_N * \overline{uu} - \bar{u}_N \bar{u}_N$ being approximated as $\tau_N = \overline{\tilde{u}_N \tilde{u}_N} - \bar{u}_N \bar{u}_N$, or it can be directly inserted into the flux function. The former approach is also called the *generalized scale-similarity model*. For the latter approach, Dunca and Epshteyn (2004) were able to prove energy stability and rigorous bounds on the modeling error.

The deconvolution equation (6.23) operates on a grid function. The support of the deconvolution kernel increases with the regularization parameter, M, even if filter kernel G has compact support. In addition, the deconvolution operator is linear with respect to the solution. Any SGS model inferred by truncating the deconvolution order to a finite M therefore is linear. It has been shown by Stolz and Adams (1999) that this linear model is sufficient for low-Reynolds-number isotropic turbulence. For larger Reynolds numbers or wall-bounded turbulence, an additional contribution to the SGS model is necessary (Stolz et al. 2001a, 2001b) where a relaxation form for this contribution was proposed. This relaxation term contains a weakly solution-dependent parameter for which a dynamic determination procedure was used. These two drawbacks of the ADM, first the increase of the support, second the linearity of the effective SGS model, can be circumvented if one allows for a nonlinear and locally defined deconvolution. This approach was proposed by Adams, Hickel, and Franz (2004) and is outlined in what follows here.

An application of the filter operation (6.16) to a function $u(x)$ returns the filtered solution in terms of a grid function u_j at x_j:

$$\bar{u}_j = G * u = \frac{1}{h} \int_{x_{j-1/2}}^{x_{j+1/2}} u(x')dx'.$$

Consistently with the finite-volume approach, we call x_j the cell centers and $x_{j\pm1/2}$ the cell faces of cell j. Filtering applied to the flux derivative, $\partial F(u)/\partial x$, in Eq. (6.1) returns Eq. (6.18), which requires an approximation of unfiltered solution $u(x)$ at the left and right faces of each cell j, which are called $u_{j-1/2}^+$ and $u_{j+1/2}^-$, respectively.

The use of a top-hat filter G_N with grid truncation according to Eq. (6.16) allows for a primitive-function reconstruction of $u(x)$ from $\bar{u}_N$ at $x_{j\pm1/2}$, as proposed by Harten et al. (1987). In this reference, a more general deconvolution reconstruction was also proposed that can be formulated for any graded filter for which sufficiently many filter moments exist. Harten et al. (1987) combined reconstruction with an interpolation-stencil selection, leading to the essentially nonoscillatory (ENO) property of the reconstructed solution. This procedure constitutes a nonlinear approximation or regularized deconvolution of the filtered solution, since the stencil selection depends on the local properties of the actual computed solution.

Following the ENO idea of adaptive deconvolution, we can formulate an *approximate local deconvolution* of the filtered grid function, $\bar{u}_N$, using local approximation polynomials. The restriction on the admissible local approximation polynomials is that their stencils interpolate at cell faces $x_{j\pm 1/2}$. Extrapolating stencils are excluded. For node j we introduce a set of interpolation polynomials of order $k = 1, \ldots, K$, for each k with shift $r = 0, \ldots, k-1$ of the left-most stencil point with respect to j. The shift also identifies the respective stencil. Admissible stencils range from from $j - r$ to $j - r + k - 1$, expressed by (k, r). On each admissible stencil, a right-face interpolant at $x_{j+1/2}$ and a left-face interpolant at $x_{j-1/2}$ of order k are given by Shu (1998):

$$p^-_{k,r}(x_{j+1/2}) = \sum_{l=0}^{k-1} c^{(k)}_{r,l}(j)\bar{u}_{j-r+l}, \qquad p^+_{k,r}(x_{j-1/2}) = \sum_{l=0}^{k-1} c^{(k)}_{r-1,l}(j)\bar{u}_{j-r+l}. \qquad (6.24)$$

Coefficients $c^{(k)}_{r,l}(j)$ contain the deconvolution, that is, the inversion of the top-hat filter on the space of admissible local interpolation polynomials, and the interpolation from the nodes (k, r) to $j \pm 1/2$. A rule for their computation is provided by Shu (1998) as

$$c^{(k)}_{r,l}(j) = h_{j-r+l} \sum_{\mu=l+1}^{k} \frac{\sum_{\substack{p=0 \\ p\neq\mu}}^{k} \prod_{\substack{v=0 \\ v\neq\mu,p}}^{k} x_{j+1/2} - x_{j-r+v-1/2}}{\prod_{\substack{v=0 \\ v\neq\mu}}^{k} x_{j-r+\mu-1/2} - x_{j-r+v-1/2}}. \qquad (6.25)$$

As indicated, this rule holds for variable mesh spacing. If $h_j = h = \text{const}$, it can be simplified and tabulated (Shu 1998). Note that the index range of $c^{(k)}_{r,l}(j)$ is $r = -1, \ldots, k-1$ and $l = 0, \ldots, k-1$ for each $k = 1, \ldots, K$.

It is well known that a main contribution to SGS dissipation can be attributed to an eddy viscosity, that is, a term $\mathcal{G}_{\text{SGS}} \sim h^2 \partial\bar{u}/\partial x$. However, a correct modeling of the local energy transfer requires additional redistribution terms. This issue is discussed, for example, by Meneveau and Katz (2000). It should be pointed out that dissipative higher-order discretizations, such as high-order upwind schemes, the spectral vanishing viscosity method, or an explicit addition of hyperviscosity, in general do not represent the SGS energy transfer correctly. These mechanisms can cause excessive dissipation in a wave number band near ξ_h, which prohibits proper energy transfer from larger scales to nonrepresented scales. Note that most of the SGS energy transfer originates from wave numbers between $\xi_h/2$ and ξ_h (Domaradzki, Liu, and Brachet 1993).

To allow for implicit model $\mathcal{G}_N$ to contain additional redistribution terms, the entire range of admissible reconstruction polynomials from degree 1 to K is considered for locally deconvolving the filtered solution. A quasi-linear combination of all possible interpolation polynomials according to Eq. (6.24) up to a certain order K can be constructed:

$$\tilde{u}^{\mp}_{j\pm 1/2} = \sum_{k=1}^{K} \sum_{r=0}^{k-1} \omega^{\mp}_{k,r}(j) p^{\mp}_{k,r}(x_{j\pm 1/2}). \qquad (6.26)$$

As a restriction, we impose that the sum of all weights, $\omega_{k,r}^{\pm}(j)$, over k and r is unity:

$$\sum_{k=1}^{K}\sum_{r=0}^{k-1}\omega_{k,r}^{\pm} = 1. \tag{6.27}$$

Equation (6.26) gives the resulting approximants for the deconvolved solution at the left and right cell faces. An additional freedom in designing the implicit SGS model originates from the choice of a suitable numerical flux function, $\tilde{F}_N$. One possible choice is a modified Lax–Friedrichs flux function,

$$\tilde{F}_N(x_{j+1/2}) = \frac{1}{2}\left[F\left(\tilde{u}_{j+1/2}^-\right) + F\left(\tilde{u}_{j+1/2}^+\right)\right] - \sigma_{j+1/2}\left(\tilde{u}_{j+1/2}^+ - \tilde{u}_{j+1/2}^-\right),$$

where $\sigma_{j+1/2}$ can be any shift-invariant functional of $\bar{u}_N$. Another possible choice is

$$\tilde{F}_N(x_{j+1/2}) = F\left(\frac{\tilde{u}_{j+1/2}^- + \tilde{u}_{j+1/2}^+}{2}\right) - \sigma_{j+1/2}\left(\tilde{u}_{j+1/2}^+ - \tilde{u}_{j+1/2}^-\right). \tag{6.28}$$

Nonlinearity by adaptivity of $\tilde{u}_{j+1/2}^{\pm}$ to the solution enters the implicit SGS model within this framework by the choices for a smoothness measure. In a manner similar to that of the weighted ENO approach (Liu, Osher, and Chan 1994), we compute weights $\omega_{k,r}^{\pm}$ by

$$\omega_{k,r}^{\pm} = \frac{1}{K}\frac{\alpha_{k,r}^{\pm}}{\sum_{\mu=0}^{k-1}\alpha_{k,\mu}^{\pm}}. \tag{6.29}$$

Coefficients $\alpha_{k,r}^{\pm}$ are defined as

$$\alpha_{k,r}^{\pm} = \gamma_{k,r}^{\pm}(\varepsilon + \beta_{k,r})^{-2}, \tag{6.30}$$

where ε is a small number to prevent singularity. Several options are available for smoothness measure $\beta_{k,r}$. A possible choice is the WENO smoothness measure (Shu 1998),

$$\beta_{k,r} = \sum_{q=1}^{k-1}\int_{x_{j-1/2}}^{x_{j+1/2}} h^{2q-1}\left(\frac{\partial^q p_{k,r}}{\partial x^q}\right)^2 dx, \tag{6.31}$$

which results in undivided differences of $\bar{u}_N$ on the respective stencils around x_j. Examples for $k = 2$ and $k = 3$ are given by Shu (1998). This smoothness measure also considers higher-order variations and was suggested as preferable for capturing interfaces and shocks with WENO schemes. For the purpose of modeling turbulent subgrid scales, the total variation of $\bar{u}_N$ on the considered stencil,

$$\beta_{k,r} = \sum_{\mu=-r}^{k-r-2} |\bar{u}_{j+\mu+1} - \bar{u}_{j+\mu}|, \tag{6.32}$$

or the variance of $\bar{u}_N$ on the considered stencil,

$$\beta_{k,r} = \sum_{\mu=-r}^{k-r-2} (\bar{u}_{j+\mu+1} - \bar{u}_{j+\mu})^2,$$

(6.33)

is suitable and computationally less expensive.

We can choose the dissipative weight in Eq. (6.28), for example, as $\sigma_{j+1/2} = |\bar{u}_{j+1} - \bar{u}_j|$. Equation (6.30) introduces free parameters for the right-face interpolant, $\gamma_{k,r}^-$, and the left-face interpolant, $\gamma_{k,k-1-r}^+$, which are constrained to be symmetric with respect to stencil center $\gamma_{k,r}^- = \gamma_{k,k-1-r}^+$. The objective of implicit SGS modeling is to determine these parameters that close the model. Since the model involves (i) locality through the use of local polynomial approximants, (ii) adaptivity through the use of a smoothness measure that renders the implicit model nonlinear, and (iii) deconvolution through an approximate inversion of the top-hat filtering, the approach is called the *adaptive local deconvolution method* (ALDM).

6.5 Modeling

As outlined in the previous section, the approximate deconvolution equation (6.26) contains model parameters $\gamma_{k,r}^\pm$. We can choose these parameters so as to maximize the order of the truncation error. This results in a WENO scheme, differing from the standard approaches only in terms of the chosen smoothness measure. However, since the objective of implicit SGS modeling is to shape the truncation error so that $\mathcal{G}_N \sim \mathcal{G}_{\text{SGS}}$ and the dominant term is $\mathcal{G}_{\text{SGS}} \sim h^2 \frac{\partial^2 \bar{u}_N}{\partial x^2}$ in most known explicit SGS models, the model parameters have to be chosen accordingly. An expansion of the filtered flux function, $F(u) = u^2/2$, which gives the inviscid Burgers equation, results in

$$G * \frac{\partial F_N(u_N)}{\partial x} = \bar{u}_N \frac{\partial \bar{u}_N}{\partial x} + \frac{1}{12} \frac{\partial \bar{u}_N}{\partial x^2} \frac{\partial^2 \bar{u}_N}{\partial x} - \frac{1}{720} \frac{\partial \bar{u}_N}{\partial x} \frac{\partial^4 \bar{u}_N}{\partial x^4} h^4$$

$$+ \frac{1}{30240} \frac{\partial \bar{u}_N}{\partial x} \frac{\partial^6 \bar{u}_N}{\partial x^6} h^6 - + \cdots,$$

(6.34)

where derivatives are to be taken at the cell centers, x_N. Error term $\mathcal{G}_1$ therefore can be expanded (for exact differentiation $D = I$ and top-hat filter G) as

$$-\mathcal{G}_1 = \frac{1}{12} \frac{\partial \bar{u}_N}{\partial x} \frac{\partial^2 \bar{u}_N}{\partial x^2} - \frac{1}{720} \frac{\partial \bar{u}_N}{\partial x} \frac{\partial^4 \bar{u}_N}{\partial x^4} h^4 + \frac{1}{30240} \frac{\partial \bar{u}_N}{\partial x} \frac{\partial^6 \bar{u}_N}{\partial x^6} h^6 - + \cdots.$$

(6.35)

A truncation of this expansion has been invoked for explicit SGS modeling, where it led to the gradient model, also called the tensor-diffusivity model, or Clark model (Clark, Ferziger, and Reynolds 1979). It is well known that these type of models without additional dissipative terms, also called *secondary regularizations*, do not properly model the SGS energy transfer. The computation of $\mathcal{G}_{\text{SGS}} = \mathcal{G}_2 + \mathcal{G}_3$ requires unavailable information, whereas a computation of $\mathcal{G}_1$ for suitable filters contains polluted or noisy but available information. The approximation of $\mathcal{G}_1$ is the *soft-deconvolution problem*, which can be handled by regularized filter inversions such as those outlined herein.

The modeling of $\mathcal{G}_{\text{SGS}}$ requires information on subgrid scales and therefore constitutes the *hard-deconvolution problem*. Approximating $\mathcal{G}_{\text{SGS}}$ by $\mathcal{G}_N$ obviously cannot be accomplished in terms of local error norms, such as by Taylor expansions. Rather, the effect of $\mathcal{G}_N$ on the SGS energy transfer should closely approximate that of $\mathcal{G}_{\text{SGS}}$. We can achieve this either by trying to represent a given explicit SGS model for $\mathcal{G}_{\text{SGS}}$ by the implicit model, or by directly adjusting the implicit-model parameters by using theoretical, experimental, or empirical data.

In the following paragraphs, we apply the modified differential equation analysis (MDEA) to the ALDM approach. MDEA is performed here for the semidiscretization and for equidistant meshes with $h_j = h$ only. Considering semi-discretizations is consistent with the spatially filtered interpretation of the LES equations, as the time step is sufficiently small for the spatial truncation error to be dominant. Adams et al. (2004) addressed the effects of varying time-step size and time-integration schemes. For implicit time integrations or larger time-step sizes, an extension of the MDEA to full discretizations should be considered. In this case, however, the LES is in effect space and time filtered (Aldama 1990). For all computations, the diffusive terms are discretized by a fourth-order central finite difference. Results of the MDEA are computed with MAPLE.*

The MDEA is applied to obtain the differential equation that has exact solution $\bar{u}_N$. Note, as we pointed out before, that if we use a grid function in a continuous sense, we mean its continuous extension by Whittaker cardinal functions. Thus $\bar{u}_N \in \mathcal{C}^\infty$, and since inversion is possible on the space of grid functions on the underlying mesh, we also have $u_N \in \mathcal{C}^\infty$. As we mentioned before, we do not intend to match expansions of $\mathcal{G}_N$ and $\mathcal{G}_{\text{SGS}}$ term by term. This alleviates us from the problem of assuming that u possesses a Taylor expansion. The central assumption for performing the MDEA is that discrete unfiltered solution u_N in a neighborhood of x_j can be represented by local approximation polynomials of degree K up to $K \le L$,

$$\bar{u}_j^{(\nu)} \doteq \sum_{\mu=\nu}^{L-1} p_j^{(\mu)} \frac{M^{(\mu-\nu)}(x_j)}{(\mu-\nu)!}, \tag{6.36}$$

for $\nu = 0, \ldots, L-1$. Here $\bar{u}_j^{(\nu)}$ and $p_j^{(\nu)}$ stand for order ν derivatives of the approximation polynomials of $\bar{u}$ and u at x_j; $M^{(\mu)}$ is the μth moment of filter kernel G:

$$M^{(\mu)}(x_j) = \int_{x_{j-1/2}}^{x_{j+1/2}} (x-x_j)^\mu G(x-x_j).$$

We can solve Eq. (6.36) for $p_j^{(\nu)}$ so that we obtain p_N in terms of the first $L-1$ derivatives of $\bar{u}_N$ and can insert it as an approximation for u_N into Eq. (6.22). As result of the MDEA, a differential equation for the continuous extension of $\bar{u}_N$ follows. We can identify the implicit SGS model by computing $\mathcal{G}_N$. Note that Eq. (6.36) holds for any graded (i.e., finite-order) filter.

---

* MAPLE 9, Waterloo Maple Inc., Ontario, Canada 2003.

Table 6.1. *Model parameters* $\gamma_{k,r}^{\pm}$ *to match the explicit Smagorinsky model*

Parameter	Value
$\gamma_{1,0}^{+}$	1
$\gamma_{2,0}^{+}$	2/3
$\gamma_{2,1}^{+}$	1/3
$\gamma_{3,0}^{+}$	3/10
$\gamma_{3,1}^{+}$	3/10
$\gamma_{3,2}^{+}$	4/10

On the example of the Smagorinsky model, we demonstrate in this Section how we can match a given explicit SGS model by adjusting K and $\gamma_{k,r}^{\pm}$ of the generic implicit SGS model. The Smagorinsky model formulated for the Burgers equation is

$$\tau_{\text{Smag}} = -C_S h^2 \left| \frac{\partial \bar{u}}{\partial x} \right| \frac{\partial \bar{u}}{\partial x}.$$

The explicit SGS model that is inserted on the right-hand side of the filtered equations is

$$\mathcal{G}_{\text{SGS}} = -\frac{\partial \tau_{\text{Smag}}}{\partial x} = 2C_S h^2 \left| \frac{\partial \bar{u}}{\partial x} \right| \frac{\partial^2 \bar{u}}{\partial x}.$$

With the implicit SGS approach, we can identify model parameters in such a way that the resulting implicit formulation matches with the explicit model for $K = 3$ up to the order of $\mathcal{O}(h^3)$ as given in Table 6.1. Choosing $\sigma_{j+1/2} = 9\,C_S\,|\tilde{u}_{j+1/2}^{-} - \tilde{u}_{j+1/2}^{+}|$ in Eq. (6.28), we see that the truncation error $\mathcal{G}_N$ follows as

$$\mathcal{G}_N = 2\,C_S \left| \frac{\partial \bar{u}}{\partial x} \right| \frac{\partial^2 \bar{u}}{\partial x^2} h^2 - \frac{1}{6}\,C_S \left| \frac{\partial \bar{u}}{\partial x} \right| \frac{\partial^4 \bar{u}}{\partial x^4} h^4 + \mathcal{O}(h^6). \tag{6.37}$$

As argued by Rider and Margolin (2003), the form of the first term in (6.37) is common for the modified equation of many nonlinear finite-volume schemes. For modeling other than reproducing a given explicit SGS model, we need to determine the model parameters on the basis of physical data for a few canonical flow cases. We then determine the quality of the proposed model by the generality of the model parameters. Different options are available. We could determine the model parameters by stochastic estimation such as proposed by Langford and Moser (1999) and applied to deriving coefficients for a finite-volume discretization with linear reconstruction by Zandonade, Langford, and Moser (2004). Another option is to employ a robust automatic optimization scheme as done by Adams et al. (2004) to reproduce DNS results. Adams et al. (2004) demonstrated that the ALDM formulation derived here is sufficiently general to recover excellent predictions for a range of forced and decaying cases of Burgers turbulence, although parameters were derived for a single case only.

6.5.1 Extension to the Navier–Stokes equations

For the extension of the aforementioned ALDM approach to the three-dimensional incompressible Navier–Stokes equations, we consider the filtered and truncated incompressible Navier–Stokes equation

$$\frac{\partial \bar{u}_{N\alpha}}{\partial t} + G_N * \frac{\partial F_{\alpha\beta}(u_N)}{\partial x_\beta} = -G_N * \frac{\partial \tau_{\alpha\beta}^{SGS}}{\partial x_\beta}, \tag{6.38}$$

where

$$\tau_{\alpha\beta}^{SGS} = F_{\alpha\beta}(u) - F_{\alpha\beta}(u_N), \tag{6.39}$$

and its ALDM discretization

$$\frac{\partial \bar{u}_{N\alpha}}{\partial t} + \tilde{G}_N * \frac{\partial \tilde{F}_{\alpha\beta}(\tilde{u}_N)}{\partial x_\beta} = 0. \tag{6.40}$$

We model the filtered divergence of the SGS-stress tensor by the truncation error,

$$\mathcal{G}_{N\alpha} = G_N * \frac{\partial F_{\alpha\beta}(u_N)}{\partial x_\beta} - \tilde{G}_N * \frac{\partial \tilde{F}_{\alpha\beta}(\tilde{u}_N)}{\partial x_\beta}. \tag{6.41}$$

We can make an approximate deconvolution to obtain $\tilde{u}_N$ from $\bar{u}_N$ by applying one-dimensional reconstructions successively in all coordinate directions. Integration over transversal directions is approximated by a Gaussian quadrature rule. The continuity equation can be satisfied by a fractional step-projection approach, where we compute the pressure from a top-hat filtered Poisson equation with the modified convection term as in Eq. (6.40). Details on the extension of ALDM for the three-dimensional Navier–Stokes equations are provided by Hickel, Adams, and Domaradzki (2005). We can take a similar approach to formulate the ALDM for the compressible Navier–Stokes equations, as described by Franz (2004).

The MDEA does not lend itself to a straightforward application to the three-dimensional Navier–Stokes equation. In addition, the numerous modification terms obtained are rather hard to interpret with respect to their physical meaning. Another option is to adjust parameters of the implicit model to make its dissipative properties consistent with the results of analytical theories of turbulence. This is most easily done for homogeneous, isotropic turbulence. Using Fourier transforms, we can write the Navier–Stokes equations for isotropic velocity field, u_n, in spectral form as

$$\left(\frac{\partial}{\partial t} + \nu k^2\right) u_n(\mathbf{k}, t) = -\frac{i}{2} P_{nlm}(\mathbf{k}) \int u_l(\mathbf{p}, t) u_m(\mathbf{k} - \mathbf{p}, t) d^3 p, \tag{6.42}$$

$$i k_n u_n = 0, \tag{6.43}$$

where i is the imaginary unit, $\mathbf{k} = (k_1, k_2, k_3)$ is the wave number, and tensor P_{nlm} is

$$P_{nlm}(\mathbf{k}) = k_m(\delta_{nl} - k_n k_l / k^2) + k_l(\delta_{nm} - k_n k_m / k^2). \tag{6.44}$$

The equation for the energy amplitudes,

$$\frac{1}{2}|u(\mathbf{k})|^2 = \frac{1}{2}u_n(\mathbf{k})u_n^*(\mathbf{k}), \tag{6.45}$$

is obtained from (6.42),

$$\frac{\partial \frac{1}{2}|u(\mathbf{k})|^2}{\partial t} = -2\nu k^2 \frac{1}{2}|u(\mathbf{k})|^2 + T(\mathbf{k}), \tag{6.46}$$

where $T(\mathbf{k})$ is the nonlinear energy transfer,

$$T(\mathbf{k}) = \frac{1}{2}\mathrm{Im}\left[u_n^*(\mathbf{k})P_{nlm}(\mathbf{k}) \int u_l(\mathbf{p})u_m(\mathbf{k} - \mathbf{p}) \right] d^3 p, \tag{6.47}$$

and the asterisk denotes a complex conjugate. The details of the derivation are given, for instance, by Pope (2000) and Lesieur (1997). Quantities $\frac{1}{2}|u(\mathbf{k})|^2$ and $T(\mathbf{k})$ are related to energy $E(k)$ and transfer spectra $T(k)$ of isotropic turbulence by the formulae

$$E(k) = 4\pi k^2 \left\langle \frac{1}{2}|u(\mathbf{k})|^2 \right\rangle, \tag{6.48}$$

$$T(k) = 4\pi k^2 \langle T(\mathbf{k}) \rangle, \tag{6.49}$$

where $\langle \ldots \rangle$ denotes averaging over thin spherical shells of radius $k = |\mathbf{k}|$. With these definitions, we obtain the classical spectral energy equation for isotropic turbulence:

$$\frac{\partial}{\partial t}E(k, t) = -2\nu k^2 E(k, t) + T(k, t). \tag{6.50}$$

The first term on the right-hand side of (6.50) accounts for the effects of the molecular viscosity on the energy decay rate, and the second is responsible for the conservative energy redistribution in wave-number space that is due to nonlinear and pressure effects.

Consider now a numerical simulation performed with a discretized Navier–Stokes equation (6.40) and a time step Δt. After mapping the velocity field from such a simulation into spectral space by using discrete Fourier transforms, we can compute the energy spectrum, viscous dissipation, and the nonlinear transfer terms appearing in (6.50) by using exact formulae (6.45) and (6.48), and (6.47) and (6.49), respectively. In contrast, the energy decay rate computed from such data,

$$\frac{\partial}{\partial t}E(k, t) \approx \frac{E(k, t + \Delta t) - E(k, t - \Delta t)}{2\Delta t}, \tag{6.51}$$

is directly affected by the truncation error (6.41), and the residual

$$\varepsilon_{\mathrm{num}}(k, t) = \frac{E(k, t + \Delta t) - E(k, t - \Delta t)}{2\Delta t} + 2\nu k^2 E(k, t) - T(k, t) \tag{6.52}$$

defines the effective numerical dissipation, attributable entirely to the discretization errors in the Navier–Stokes solver. In practice we compute $\varepsilon_{\mathrm{num}}$ by first storing velocity fields at three subsequent time steps in a simulation, computing energy spectra from

(6.45) and (6.48) and the transfer spectrum from (6.47) and (6.49), and finally employing Eq. (6.52). By analogy with the viscous dissipation

$$\varepsilon_v = 2vk^2 E(k), \tag{6.53}$$

the numerical dissipation (6.52) allows us to define the k-dependent numerical viscosity:

$$v_{num}(k) = \frac{\varepsilon_{num}(k)}{2k^2 E(k)}. \tag{6.54}$$

This approach was first used to quantify dissipative properties of a specific numerical code, MPDATA (Domaradzki, Xiao, and Smolarkiewicz 2003).

Varying ALDM parameters will result in different numerical eddy viscosities. As shown by Hickel et al. (2005), a set of model parameters can be derived that provides an optimum agreement with the spectral eddy viscosities computed from the analytical theories of turbulence.

Analytical theories of turbulence provide expressions for the energy transfer, $T(k | p, q)$ to or from scales with wave number k as a result of interactions with scales of motion corresponding to wave numbers p and q such that all three wave numbers form a closed triangle. The integral of $T(k | p, q)$ over all allowable values of wave numbers p and q for a given k is the classical energy transfer, $T(k)$, given by formula (6.49). We obtain SGS energy transfer $T_{SGS}(k | k_c)$ by integrating $T(k | p, q)$ over wave numbers p and q subject to the condition that at least one of them is greater than k_c, where k_c is the cutoff wave number separating the range of resolved scales, $k < k_c$, from the subgrid scales, $k > k_c$. The theoretical eddy viscosity is defined as

$$v_{th}(k | k_c) = -\frac{T_{SGS}(k | k_c)}{2k^2 E(k)}. \tag{6.55}$$

Kraichnan (1976) and later Chollet and Lesieur (1981) computed such a spectral eddy viscosity by using test field model and eddy-damped quasi-normal Markovian approximation, respectively. A good overview of these approaches can be found in Lesieur (1997).

Frequently, we normalize the eddy viscosity by using values of the energy spectrum at the cutoff

$$v_{th}^+(k | k_c) = \frac{v_{th}(k | k_c)}{\sqrt{E(k_c)/k_c}}. \tag{6.56}$$

Assuming infinite inertial range spectrum $k^{-5/3}$, we can compute integrals leading to $T_{SGS}(k|k_c)$ numerically; the normalized eddy viscosity (6.56) is well fitted by the expression given by Chollet (1984),

$$v_{th}^+(k | k_c) = C_K^{-3/2}(0.441 + 15.2e^{-3.03k_c/k}), \tag{6.57}$$

where C_K is the Kolmogorov constant taken usually as 1.4. This expression has been used by Hickel et al. (2005) as a target to optimize the model parameters in the ALDM approach.

It is often observed that the dissipative properties of discretization schemes proposed for implicit LES depend on the grid resolution (Garnier et al. 1999) and on the time step (Domaradzki and Radhakrishnan 2005). The ALDM, however, employs a numerical flux function that compensates for these resolution effects; see Eqs. (33 and 34) and Figure 6.1 in Hickel et al. (2005). It is therefore possible to optimize parameters of a numerical scheme only once and then apply the same values to varying numerical resolutions and flows. The validation of the model for a range of different cases of decaying and forced isotropic turbulence showed an excellent agreement with direct numerical simulation (DNS) and experimental data (Hickel et al. 2005), without a need to adjust the parameters of the scheme.

6.5.2 Transition in the three-dimensional Taylor–Green vortex

Here, we illustrate the performance of the ALDM on one of the most demanding test cases for SGS models, that is, on laminar–turbulent transition. For the onset of transition, the SGS model must not affect the instability modes of the laminar flow. Most eddy-viscosity models, for instance the Smagorinsky model and the unmodified structure-function model of Métais and Lesieur (1992), do not satisfy this requirement (Lesieur 1995, 1997).

A suitable test scenario for a periodic computational domain is the three-dimensional Taylor–Green vortex (TGV); also see Chapter 4a. This flow is characterized by the initial data

$$\bar{u}(t=0) \approx u(t=0) = \begin{bmatrix} 0 \\ \cos(x)\sin(y)\cos(z) \\ -\sin(x)\cos(y)\sin(z) \end{bmatrix}. \tag{6.58}$$

At $t = 0$, the entire kinetic energy is contained within eight Fourier modes on wavenumber shell $k = \sqrt{3}$. At early times the TGV evolution is laminar and strongly anisotropic. Later, the energy is transferred to larger wave numbers by vortex stretching, and eventually the flow becomes fully turbulent. In the final steps the small scales are nearly isotropic and exhibit a $k^{-5/3}$ inertial range of the kinetic-energy spectrum.

We compare our LES with the DNS of Brachet et al. (1983), which was originally performed on a grid of 256^3 modes in 1983 and repeated with 864^3 modes about 10 years later (Brachet et al. 1992). These spectral simulations exploit the spatial symmetries of the TGV to reduce the effective computational cost by a factor of 8. It was therefore possible to resolve Reynolds numbers up to Re $= 3000$ in Brachet et al. (1983).

For our purposes it is not necessary to exploit spatial symmetries. The computational domain is a periodic $(2\pi)^3$ box, discretized with 64^3 cells, that initially contains eight counterrotating vortices. To assess the quality of the LES using the ALDM, we compare the time evolution of the dissipation rate with the DNS of Brachet et al. (1983); see Figure 6.1. The considered Reynolds numbers range from Re $= 100$ to Re $= 3000$. For an assessement of the ALDM with respect to standard LES, we also show results for the

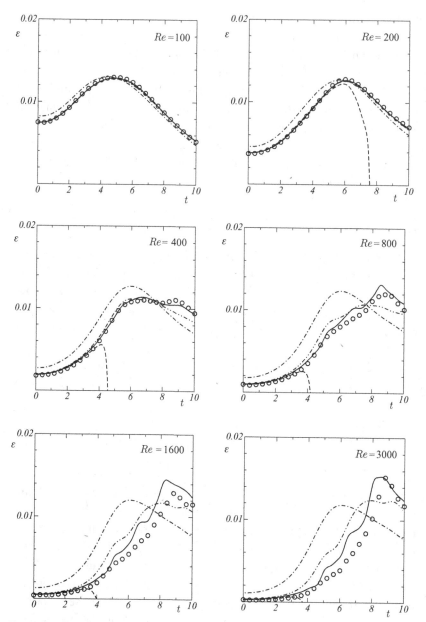

Figure 6.1. Rate of energy dissipation for LES of the TGV: - - - -, without SGS model; ----, Smagorinsky model; ------, dynamic Smagorinsky model; —, ALDM; o, DNS Data from Brachet et al. (1983).

conventional ($C_S = 0.18$) and the dynamic Smagorinsky model at the same resolution. To demonstrate the effect of the SGS models, we also performed a simulation without SGS model. This simulation became unstable as soon as the energy transfer reached the highest resolved wave numbers; see Figure 6.1.

The Smagorinsky model with constant parameter C_S is obviously not well suited for transitional flows. Even in the fully resolved Re $= 100$ case, excessive dissipation

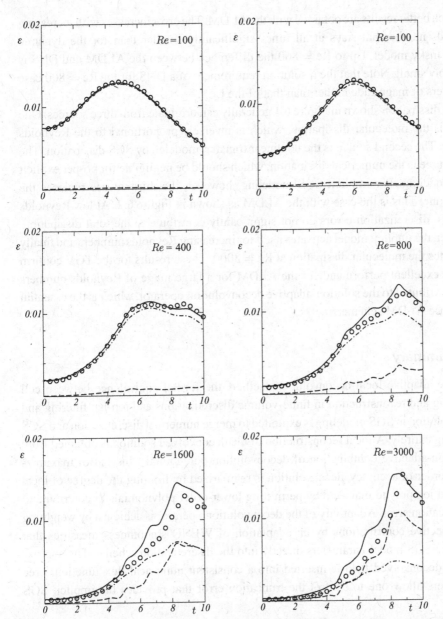

Figure 6.2. Contributions to energy dissipation in the ALDM for LES of the TGV: - - -, molecular dissipation; ·-·-·, implicit SGS dissipation; —, total dissipation $\partial K / \partial t$; o, DNS data from Brachet et al. (1983).

affects the flow evolution. At larger Reynolds numbers the conventional Smagorinsky model gives incorrect dissipation rates and a completely wrong flow structure. If the Smagorinsky parameter is adjusted dynamically, the behavior improves significantly. The SGS viscosity vanishes while the flow is laminar. The prediction is good for the two lowest Reynolds numbers throughout the entire time interval and up to $t = 8$ for Re = 400.

Much better results are obtained with the ALDM. The error increases with increasing Reynolds number but stays at all times significantly smaller than for the dynamic Smagorinsky model. Up to Re = 800 the difference between the ALDM and DNS is negligibly small. Note that the resolution requirement of a DNS for the Re = 800 case is 2 orders of magnitude higher than that of the LES.

The dissipation shown in Figure 6.1 generally can originate from three sources. One source is the molecular dissipation, which is inversely proportional to the Reynolds number. The second source is the nonlinear transfer, modeled by SGS dissipation. The third source is the numerical dissipation, which should be negligible for proper explicit SGS modeling. For implicit SGS modeling, however, the third source replaces the second one. This is the case with the ALDM as shown in Figure 6.2. At low Reynolds numbers, discretization errors do not significantly contribute to the total dissipation. However, the implicit model activates itself for increasing Reynolds numbers and finally dominates the molecular dissipation at Re = 3000. These results for the TGV confirm that the excellent performance of the ALDM for a large range of Reynolds numbers can be attributed to the solution-adaptive deconvolution operator, which is the essential part of the ALDM scheme.

6.6 Summary

With the adaptive local deconvolution method, the formal equivalence between cell averaging and reconstruction in finite-volume discretizations and top-hat filtering and deconvolution in SGS modeling is exploited to merge numerical discretization and SGS modeling entirely. A local reconstruction of the deconvolved solution is obtained from a solution-adaptive combination of deconvolution polynomials. Instead of maximizing the order of accuracy, deconvolution is regularized by limiting the degree of local interpolation polynomials and by permitting lower-order polynomials to contribute to the truncation error. Adaptivity of the deconvolution operator is achieved by weighting the respective contributions by an adaptation of WENO smoothness measures that introduce SGS model parameters directly into the discretization scheme. The approximately deconvolved field is inserted into a consistent numerical flux function. Free parameters allow one to control the truncation error that provides the implicit SGS model.

By an analysis of the modified differential equation, the implicit SGS model can be determined analytically. It is observed that the truncation error of the nonlinear discretization scheme contains functional expressions that are similar to explicit SGS models. By adjusting the truncation error appropriately, one can match the implicit model with an explicit SGS model. This is demonstrated for the one-dimensional Burgers equation and the Smagorinsky SGS model.

For the three-dimensional Navier–Stokes equation, a spectral-space analysis of the modified differential equation is employed to compare the effective spectral numerical viscosity of the ALDM with theoretical predictions for isotropic turbulence. By means of an evolutionary optimization algorithm, a set of parameters is found that gives an excellent match with the spectral eddy viscosity predicted by eddy-damped quasi-normal Markovian theory.

Computational results for different large-scale forced and decaying fully turbulent flow configurations show that the ALDM performs as well as established explicit models (Hickel et al. 2005). The generality of the derived model becomes evident from an application to the three-dimensional TGV. Instability growth and transition to developed turbulence show an excellent agreement with DNS data.

REFERENCES

Adams, N. A., Hickel, S., & Franz, S. 2004 Implicit subgrid-scale modeling by adaptive deconvolution. *J. Comput. Phys.* **200**, 412–431.

Adams, N. A., & Stolz, S. 2002 A deconvolution approach for shock-capturing. *J. Comput. Phys.* **178**, 391–426.

Aldama, A. A. 1990 *Filtering Techniques for Turbulent Flow Simulation*, Vol. 56 of Lecture Notes in Engineering. New York: Springer.

Bardina, J., Ferziger, J. H., & Reynolds, W. C. 1983 *Improved Turbulence Models Based on Large Eddy Simulation of Homogeneous, Incompressible, Turbulent Flows*. Stanford University: Technical Report TF-19.

Brachet, M., Meiron, D., Orszag, S., Nickel, B., Morf, R., & Frisch, U. 1983 Small-scale structure of the Taylor–Green vortex. *J. Fluid Mech.* **130**, 411–452.

Brachet, M., Meneguzzi, M., Vincent, A., Politano, H., & Sulem, P.-L. 1992 Numerical evidence of smooth self-similar dynamics and possibility of subsequent collapse for three-dimensional ideal flow. *Phys. Fluids* **4**, 2845–2854.

Chollet, J. P. 1984 Two-point closures as a subgrid-scale modeling tool for large-eddy simulations, in *Turbulent Shear Flows IV*, ed. F. Durst & B. E. Launder. Heidelberg: Springer, 62–72.

Chollet, J., & Lesieur, M. 1981 Parameterization of small scales of three-dimensional isotropic turbulence utilizing spectral closures. *J. Atmos. Sci.* **38**, 2767.

Clark, R., Ferziger, J., & Reynolds, W. 1979 Evaluation of subgrid-scale models using an accurately simulated turbulent flow. *J. Fluid Mech.* **91**, 1–16.

Domaradzki, J. A., & Adams, N. A. 2002 Modeling subgrid scales of turbulence in large-eddy simulations. *J. Turb.* **3**, 24.

Domardzki, J. A., Liu, W., & Brachet, M. 1993 An analysis of subgrid-scale interactions in numerically simulated isotropic turbulence. *Phys. Fluids A* **5**, 1747–1759.

Domaradzki, J. A., & Radhakrishnan, S. 2005 Effective eddy viscosities in implicit modeling of decaying high Reynolds number turbulence with and without rotation. *Fluid Dyn. Res.* **36**, 385–406.

Domaradzki, J. A., Xiao, Z., & Smolarkiewicz, P. 2003 Effective eddy viscosities in implicit large eddy simulations of turbulent flows. *Phys. Fluids* **15**, 3890–3893.

Dunca, A., & Epshteyn, Y. 2004, On the Adams–Stolz deconvolution LES models. *SIAM J. Math. Anal.* **37**, 1890–1902.

Franz, S. 2004 *Entwicklung eines impliziten LES Modells für kompressible Turbulenz*. Institut für Strömungsmechanik: Technical Report D 1619.

Garnier, E., Mossi, M., Sagaut, P., Comte, P., & Deville, M. 1999 On the use of shock-capturing schemes for large-eddy simulation. *J. Comput. Phys.* **153**, 273–311.

Geurts, B. 1997 Inverse modeling for large-eddy simulation. *Phys. Fluids* **9**, 3585–3587.

Ghosal, S. 1996 An analysis of numerical errors in large-eddy simulations of turbulence. *J. Comput. Phys.* **125**, 187–206.

Harten, A., Engquist, B., Osher, S., & Chakravarthy, S. 1987 Uniformly high order essentially non-oscillatory schemes, III. *J. Comput. Phys.* **71**, 231–275.

Hickel, S., Adams, N. A., & Domaradzki, J. A. 2005 An adaptive local deconvolution method for implicit LES. *J. Comput. Phys.* **213**, 413–436.

Kraichnan, R. 1976 Eddy viscosity in two and three dimensions. *J. Atmos. Sci.* **33**, 1521.

Langford, J. A., & Moser, R. D. 2001 Optimal LES formulations for isotropic turbulence. *J. Fluid Mech.* **398**, 321–346.

Lele, S. K. 1992 Compact finite difference schemes with spectral-like resolution. *J. Comput. Phys.* **103**, 16–42.

Leonard, A. 1974 Energy cascade in large eddy simulations of turbulent fluid flows. *Adv. Geophys.* **18A**, 237–248.

Lesieur, M. 1995 New trends in large-eddy simulations of turbulence. *Annu. Rev. Fluid Mech.* **28**, 45–82.

Lesieur, M. 1997 *Turbulence in Fluids*. Dordrecht: Kluwer Academic, 3rd ed.

LeVeque, R. J. 1992 *Numerical Methods for Conservation Laws*. Basel: Birkhäuser, 2nd ed.

Liu, X.-D., Osher, S., & Chan, T. 1994 Weighted essentially non-oscillatory schemes. *J. Comput. Phys.* **115**, 200–212.

Margolin, L. G., & Rider W. J. 2002 A rationale for implicit turbulence modeling. *Int. J. Numer. Meth. Fluids* **39**, 821–841.

Margolin, L. G., & Rider, W. J. 2005 The design and construction of implicit LES models. *Int. J. Numer. Meth. Fluids* **47**, 1173–1179.

Mellado, J., Sarkar, S., & Patano, C., 2003 Reconstruction subgrid models for nonpremixed combustion. *Phys. Fluids* **15**, 3280–3307.

Meneveau, C., & Katz, J. 2000 Scale-invariance and turbulence models for large-eddy simulation. *Annu. Rev. Fluid. Mech.* **32**, 1–32.

Métais, O., & Lesieur, M. 1992 Spectral large-eddy simulations of isotropic and stably-stratified turbulence. *J. Fluid Mech.* **239**, 157.

Pope, S. 2000 *Turbulent Flows*. New York: Cambridge University Press.

Rider, W., & Margolin, L. 2003 From numerical analysis to implicit turbulence modeling. Paper AIAA 2003-4101.

Sagaut, P. 2005 *Large-Eddy Simulation for Incompressible Flows*. New York: Springer, 3rd ed.

Schumann, U. 1975 Subgrid scale model for finite-difference simulations of turbulence in plance channels and annuli. *J. Comput. Phys.* **18**, 376–404.

Shah, K. B., & Ferziger, J. 1995 A new non-eddy viscosity subgrid-scale model and its application to channel flow, in *CTR Annual Research Briefs 1995*. Stanford, CA: NASA Ames/Stanford University.

Shu, C.-W. 1998 Essentially non-oscillatory and weighted essentially non-oscillatory schemes for hyperbolic conservation laws, in ed. B. Cockburn, C. Johnson, C.-W. Shu, E. Tadmor, & A. Quarteroni. *Advanced Numerical Approximation of Nonlinear Hyperbolic Equations*, Vol. 1697 of Lecture Notes in Mathematics. Berlin: Springer 325–432.

Stolz, S., & Adams, N. A. 1999, An approximate deconvolution procedure for large-eddy simulation. *Phys. Fluids* **11**, 1699–1701.

Stolz, S., Adams, N. A., & Kleiser, L. 2001a An approximate deconvolution model for large-eddy simulation with application to incompressible wall-bounded flows. *Phys. Fluids* **13**, 997–1015.

Stolz, S., Adams, N. A., & Kleiser, L. 2001b The approximate deconvolution model for LES of compressible flows and its application to shock-turbulent-boundary-layer interaction. *Phys. Fluids* **13**, 2985–3001.

Vichnevetsky, R., & Bowles, J. B. 1982 *Fourier Analysis of Numerical Approximations of Hyperbolic Equations*. Philadelphia: SIAM.

Zandonade, P. S., Langford, J. A., & Moser, R. D. 2004 Finite-volume optimal large-eddy simulation of isotropic turbulence. *Phys. Fluids* **16**, 2255–2271.

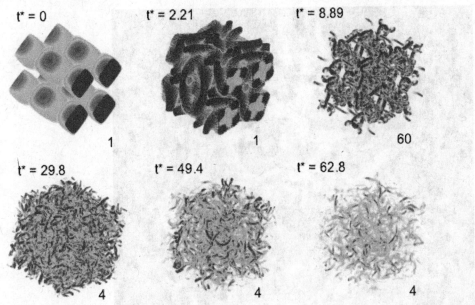

Plate 1. (Fig. 4a.6) Flow visualizations of the TGV flow, using volume renderings of λ_2. The results shown here (representative of all methods discussed) are from the fourth-order 3D monotone FCT on the 128^3 grid.

These plates are available for download in color from www.cambridge.org/9780521172721

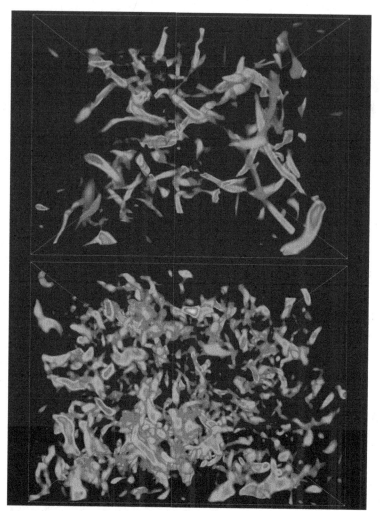

Plate 2. (Fig. 7.1) Two contrasting visualizations of homogeneous, compressible turbulence as computed on a 1024^3 grid with the PPM scheme in 1993. In the upper panel, vorticity structures near the dissipation range of length scales are shown in a small region of the fully developed turbulent flow. In the lower panel, the data have been filtered before the vorticity image was rendered. These vorticity structures are in the Kolmogorov inertial range of length scales. Each region shown is the same width relative to that of the vortex tubes it contains. The two images here were rendered for direct and unbiased comparison. In each subvolume shown, the same volume fraction is occupied by each opacity–color level. The very different appearance of these images therefore reflects real differences in their dynamics. The relatively straight vortex tubes near the dissipation range do not readily kink to form still smaller structures.

Plate 3. (Fig. 7.4) Volume rendering of a thin diagonal slice through the cube of turbulence computed at NCSA in 2003 with our PPM–Euler code on a grid of 2048^3 cells. The logarithm of the vorticity magnitude is shown. The flow is shown at 1.15 sound or flow-crossing times (the initial rms velocity is Mach 1) of the energy-containing scales, which are half the size of the cubical computational domain. At this stage it is clear that small-scale turbulence is developing more rapidly in some regions of this flow than in others, despite the statistical homogeneity imposed on the initial condition for the problem. Analysis of the flow on larger scales in these regions can reveal why the turbulence is developing rapidly there and not elsewhere.

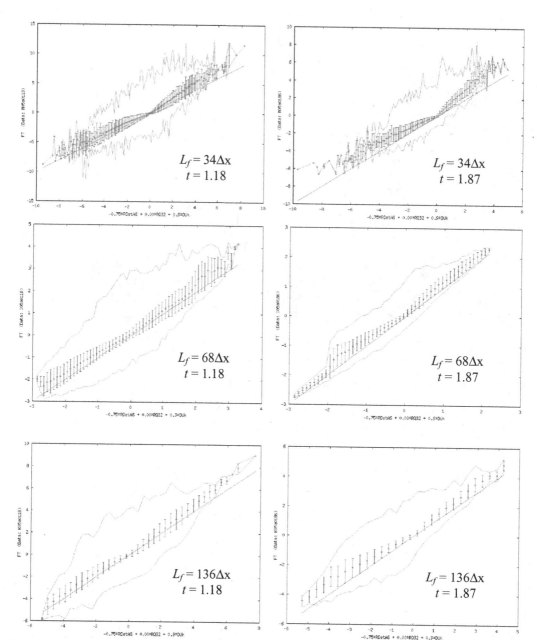

Plate 4. (Fig. 7.5) Data from a PPM simulation of homogeneous, Mach 1, decaying turbulence on a grid of 2048^3 cells (Woodward et al. 2004) was used to evaluate the term $F_{\text{SGS}} = -\tau_{ij}\,\partial_j\,\partial\bar{u}_i$ at two different times and using three different Gaussian filter widths. F_{SGS} is plotted as the ordinate, and the corresponding values of the model equation discussed in the text are plotted as the abcissa. The two quantities are seen to be well correlated for this run.

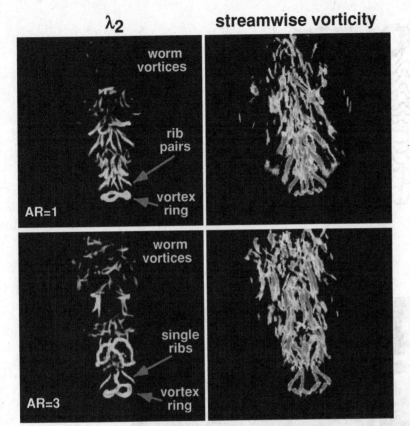

Plate 5. (Fig. 8.5) Comparative instantaneous visualizations based on volume visualizations of λ_2, the second-largest eigenvalue of tensor $\mathbf{S}^2 + \Omega^2$, where $\mathbf{S}$ and Ω are the symmetric and antisymmetric components of the velocity gradient tensor, are shown on the left. Corresponding visualizations of the streamwise vorticity are shown on the right.

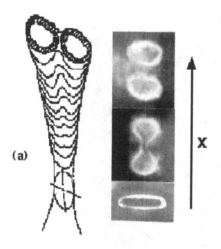

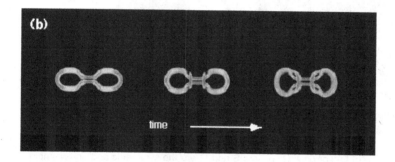

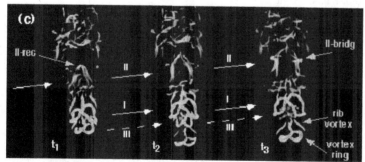

Plate 6. (Fig. 8.6) (a) Characteristic axis-switching and bifurcation phenomena for AR = 4 from visualizations of laboratory elliptical jets subject to strong excitation at the preferred mode (Hussain and Husain 1989); (b) visualizations of an AR = 4 rectangular jet based on ILES data (Grinstein 1995); (c) ILES simulation of an AR = 3 rectangular jet (Grinstein 2001).

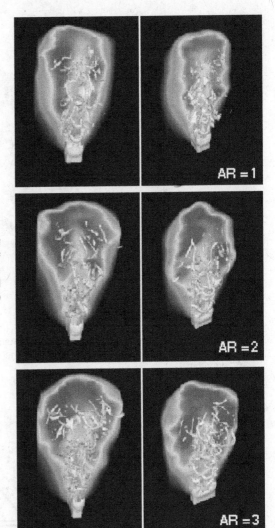

Plate 7. (Fig. 8.9) Instantaneous visualizations of non-premixed-combustion regions as a function of AR at two selected times (Grinstein 2001). Temperature distributions in the back half of the visualized subvolume are superimposed to isosurfaces of the vorticity magnitude (gray).

AR = 1

AR = 2

AR = 3

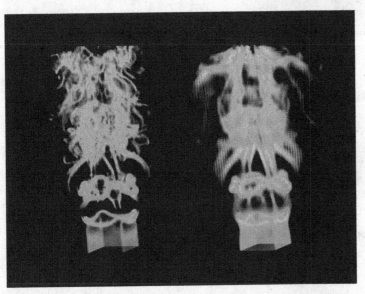

Plate 8. (Fig. 8.10) Comparative instantaneous volume visualization of the vorticity magnitude based on the database of square jets at the same time simulations on the finest grid, left, and coarsest grid, right (from Grinstein 2001).

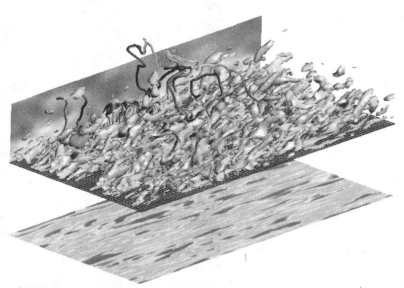

Plate 9. (Fig. 10.1) Fully developed turbulent channel flow: Perspective of a fully developed turbulent channel flow at $Re_\tau = 1800$ from ILES, together with the wall model (ILES + WM) using the FCT algorithm.

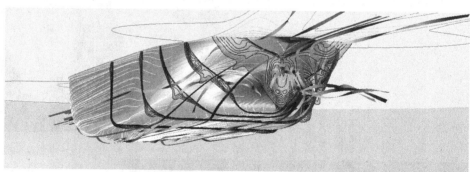

Plate 10. (Fig. 10.14) Ship flow: Perspective view from the stern of the flow past the KVLCC model, showing the time-averaged boundary layer profiles, surface streamlines, and streamlines released at the bow.

Plate 11. (Fig. 14.1) Geophysical turbulence. Scales of motion $\mathcal{O}(10^7)$, $\mathcal{O}(10^4)$, and $\mathcal{O}(10^{-2})$ m, from left to right, respectively.

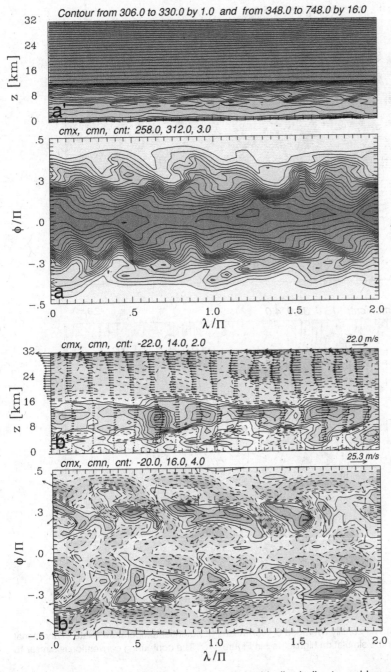

Plate 12. (Fig. 14.7) Instantaneous solutions of the idealized climate problem after 3 years of simulation.

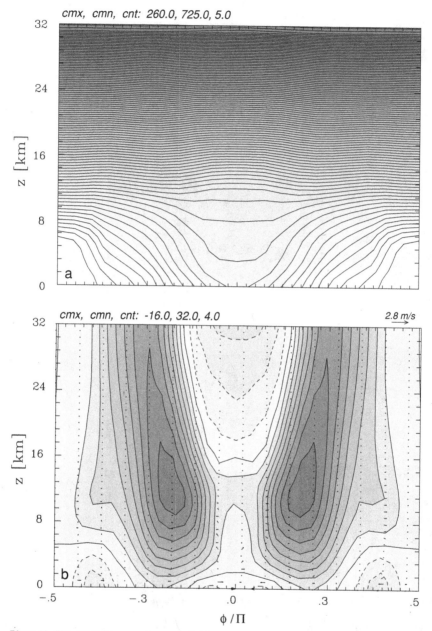

Plate 13. (Fig. 14.8) The zonally averaged 3-year means of potential temperature (plate a) and zonal velocity (plate b) for the simulation highlighted in Figure 14.7. The contouring convention is similar to that used in Figure 14.7.

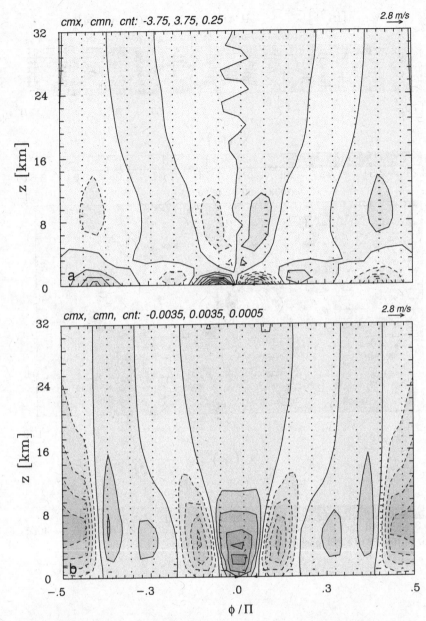

Plate 14. (Fig. 14.9) The zonally averaged 3-year means of meridional and vertical velocities (plates a and b, respectively) for the simulation highlighted in Figure 14.7.

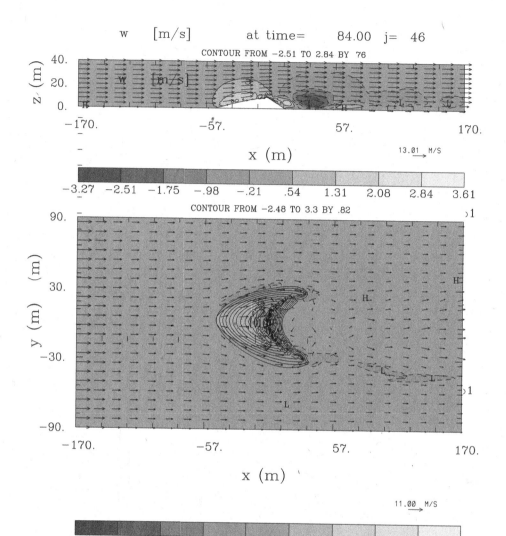

Plate 15. (Fig. 14.10) LES of a PBL past a rapidly evolving sand dune.

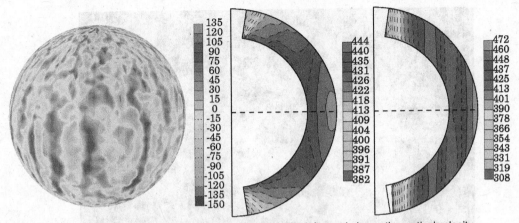

Plate 16. (Fig. 14.12) NFV simulations of solar convection. The left panel shows the vertical velocity (ms⁻¹) on a horizontal surface near the middle of the domain for the ILES run. Central and right panels show the time-averaged angular velocity (nHz) for, respectively, DNS and ILES runs.

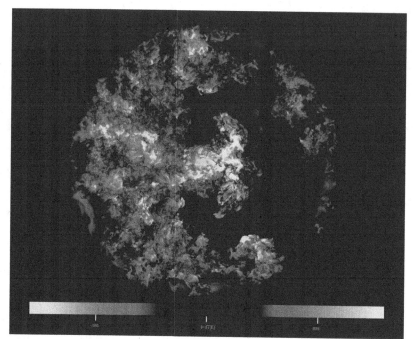

Plate 17. (Fig. 15.14) Temperature fluctuations in a thick section from a simulation of convection in spherical geometry. this section passes through the center of the model and is aligened to cut through the principle down flow plume, which is seen as cool gas (blue and aqua hues) on the left side of the figure.

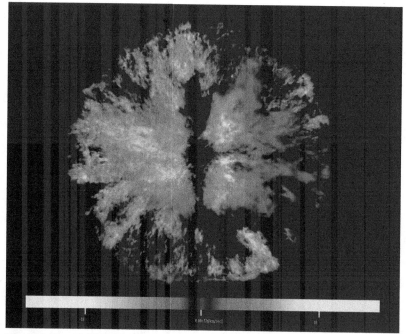

Plate 18. (Fig. 15.16) Radial velocity in a thick section. The predominantly negative (crimson hues) and positive (blue and aqua) radial velocities on the left and right sides of this figure correspond to a dipolar flow.

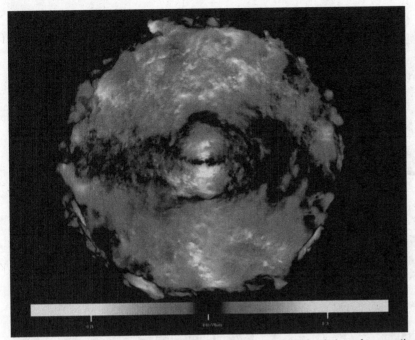

Plate 19. (Fig. 15.17) Angular component of velocity from a simulation of convection in spherical geomety. The view, orientation, and section shown here are the same as those used in figures 15.14 and 15.16. Negative (crimson hues) and positive (aqua hues) values of angular velocity correspond to counterclockwise and clock wise motion about the axis along the line of sight through the center of the model. The predominant pattern of angular motion indicates a dipolar flow filling the interior of the star.

Plate 20. (Fig. 17.3) Contaminant dispersion from an instantaneous release in Times Square, New York City, as predicted by the FAST3D-CT MILES model. The frames show concentrations at 3, 5, 7, and 15 min after release.

SECTION C

VERIFICATION AND VALIDATION

7 Simulating Compressible Turbulent Flow with PPM

David H. Porter and Paul R. Woodward

7.1 Introduction

The use of the piecewise parabolic method (PPM) gas dynamics simulation scheme is described in detail in Chapter 4b and used in Chapter 15 (see also Woodward and Colella 1981, 1984; Collela and Woodward 1984; Woodward 1986, 2005). Here we review applications of PPM to turbulent flow problems. In particular, we focus our attention on simulations of homogeneous, compressible, periodic, decaying turbulence. The motivation for this focus is that if the phenomenon of turbulence is indeed universal, we should find within this single problem a complete variety of particular circumstances. If we choose to ignore any potential dependence on the gas equation of state, choosing to adopt the gamma law with $\gamma = 1.4$ that applies to air, we are then left with a one-parameter family of turbulent flows. This single parameter is the root-mean-square (rms) Mach number of the flow. We note that a decaying turbulent flow that begins at, say, Mach 1 will, as it decays, pass through all Mach numbers between that value and zero. Of course, we will have arbitrary possible entropy variations to deal with, but turbulence itself will tend to mix different entropy values, so that these entropy variations may not prove to be as important as we might think. In all our simulations of such homogeneous turbulence, we begin the simulation with a uniform state of density and sound speed unity and average velocity zero. We perturb this uniform state with randomly selected sinusoidal velocity variations sampled from a distribution peaked on a wavelength equal to half that of our periodic cubical simulation domain. The PPM simulations of this type that are reviewed here have the amplitudes of these velocity variations chosen in order to achieve an initial rms Mach number of either 1/2 or unity.

We exploit the numerical dissipation of PPM, which is strongly targeted at the very smallest scales that can be described on our grid. This type of viscosity allows us to have in our simulations the largest possible dynamic range of scales of motion that is essentially unaffected by viscosity of any kind, either numerical or real. The idea here is that information will pass preferentially down the spectrum from larger to smaller scales of motion. Therefore, as long as we dissipate any small-scale motions into the proper amount of heat at the proper places and the proper times, we should be able to compute the dynamics of the larger scales correctly and accurately. In essence, were this not so, then there would be no hope to simulate the dynamics of a turbulent flow without computing all the motions on all the scales, however small. Nevertheless, we demand a consistency with this assumption from the results of our simulations, and this consistency is, thankfully, observed.

7.2 Large-scale 3D simulations of turbulent flows

Together with our collaborator, Annick Pouquet, we decided in 1991 to exploit the high resolving power of our PPM gas dynamics scheme and the power of the University of Minnesota's Cray-2 supercomputer to make a three-dimensional (3D) simulation of decaying, homogeneous, compressible turbulence (Porter, Pouquet, and Woodward 1991). We initialized the flow on a periodic, cubical domain of 256^3 grid cells so that the rms Mach number of the flow was unity. Because of Kolmogorov's compelling arguments based upon dimensional analysis and self-similarity (Kolmogorov 1941), we expected to see a $k^{-5/3}$ velocity power spectrum develop. Instead, what we observed was a k^{-1} behavior between the energy-containing scales and the scales on which the numerical dissipation of PPM came into play (Porter, Pouquet, and Woodward 1992a, 1992b). The Kolmogorov behavior was nevertheless present as the spectrum decayed, since the spectra at different times, when plotted together on the same graph, traced out a clear envelope with a $k^{-5/3}$ slope. This result was a surprise, since we knew from separate measurements of the effective viscosity of PPM (Edgar and Woodward 1993; Porter and Woodward 1994) that all scales above 12, or at the most 16, grid-cell widths were essentially inviscid. This was our first introduction to a physical effect, which we later clearly showed occurs with the Navier–Stokes viscosity as well, of a pileup of kinetic energy near the end of the turbulent cascade. An argument offered by Jack Herring at the time was that the removal of kinetic energy on the very smallest scales supported by the computational grid inhibited scales just larger than this from transferring their energy through nonlinear interactions with smaller scales. In earlier two-dimensional (2D) simulations of homogeneous, compressible turbulence with PPM (Porter, Pouquet, and Woodward 1992a), we had observed no such effect near the end of the spectrum. The flattening of the near-dissipation-range spectrum occurs only in three dimensions, and only in the solenoidal component of the velocity field.

In 1992, we were able to clarify this process of flattening of the velocity power spectrum just before the dissipation range of scales is reached. Using the University of Minnesota's new Connection Machine at the Army High Performance Computing Research Center, we were able to expand the grid of our earlier 256^3 simulation to 512^3 cells. At last a short segment of the Kolmogorov inertial range emerged (Porter, Pouquet, and Woodward 1994; Woodward 1994; Woodward et al. 1995). We were able to see that, for the PPM dissipation, the Kolmogorov inertial range extends to a shortest disturbance wavelength of about 32 grid-cell widths. This scale marks the end of the indirect effects of PPM dissipation rather than of its direct effects, since the propagation with PPM of a 16-cell sinusoidal velocity disturbance is very accurate over the brief duration of any of these turbulence simulations. From this work we concluded that, to study the statistics of the inertial range of turbulence, we needed a truly enormous grid, since with only 512^3 cells our inertial range was only about a factor of 3 in length.

With a dissipation such as the Navier–Stokes dissipation, which acts significantly on much larger scales than the PPM dissipation, the inertial range on a 512^3 grid would be even smaller. Indeed, several years later we performed a resolution study with both PPM–Euler and PPM–Navier–Stokes methods. This study used resolved Navier–Stokes simulations, for which the velocity power spectrum does not change upon a grid refinement if the coefficient of viscosity is held constant. We discovered that on a 512^3 grid such a Navier–Stokes simulation of decaying, Mach 1/2 turbulence reveals no inertial range at all (Sytine et al. 2000). The flattening of the spectrum before the dissipation range in Navier–Stokes flows was confirmed in 2002 by a Japanese group who used the Earth Simulator to perform Navier–Stokes simulations of decaying homogeneous turbulence on grids up to 4096^3 (Yokokawa et al. 2002).

In 1993 our quest for a machine that could simulate turbulence on a 1024^3 grid led us to collaborate with Silicon Graphics (Silicon Graphics and the University of Minnesota 1993). A cluster of multiprocessor SGI machines was built for us temporarily in the Silicon Graphics manufacturing facility. The combined cluster gave a sustained computational performance of 5 Gflops, which was not a record at the time. However, the key feature of this system, aside from its very low cost (it cost us nothing), was its tremendous memory capacity, which made it the only system in the world at the time that was capable of carrying out a turbulence simulation on a 1024^3 grid. Some results of this simulation are shown in Figure 7.1 (see also Porter, Pouquet, and Woodward 1994; Woodward 1994; Woodward et al. 1995; Porter, Pouquet, and Woodward 1998a, 1998b; Porter et al. 1999). Unfortunately, the SGI personnel had to dismantle this system and distribute its components to customers before we could compute long enough to let the velocity power spectrum completely settle. It was not until 1997, using a new ASCI computer system at Los Alamos, in collaboration with Karl-Heinz Winkler and Steve Hodson, that we were at last able to calculate this flow completely. Using a new Itanium-based cluster system at NCSA in 2001, we were able to do a similar computation, this time at Mach 1, and to bring a complete, detailed set of data over the Internet to our lab in Minnesota (Green 2001; Woodward et al. 2001; Woodward, Porter, and Jacobs 2003). In 2003, we were able, on a new Itanium cluster at NCSA, to simulate Mach 1 homogeneous turbulence with PPM on a grid of 2048^3 cells (Woodward et al. 2004; Porter, Woodward, and Iyer 2005).

The images in Figure 7.1 show the difference in the structures that develop in the near-dissipation range and in the Kolmogorov inertial range. Since the Fourier transform of a line vortex has a k^{-1} spectrum, it is not surprising to find spaghetti-like structures in the near-dissipation range. In order to get a clear view of the structures in the inertial range, we applied a Gaussian filter to the data from this run, thus removing the spaghetti structures and letting the macaroni-like structures of the inertial range come clearly into view. Here we still see vortex tubes, but they are shorter and they appear to kink much more easily.

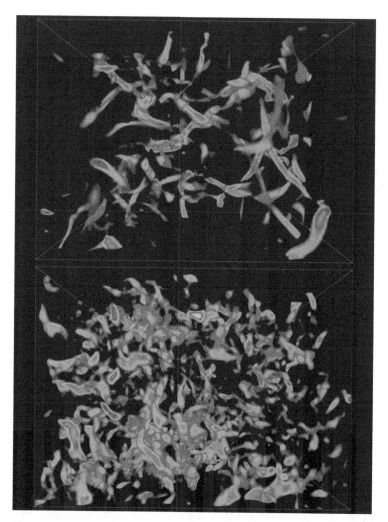

Figure 7.1. Two contrasting visualizations of homogeneous, compressible turbulence as computed on a 1024^3 grid with the PPM scheme in 1993. In the upper panel, vorticity structures near the dissipation range of length scales are shown in a small region of the fully developed turbulent flow. In the lower panel, the data have been filtered before the vorticity image was rendered. These vorticity structures are in the Kolmogorov inertial range of length scales. Each region shown is the same width relative to that of the vortex tubes it contains. The two images here were rendered for direct and unbiased comparison. In each subvolume shown, the same volume fraction is occupied by each opacity–color level. The very different appearance of these images therefore reflects real differences in their dynamics. The relatively straight vortex tubes near the dissipation range do not readily kink to form still smaller structures. (See color inset, Plate 2).

7.3 Purpose of the simulations: Validation and testing of turbulence models

The principal purpose of our turbulence simulations has been to create data sets that can be regarded in much the same way as experimental data, except that the simulations make possible the capture of data that are more complete and more detailed than is now possible from laboratory experiments. From our recent PPM simulation of Mach 1 homogeneous turbulence on a 2048^3 grid, for example, we get values of all fluid state variables sampled at 8.6 billion locations at regular time intervals; see Figure 7.4 and also our description of the data handling in (Porter, Woodward, and Lyer 2005). From this data set we can build grids of macrocells, each a cube of 32, 64, or even 128 grid cells on a side, in which we have computed in detail the subgrid-scale turbulence. From this data set we can correlate the amount of subgrid turbulence and its time rate of change with the local character of the flow on larger scales and compare these results with the predictions or assumptions of turbulence closure models. In order to do this with confidence, we need to establish the range of scales on which the dissipation of the PPM gas dynamics scheme has a significant effect.

If our macrocells are large enough, the effects of this dissipation on the flow on scales resolved by the macrocells should be minimal. On a fine enough grid, we will be able to have the flow on the macrogrid be essentially inviscid, while at the same time the flow we have computed inside each macrocell may still be essentially inviscid on the largest submacrogrid scales. A test of a turbulence model could then be constructed, in which the model is used in a large eddy simulation on the macrogrid, or perhaps on a grid finer than this by a small factor (2 or 4), and the results are compared with the macrogrid data from the 2048^3 PPM run plus the statistical properties of the submacrogrid-scale turbulence directly computed in that run. To be of value, such a large eddy simulation would have to come closer to the data from the 2048^3 run than a simple Euler calculation with a code like PPM using the same, coarser grid as the run with the turbulence model. Such a test remains to be performed, but now we have the data that make such testing possible.

In order to determine the range of scales in our simulations with PPM that are essentially unaffected by the dissipation of this numerical scheme, we have performed both 2D (Porter, Pouquet, and Woodward 1992a) and 3D (Sytine et al. 2000) grid-resolution studies. In these studies we have taken a single initial condition and computed its evolution with both the standard PPM–Euler scheme described in this book and with this scheme to which the Navier–Stokes dissipation terms, for a Prandtl number of unity, have been added. For each numerical scheme, we have progressively refined the grids and computed to the same final time level. For the Navier–Stokes runs, in which there is a coefficient of viscosity to be chosen, we have used the smallest values for which a grid refinement, keeping this coefficient constant, will yield essentially the same velocity power spectrum in the computed result on the finer grid. As the grid is progressively refined, higher and higher Reynolds numbers can be reached.

It is the purpose of an Euler scheme to produce an approximation to the limit as the Reynolds number goes to infinity of well-resolved Navier–Stokes simulations. On any particular grid, one may compare the closeness to this limit solution, to the extent that it can be known, of either the highest-Reynolds-number flow that can be correctly computed from the Navier–Stokes equations on that grid or of the Euler flow that can be computed on that grid. Our convergence studies essentially yield this comparison. If we take either the highest-resolution Euler or Navier–Stokes solution as our best approximation to the desired infinite-Reynolds-number limit, then we need only compare Euler and Navier–Stokes runs at any of the coarser grid levels. By this measure we find that the Euler solutions are better approximations to the high-Reynolds-number limit on a given grid than are the Navier–Stokes solutions. This of course is no surprise, since the Euler scheme is designed to approximate this limit, while the Navier–Stokes solution is not. It is also apparent from these resolution studies that both limit sequences, that of the progressively finer Euler simulations and that of the progressively higher Reynolds-number Navier–Stokes flows, approach the same limit, at least as far as the velocity power spectrum is concerned. The PPM simulation sequence approaches the limit faster, and for any given grid resolution has a larger range of scales in the velocity power spectrum that have actually converged to the limit to within a given tolerance.

The manner in which the velocity power spectra for the sequence of PPM runs in Sytine et al. (2000) converge is shown in Figure 7.2. The top panel of the figure gives the spectra for the solenoidal component of the velocity field, while the bottom panel gives spectra for the compressional component, which contains only about a tenth of the kinetic energy. The Kolmogorov trend is indicated in each panel. The flattening of the power spectrum for the solenoidal component in the near-dissipation regime is clearly seen; such a flattening is not seen in our 2D resolution study (Porter, Pouquet, and Woodward 1992a). As the grid is refined, this feature in the spectrum simply translates to smaller scales, while the portion of the spectrum above this feature at each grid resolution (scales larger than about 32 cell widths) is essentially converged. This same sort of behavior is seen in the compressional spectrum, except that the converged portion of the spectrum extends all the way down to a scale of about 10 to 12 cell widths. The sequence of Navier–Stokes simulations whose spectra are shown in Figure 7.3, also taken from Sytine et al. (2000), clearly show that a much broader portion of the spectrum on any particular grid is affected by the Navier–Stokes viscosity than is affected on that same grid by the PPM numerical viscosity. Of course, were this not so PPM would be a terrible Euler scheme, so this is no surprise. However, the Navier–Stokes spectra for the solenoidal component of the velocity field also show the flattening in the near-dissipation range, so that it is clear that this feature of the turbulent velocity spectrum is physically real and not a numerical artifact. In the PPM simulations, this feature is no doubt different in detail from that of the Navier–Stokes flows; but since it is not a feature of interest for scales of reasonable size in the very-high-Reynolds-number limit we seek to approximate, we do not regard its detailed shape as important. In fact, we can clearly see that were we to continue to refine our grid, this feature would move on down the spectrum. In this respect, it is a numerical error feature, since for the infinite-Reynolds-number limit it occurs at infinite, rather than any finite, wave number. Recent work by Yokokawa et al. (2002) on the Earth Simulator shows this feature as well in Navier–Stokes flows that are unresolved in our sense here – that is, the coefficient of viscosity is too small for the grid resolution used – even though they are computed on grids up to 4096^3.

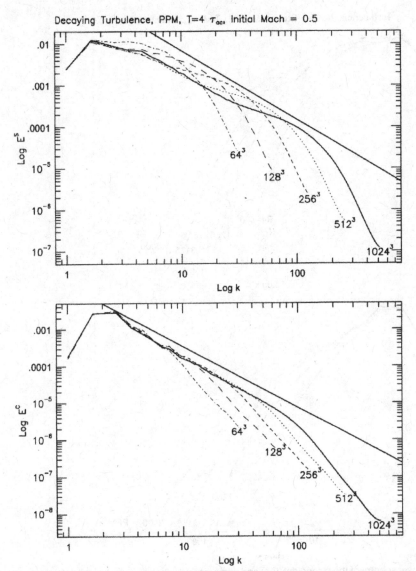

Figure 7.2. Power spectra of the solenoidal (top) and compressional velocity fields in a series of PPM simulations of the same Mach 1/2 (rms) decaying, homogeneous, compressible turbulence problem carried out on grids of 64^3, 128^3, 256^3, 512^3, and 1024^3 cells.

7.4 Potential role of turbulence models

The aforementioned results of our grid-resolution studies show that motions on scales between about 12 and 32 grid-cell widths are being falsified, with respect to the infinite-Reynolds-number limit, by the pileup of energy toward the bottom end of the turbulent cascade. The damping of motions shorter than 12 grid-cell widths is a necessary feature of numerical simulation, at least for compressible flows like these that contain shocks. However, our simulations could perhaps be improved in the range of scales from

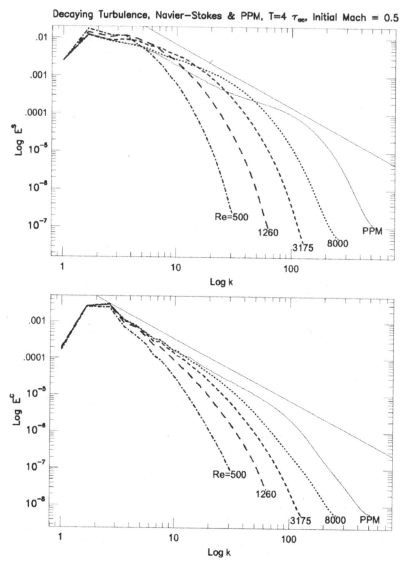

Figure 7.3. Power spectra of the solenoidal (top) and compressional (bottom) velocity fields in a series of PPM–Navier–Stokes simulations of the same Mach 1/2 (rms) decaying, homogeneous, compressible turbulence problem carried out on grids of 64^3, 128^3, 256^3, and 512^3 cells at Reynolds numbers of 500, 1260, 3175, and 8000. A PPM run with 1024^3 cells is shown for comparison.

12 to 32 cell widths. If indeed the reason for the excess kinetic energy in the simulation on these scales is the difficulty of sending this energy further down the spectrum, then perhaps removing the excess on these scales by means of a turbulence model and placing it into a reservoir of subgrid-scale turbulence could improve the ability of the simulated flow to approximate the infinite-Reynolds-number limit. It is not clear that the application of a turbulence model will in fact improve the solution, since it may well matter precisely how the excess kinetic energy is removed and precisely what is

Figure 7.4. Volume rendering of a thin diagonal slice through the cube of turbulence computed at NCSA in 2003 with our PPM–Euler code on a grid of 2048^3 cells. The logarithm of the vorticity magnitude is shown. The flow is shown at 1.15 sound or flow-crossing times (the initial rms velocity is Mach 1) of the energy-containing scales, which are half the size of the cubical computational domain. At this stage it is clear that small-scale turbulence is developing more rapidly in some regions of this flow than in others, despite the statistical homogeneity imposed on the initial condition for the problem. Analysis of the flow on larger scales in these regions can reveal why the turbulence is developing rapidly there and not elsewhere. (See inset, Plate 3).

done with it. For example, a simple dissipation of this excess energy directly into heat may not be helpful, since the proper flow structures on these scales of 12 to 32 cell widths are turbulent vorticity structures as shown in the lower panel of Figure 7.1; however, dissipation of excess kinetic energy into heat may simply leave the spaghetti-like structures of the upper panel of that figure intact, while only weakening them a bit. The convergence studies already discussed indicate that, for a given turbulent flow, if we provide enough grid cells to resolve any desired scale with 32 or more cell widths, then a PPM–Euler simulation should suffice to give a good approximation to the high-Reynolds-number limit on those scales or larger ones. To be useful, augmentation of a scheme like PPM with a turbulence model must do better than this by giving us correct statistical flow behavior on scales smaller than 32 cell widths. That this is possible remains to be established. In short, we know that the velocity power spectrum we obtain in the near-dissipation range is wrong, in the sense that the spectrum at these same wave numbers will change upon a grid refinement, but we do not know that a turbulence model can make the description of the flow in this range right.

Not only can the data from our PPM simulations of turbulence be used to validate proposed turbulence models, the data can also point the way to formulations of such models. A key element of any such model is an equation for the rate of generation of small-scale turbulent kinetic energy in terms of the local larger-scale flow. Our turbulence simulations have such fine grids that they allow us to construct estimates of this energy transfer to small-scale motions and to visualize those estimates. Then we may seek characteristics of the regions where turbulence is growing in strength and to establish that these characteristics are indeed well correlated with growth of turbulence (i.e., they do not tend to occur elsewhere). Following this approach, we first noticed such a correlation in data from a simulation of a Richtmyer–Meshkov instability experiment (Vetter and Sturtevant 1995), which we carried out as members of a large collaboration (Mirin et al. 1999; Cohen et al. 2002); see also Dannevik et al. (1996), Mirin et al. (1997), Schilling et al. (1997), Cohen, Dannevik, et al. (1997), and Cohen, Dimits, et al. (1997) centered on the ASCI turbulence team at Livermore. Analyzing the 3 Terabyte data set from this run, which was carried out on an 8-billion-cell grid using our sPPM code, we noticed that the local topology of the flow was well correlated with the regions from which growing turbulent motions emerged, transported along with the local, large-scale fluid velocity (Cohen et al. 2002; Woodward et al. 2000; Woodward, Porter, and Jacobs 2003). These regions turned out to be those where the flow was compressing in one dimension and expanding in the other two. The time reversal of this sort of flow – compression in two dimensions and expansion in the remaining one – was correlated with the decay of small-scale turbulent motions. The first sort of flow results when you clap your hands. The compression magnifies diffuse shear, and the squirting out in the other two dimensions creates shear in a thin layer even if none was originally present. The fluid instabilities that lead to turbulence are then secondary consequences of this large-scale organization of the flow. In the time-reversed case, which is a flow like the squirting of toothpaste from a tube or like the flow in a tornado, the vorticity becomes organized into tubes that tend to be aligned, so that they naturally merge into larger structures. This process leads to the transport of energy up the spectrum, from small scales to larger ones. After noticing these correlations in the Richtmyer–Meshkov flow, we realized that we had been seeing them for years, without understanding their significance in this regard, in our simulations of compressible convection (see description elsewhere in this volume). In those convection flows we also saw that the intensity of turbulence was correlated with the compression of the gas in all three dimensions, which amplifies the vorticity and takes large-scale motions directly into smaller-scale ones by simple compression. We also saw that expansion of the convection flow in local upwellings is correlated with diminishing turbulent intensity (Woodward, Porter, and Jacobs 2003).

7.5 Correlation of the action of a turbulent cascade with the local flow topology

To see the relation between the rate of generation of subgrid-scale (SGS) kinetic energy, which we write as F_{SGS}, and the topology of the larger-scale flow, we follow the classic Reynolds-averaging approach. We apply a Gaussian filter with a prescribed full width at half-maximum, L_f, to our simulation data to arrive at a set of filtered, or "resolved," variables and a set of "unresolved" state variables that fluctuate rapidly in space. For any particular state variable, Q, we define the filtered (or resolved) value, $\overline{Q}$, by

$$\overline{Q}(x) = \int e^{-[k_f(x-x_1)]^2} Q(x_1) \, d^3x \bigg/ \int e^{-[k_f(x-x_1)]^2} \, d^3x,$$

where the wave number of the filter, k_f, is related to the full width at half-maximum, L_f, by $L_f = 1.6688/k_f$, and where the integral in the denominator is, of course, equal to $(2\pi)^{3/2}/(2k_f)^3$. The mass-weighted, or Favre, average of a state variable, Q, is denoted by $\tilde{Q}$. Manipulating the Euler equations by using these definitions in order to arrive at the time rate of change of the kinetic energy in a frame moving with the filtered flow velocity, we get

$$\frac{\partial k_{\text{SGS}}}{\partial t} + \partial_j(\tilde{u}_j k_{\text{SGS}}) = \overline{\rho} \, \frac{D}{Dt} \left(\frac{k_{\text{SGS}}}{\overline{\rho}} \right)$$

$$= \left(\overline{p \, \partial_i u_i} - \bar{p} \, \partial_i \tilde{u}_i \right) - \tau_{ij} \, \partial_j \tilde{u}_i$$

$$- \partial_j \left(\overline{u_j p} - \tilde{u}_j \bar{p} - \tilde{u}_i \tau_{ij} + \tfrac{1}{2} \overline{\rho \, u_i^2 \, u_j} - \tfrac{1}{2} \overline{\rho \, u_i^2} \, \tilde{u}_j \right).$$

Here k_{SGS} is the SGS kinetic energy, D/Dt denotes the co-moving time derivative, and τ_{ij} is the SGS stress tensor,

$$\tau_{ij} = \overline{\rho \, u_i u_j} - \bar{\rho} \, \tilde{u}_i \tilde{u}_j.$$

Using our simulation data, we can establish the relative importance of the various terms grouped on the right in this expression for the time rate of increase of SGS kinetic energy per unit mass in the co-moving frame. A preliminary analysis indicates that, statistically, the divergence terms tend to average out and that the first terms in parentheses on the right, the $p \, DV$ work terms, also tend to have little effect on the average. However, term $-\tau_{ij} \, \partial_j \tilde{u}_i$ has systematic behavior that tends to make it dominant over space and over time. We refer to this term as the *forward energy transfer* to the SGS, or F_{SGS}. By analysis of several detailed data sets from PPM simulations of turbulent flows, we have correlated this F_{SGS} term to the topology of the filtered flow field, expressed in terms of the determinant of the deviatoric symmetric rate of strain tensor, given by

$$(S_D)_{ij} = \frac{1}{2} \left(\frac{\partial \tilde{u}_i}{\partial x_j} + \frac{\partial \tilde{u}_j}{\partial x_i} - \frac{2}{3} \delta_{ij} \nabla \cdot \tilde{u} \right).$$

There is, of course, also a correlation with the divergence of this velocity field, so that a model for this forward energy transfer term results in the following form:

$$FT_{\text{MODEL}} = A \, L_f^2 \overline{\rho} \, \det(S_D) + C \, k_{\text{SGS}} \, \nabla \cdot \tilde{u}.$$

This model equation is intended for use in a large eddy simulation in which the SGS kinetic energy, k_{SGS}, is carried as an additional independent variable, so that it is available

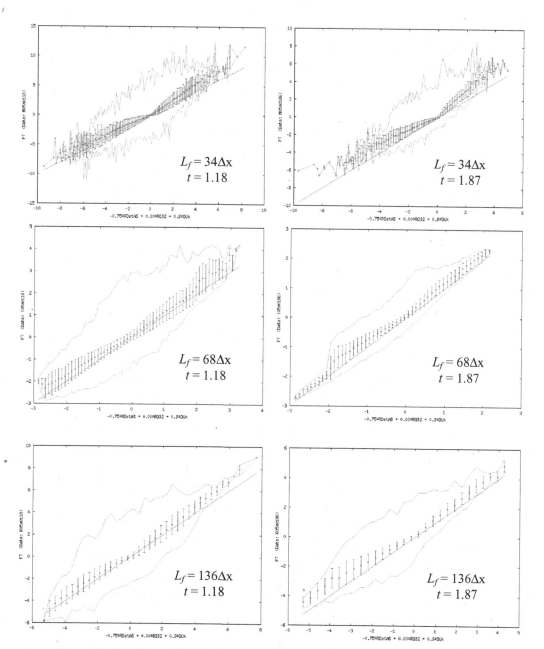

Figure 7.5. Data from a PPM simulation of homogeneous, Mach 1, decaying turbulence on a grid of 2048^3 cells (Woodward et al. 2004) was used to evaluate the term $F_{\text{SGS}} = -\tau_{ij} \ \partial_j \ \partial \bar{u}_i$ at two different times and using three different Gaussian filter widths. F_{SGS} is plotted as the ordinate, and the corresponding values of the model equation discussed in the text are plotted as the abcissa. The two quantities are seen to be well correlated for this run. (See color inset, Plate 4).

for use in the second, compressional term here. We find that the best fits for coefficients A and C in the model are

$$A = -0.75, \quad C = -0.90.$$

The best-fit coefficient for a term in the norm of the rate of strain tensor is zero.

The results of the aforementioned set of coefficients are shown in Figure 7.5 for data from our 2048^3 PPM simulation at two times during the run and for three choices of the filter width (the model values are along the x axes and the actual ones along the y axes). The fits are very good in all cases, so that we are encouraged to construct a SGS turbulence model by using this model equation. Whether such a model can deliver improved simulation capabilities for our PPM code remains to be established.

7.6 Visual evidence for the correlation of F_{SGS} with det(S)

The aforementioned correlation diagrams provide quantitative confirmation of ideas that occurred to us from visual explorations of many turbulent flows. Their importance depends upon our association of the term $F_{SGS} = -\tau_{ij}\,\partial_j\tilde{u}_i$ with a useful approximation to the actual transfer of turbulent kinetic energy from the larger to the smaller scales in the flow. We therefore present in Figure 7.6 a few visual representations of fairly thin slices through a selected region of our PPM simulation of decaying Mach 1 turbulence on a 2048^3 grid. These provide qualitative, visual support for our identification of $F_{SGS} = -\tau_{ij}\,\partial_j\tilde{u}_i$ as a measure of the rate of turbulent kinetic energy transfer, and they also support the correlation of F_{SGS} with the determinant, $\det(\tilde{S})$, of the local rate of strain tensor for the filtered velocity field. Here we do not use the deviatoric tensor, since that becomes large in shocks. In arriving at the quantitative correlations presented earlier, we sensed strong shocks in the flow and rejected results generated inside shock structures. The results presented here have been generated from data that were blended over bricks of 4^3 cells. As a result of an inadvertant erasure of the data from this large run by the San Diego Supercomputer Center in late 2003, a handful of such blended data dumps is all the quantitative data that now remain (140 detailed dumps of the vorticity distributions, sent to our lab during the run, also remain), and we are therefore forced to use this reduced resolution data here.

At each of five times – 0.30, 0.51, 0.71, 0.90, and 1.10 – we have plotted the magnitude of the vorticity, $|\nabla \times \vec{u}|$, generated from the averaged velocities in the bricks of 4^3 cells, in the upper-left quadrant of our image panel. This quantity allows easy visual recognition of regions where small-scale turbulence is developing. In the upper-right quadrant of each image panel, we show a different measure of small-scale turbulent kinetic energy, namely

$$\text{PKE} = \rho\left(u_x'^2 + u_y'^2 + u_z'^2\right)/2,$$

where the primes denote the difference between velocity component u_i averaged over a brick of 4^3 cells and $\tilde{u}_i$, the average of this velocity component over a brick of 64^3 cells centered on the 4^3 brick. This perturbed kinetic energy, PKE, is constructed so that smooth variations of $\tilde{u}_i$ do not contribute. In the lower two quadrants of each image panel, we show $\det(\tilde{S})$ at the left and $\nabla \cdot \vec{u}$ at the right. Here the divergence is constructed from the velocities, $u_i(\ )$, averaged over the 4^3-cell bricks, so that it gives a good representation of the shocks in this flow. The rate of strain tensor, $\tilde{S}$, defined in

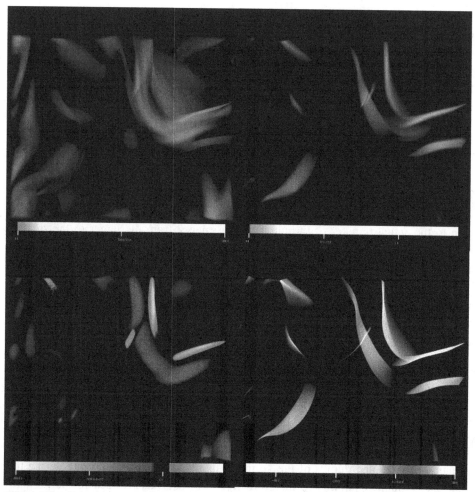

Figure 7.6a. Data from a PPM simulation of homogeneous, Mach 1, decaying turbulence on a grid of 2048^3 cells (Woodward et al. 2004) blended over bricks of 4^3 (velocity components u_i) and 64^3 cells (velocity components $\bar{u}_i$), was used to create visualizations of $|\nabla \times \bar{u}|$ (upper left), $\rho[(u_x - \bar{u}_x)^2 + (u_y - \bar{u}_y)^2 + (u_z - \bar{u}_z)^2]/2$ (upper right), det($\tilde{S}$) (lower left), and $\nabla \cdot \bar{u}$ (lower right) in a thin slice of the flow at time 0.30 (see text).

the following equation, is constructed from the velocities, $\tilde{u}_i$, averaged over 512^3 bricks of 64^3 cells each (i.e., we use a moving average on 64^3 cells):

$$\tilde{S}_{ij} = \left(\partial_i \tilde{u}_j + \partial_j \tilde{u}_i\right)/2.$$

This plot of det($\tilde{S}$) gives us a good idea of where a turbulence model might suggest that energy is being transferred to motions on the smallest scales. If such a suggestion were to be correct, we would expect to notice this turbulent energy appearing at a later time in the plots of PKE at the upper right in each panel. We would also expect to see the tangles of small vortex tubes characteristic of fully developed turbulence appearing in the corresponding plots of the magnitude of the vorticity at the upper left.

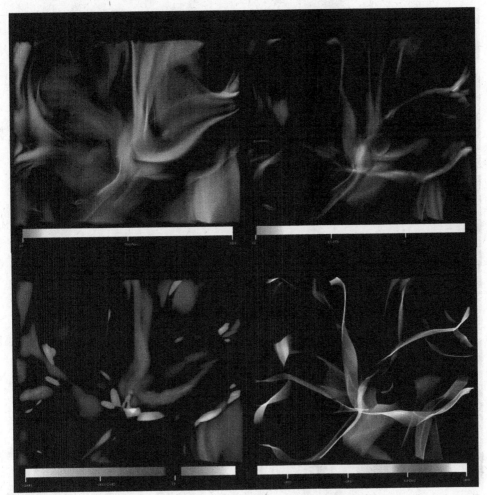

Figure 7.6b. Same as Figure 7.6a, but at time 0.51 (see discussion in the text).

Color versions of these figures can be found on the LCSE Web site, but even the black-and-white versions here are fairly easily interpreted, once we realize that the positive values of the determinant $\det(\tilde{S})$ that are rendered as blue in the color plots show up as the whitest features in black and white. Since the locations of shocks are determined during the standard PPM gas dynamics computation, it is an easy matter to set $\det(\tilde{S})$ to zero in those locations in an implementation of a turbulence model while keeping the term in FT_{MODEL} involving the divergence of the filtered velocity active. Recognizing that, to first order, the vorticity tends to be advected with the large-scale flow, we can see from the upper-left quadrants of Figure 7.6 that the flow in this slice of the volume involves a stream coming from the upper right and eventually traveling through to the lower left. At time 0.30, in Figure 7.6a, we see that our measure, PKE, of the turbulent kinetic energy is picking up false signals in strong shock fronts, which show up clearly in the plot of $\nabla \cdot \vec{u}$. At this early time, there is no small-scale turbulence in evidence in the vorticity plot, but we see that $\det(\tilde{S})$ is large and negative in the region

Figure 7.6c. Same as Figure 7.6a, but at time 0.71 (see discussion in the text).

between the two roughly parallel shock fronts, where the gas is being squeezed in the direction of shock propagation but where the flow is roughly divergence free. In this same region, the shear causes a low-level, smooth feature to appear in the vorticity plot. This is not turbulence but a precursor to turbulence that is being signaled by large negative values of $\det(\tilde{S})$. The picture at time 0.51 is much more confused. Several strong shocks are evident. While the shearing region between the shocks at time 0.30 has now traveled into the middle of the picture, and its vorticity has intensified, there is still no small-scale turbulence. The PKE plot again mainly picks up false signals from the shocks, but the strong feature in $\det(\tilde{S})$ extending from the top middle to the center of the plot and then off to the right middle is located well behind the shocks. This feature, marking gas in which conditions are ripe for the development of small-scale turbulence, is reflected in the vorticity plot, where a strong region of intensifying shear can clearly be discerned. By time 0.71, shock features no longer dominate the PKE plot, because small-scale turbulent features are at last beginning to emerge. The slip

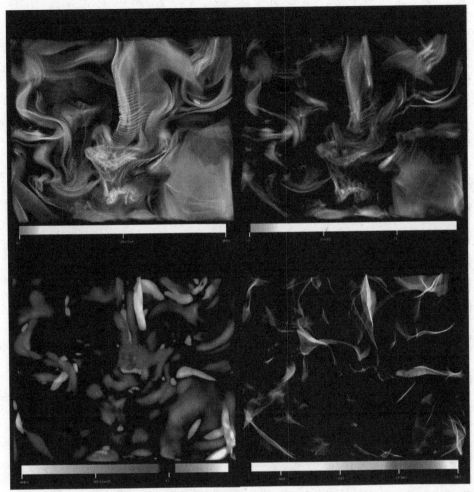

Figure 7.6d. Same as Figure 7.6a, but at time 0.90 (see discussion in the text).

surface that is being driven downward and to the left by the stream of gas from the upper right is now very thin and is beginning to display a ribbed appearance from developing small-scale folds. These give rise to the strongest features in the vorticity plot, the PKE plot, and in the plot of $\det(\tilde{S})$, but there are no corresponding features at all in the plot of $\nabla \cdot \vec{u}$, so we can be assured that we are not observing effects caused directly by the compressibility of the gas.

From time 0.71 onward, the large-scale flow nearly stagnates in the region of the strongest vorticity features, near the head of the stream plunging from the upper right toward the lower left. Clear evidence in the vorticity plot of small-scale turbulence is seen along the edges of this plunging flow, and corresponding features unrelated to shocks stand out in the PKE plots. The plots of $\det(\tilde{S})$ indicate energy transfer to turbulence in these regions, but they also identify regions, less noticeable in the vorticity plots, where small-scale turbulence is destined later to appear. By time 1.10, several regions of positive $\det(\tilde{S})$ have appeared. These are shown in blue in the color versions

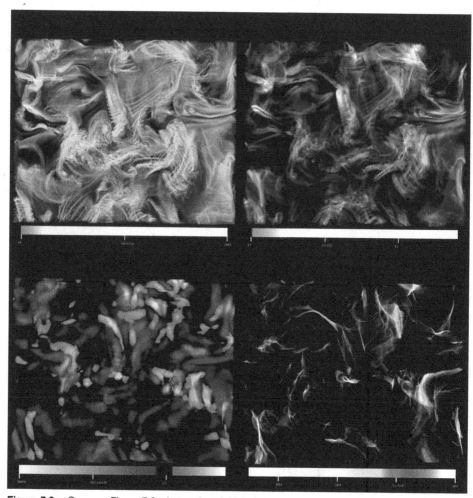

Figure 7.6e. Same as Figure 7.6a, but at time 1.10 (see discussion in the text).

on the LCSE Web site, and they show up here as the whitest regions. Unlike the situation at earlier times, many of these regions are not correlated with shock fronts. It is natural to argue that the strength of the local shear in the filtered velocity field should be a good indicator of energy transfer to small-scale turbulence, and there is support for this view in the plots of Figure 7.6. However, such arguments cannot locate regions where the energy transfer runs in the opposite direction, namely from small-scale to larger-scale motions. This, we believe, is a major advantage of $\det(\tilde{S})$ as an indicator of turbulent energy tranfer.

7.7 Summary

Together with many collaborators, we have used the high resolving power and low numerical viscosity of the PPM gas dynamics scheme to simulate homogeneous, compressible, decaying turbulence in great detail. We have shown that such Euler simulations converge more rapidly to the high-Reynolds number limit of viscous flows than do simulations based upon the Navier–Stokes equations. The simulations can provide high-quality data for use in understanding turbulence and in guiding the development of statistical models of turbulence. In this respect, such simulations can play a role similar to experiments; but, unlike experiments, they can produce hundreds of snapshots of the flow, with all the state variables sampled at billions of locations in each one. Processing of these rich data sets is producing insights useful in the design of turbulence models.

REFERENCES

Cohen, R. H., Dannevik, W. P., Dimits, A. M., Eliason, D. E., Mirin, A. A., Porter, D. H., Schilling, O., & Woodward, P. R. 1997 Three-dimensional high-resolution simulations of the Richtmyer–Meshkov mixing and shock-turbulence interaction, in Proceedings of the 6th International Workshop on the Physics of Compressible Turbulent Mixing, June 18–21, Marseille, France; LLNL Report UCRL-JC-125309.

Cohen, R. H., Dannevik, W. P., Dimits, A. M., Eliason, D. E., Mirin, A. A., Zhou, Y., Porter, D. H. & Woodward, P. R. 2000 Three-dimensional simulation of a Richtmyer–Meshkov instability with a two-scale initial perturbation. Phys. Fluids 14, 3692–3709; also available as LLNL Preprint UCRL-JC-144836.

Cohen, R. H., Dimits, A. M., Mirin, A. A., Dannevik, W. P., Eastman, R. G., Eliason, D. E., Schilling, O., Porter, D. H., & Woodward, P. R. 1997 Three-dimensional PPM simulations of re-shocked Richtmyer–Meshkov instability. Lawrence Livermore National Laboratory: LLNL Report UCRL-MI-128783; VHS video.

Colella, P., & Woodward, P. R. 1984 The Piecewise-Parabolic Method (PPM) for gas dynamical simulations, J. Comput. Phys. 54, 174–201.

Dannevik, W. P., Dimits, A. M., Cohen, R. H., Eliason, D., Schilling, O., Porter, D. H., & Woodward, P. R. 1996 Three-Dimensional Compressible Rayleigh–Taylor Simulation on the ASCI Blue-Pacific ID System. Lawrence Livermore National Laboratory: LLNL Report UCRL-MI-125449; VHS video.

Edgar, B. K., & Woodward, P. R. 1993 Diffraction of a shock wave by a wedge: Comparison of PPM simulations with experiment. Paper AIAA 91-0696. Video J. Eng. Res. 3, 25–33.

Green, K. 2001 Going with the flow. NCSA Access Magazine, Fall; available at www.ncsa.uiuc.edu/News/Access/Stories/itaniumflow.

Kolmogorov, A. 1941 The local structure of turbulence in incompressible viscous fluid for very large Reynolds numbers. Dokl. Akad. Nauk SSSR 30, 301. [reprinted in Proc. R. Soc. London Ser. A 434, 9 (1991)]

LCSE movies from a variety of turbulent fluid flow simulations can be downloaded and/or viewed at www.lcse.umn.edu/MOVIES.

Mirin, A. A., Cohen, R. H., Curtis, B. C., Dannevik, W. P., Dimits, A. M., Duchaineau, M. A., Eliason, D. E., Schikore, D. R., Anderson, S. E., Porter, D. H., Woodward, P. R., Shieh, L. J., & White, S. W., 1999 Very High Resolution Simulation of Compressible Turbulence on the IBM-SP System. Lawrence Livermore National Laboratory: Report UCRL-MI-134237.

Mirin, A. A., Cohen, R. H., Dannevik, W. P., Dimits, A. M., Eliason, D. E., Porter, D. H., Schilling, O., & Woodward, P. R. 1997 Three-Dimensional Simulations of Compressible Turbulence on High-Performance Computing Systems. Lawrence Livermore National Laboratory: LLNL Report UCRL-JC-125949.

Porter, D. H., Pouquet, A., Sytine, I., & Woodward, P. R. 1999 Turbulence in compressible flows. Physica A 263, 263–270.

Porter, D. H., Pouquet, A., & Woodward, P. R. 1991 Supersonic homogeneous turbulence. Lecture Notes Phys. 392, 105–125.

Porter, D. H., Pouquet, A., & Woodward, P. R. 1992a A numerical study of supersonic turbulence. Theor. Comput. Fluid Dyn. 4, 13–49.

Porter, D. H., Pouquet, A., & Woodward, P. R. 1992b Three-dimensional supersonic homogeneous turbulence: A numerical study. Phys. Rev. Lett. 68, 3156–3159.

Porter, D. H., Pouquet, A., & Woodward, P. R. 1994 Kolmogorov-like spectra in decaying three-dimensional supersonic flows. *Phys. Fluids A* **6**, 2133–2142.

Porter, D. H., Pouquet, A., & Woodward, P. R. 1995 Compressible flows and vortex stretching, in *Small-Scale Structures in Three-Dimensional Hydrodynamic and MHD Turbulence*, ed. A., Pouquet, M. Meniguzi, & P-L. Sulem. New York: Springer-Verlag, 51–58.

Porter, D. H., Pouquet, A., & Woodward, P. R. 1998 Intermittency in compressible flows, in Advances in Turbulence VII, pp. 255–258, Fluid Mechanics and Its Application, U. Frisch ed., Kluwer: Dordrecht.

Porter, D. H., Pouquet, A., & Woodward, P. R. 2002 Measures of intermittency in driven supersonic flows. *Phys. Rev. E* **66**, 026301.

Porter, D. H., & Woodward, P. R. 1994 High resolution simulations of compressible convection with the piecewise-parabolic method (PPM). *Astrophys. J. Suppl.* **93**, 309–349.

Porter, D. H., Woodward, P. R. & Iyer, A. 2005 Initial experiences with grid-based volume visualization of fluid flow simulations on PC clusters. *Proc. Visualization and Data Analysis*, accepted for publication.

Porter, D. H., Woodward, P. R., & Pouquet, A. 1998 Inertial range structures in decaying turbulent flows. *Phys. of Fluids* **10**, 237–245.

Schilling, O., Cohen, R. H., Dannevik, W. P., Dimits, A. M., Eliason, D. E., Mirin, A. A., Porter, D. H., & Woodward, P. R. 1997 *Three-Dimensional High-Resolution Simulations of Compressible Rayleigh–Taylor Instability and Turbulent Mixing*. Lawrence Livermore National Laboratory: LLNL Report UCRL-JC-125308.

Silicon Graphics & the University of Minnesota. 1993 *Grand Challenge Computing, A No-Frills Technical Documentary*. Video documentary.

Sytine, I. V., Porter, D. H., Woodward, P. R., Hodson, S. W., & Winkler K.-H. 2000 Convergence tests for piecewise parabolic method and Navier–Stokes solutions for homogeneous compressible turbulence. *J. Comput. Phys.* **158**, 225–238.

Vetter, M., & Sturtevant, B. 1995 Experiments on the Richtmyer–Meshkov instability of an air SF6 interface. *Shock Waves* **4**, 247–252.

Woodward, P. R. 1986 Numerical methods for astrophysicists, in *Astrophysical Radiation Hydrodynamics*, ed. Winkler, K.-H. & M. L., Norman. Dordrecht: Reidel, 245–326.

Woodward, P. R. 1994 Superfine grids for turbulent flows. *IEEE Comput. Sci. Eng.* **1**, 4–5.

Woodward, P. R. 2005 A complete description of the PPM compressible gas dynamics scheme. LCSE internal report available from the main LCSE page at www.lcse.umn.edu.

Woodward, P. R., & Colella, P. 1981 High-resolution difference schemes for compressible gas dynamics, *Lecture Notes Phys.* **141**, 434.

Woodward, P. R., & Colella, P. 1984 The numerical simulation of two-dimensional fluid flow with strong shocks. *J. Comput. Phys.* **54**, 115–173.

Woodward, P. R., Porter, D. H., Edgar, B. K., Anderson, S. E., & Bassett, G., 1995 Parallel computation of turbulent fluid flow. *Comput. Appl. Math* **14**, 97–105.

Woodward, P. R., Porter, D. H., Sytine, I., Anderson, S. E., Mirin, A. A., Curtis, B. C., Cohen, R. H., Dannevik, W. P., Dimits, A. M., Eliason, D. E., Winkler, K.-H., & Hodson, S. W. 2001 Very high resolution simulations of compressible, turbulent flows, in *Computational Fluid Dynamics*, ed. E. Ramos, G. Cisneros, R. Fernández-Flores, & A. Santillan-González. A Singapore: World Scientific; available at www.lcse.umn.edu/mexico.

Woodward, P. R., Porter, D. H., & Jacobs, M. 2003 3-D simulations of turbulent, compressible stellar convection, in *3D Stellar Evolution*, ed. S. Turcotte, S. C. Keller, & R. M. Cavallo. Vol. **293** of the *ASP Conference Series*, 45–63; available at www.lcse.umn.edu/3Dstars.

Woodward, P. R., Anderson, S. E., Porter, D. H., & Iyer, A. 2004 *Cluster Computing in the SHMOD Framework on the NSF TeraGrid*. LCSE internal report, available at www.lcse.umn.edu/turb2048.

Yokokawa, M., Itakura, K., Uno, A., Ishihara, T., & Kaneda, Y. 2002 16.4 Tflops direct numerical simulation of turbulence by a Fourier spectral method on the earth simulator, presented at Supercomputing 2002, Novembar, Baltimore, MD.

8 Vortex Dynamics and Transition to Turbulence in Free Shear Flows

Fernando F. Grinstein

8.1 Introduction

Shear flows driven by Kelvin–Helmholtz instabilities such as mixing layers, wakes, and jets are of great interest because of their crucial roles in many practical applications. The simulation of shear flows is based on the numerical solution of the Navier–Stokes (NS) or Euler (EU) equations with appropriate boundary conditions. The important simulation issues that have to be addressed relate to the appropriate modeling of (1) the required open boundary conditions for flows developing in both space and time in finite-size computational domains, and (2) the unresolved subgrid-scale (SGS) flow features.

Appropriate boundary condition modeling is required because, in studying spatially developing flows, we can investigate only a portion of the flow – as in the laboratory experiments, where finite dimensions of the facilities are also unavoidable. We must ensure that the presence of artificial boundaries adequately bounds the computational domain without *polluting* the solution in a significant way: numerical boundary condition models must be consistent numerically and with the physical flow conditions to ensure well-posed solutions, and emulate the effects of virtually assumed flow events occurring outside of the computational domain. SGS models are needed that ensure the accurate computation of the inherently three-dimensional (3D) time-dependent details of the largest (grid-scale) resolved motions responsible for the primary jet transport and entrainment. At the high Reynolds number of practical interest, direct numerical simulation (DNS) cannot be used to resolve all scales of motion, and some SGS modeling becomes unavoidable to provide a mechanism by which dissipation of kinetic energy accumulated at high wave numbers can occur.

In what follows I first motivate the implicit large eddy simulation (ILES) approach to simulate free shear flows, and then I highlight the basic components involved in a typical simulation model. Next, I use examples from the ILES jet studies to address major aspects of the transition to turbulence from laminar conditions in free shear flows, namely, the occurrence of global instabilities, the complex dynamics of 3D vortical geometries, and their impact on jet entrainment and combustion. I then make quantitative statements on the properties of the small-scale modes in the turbulent portions to which the simulated jet has transitioned from the initial laminar conditions at the nozzle exit; I address convergence issues of ILES in this context. Finally, I discuss aspects of sensitivity of ILES to the implicit SGS model specifics, and I note unresolved issues.

8.2 ILES approach

8.2.1 Motivation

Experimental investigations have shown that large-scale vortical coherent structures (CS) dominate the entrainment and mixing processes in free shear flows (Brown and Roshko 1974; Hussain 1986). By design, large eddy simulation (LES) approaches focus on the modeling of the large-scale-driven flow features, and thus they are ideally suited to model these CS. Conventional LES involves low-pass filtering of the NS equations followed by explicit SGS modeling (Sagaut 2005; also see Chapter 3, this volume). LES seeks to resolve most of the entrainment-dominating CS by choosing the cutoff wavelength within the inertial subrange. The main challenge is to appropriately emulate the fluid dynamics near the cutoff, to ensure that proper interactions between resolved and SGS scales are represented.

The existence of intermittent CS, implied by early laboratory studies of turbulent flows (Kuo and Corrsin 1972) and numerical simulations (Siggia and Patterson 1988), indicate that even isotropic homogeneous flows are not completely random at the smallest scale. The vorticity field tends to be highly organized, and intense vorticity concentrates in elongated filaments. These are the so-called worm vortices, characterizing the smallest CS of turbulent flows, with typical cross-sectional diameters in the range between 4η and 10η, where η is the Kolmogorov scale (e.g., Jiménez et al. 1993). The existence of worm vortices can be traced to an inherently anisotropic feature of the small-scale organization of turbulent flows: High-magnitude vorticity is preferentially aligned with the eigenvector corresponding to the intermediate (weakest) eigenvalue of the rate of strain tensor, while there is very little such preferential alignment for the lower-magnitude vorticity (Ashurst et al. 1987; Kerr 1985) . This alignment is a kinematical property completely independent of the particular dynamical mechanism involved in the vorticity generation (Jiménez 1992), and thus it is a key feature to to be emulated at the smallest scales (Fureby and Grinstein 2002).

From the numerical simulation point of view, we recognize two crucial aspects of the high-Reynolds number turbulent flows to be captured: (1) the dominant vortex interaction mechanisms underlying the cascade dynamics in the inertial subrange occur on convective time scales much shorter than the diffusive time scales, and are thus essentially inviscid; (2) when worm vortices are much thinner than the main flow scales, the details of their internal structure (and actual core diameters) may no longer be significant, and the strengths and positions of the centerlines of such characteristic regions of intense vorticity may be the most important features to be captured. These observations suggest that nearly inviscid methods capable of handling vortices in a manner similar to that of shock-capturing schemes might provide an efficient computational framework in such flow regimes. Similar ideas have been also envisaged as the premise of, for example, contour dynamics (Oberman and Zabusky 1982) and of the vorticity confinement method (Steinhoff and Underhill 1994, also see Chapter 4e this volume). An additional important aspect to keep in mind as we seek an appropriate simulation framework is the fact that relevant laboratory objects to be studied (observables) are inherently discrete in nature: measurements always involve finite-fluid parcels transported over finite periods of time (Margolin and Rider 2002; Margolin, Rider, and Grinstein 2006; also see Chapters 2 and 5, this volume).

We can use these considerations to propose an ILES framework based on finite-volume (FV) numerics incorporating a thin-vortex-capturing capability operating at the smallest resolved scales, emulating near and above the small-scale cutoff the expected behavior in the high-wave-number end of the inertial subrange region. The idea that a suitable SGS reconstruction might be implicitly provided by discretization in a particular class of numerical schemes led to the proposal of the monotone integrated LES (MILES) approach (Boris 1990, Boris et al. 1992; also see Chapter 1, this volume) – mostly referred to hereafter as (monotonicity-preserving-based) ILES.

Early MILES studies were able to reproduce the large-scale features of the flow observed in laboratory experiments, such as the asymmetric entrainment (Grinstein, Oran, and Boris 1986), the distribution of merging locations (Grinstein, Oran, and Boris 1987), the spreading rate of the mixing layers (Grinstein, Oran, and Hussain 1989; Grinstein, Hussain, and Oran 1990), and the base-pressure and vortex-shedding dynamics in bluff-body near-wakes (Grinstein, Boris, and Griffin 1991). Important milestones to mention here were the first demonstration of physical global instabilities in free mixing layers (Grinstein, Oran, and Boris 1990), and the first 3D ILES studies of transition from laminar initial conditions in mixing layers (Grinstein, Oran, and Hussain 1989) and jets (Grinstein and DeVore 1992) – some of which are reviewed in the following paragraphs (also see the tests in a canonical case study in Chapter 4a).

Later theoretical studies (Fureby and Grinstein 1999, 2002) showed that a certain class of flux-limiting algorithms with dissipative leading-order terms provide effective built-in (implicit) SGS models of a mixed tensorial (generalized) eddy-viscosity type. Key features in the early comparisons with classical LES that led to the identification of this implicit SGS model were the use of the modified equation framework and the FV formulation, which readily allowed the recasting of the leading-order truncation terms in divergence form (Chapter 4a).

ILES based on flux-corrected transport (FCT) algorithms has been extensively used in the study of free shear flow phenomena to investigate the vortex dynamics and mechanisms of transition to turbulence in moderately high-Reynolds-number planar mixing layers (Grinstein, Oran, and Boris 1986, 1990, 1991), in planar wakes (Grinstein, Hussain, and Boris 1991; Jeong, Grinstein, and Hussain 1994), in jets emerging from round nozzles (Grinstein, Glauser, and George 1995; Grinstein et al. 1996) and rectangular nozzles (Grinstein 1995, 2001; Grinstein and DeVore 1996; Grinstein, Gutmark, and Parr 1995), and in swirling jet flows (Grinstein et al. 2002). ILES has also been extensively used to investigate convectively dominated chemically reacting mixing layers and jets in nonpremixed regimes (Grinstein 2001, 2002; Grinstein and Kailasanath 1992), as well as in premixed regimes (Fureby, Grinstein, and Kailasanath 2000; Grinstein and Fureby 2005).

8.2.2 FCT-based model

In the compressible free shear flow simulations discussed in the following paragraphs, complex vortex dynamics and its associated acoustic production must be distinctly captured, and spurious boundary condition effects have to be avoided (or, at the very least, clearly identified and controlled). ILES is based on the use of fourth-order FCT algorithms in the convection stage (Chapter 4a) to solve the time-dependent, compressible flow conservation equations for mass, energy, and momentum density and chemical species (when appropriate) with an ideal gas equation of state (see model discussion in Fureby and Grinstein 1999). In the present work I model nominally inviscid conditions (when EU based), denoted ILES–EU, or a linear viscous fluid (when NS based); I denote the latter as ILES when SGS effects are neglected, or as ILES–SMG when an explicit Smagorinsky-type viscosity is added.

For the two-dimensional (2D) simulations I used the Boris and Book one-dimensional (1D) FCT algorithm in conjunction with directional and time-step splitting techniques. The transitional 3D simulations are based on splitting the integrations for streamwise convection, cross-stream convection, and other local processes (e.g., viscosity, thermal conduction, molecular diffusion, and finite-rate chemistry, as appropriate). I performed convective spatial integrations by utilizing the 1D FCT for the streamwise direction, DeVore's 2D monotone FCT on cross-stream planes, and 3D central differences otherwise. I implemented time integrations by using a second-order predictor–corrector scheme.

I imposed open boundary conditions at boundaries in the streamwise direction (inflow and outflow), and I imposed free-stream flow conditions in the cross-stream directions. Inflow boundary conditions specify the values for the mass density and velocities. The inflow pressures are obtained as solutions of a one-sided finite-difference expression based on the 1D, unsteady, inviscid pressure equation (Grinstein 1994). A nonreflecting 1D boundary condition with or without explicit asymptotic relaxation to ambient conditions (Poinsot and Lele 1992) on the pressure is specified at the downstream (outflow boundary), where I introduced the additional numerical conditions required for closure of the discretized equations by requiring that the mass and momentum densities be advected with the local streamwise velocity at the boundary. The structured Cartesian computational grids were held fixed in time, and used evenly spaced cells in the shear flow region of interest. I frequently used geometric stretching in the cross-stream and downstream directions outside of the latter region to implement the open boundary conditions there. More specific details can be found in the cited references.

8.3 Transition from laminar conditions

A crucial feature of the shear flow development in transitioning from laminar conditions to turbulence is the rate at which fluid elements from the free streams become entangled or mixed as they join at the mixing layers. This information is given by the *entrainment* rates, which define the rate of propagation of the interface between rotational and irrotational fluids. The viscous diffusion of vorticity, acting primarily at the smallest scales of motion where the gradients are the largest, acts to propagate vorticity into the irrotational fluid. However, the production of this small-scale vorticity is largely controlled by the straining provided at much larger scales of motion. Thus the entrainment rate is controlled by the speed at which the interface contortions with the largest scales move into the surrounding fluid (Tennekes and Lumley 1972). These controlling large-scale vortices tend to be coherent and easily recognizable features; hence their name, CS. Control of the shear flow development in practical applications is strongly dependent on understanding the dynamics and topology of CS, particularly how their properties can be affected through control of the formation, interaction, merging, and breakdown of CS. I use selected examples from the MILES jet studies in the following paragraphs to demonstrate their important contributions in representative applications. A detailed analysis of ILES transition to turbulence in the case of a canonical case study (the Taylor–Green vortex) can be found in Chapter 4a.

The discussion that follows illustrates three crucial aspects of the development of the Kelvin–Helmholtz instability in subsonic free jet flows, namely global instabilities, flow self-organization, and complex 3D vortex dynamics. I examine the impact of the latter on jet entrainment and combustion dynamics in convectively dominated flow regimes. Finally, I comment on the trends of the population of the small-scales vortices in the downstream portion of the simulated jets, where the flow is characterized by filament vortices similar to those observed in fully developed turbulent flows. I test quantitatively how the latter features of the *high end* of the physical inertial subrange region are effectively emulated near the ILES cutoff; relevant issues of ILES convergence are addressed in this context.

8.3.1 Global instabilities

Characterizing the local nature of free shear flow instabilities and their global nonlinear development in space and time is of fundamental importance for practical shear flow control. An important question is whether a free mixing layer can be globally unstable with the self-excitation upstream induced by pressure disturbances generated by means of finite-amplitude fluid accelerations downstream. From the point of view of linear inviscid stability analysis, the spatially evolving subsonic mixing layer is convectively unstable with respect to vortical fluctuations (Huerre and Monkewitz 1990).

As a consequence, turbulent mixing layers are expected to be driven by environmental disturbances (Morkovin 1988) – except in rare configurations with global-absolute instabilities – and self-excitation effects in free shear flows are not generally expected to be significant.

The idea of a feedback mechanism through which downstream flow events can influence the upstream flow was first proposed by Dimotakis and Brown (1976) to explain unusually long autocorrelation times of the streamwise velocity fluctuations observed in their planar shear flow experiments. Subsequently, Laufer and Monkewitz (1980) observed that the unstable shear layer close to the nozzle of a jet was modulated with a low frequency corresponding to the passage frequency of the large-scale structures at the end of the potential core.

Mechanisms that trigger instabilities and transition to turbulence in free shear flows (e.g., vortex roll-up and merging) are the following: disturbances in the free streams, disturbances that are due to boundary layers, wakes, and small recirculation zones, acoustic environmental disturbances, and disturbances that feed back from downstream events in the flow. Isolating these mechanisms is particularly difficult in laboratory studies, because turbulence in free streams and boundary layers cannot be eliminated. Numerical simulations of spatially evolving shear flows can eliminate the first two types of disturbances, and they can address the third through careful implementation of the prescribed boundary conditions. Grinstein, Oran, and Boris (1990, 1991) successfully addressed these questions with FCT-based ILES-EU of a spatially evolving mixing layer in which the effects of boundary conditions could be neglected. The studies gave direct evidence for the existence of a global-instability mechanism for self-excitation of the mixing layer, and thus for self-sustaining instabilities, which are due to upstream feedback from the fluid accelerations originated by the vortex roll-ups and vortex pairings downstream (Figure 8.1). The nonlinear, global-instability mechanism in initially laminar, subsonic, free mixing layers depends on having significantly different propagation velocities for pressure waves on each side of the shear layer, and thus, in particular, depends on a finite sound speed. As a consequence, this isolated acoustic reinitiation mechanism becomes less pronounced as the acoustic propagation velocities on the sides of the mixing layer tend to be the same, that is, (a) as the incompressible limit is approached, and (b) as free-stream velocity ratios approach unity. Later studies demonstrated similar self-excited global instabilities in supersonic jets, based on upstream feedback mechanisms acting on the subsonic outer jet regions (Grinstein and DeVore 2002).

A key computational capability used in the global-instability studies was the ability to isolate the generation and propagation of acoustic disturbances correlated with the large-scale vortex dynamics. Relevant features accurately captured with ILES included the quadrupole pattern of acoustic production associated with vortices and the significantly more intense dilatation and pressure fluctuations associated with vortex pairing (Grinstein, Oran, and Boris 1991), as well as the very low fluctuation levels involved – for example, four orders of magnitude smaller than ambient values.

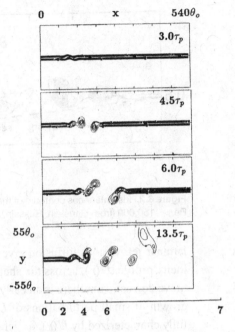

Figure 8.1a. Global instabilities in the convectively unstable planar mixing layer (from Grinstein, Oran, and Boris 1990).

8.3.2 Self-organization

Let us focus on the role of feedback and self-induction phenomena in organizing the spatial and temporal development of vortical structures in axisymmetric subsonic transitional air jets emerging into quiescent air at the same temperature (Grinstein, Oran, and Boris 1987; Grinstein, Hussain, and Oran, 1990). In the early stages of jet development immediately downstream of the nozzle, a shear layer is formed between the jet stream and the surroundings. In the simplest possible conceptual picture of a circular

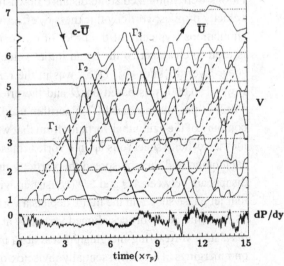
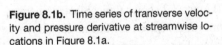

Figure 8.1b. Time series of transverse velocity and pressure derivative at streamwise locations in Figure 8.1a.

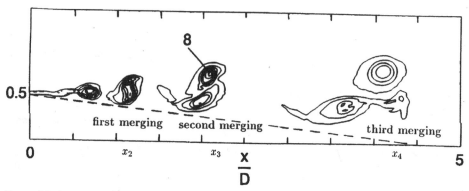

Figure 8.2. Instantaneous contours of the azimuthal vorticity from round jet ILES studies, $D/\theta_o = 140$, $Re_D \sim 150{,}000$ (from Grinstein, Hussain, and Oran 1990).

laminar jet, viscous diffusion gives rise to a weakly diverging steady streamwise velocity profile, $U(r)$, across the shear layer, having initial vorticity thickness δ_o. As we
move downstream, there is an early *linear* instability jet regime, involving exponential
growth of small perturbations of $U(r)$, introduced at the jet exit. This regime can be
fully characterized by $U(r)$, δ_o, jet exit velocity U_o, Mach number M, jet diameter D,
and Reynolds number $Re_D = U_o D/\nu$, where ν is the kinematic viscosity (Michalke
1984). Beyond this development stage, in the *nonlinear* Kelvin–Helmholtz instability
regime, large-scale vortex rings roll up, and their dynamics of formation and merging
become the defining feature of the transitional shear flow. Inherently 3D (nonaxisym-
metric) effects become increasingly more important as we move downstream away from
the nozzle.

The initial jet shear-layer thickness and D, provide the two length scales for describing the circular jet flow. In the nominally inviscid round jet MILES studies by
Grinstein, Oran, and Boris (1987) and Grinstein, Hussain, and Oran (1990), the simulations were initialized some distance downstream from the nozzle with top-hat axial
velocity profiles; vorticity thickness δ_o effectively equal to the rim of the nozzle in the
simulations (typically of the order of one computational cell across the shear layer)
is used to characterize the mean initial velocity conditions. The ratio of D to initial
shear-layer thickness δ_o, D/δ_o, was in the range 70–140, and the Reynolds number,
$Re_D \sim 150{,}000$, was based on D and the effective viscosity (Grinstein and Guirguis
1992) of the fourth-order FCT algorithm used. Figure 8.2 shows typical instantaneous
contours of the azimuthal vorticity from that work for $D/\delta_o = 140$. The initial axisymmetric vorticity sheet rolls up into vortex rings, which move downstream, interact with
each other, and thereby spread the vorticity until the central, potential core of the jet
disappears between $4D$ and $5D$ (consistent with laboratory observations). The particular jet phase shown in the figure depicts a vortex ring resulting from a *first* merging at
x_2, two (merged) vortices aligned vertically at x_3 undergoing a *second* merging, and a
third such vortex-ring-merging process at x_4 involving vortices that resulted from second mergings. Through essentially inviscid vortex-pairing interactions, the transverse
cross section of the ring structures nearly doubles at each merging. The vortex-merging

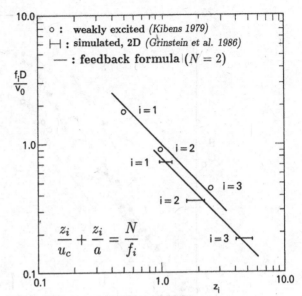

Figure 8.3. Axisymmetric jet merging locations (Grinstein, Oran, and Boris 1986.)

locations tend to be distributed in fairly regular patterns (e.g., Kibens 1979), and the number of observed mergings depends on ratio D/δ_o (Grinstein, Hussain, and Oran 1990).

Vortex ring roll-up and merging are major mechanisms for the spread of the jet shear layer for high-Reynolds-number convectively dominated flows for which diffusion processes are much less important. Vortex interactions are promoted by small displacements of the newly formed structures from the center of the shear layer as they are convected downstream. These displacements may be caused directly or indirectly by pressure and velocity fluctuations originating from perturbations downstream at vortex-ring mergings and at the end of the potential core. When the displacements of a pair of close structures are properly phased, the structures roll around each other by mutual induction, and this leads in turn to the vortex merging. As is the case with other fundamental vortex interactions (including, e.g., reconnection, which is subsequently examined), once triggered, vortex pairing is an essentially inviscid process.

In order to explain the observed jet self-organization, a feedback mechanism for free and forced shear flows was proposed in which the cross-stream perturbations required for the vortex-pairing process are provided by the pressure field induced by the downstream (already paired) vortices (Ho and Huerre 1984; Ho and Nosseir 1981; Laufer and Monkewitz 1980). The feedback loop from the trailing edge to the merging location involves the unstable vortex roll-ups convecting downstream and pressure waves from these changing, predominantly rotational flows, propagating upstream. The resulting *feedback formula* implied by modeling this mechanism describes fairly well the distribution of merging locations found in weakly excited experimental jets (Laufer and Monkewitz 1980) and also captured in the ILES studies (Figure 8.3) by Grinstein, Oran, and Boris (1986). Extensive experimental investigations were unable to conclusively establish the existence of this feedback loop for laboratory free jets as a

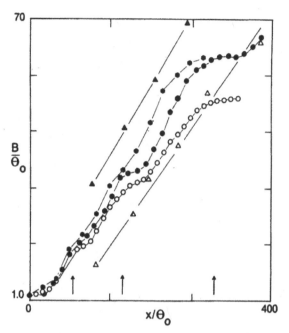

Spread of the mixing layer B/θ_0. ●: M = 0.57, $D/\theta_0 = 140$; ○: M = 0.57, $D/\theta_0 = 70$;
◖: M = 0.43, $D/\theta_0 = 140$; ▲: Husain (1981); △: Hussain and Zedan (1978).

Figure 8.4. Time-averaged mixing-layer growth results for unexcited jets (Grinstein, Hussain, and Oran 1990). The figure shows the approximately linear growth in the axial direction for both calculated and experimental results. The normalized width of the shear layer, B/θ_0, is shown as a function of normalized streamwise distance x/θ_0 and Mach number.

result of so-called unavoidable facility-dependent background disturbances (Gutmark and Ho 1983) – reflecting the fact that different facility boundary conditions prescribe different jet flows. The feedback loop was first isolated and demonstrated in the ILES studies of global instabilities described herein (Grinstein, Oran, and Boris 1990, 1991).

An important jet modeling issue deserves to be noted here, namely, the physical effectiveness of initiating the jet simulation some distance downstream of the nozzle with appropriate initial (inflow) conditions. As discussed by Grinstein, Hussain, and Oran (1990), a crucial ingredient was choosing an effective initial momentum thickness, θ_o, equal to δ_o. Consistency tests of this initial length scale "calibration" thus tied with the assumed nozzle thickness (and thus also to grid size) were provided as follows. First, with this choice, the simulated initial shear-layer instability Strouhal-frequency peak lies at $St_{\delta o} = f\delta_o/U_o \approx 0.014$, within the expected range 0.0125–0.0155 (Husain and Hussain 1983). Moreover, the said choice, $\theta_o = \delta_o$, provided a suitable length scale to normalize the time-averaged mixing-layer growth results as presented in Figure 8.4 for unexcited jets (from Grinstein, Hussain et al. 1990), showing the normalized width of the shear layer, B/θ_o, as a function of x/θ_o and Mach number. The figure depicts the approximately linear growth of the mixing layer in the axial direction for both the calculated and the experimental results and the good agreement among growth rates dB/dx, particularly in the initial streamwise portion governed by vortex roll-up phenomena in all cases. Agreement with the experimental data for dB/dx was found to be better when the conditions in the simulations were closer to those in the experiments, namely, for lower Mach number and larger ratio D/θ_o. For $\theta_o/D \ll 1$ (as we have here), the properties of the axisymmetric shear layer in the near field of the jet approach those of the plane mixing layer, and they are also consistent with the predictions of linear stability theory (Michalke 1965).

8.3.3 Complex vortex dynamics

Mixing between free streams occurs in two stages, an initial stage of bringing relatively large amounts of the fluids together (large-scale stirring), and a second stage promoted by the small-scale velocity fluctuations that accelerate mixing at the molecular level. Large-scale 2D spanwise rollers (in plane mixing layers or wakes) or vortex rings (in jets such as just discussed) and their interactions play a crucial role in controlling transitional shear-layer growth and entrainment at moderately high Reynolds numbers. For higher Reynolds numbers, or when the azimuthal or spanwise symmetry is broken or not present, three-dimensionality becomes the crucially important feature. The dynamics of streamwise vortices becomes dominant, and mechanisms such as self-induction, vortex stretching, vorticity production due to baroclinicity and expansion, as well as vortex reconnection become major fluid-dynamical processes involved in the transition from laminar to turbulent regimes.

Practical control of jet entrainment involves manipulating the natural development of CS and their breakdown into turbulence to enhance three-dimensionality and thus mixing. Popular passive control approaches are based on geometrical modifications of the jet nozzle that directly alter the flow development downstream relative to using a conventional circular nozzle; research on flow control with noncircular jets was reviewed by Gutmark and Grinstein (1999). Rectangular jet configurations are of particular interest because they offer passively improved mixing at both ends of the spectrum: enhanced large-scale entrainment caused by axis-switching, and enhanced small-scale mixing near corner regions and farther downstream caused by faster breakdown of vortex-ring coherence and hence faster transition to turbulence.

Rectangular jet ILES studies (e.g., Grinstein 2001; Grinstein and DeVore 1996) with aspect ratio AR = 1–4, were carried out in various nonreactive and reactive regimes. As in the case of simulated round jets discussed herein, for computational efficiency, the focus was on initially laminar jets developing from thin rectangular vortex sheets (with slightly rounded-off corner regions here), and uniform momentum thickness θ enforced in terms of appropriate top-hat velocity profiles. Characteristic velocity scales for the rectangular jet simulations were U_j and the peak rms velocity fluctuation value u', where U_j denotes the free-stream jet velocity. The gas jets investigated were assumed to emerge into a quiescent background with Mach number $M = 0.3$–0.6. Characteristic length scales are the initial shear layer thickness, the jet equivalent diameter, D_e, and the Taylor microscale, $L_T = L/\sqrt{(\mathrm{Re}_T)}$, where $\mathrm{Re}_T = u'L/\nu$, and integral scale L is the full core width of the initial vortex rings (size of the largest eddies in the flow). The typical Reynolds number for the nominally inviscid jets was $Re = U_j D_e / \nu > 78{,}000$ ($Re_T > 90$) – based on estimated upper bounds for the effective numerical viscosity of the FCT algorithm (Grinstein and Guirguis 1992) or comparisons with DNS turbulence data (Fureby and Grinstein 2002; Grinstein 2001; also see subsequent information); otherwise, typical DNS studies had $Re = 3200$, based on air viscosity at STP.

The jet simulations demonstrated the dynamical vorticity geometries characterizing the near jet, involving (1) self-deforming vortex rings, (2) interacting ring and braid (rib) vortices, and (3) a more disorganized flow regime in the far jet downstream, where the rotational fluid volume is occupied by a relatively weak vorticity background with strong, slender tube-like filament vortices filling a small fraction of the domain. Figure 8.5 highlights the characteristic observed differences as a function of AR based on representative instantaneous volume visualizations. Two qualitatively different ring–rib coupling geometries are shown: jets with $AR \geq 2$ are characterized by single ribs aligned with corner regions; square jets are characterized by pairs of counterrotating ribs aligned with the corners.

Other distinct geometrical features are associated with the occurrence of vortex-ring splittings caused by reconnection, which are observed for the larger values of AR. Figure 8.6(a) illustrates characteristic axis-switching and bifurcation phenomena for $AR = 4$ from visualizations of laboratory elliptical jets subject to strong excitation at the preferred mode (Hussain and Husain 1989), compared to carefully developed simulations, Figure 8.6(b), designed to address unresolved issues of the vortex dynamics. Underlying aspects of the vortex-ring bifurcation process – namely, reconnection, bridging, and threading – had been conjectured (Hussain and Husain 1989) but could not be captured in the laboratory studies, and were first demonstrated by the simulations (Grinstein 1995). Indeed, although vortex bridging and threading had been recognized as basic ingredients of the vortex reconnection dynamics in DNS studies of curved antiparallel vortex tubes (Melander and Hussain 1989), evidence of their involvement was not actually suggested in laboratory flow visualizations of vortex reconnection phenomena until later (Leweke and Williamson 1998). Various vortex interaction mechanisms contribute in a crucial way to the breaking of large-scale CS into increasingly smaller ones in fully developed jets; this includes further complex reconnection and bridging mechanisms. See, for example, Figure 8.6(c), from Grinstein (2001), where this is indicated by arrows and labels "II-rec" and "II-bridg." Likely cascade scenarios were demonstrated (Grinstein 1995, 2001), suggesting pathways for transition to turbulence based on successive vortex self-inductions, stretching, and reconnections. ILES has proven to be a capable simulation tool in this context.

In terms of characteristic ring–rib coupling geometry, significantly larger jet spreading and streamwise vorticity (ω_1) production in the jet for $AR = 1$ are shown in Figure 8.5, reflecting the appearance of rib pairs aligned with corner regions – rather than single ribs for $AR = 3$. However, also associated with the larger near-jet ω_1 production for $AR = 1$, the vortex rings tend to be more unstable azimuthally and break down closer to the jet exit. As a consequence, Figure 8.5 presents larger jet spreading for $AR = 3$ further downstream (say for $x > 6 D_e$), where better entrainment properties are suggested by the more intense distributions of streamwise vorticity (much more intense

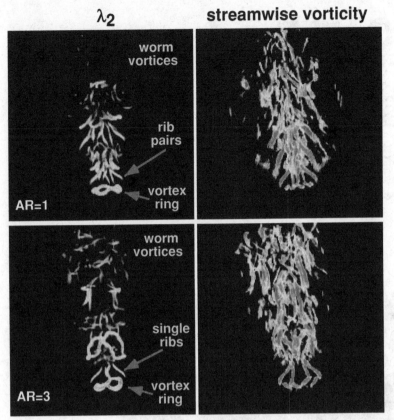

Figure 8.5. Comparative instantaneous visualizations based on volume visualizations of λ_2, the second-largest eigenvalue of tensor $\mathbf{S}^2 + \Omega^2$, where $\mathbf{S}$ and Ω are the symmetric and antisymmetric components of the velocity gradient tensor, are shown on the left. Corresponding visualizations of the streamwise vorticity are shown on the right. (See color inset, Plate 5).

for AR = 3 than for AR = 2). Furthermore, vortex self-deformation and axis-switching can occur (and promote mixing better) closer to the jet exit for the AR = 1 case because of significantly smaller characteristic vortex-ring axis-switching times. Typically these are half as large and less than one-third as large, respectively, compared with those for AR = 2 and AR = 3 (e.g., Grinstein 1995).

Thus, there are different interesting possibilities for how the jet flow develops, depending on the particular jet initial conditions and specific type of unsteady vortex interactions initiated, the nozzle geometry, and modifications introduced at the jet exit. This is of interest in the context of improving the mixing of a jet (or plume) with its surroundings in many practical applications. Taking advantage of these flow control possibilities is of interest to improve the mixing of a jet, or plume, with its surroundings in practical applications. For example, we may seek enhanced combustion between injected fuel and a background oxidizer (subsequently discussed), rapid initial mixing and submergence of effluent fluid, less intense jet noise radiation (see, e.g., Grinstein and Fureby 2002), or reduced infrared plume signature. The crucial feature to control here is the jet entrainment rate – the rate at which fluid from the jet becomes entangled or mixed with its surroundings, determined by the characteristic large-scale vortex dynamics.

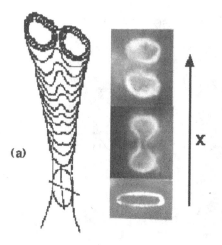

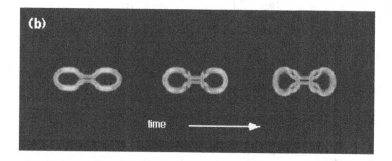

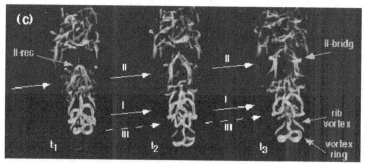

Figure 8.6. (a) Characteristic axis-switching and bifurcation phenomena for AR = 4 from visualizations of laboratory elliptical jets subject to strong excitation at the preferred mode (Hussain and Husain 1989); (b) visualizations of an AR = 4 rectangular jet based on ILES data (Grinstein 1995); (c) ILES simulation of an AR = 3 rectangular jet (Grinstein 2001). (See color inset, Plate 6).

8.3.4 Entrainment and combustion dynamics

Jet entrainment measurements are often based on evaluating the streamwise mass flux Q, which is normalized with its value at the jet exit, Q_o, and plotted as a function of x/Q_e, where D_e is the circular-equivalent diameter (e.g., Figure 8.7 – comparing Q for

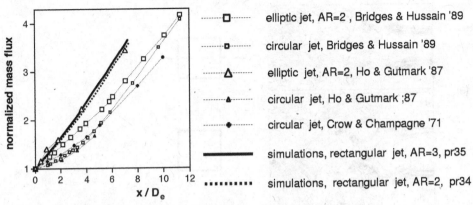

Figure 8.7. Entrainment measurements for laboratory and simulated jets.

rectangular, elliptic, and round jets with various different initial conditions at the nozzle exit). With other approaches – requiring more detailed velocity data not always available in the laboratory experiments – only vorticity-bearing fluid elements are allowed to contribute when Q is evaluated (e.g., Grinstein, Gutmark, and Parr 1995; Husain and Hussain 1993), or the entrainment rate (dQ/dx) is evaluated directly based on the radial volumetric flux (Liepmann and Gharib 1992). Discussion and examples of the inherent uncertainties in evaluating Q can be found in Grinstein (2001) and references therein.

A database of ILES reactive jet simulations was used to obtain insights on the close relationship between unsteady fluid and non-premixed-combustion jet dynamics (Grinstein 2001, 2002). The combustion of propane-nitrogen (or hydrogen-nitrogen) rectangular jets emerging into a quiescent gaseous oxygen-nitrogen background was investigated. The focus was on (1) regimes with a moderately high Reynolds number, for which chemical exothermicity affects the jet development mainly through inviscid volumetric expansion and baroclinic vorticity production mechanisms; and (2) regimes with a sufficiently lower Reynolds number, for which local temperature changes in the flow that are due to chemical energy release and compressibility combined with temperature-dependent viscous effects play a major role in controlling the flow dynamics. Investigation of the effects of density differences, exothermicity, relaminarization, and preferential diffusion on the jet dynamics produced the first detailed documentation of temperature and mixedness distributions in reactive square jets, and the relation between combustion and underlying fluid dynamics.

Multispecies temperature-dependent diffusion and thermal conduction processes (Grinstein and Kailasanath 1996) were calculated explicitly by use of central difference approximations and coupled to convection and global finite-rate models for the chemical kinetics, using time-step splitting techniques (Oran and Boris 2001). Details of the simulated transport properties and their validation are discussed in Grinstein and Kailasanath (1996) and references therein. Initial non-premixed conditions at the inflow were modeled in terms of appropriate step-function profiles for the reacting species concentrations. Convectively dominated, reactive flow regimes were considered – for which

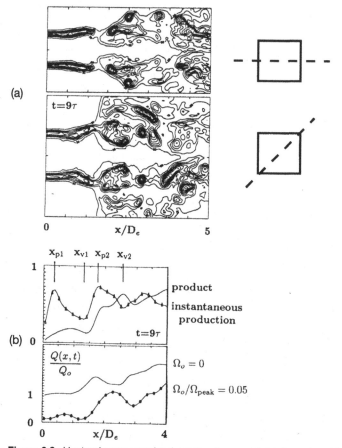

Figure 8.8. Unsteady non-premixed-combustion and fluid dynamics: (a) contours of vorticity magnitude Ω in planes indicated; (b) cross-sectional averaged measures of instantaneous chemical product, product formation, unconstrained, and vorticity-bearing ($\Omega > 5\%$ peak value) streamwise mass flux Q (Grinstein 2002).

an ILES approach can be expected to be meaningful. For typical propane-nitrogen jets studied, instantaneous peak values for the Damköhler number were Da $\sim$ 1,000, where Da measures the characteristic ratio between chemistry and convection times. SGS fluctuations were neglected in the ILES of reactive jets, and instantaneous evaluations of relevant combustion quantities such as diffusivities, thermal conductivities, and fuel burning rates were performed directly in terms of unfiltered variables. Convergence and chemistry modeling issues relevant to this simulation approach are discussed elsewhere (e.g., Fureby, Grinstein, et al. 2000; Grinstein and Fureby 2005).

Instantaneous chemical product formation rates were shown to be closely correlated with entrainment rates $(D_e/Q_0)dQ/dx$ (Grinstein and Kailasanath 1995) – found to be first significant in the regions of roll-up and initial self-deformation of vortex rings, and then farther downstream, where fluid and momentum transport between jet and surroundings are considerably enhanced by the presence of braid vortices (Figure 8.8). The jet simulations also addressed the potential impact of nozzle AR on non-premixed

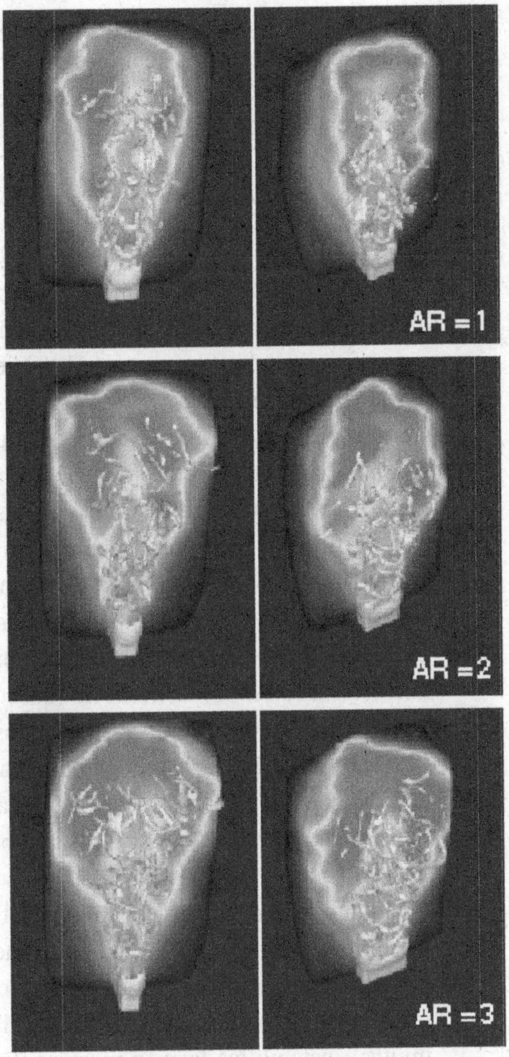

Figure 8.9. Instantaneous visualizations of non-premixed-combustion regions as a function of AR at two selected times (Grinstein 2001). Temperature distributions in the back half of the visualized subvolume are superimposed to isosurfaces of the vorticity magnitude (gray). (See color inset, Plate 7).

rectangular-jet combustion (Grinstein 2001), an issue of great importance in the context of passive jet entrainment control strategies (Gutmark and Grinstein 1999). By design, the simulated jets with AR = 1, 2, and 3 compared in Figure 8.9 only differ in the actual shape of the initial jet cross section but have otherwise identical initial conditions, for example, including the same cross-sectional jet outlet areas. The near-jet entrainment and non-premixed-combustion properties were largely determined by the characteristic braid-vortex topology and vortex-ring axis-rotation times. Because of the initial enhanced entrainment associated with rib pairs aligned with the corner regions, the square jet non-premixed combustion turns out to be more effective immediately downstream of the nozzle; on the other hand, the jet with AR = 3 exhibits better combustion farther away from the jet exit, reflecting better entrainment there as a result of the azimuthally stabler vortex rings and more intense streamwise vorticity distributions (cf. Figure 8.5) and on the axis-switching process being completed farther downstream – since vortex-ring axis-rotation times increase with AR (Grinstein 1995). Thus, axis-switchings occur closer to the jet exit and can be more frequent in the near jet for AR = 1. The effectiveness of the non-premixed jet combustion as a function of AR can be consistently related to the AR-dependent CS dynamics.

8.4 Small-scale emulation

Figure 8.10 shows instantaneous visualizations of a developed square jet from my ILES–EU studies (Grinstein 2001; Grinstein and DeVore 1996). To facilitate the analysis of the results, we organized the jet flow by using weak forcing with a (single) Strouhal frequency, $St = fD_e/U_j = 0.55$, chosen within the range of observed laboratory preferred (large-scale) jet frequencies, and rms level 2.5% of U_j. We chose a forcing amplitude to be low enough to ensure good agreement of the initial shear-layer growth rate of the simulated jets with those of comparable unforced laboratory orifice jets with nearly laminar initial conditions. Figure 8.10 compares unfiltered data at selected representative times from simulations carried out with identical initial and boundary conditions in grids denoted F and C, with nominal smallest cell spacings Δ and 2Δ. We carried the simulations out with fixed Courant number 0.4, and estimated Reynolds numbers (based on U_j, D_e, and effective algorithm viscosity) were Re > 220,00 and Re > 78,000, respectively. The smallest grid size we used was $\Delta = D/42$, where D is the length of the minor side of the rectangular nozzle. These comparisons and further subsequent analysis are used to examine dependence on spatial resolution (effective Reynolds number) of the simulations as well as issues of ILES convergence.

Figure 8.10 depicts the rolling up of the square vortex sheet into vortex rings near the inflow, self-induced distortion of the nominally square vortex rings, formation of hairpin (braid) vortices between distorted rings, strong coupling of these vortices into ring–hairpin bundles, and merging of the vortex bundles. Vortex interactions and azimuthal instabilities lead to more contorted vortices; vortex stretching, kinking, and reconnection lead to their breakdown, and to a more disorganized flow regime farther downstream, characterized by elongated vortices resembling those typical of fully developed turbulent flows (e.g., Jiménez et al. 1993; Porter, Pouquet, and Woodward 1994; also see Chapter 7, this volume). The figure shows quite good visual agreement on the large-scale dynamics of ring and hairpin vortices near the jet exit, but their faster coherence breakdown downstream is apparent with increasing resolution (effective Reynolds number), *as finer dynamical features can contribute* and affect the larger scales.

The focus is now on making quantitative statements on the trends of the population of the small-scales vortices in the downstream portion of the simulated jets, where the flow is characterized by thinner filament vortices similar to those observed in fully developed turbulent flows. The goal is to quantify characteristic features of the "turbulence" to which the simulated flow transitions from laminar initial conditions to what is regarded as its established metrics.

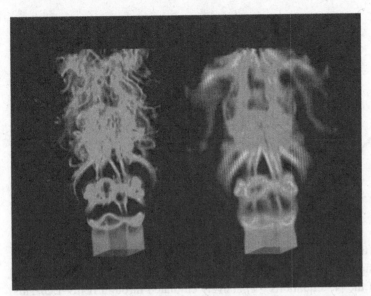

Figure 8.10. Comparative instantaneous volume visualization of the vorticity magnitude based on the database of square jets at the same time simulations on the finest grid left, and coarsest grid right (from Grinstein 2001). (See color inset, Plate 8).

Spectral analysis concentrating on the latter portion of the simulated developed jets can be used to investigate the small-scale jet behavior captured by the simulations. Analysis is based on the instantaneous velocity databases for the grid resolutions considered on (uniformly spaced) downstream subvolumes. These volumes include only the appropriate downstream portions of the jets chosen symmetrically around the centerline (Fureby and Grinstein 1999; Grinstein 2001; Grinstein and DeVore 1996).

In order to improve the basis for comparison of the current compressible (subsonic) jet data with available DNS data of incompressible homogeneous isotropic turbulence, it is helpful to base the analysis on the solenoidal component of the jet velocity data. We can decompose the instantaneous velocity fluctuation into its solenoidal and compressible components according to $v = v^s + v^c$, with the condition $\nabla \cdot (v^s) = 0$ in physical space translating into the condition $\mathbf{k} \cdot \widehat{\mathbf{v}}^s = 0$ in Fourier space. We use this condition explicitly to obtain the solenoidal component of the Fourier velocity transform in the form $\widehat{\mathbf{v}}^s = \widehat{\mathbf{v}} - (\widehat{\mathbf{v}} \cdot \mathbf{k})\mathbf{k}/|\mathbf{k}|^2$, whereby $\mathbf{v}^s$ is evaluated by means of an inverse Fourier transform. Thus we can also obtain the solenoidal part of the turbulent kinetic energy, E^S.

Figure 8.11a shows time-averaged plots of the turbulent kinetic energy spectra in the case AR = 1. In each case, spatial fast fourier transform analysis is based on data sets for 40 successive times separated by a time interval, $0.1/f$, on 140^3 and 70^3 downstream grid subvolumes, for the two resolutions. The largest wave number for which spectral amplitudes are plotted corresponds to a wavelength of four computational cells; the spectra depict short simulated inertial subranges consistent with the $k^{-5/3}$ inviscid subrange of the Kolmogorov K41 theory, and a longer inertial range for higher-resolution cases.

Figure 8.11b shows spectra as a function of AR for intermediate resolution (1.5D), and as based on 110^3 downstream subvolumes; the figure suggests the experimentally

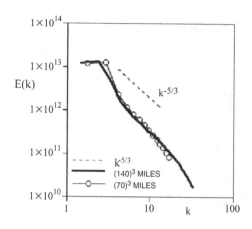

Figure 8.11a. Compressible (subsonic) jet flow: Time-averaged plots of the turbulent kinetic energy spectra in the square jet case.

observed shallower (than the $k^{-5/3}$) slope in the near-dissipation region (e.g., Saddoughi 1992, 1997), which appears to be a characteristic feature of high- (but finite) Reynolds-number solutions of the NS equations – *effectively* emulated through ILES–EU. This so-called near-dissipation *bump* was also noted in turbulent decay ILES studies using fourth-order, piecewise parabolic method (Porter, Pouquet, and Woodward 1994; also see Chapter 7, this volume). The emulation of the bump indicates that, in spite of relying on an inherently dissipative-type of SGS modeling, ILES can implicitly incorporate some degree of desirable backscatter that is dependent on the existence of a simulated dissipation region – but mostly independent of its precise nature. The simulated inertial range is followed by the faster decay of the amplitudes as a result of the FCT dissipation for wavelengths $\lambda < 10\Delta$. This limiting length scale corresponds to approximately twice the smallest characteristic (full-width) cross-sectional length scales of the elongated vortex tubes in the transitional region of the jet (Grinstein and DeVore 1996).

The distribution of vorticity intensities can be examined on the basis of the cumulative distribution function (CDF), defined as the volume fraction occupied by vorticity magnitude values above a given threshold level. Figures 8.12 focuses again on the jet with AR = 1, showing the CDFs of the vorticity magnitude $|\omega|$ for $|\omega| \geq 1$ based on

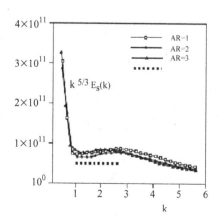

Figure 8.11b. Compressible (subsonic) rectangular jet flow: Compensated spectra as a function of AR = 1–3 for intermediate resolution (1.5D), based on 110^3 downstream subvolume data.

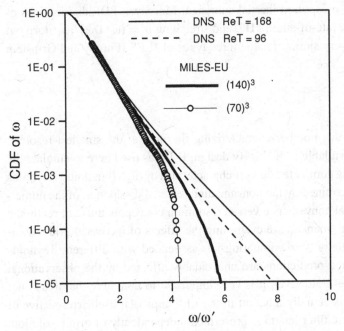

Figure 8.12. Cumulative distribution function of the vorticity magnitude.

downstream subvolume data for the two extreme grid resolutions F and C, superimposed to CDFs based on DNS turbulence data by Jiménez et al. (1993). The vorticity magnitude is scaled with its rms value ω' for each jet subvolume data. The figure shows good agreement of the grid-F and DNS data even for fractional volumes of less than 1%, while differences between DNS and the grid-C data are significant much before, for fractional volumes between 10% and 1%. The structure of the resolved vorticity implies that high-intensity regions tend to be organized mostly in elongated tubes. Because "worm" vortices typically involve fractional volumes of the order of 1% or less (Jiménez et al. 1993), of these two cases, only the grid-F simulation resolves these vortices, and this is supported by the volume visualizations of Figure 8.10.

The possible dependence of the resolved small-scale jet features on whether ILES was based on the Euler or NS equations (with or without SMG viscosity) was assessed by carrying out separate appropriate runs (Fureby and Grinstein 1999, 2002). The simulations were performed by using the intermediate resolution grid (cell size 1.5Δ) with identical initial conditions, and our small-scale statistical analysis was based on instantaneous velocity data on 90^3 grid downstream subvolumes. Comparative flow analysis depicted virtually identical initial larger-scale vortex dynamics; globally similar but distinctly different local features were apparent downstream, as the flow regime becomes disorganized and dominated by the presence of thin elongated vortices. The velocity-fluctuation spectra was found to be very similar for all three ILES approaches, indicating (1) somewhat smaller amplitudes for lower wave numbers associated with ILES and ILES–SMG, depicting viscous damping of resolved features; (2) essentially coincident high-wave-number amplitudes – reflecting the unresolved small-scale viscous effects; and (3) a captured inertial subrange. Moreover, PDF distributions of vorticity magnitude $|\omega|$, rate-of-strain $\|\mathbf{D}\|$, and stretching $\sigma = (\omega \cdot \mathbf{D}\omega)/|\omega|^2$ turn out to be correspondingly very similar for the three types of ILES (Fureby and Grinstein 1999, 2002).

8.4.1 Convergence

Actual values of Reynolds number characterizing the flow at the smallest resolved scales are not a priori available in ILES. By design, ILES as used here to emulate the dynamics of convectively dominated flows is characterized by high (but finite) Reynolds numbers ultimately determined by the nonvanishing residual dissipation of the numerical algorithm. The usual convergence versus resolution concept is thus not really the relevant one to consider when we are examining the effects of increased grid resolution, since we have actually been comparing jets associated with different Reynolds numbers. Thus, simulated predictions are unavoidably affected by the observational (simulation) process itself. However, it is very important to note here that this is not specific to ILES, but it is actually inherent in the LES approach itself irrespective of whether the characteristic filter length is prescribed independently of grid resolution, and in fact, it is inherent in making any (computational or laboratory) observation (Grinstein 2006).

We can still legitimately ask about possible convergence of the larger scale jet features and dynamics versus Reynolds number, that is, those scales much larger than characteristic resolution cutoff scales. A resolution instance in which we can make a clear statement regarding LES versus laboratory truths occurs when, for the conditions of the problem under study (1) the Reynolds number is sufficiently large that an inertial subrange exists, and (2) the LES resolution cutoff can be selected within that said inertial subrange. This is the case where (by design) LES can capture most of the energy-containing scales, modeling of the SGS can be achieved in a fairly universal robust fashion without considerably affecting the large-scale dynamics, and significant overlap with the laboratory observations can thus be attained on the large-scale driven flow features. These conditions are frequently realized with free shear flows, and we have exemplified this situation with the ILES jet simulations herein.

8.4.2 Sensitivity to multidimensional SGS model specifics

Figure 8.13 from Grinstein, Fureby, and DeVore (2005) is used in what follows to address issues of possible sensitivity to the particular choice of the multidimensional FCT limiter used in the rectangular jet studies discussed herein. Temporal sequences of visualizations of the vorticity magnitude were obtained by using either the positivity-preserving, Figure 8.13(a), or monotonicity-preserving, Figure 8.13(b) versions of multidimensional FCT (see Chapter 4a) in the cross-stream integrations. It is important to note here that the visualizations in Figure 8.6(b) correspond to the vortex-ring bifurcation phases in the top row of Figure 8.13(b), both being generated with the same unsteady

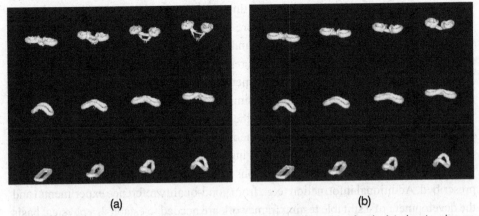

(a)   (b)

Figure 8.13. Self-deformation, reconnection, and subsequent bifurcation of an isolated vortex ring are shown, using (a) Zalesak's flux limiter (Zalesak 1979), which preserves positivity but not monotonicity, and (b) the monotone flux limiter (DeVore 1998), which involves a prelimiting step using monotonicity-preserving 1D FCT in each cross-stream direction.

ILES data by use of monotonicity-preserving FCT modules – as are all other jet simulations discussed herein.

An initially laminar air jet at STP issues from a rectangular nozzle of AR = 4:1 at Mach 0.6. The quiescent background gas also is at STP. In order to focus on the dynamics of the individual vortex rings of the jet, we puff out an isolated ring by turning off the jet inflow after a suitable finite time (chosen as the ratio of the jet equivalent diameter to the flow speed). The evolution of the ring is then followed as it convects downstream. In the panels of Figure 8.13, time increases from bottom left to top right; the flow direction is from bottom to top in each frame.

The bottom-most six frames in each layout in Figure 8.13 show the self-induced deformation and axis-switching of the vortex ring. The highly curved corners accelerate ahead of the ring sides and toward the centerline, pulling the minor axis-sides along with them. This process bends the ring along its major axis. The increasing curvature at the midpoints of the major sides accelerates those portions streamwise toward the leading minor sides but away from the jet centerline. This results in a nearly planar, axis-switched configuration of the vortex ring at frame "8" in both simulations. This early evolution is essentially identical in the two cases, save for some intermittent, small-scale, numerical features evident with Zalesak's limiter.

Subsequently, the ring's new major sides pinch together and reconnect, forming a pair of vortex rings linked on the underside by two thin threads. This is shown in the following (top) four panels of Figures 8.13. While both simulations clearly capture the laboratory-observed bifurcation of the ring, Figure 8.6(a), the predicted fine structure dynamics (*not captured* in the laboratory visualizations) associated with the threads bridging the two daughter rings increasingly differ with the two limiters. Zalesak's limiter (Zalesak 1979) effectively allows fluctuations on the threads of vorticity, which show up as spikes attached to the isosurfaces and lead in Frame 12 to the fragmenting of the threads, and, as suggested but not shown in Figure 8.13(a), the two daughter vortex rings irrevocably separate at later times. In contrast, Figure 8.13(b), with the locally monotone (and thus inherently more diffusive) 2D limiter (DeVore 1998), the fine structure appears *cleanly* represented, and the daughter rings stay in close proximity to one another, leading to significantly different later vortex dynamics (Grinstein 1995). However, more detailed laboratory experiments for the later-time phases are necessary to establish the relative merits of the two simulated predictions.

The latter example demonstrates that the detailed transition dynamics can be significantly influenced by the specifics of the implicitly built SGS model. By choosing among positivity- or monotonicity-preserving FCT schemes, different SGS (backscatter) models for how virtual SGS fluctuations affect the large scale dynamics are being prescribed. Additional information (e.g., from good-quality reference experiments) and the development of a suitable testing framework are needed to establish a physical basis for improved limiting choices. However, it is also important to note here that tests of various different (monotonic) flux-limiting approaches in the context of the Taylor–Green vortex problem (Chapter 4a) indicate that, while transition details can depend on the limiting specifics, features such as global characteristic transition times and peak dissipation rates associated with inherently inviscid instabilities can be captured in a fairly robust fashion with these numerical schemes. Extending such canonical tests in the future to include nonmonotonic nonoscillatory finite-volume schemes will be important to complete our understanding of this picture. Further studies addressing ILES performance versus numerics are discussed elsewhere in this volume. For example, details on the effective choice of implicit SGS associated with popular nonoscillatory finite-volume methods are given in Chapters 4a and 5, and specific modifications of the low-order component of FCT for studies of flow and dispersion in urban canyons are given in Chapter 17.

REFERENCES

Ashurst, W. T., Kerstein, A. R., Kerr, R. M., & Gibson, C. H. 1987 Alignment of vorticity of scalar gradients with strain rate in simulated Navier Stokes turbulence. *Phys. Fluids* **30**, 2343–2353.

Boris, J. P. 1990 On large eddy simulation using subgrid turbulence models, in *Whither Turbulence? Turbulence at the Crossroads*, ed. J. L. Lumley. New York: Springer, 344–353.

Boris, J. P., Grinstein, F. F., Oran, E. S., & Kolbe, R. J. 1992 New insights into large eddy simulation, *Fluid Dyn. Res.* **10**, 199–228.

Brown, G., & Roshko, A. 1974 On density effects and large structure in turbulent mixing layers. *J. Fluid Mech.* **64**, 775–816.

DeVore, C. R. 1998 *An Improved Limiter for Multidimensional Flux-Corrected Transport.* U.S. Naval Research Laboratory, Washington, DC: Report NRL-MR-6440-98-8330.

Dimotakis, P. E., & Brown, G. L. 1976 The mixing layer at high Reynolds number: Large-structure dynamics and entrainment. *J. Fluid Mech.* **78**, 535–560.

Fureby, C., & Grinstein, F. F. 1999 Monotonically integrated large eddy simulation of free shear flows. *AIAA J.* **37**, 544–556.

Fureby, C., & Grinstein, F. F. 2002 Large eddy simulation of high Reynolds number free and wall bounded flows. *J. Comput. Phys.* **181**, 68–97.

Fureby, C., Grinstein, F. F., & Kailasanath, K. 2000 Large eddy simulation of premixed turbulent flow in a rearward-facing-step combustor. Paper AIAA 00-0863.

Grinstein, F. F. 1994 Open boundary conditions in the simulation of subsonic turbulent shear flows. *J. Comput. Phys.* **115**, 43–55.

Grinstein, F. F. 1995 Self-induced vortex ring dynamics in subsonic rectangular jets. *Phys. Fluids* **7**, 2519–2521.

Grinstein, F. F. 2001 Vortex dynamics and entrainment in regular free jets. *J. Fluid Mech.* **437**, 69–101.

Grinstein, F. F. 2002 Vortex dynamics, entrainment, and non-premixed combustion in rectangular jets, in *Advances in Chemical Propulsion,* ed. G. Roy. Boca Raton, FL: CRC Press, 215–230.

Grinstein, F. F. 2006 On integrating numerical and laboratory turbulent flow experiments, Paper AIAA 2006–3048.

Grinstein, F. F., Boris, J. P., & Griffin, O. M. 1991 Passive pressure-drag control in a plane wake. *AIAA J.* **29**, 1436–1442.

Grinstein, F. F., & DeVore, C. R. 1992 Coherent structure dynamics in spatially developing square jets. Paper AIAA 92-3441 .

Grinstein, F. F., & DeVore, C. R. 1996, Dynamics of coherent structures and transition to turbulence in free square jets. *Phys. Fluids* **8**, 1237–1251.

Grinstein, F. F., & DeVore, C. R. 2002 On global instabilities in countercurrent jets. *Phys. Fluids* **14**, 1095–1100.

Grinstein, F. F., & Fureby, C. 2002 Recent progress on MILES for high Reynolds-number flows. *J. Fluids Eng.* **124**, 848–886.

Grinstein, F. F., & Fureby, C. 2005 LES Studies of the flow in a swirl gas combustor, in *Proceedings of the Combustion Institute.* New York: Elsevier **30**, 1791–1798.

Grinstein, F. F, Fureby, C., & DeVore, C. R. 2005 On MILES based on flux-limiting algorithms. *Int. J. Numer. Meth. Fluids* **47**, 1043–1051.

Grinstein, F. F. & Guirguis, R. H. 1992 Effective viscosity in the simulation of spatially evolving shear flows with monotonic FCT models *J. Comput. Phys.* **101**, 165–175.

Grinstein, F. F., Gutmark, E., & Parr, T. P. 1995 Near-field dynamics of subsonic, free square jets. A computational and experimental study. *Phys. Fluids* **7**, 1483–1497.

Grinstein, F. F., Gutmark, E. J., Parr, T. P., Hanson-Parr, D. M., & Obeysekare, U. 1996 Streamwise and spanwise vortex interaction in an axisymmetric jet. A computational and experimental study. *Phys. Fluids* **8**, 1515–1524 .

Grinstein, F. F., Hussain, F., & Oran, E. S. 1990 Vortex-ring dynamics in a transitional subsonic free jet. A numerical study. *Eur. J. Mech. B Fluids* **9**, 499–525.

Grinstein, F. F., Hussain, F., & Boris, J. P. 1991 Dynamics and topology of coherent structures in a plane wake, in *Advances in Turbulence* 3, ed. A. V. Johansson & P. H. Alfredsson. Berlin: Springer, 34–41.

Grinstein, F. F., Glauser, M. N., & George, W. K. 1995 Vorticity in jets, in fluid vortices, ed. S. I. Green, Dordrecht: Kluwer, 65–94.

Grinstein, F. F., & Kailasanath, K. 1992 Chemical energy release and dynamics of transitional reactive, free shear flows. *Phys. Fluids* **A4**, 2207–2221.

Grinstein, F. F., & Kailasanath, K. 1995 Three-dimensional numerical simulations of unsteady reactive square jets. *Combust. Flame,* **100**, 2–10 erratum, **101**, 192.

Grinstein, F. F., & Kailasanath, K. 1996 Exothermicity and relaminarization effects in reactive square jets. *Combust. Sci. Tech.* **113–114**, 291–312.

Grinstein, F. F., Oran, E. S., & Boris, J. P. 1986 Numerical simulations of asymmetric mixing in planar shear flows. *J. Fluid Mech.* **165**, 201–220.

Grinstein, F. F., Oran, E. S., & Boris, J. P. 1987 Direct numerical simulation of axisymmetric jets. *AIAA J.* **25**, 92–98.

Grinstein, F. F. , Oran, E. S., & Boris, J. P. 1990 Reinitiation and feedback in global instabilities of spatially developing mixing layers. *Phys. Rev. Lett.* **64**, 870–873.

Grinstein, F. F., Oran, E. S., & Boris, J. P. 1991 Pressure field, feedback and global instabilities of subsonic spatially developing mixing layers. *Phys. Fluids* **A3**, 2401–2409.

Grinstein, F. F., Oran, E. S., & Hussain, A. K. M. F. 1989 A Numerical study of mixing control in spatially evolving shear flows. Paper AIAA 89-0977.

Grinstein, F. F., Young, T. R., Gutmark, E. J., LI, G., Hsiao, G., & Mongia, H. C. 2002 Flow dynamics in a swirl combustor. *J. Turb.* **3**, 030, 1–19.

Gutmark, E. J., & Grinstein, F. F. 1999 Flow control with noncircular jets. *Annu. Rev. Fluid Mech.* **31**, 239–272.

Gutmark, E. J., & Ho, C. M. 1983 Preferred modes and the spreading rates of jets. *Phys. Fluids* **26**, 2932–2938.

Ho, C. M., & Huerre, P. 1984 Perturbed free shear layers. *Annu. Rev. Fluid Mech.* **16**, 365–424.

Ho, C. M., & Nosseir, N. S. 1981 Dynamics of an impinging jet. Part 1. The feedback phenomenon, *J. Fluid Mech.* **105**, 119–142.

Huerre. P., & Monkewitz, P. A. 1990 Local and global instabilities in spatially developing flows. *Annu. Rev. Fluid Mech.* **22**, 473–537.

Husain, H. S., & Hussain, A. K. M. F. 1983 Controlled excitation of elliptic jets, *Phys. Fluids* **26**, 2763–2765.

Husain, H. S., & Hussain, F. 1991 Elliptic jets. Part 2. Dynamics of coherent structures: Pairing. *J. Fluid Mech.* **233**, 439–482.

Husain, H. S., & Hussain, F. 1993 Elliptic jets. Part 3. Dynamics of preferred mode coherent structure. *J. Fluid Mech.* **248**, 315–361.

Hussain, A. K. M. F. 1986 Coherent structures and turbulence. *J. Fluid Mech.*, **173**, 303–356.

Hussain, F., & Husain, H.S. 1989 Elliptic jets. Part I. Characteristics of unexcited and excited Jets. *J. Fluid Mech.* **208**, 257–320.

Jeong, J., Grinstein, F. F., & Hussain, F. 1994 Eduction of coherent structures in a numerically simulated plane wake. *Appl. Sci. Res.* **53**, 227–236.

Jeong, J., & Hussain, F. 1995 On the identification of a vortex. *J. Fluid Mech.* **285**, 69–94.

Jiménez, J. 1992 Kinematic alignment effects in turbulent flows. *Phys. Fluids* **A4**, 652–654.

Jiménez, J., Wray, A., Saffman, P., & Rogallo, R. 1993 The structure of intense vorticity in isotropic turbulence. *J. Fluid Mech.* **255**, 65–90.

Kerr, R. M. 1985 Higher order derivative correlation and the alignment of small scale structures in numerical turbulence. *J. Fluid Mech.* **153**, 31–58.

Kibens, V. 1979 Discrete noise spectrum generated by an acoustically excited jet. *AIAA J.* **18**, 434–441.

Kuo, A. Y., & Corrsin, S. 1971 Experiment on the geometry of the fine-structure regions in fully developed turbulent fluid. *J. Fluid Mech.* **56**, 447–479.

Laufer, J., & Monkewitz, P. A. 1980 On turbulent jet flows: A new perspective. Paper AIAA 80-0962.

Leweke, T., & Williamson C. H. K. 1998 reconnection of a counterrotating vortex pair, in *Advances in Turbulence VII*, ed. U. Frisch. Dordrecht: Kluwer, 55–58.

Liepmann, D., & Gharib, M. 1992 The role of streamwise vorticity in the near-field entrainment of round jets. *J. Fluid Mech.* **245**, 643–668.

Margolin, L. G., & Rider W. J. 2002 A rationale for implicit turbulence modeling. *Int. J. Numer. Meth. in Fluids* **39**, 821–841.

Margolin, L. G., Rider W. J., & Grinstein, F. F. 2006 Modeling turbulent flow with implicit LES. *J. Turb.* **7** (15), 1–27.

Melander, M. V., & Hussain F. 1989 Cross-linking of two antiparallel vortex tubes. *Phys. Fluids A* **1**, 633–636.

Michalke, A. 1965 On spatially growing disturbances in an inviscid shear layer. *J. Fluid Mech.* **23**, 521–544.

Michalke, A. 1984 Survey of jet instability theory. *Prog. Aerospace Sci.* **21**, 159–199.

Morkovin, M. 1988 *Recent Insights into Instability and Transition to Turbulence in Open-Flow Systems.* NASA Langley Research Center, Institute for Computer Applications in Science and Engineering: ICASE Report 88-44.

Oberman, E. A., & Zabusky, N. J. 1982 Evolution and merger of isolated vortex structures. *Phys. Fluids* **25**, 1297–1305.

Oran, E. S., & Boris, J. P. 2001 *Numerical Simulation of Reactive Flow.* New York: Cambridge University Press, 2nd ed.

Poinsot, T. J., & Lele, S. K. 1992 Boundary conditions for direct simulations of compressible viscous flows. *J. Comput. Phys.* **101**, 104–129.

Pope, S. B. 2004 Ten questions concerning the large eddy simulation of turbulent flows. *New J. Phys.* **6**, 35.

Porter, D. H., Pouquet, A., & Woodward, P. R. 1994 Kolmogorov-like spectra in decaying three-dimensional supersonic flows. *Phys. Fluids* **6**, 2133–2142.

Saddoughi, S. G., 1992 Local isotropy in high Reynolds number turbulent shear flows, in *CTR Annual Research Briefs*. Stanford, CA: NASA Ames/Stanford University, 237–262.

Saddoughi, S. G. 1997 Local isotropy in complex turbulent boundary layers at high Reynolds number. *J. Fluid Mech.* **348**, 201–245.

Sagaut, P. 2005 *Large Eddy Simulation for Incompressible Flows*. New York: Springer, 3rd ed.

Siggia, E. D., & Patterson, G. S. 1978 Intermittency effects in a numerical simulation of stationary three dimensional turbulence. *J. Fluid Mech.* **86**, 567–592.

Steinhoff, J., & Underhill, D. 1994 Modification of the Euler equations for 'Vorticity Confinement': Application to the computation of interacting vortex rings. *Phys. Fluids* **6**, 2738–2744.

Tennekes, H., & Lumley, J. L. 1972 *A First Course in Turbulence*. Cambridge, MA: MIT Press.

Zalesak, S. T. 1979 Fully multidimensional flux-corrected transport algorithms for fluids. *J. Comput. Phys.* **31**, 335–362.

9 Symmetry Bifurcation and Instabilities

Dimitris Drikakis

9.1 Introduction

The importance of investigating nonlinear bifurcation phenomena in fluid mechanics lies in enabling a clearer understanding of hydrodynamic stability and the mechanism of laminar-to-turbulent flow transition. Bifurcation phenomena have been observed in a number of laboratory flows, with incompressible flow in sudden expansions being one of the classical examples. At certain Reynolds numbers, these flows present instabilities that may lead to bifurcation, unsteadiness, and chaos (Mullin 1986).

For example, the existence of symmetry-breaking bifurcation in suddenly expanded flows has been demonstrated (Chedron, Durst, and Whitelaw 1978; Fearn, Mullin, and Cliffe 1990). This is manifested as an asymmetric separation that occurs beyond a certain value of Reynolds number. Similarly, Mizushima et al. (Mizushima, Okamoto, and Yamaguchi 1996; Mizushima and Shiotani 2001) have conducted experimental investigations to extend suddenly expanded flows to suddenly expanded and contracted channel flow. They found that this type of geometry exhibits similar flow effects to the simpler suddenly expanded channel, with instabilities manifesting as asymmetric separation at Reynolds numbers within a critical range. In the experiments, the instabilities were triggered by geometrical imperfections and asymmetries in the inflow conditions upstream of the expansion. In a symmetric numerical setup, however, these asymmetries can only be generated by the numerical scheme and are associated with dissipation and dispersion properties of the numerical method employed. In the past, computational investigations have been conducted for unstable separated flows through sudden expansions (Alleborn et al. 1997; Drikakis 1997). In particular, numerical experiments by Patel and Drikakis (2004) using explicit (symmetric) solvers and different high-resolution schemes were conducted to show that symmetry breaking depends solely on the details of the numerical scheme employed for the discretization of the advective terms. The details of the method used here are described in Chapter 4a.

9.2 Results

9.2.1 Symmetry breaking in a suddenly expanded–contracted channel

An example showing the effects of the high-resolution method on symmetry breaking is presented here. The phenomenon of asymmetric separation in a suddenly expanded and contracted channel has been discussed by Mizushima et al. (1996; also see Mizushima and Shiotani 2001) both experimentally and numerically. At very low Reynolds numbers, the flow remains symmetric, with separation regions of equal length on both channel walls directly after the sudden expansion and before the sudden contraction. As the Reynolds number is increased, the separation length is also increased. A critical value for asymmetries is reached when one recirculation region grows at the expense of the other. A further increase in the Reynolds number makes the asymmetry between the two recirculation regions become more prominent to the extent that both bubbles can reach the size of the entire length of the expanded part. A second critical value is reached as the Reynolds number is further increased where the separation returns to a stable symmetric solution.

Patel and Drikakis (2004) performed a numerical investigation of high-resolution methods for the aforementioned experimental setup. They used different high-resolution (Godunov-type) schemes (Drikakis and Rider 2004), namely the characteristics-based scheme (Drikakis, Govatsos, and Papatonis 1994) and Rusanov's scheme (Rusanov 1961), for solving the incompressible flow equations. They used a sudden expansion–contraction geometry with a 1:3 expansion ratio in the simulations. They obtained a comparison among the various numerical schemes for several flow cases on two different grids in order to test whether the asymmetric separation was affected by the numerical scheme. Furthermore, they performed computations by using first-, second-, and third-order interpolation schemes for the intercell variables (see Chapter 4a). The characteristics-based scheme led to an asymmetric flow for both grid-independent and underresolved grid conditions whereas the Rusanov scheme (similar results were also obtained with the HLLE scheme–Einfeldt 1988) led to symmetric solutions for the same Reynolds number (Re = 120). With the use of higher-order interpolations (second and third order) at the same Reynolds number, all the numerical schemes investigated led to asymmetric separation for both of the grid sizes investigated. Figure 9.1 shows grid-independent results for the characteristics-based scheme and the Rusanov scheme (based on first-order Godunov-type interpolation) for a Re = 120. The results show the sensitivity of symmetry-breaking computations to the high-resolution numerical variant employed.

9.2.2 Shock propagation in an enclosure

A second problem that shows the dependence of transition and symmetry breaking on the high-resolution method employed for solving the equations is blast-wave propagation in an enclosure (a square box as seen in Figure 9.2). The problem features

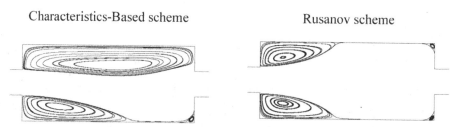

Figure 9.1. Symmetric and asymmetric solutions as obtained by different high-resolution schemes. The schemes are accurate to second order and have been combined with first- and second-order Godunov-type interpolation for the cell-face variables (Drikakis and Rider 2004).

several shock interactions and Richtmyer–Meshkov instabilities and has been investigated numerically by a few researchers (Bagabir and Drikakis 1999, 2004; Marconi 1994; Sakurai 1965). The initial condition is a one-dimensional symmetric blast, before the blast reflects from the walls (Marconi 1994; Sakurai 1965), which can be obtained by solving the one-dimensional equations of motion for an arbitrary strong shock running into a gas at rest. There is a very hot region in the center of the explosion with very low density and finite pressure, which slowly expands as time goes on. The radial velocity is linearly decreased from the shock value to zero at the origin. After an initial start-up period, all variables approach a self-similar variation between the origin and the shock.

It is known that when two fluids of different density are impulsively accelerated into each other by a shock wave, and when at the same time the interface separating them is not perfectly flat, a shock excitation occurs; this is known as a Richtmyer–Meshkov (or impulsive Rayleigh–Taylor) instability (Meshkov 1970; Richtmyer 1960). The computations of Marconi (1994) and Bagabir and Drikakis (2004) revealed that a similar instability occurs in the case of blast propagation in an enclosure. The instability

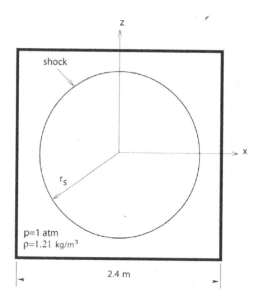

Figure 9.2. Schematic of the shock-propagation problem in an enclosure.

initiates after the reflection of the blast from the walls and its interaction with the low-density region around the center of the enclosure. The instability manifests itself as asymmetric flow, initially located around the center of the enclosure, but spreading quickly throughout the flowfield. The occurrence of the instability, however, depends strongly on the numerical scheme employed. Bagabir and Drikakis (2004) carried out computations for six high-resolution methods; these showed a variety of results, depending on the advective scheme employed. We note that they also investigated time-integration (both explicit and implicit) schemes, but they found that the manifestation of the instability and transition to symmetry breaking depends solely on the advective scheme employed.

The high-resolution methods used were as follows: (i) the Harten–Lax–van Leer (HLL) Riemann solver (Harten et al. 1983); (ii) the HLLC Riemann solver (C stands for contact discontinuity) by Toro, Spruce, and Speares (1994); (iii) Roe's Riemann solver (Roe 1981; Roe and Pike 1984); (iv) a variant of the Steger–Warming (1981) flux-vector splitting (FVS) scheme (Drikakis and Tsangaris 1993; Zoltak and Drikakis 1998); (v) the van Leer FVS (VL-FVS) scheme (van Leer 1982); the Rusanov scheme (Rusanov 1961); and (vi) a hybrid total variation diminishing scheme (CBM-FVS) based on the combination of the characteristic-based method of Eberle, Rizzi, and Hirschel (1992) and the SW-FVS (Drikakis and Tsangaris 1993; Zoltak and Drikakis 1998). All of these schemes were implemented in conjunction with third-order interpolation (reconstruction) of the conservative variables at the cell faces. This is achieved through the monotone-upstream scheme for conservation laws, based on the formulation proposed by Thomas, van Leer, and Walters (1985). Results obtained by these schemes are shown in Figure 9.3 by means of iso-Mach contours. The asymmetric flow as predicted by the CBM-FVS, Roe, and HLLC schemes has already been initiated at the center of the enclosure at $t = 6$ ms and spreads further as times goes by. Computations using an explicit fourth-order Runge–Kutta scheme led to similar asymmetric flow patterns. Therefore, the prediction of the instability is not a numerical artifact arising from the implicit solver. Computations using the SW-FVS, VL-FVS, and HLL schemes did not lead to flow asymmetries. The flow remains symmetric throughout the computation. Although there is no experimental evidence about the occurrence of instability for the flow in question, we speculate that the symmetric flow obtained by the SW-FVS, VL-FVS, and HLL schemes is the result of excessive numerical dissipation.

9.2.3 Interaction of a shock wave with a bubble

The effects of numerical dissipation on flows featuring instabilities can also be demonstrated through simulations for the interaction of a shock wave with an inhomogeneous surface, such as a bubble. When two motionless gases are brought into contact, they mix by diffusion at a rate that is proportional to the product of the contact area and the concentration gradient at the interface. If the gases are set into motion by a passing shock wave, the mixing rate will increase because the motion produces an increase in

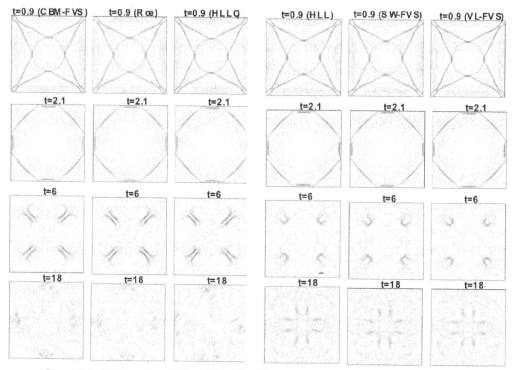

Figure 9.3. Stable and unstable solutions (iso-Mach line contours) for shock-wave propagation in an enclosure using high-resolution schemes.

the contact area. The physical problem can be modeled by a planar shock wave propagating in a rectangular shock tube, interacting with a bubble (Figure 9.4). The bubble may be lighter or denser than the surrounding medium. The shock–bubble interaction results in refraction, diffraction, and reflection of the shock wave. In Haas and Sturtevant (1987), the interaction of a planar shock wave with a bubble as a model problem was proposed for studying vorticity and turbulence generation in compressible flows with shock waves.

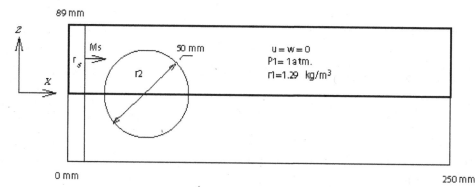

Figure 9.4. Computational setup for simulations of shock–bubble interaction in a shock tube.

Bagabir and Drikakis (2004) chose the aforementioned setup to examine properties of high-resolution methods. A planar shock wave propagates through air from the left to the right of a tube and impinges on a cylindrical bubble. The cylindrical bubble has a radius of 25 mm and the vertical dimension of the shock tube is 89 mm. The horizontal dimension of the shock tube is 10 times the radius of the bubble. The incident shock Mach number is $M_s = 1.22$. In the experiment by (Haas and Sturtevant (1987), the cylindrical bubble was produced by a very thin nitrocellulose membrane. Therefore, good control was exercised over the shape of the bubble, and the flow was almost two dimensional (Picone and Boris, 1988; Quirk and Karni, 1996). The present numerical model simulates a two-dimensional cross section of the cylindrical bubble using Cartesian coordinates. Similar to previous studies (Picone and Boris, 1988; Quirk and Karni, 1996), the flowfield is assumed to be symmetric about the axis of the shock tube; therefore, only the top half of the flowfield has to be computed. The numerical experiments were performed by use of a 500×100 grid corresponding to $Dx = Dz = 0.445$ mm, thus yielding that one bubble radius is covered by about 56 cells. The upper and lower boundaries of the domain are the shock-tube wall and axis of symmetry, respectively. The left boundary corresponds to the region behind the shock wave. The Rankine–Hugoniot relations obtain the flow conditions at this boundary. Smooth outflow conditions are assumed for the reflected waves at the left boundary, and all variables are extrapolated from the interior of the domain. The right boundary is also considered to permit smooth outflow of any rightward-moving waves by maintaining a zero-gradient condition for all variables. The bubble is assumed to be both in thermal and mechanical equilibrium with the surrounding air. The time, t, has become dimensionless using the shock velocity and radius of the bubble, that is, $t = T/(r/sM_s)$, where s, r, and T are the sound-speed of ambient air, radius of the bubble, and real time, respectively. The dimensionless time, $t = 0$, corresponds to the first collision of the shock wave with the bubble.

The numerical experiments in Bagabir and Drikakis (2004) were performed for a helium bubble, which is lighter than the surrounding air (the density ratio of the bubble to the ambient air is 0.166). For the present grid resolution, the results obtained by different high-resolution schemes (the methods are described in Chapter 4a and Section 2.2) are shown in Figure 9.5. The high-resolution schemes employed here represent discontinuities as being smeared over two to three cells. The computed flow patterns were found to be in good agreement with the experimental flow visualizations (Haas and Sturtevant 1987). At $t = 0.9$, the incident shock is reflected on the surface of the bubble and is transmitted (refracted) through the bubble. At later time instants, the bubble interface takes a kidney-like shape. This is due to a high-speed jet formed at the upstream interface. At $t = 3$ onward, the solutions obtained by the CBM-FVS, HLLC, and Roe schemes exhibit ripples at the bubble interface. These ripples are not a numerical defect; they are a manifestation of the Richtmyer–Meshkov instability. The SW-FVS, VL-FVS, HLL, and Rusanov schemes result in slightly smoother and thicker interfaces, but what is most important is that they do not capture the Richtmyer–Meshkov instability (note that this does not, however, affect the results for the position of the bubble

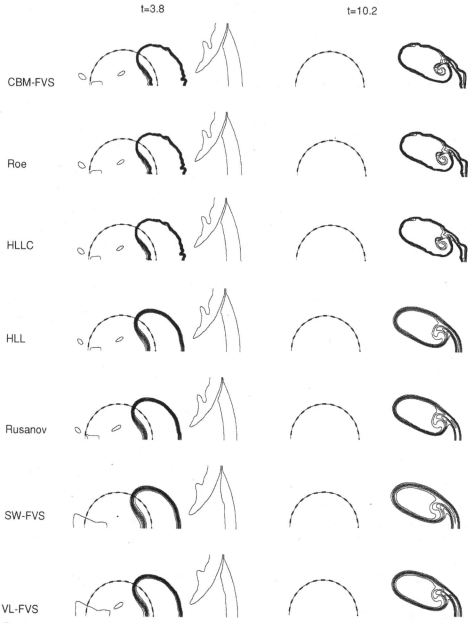

Figure 9.5. Results for the shock-bubble interaction using different high-resolution schemes. The circle line shows the initial position of the bubble and the left and right columns show the evolution of interaction at two different time instants.

interface). The CBM-FVS, HLLC, and Roe schemes capture the discontinuities within two computational cells, whereas the SW-FVS, VL-FVS, HLL, and Rusanov schemes smear it over four cells.

During the investigation, we also carried out numerical experiments using a longer domain in order to allow the deformation of the bubble to continue further. In a manner similar to the experiment of Jacobs (1992), we found that the upstream portion of the bubble gradually shrinks and eventually disappears. However, when considering these observations, bear in mind the uncertainties associated with the numerical modeling. In the case of the present inviscid solution, the shrinking or disappearance of the bubble material is entirely due to numerical diffusion. Therefore, a direct comparison between such simulations and real-world effects of molecular diffusion and turbulent mixing should be considered with extreme prudence.

REFERENCES

Alleborn, N., Nanakumar, K., Raszillier, F., &. Durst, F. 1997 Further contributions on the two-dimensional flow in a sudden expansion. *J. Fluid Mech.* **330**, 169–188.

Bagabir, A., & Drikakis, D. 1999 On the Richtmyer–Meshkov instability produced by blast-wave propagation in an enclosure, in *Proceedings of the 22nd Symposium on Shock Waves,* ed. G. J. Ball, G. T. Roberts, and R. Hillier. Fyshwick, Australia: Panther 865–870.

Bagabir, A., & Drikakis, D. 2004 Numerical experiments using high-resolution schemes for unsteady, inviscid, compressible flows. *Comput. Meth. Appl. Mech. Eng.* **193/42–44**, 4675–4705.

Battaglia, F., Tavener, S. J., Kulkarni A. K., & Merkle, C. L. 1997 Bifurcation of low Reynolds number flows in symmetric channels. *AIAA J.* **35**, 99–105.

Chedron, W., Durst, F., and Whitelaw, J. H. 1978 Asymmetric flows and instabilities in symmetric ducts with sudden expansions. *J. Fluid Mech.* **84**, 13–31.

Drikakis, D. 1997 Bifurcation phenomena in incompressible sudden expansion flows. *Phys. Fluids* 9, 76–87.

Drikakis, D., Govatsos, P. A., & Papatonis, D. E. 1994 A characteristic-based method for incompressible flows. *Int. J. Numer. Meth. Fluids* **19**, 667–685.

Drikakis, D., & Rider, W. J. 2004 *High-Resolution Methods for Incompressible and Low-Speed Flows.* New York: Springer.

Drikakis, D., & Tsangaris, S. 1993 On the accuracy and efficiency of CFD methods in real gas hypersonics. *Int. J. Numer. Meth. Fluids* **16**, 759–775.

Eberle, A., Rizzi, A., & Hirschel, E. H. 1992 *Numerical Solutions of the Euler Equations for Steady Flow Problems.* Vol. 34 of Notes on Numerical Fluid Mechanics. Wiesbaden: Vieweg Verlag.

Einfeldt, B. 1988 On Godunov-Type methods for gas dynamics. *SIAM J. Numer. Anal.* **25**, 294–318.

Fearn, R. M., Mullin T., & Cliffe, K. A. 1990 Nonlinear flow phenomena in a symmetric sudden expansion. *J. Fluid Mech.* **211**, 595–608.

Harten, A., Lax, P. D., & Van Leer, B. 1983 On upstream differencing and Godunov-type schemes for hyperbolic conservation laws. *SIAM Rev.* **25**, 35–61.

Haas, J. F., & Sturtevant, B. 1987 Interaction of weak shock waves with cylindrical and spherical gas inhomogeneities. *J. Fluid Mech* **181**, 41–76.

Jacobs, J. W. 1992 Shock-induced mixing of a light-gas cylinders. *J. Fluid Mech.* **234**, 629–649.

Marconi, F. 1994 Investigation of the interaction of a blast wave with an internal structure. *AIAA J.* **32**, 1561–1567.

Meshkov, E. E. 1970 *Instability of a Shock Wave Accelerated Interface Between Two Gases.* NASA: NASA Translation TT F-13, 074 R. F., 151–158.

Mizushima, J., Okamoto, H., & Yamaguchi, H. 1996 Stability of flow in a channel with a suddenly expanded part. *Phys. Fluids* **8**, 2933–2942.

Mizushima, J., & Shiotani, Y. 2001 Transitions and instabilities of flow in a symmetric channel with a suddenly expanded and contracted part. *J. Fluid Mech.* **434**, 355–369.

Mullin, T., & Cliffe, K. A. 1986 Symmetry-breaking and the onset of time dependence in fluid mechanical systems, in *Nonlinear Phenomena and Chaos*, ed. S. Sarkar. Adam Hilger, 96–112.

Patel, S., & Drikakis, D. 2004 *Prediction of Flow Instabilities and Transition Using High-Resolution Methods*, CD-Rom Proceedings of the ECCOMAS Congress.

Picone, J. M., & Boris, J. P. 1988 Vorticity generation by shock propagation through bubbles in a gas. *J. Fluid Mech.* **189**, 23–51.

Quirk, J. J., & Karni, S. 1996 On the dynamics of a shock-bubble interaction. *J. Fluid Mech.* **318**, 129–163.

Richtmyer, R. D. 1960 Taylor instability in shock acceleration of compressible fluids. *Commun. Pure Appl. Math.* **13**, 297–319.

Rusanov, V. V. 1961 Calculation of interaction of non-steady shock waves with obstacles. *J. Comput. Math. Phys. USSR* **1**, 267–279.

Roe, P. L. 1981 Approximate Riemann solvers, parameter vectors and difference schemes. *J. Comput. Phys.* **43**, 357–372.

Roe, P. L., & Pike, J. 1984 Efficient construction and utilisation of approximate Riemann solutions, in *Computing Methods in Science and Engineering*. New York: North-Holland.

Sakurai, A. 1965 *Blast Wave Theory: Basic Developments in Fluid Dynamics*, ed. M. Holts. New York: Academic Press.

Steger, J. L., & Warming, R. F. 1981 Flux vector splitting of the inviscid gasdynamic equations with applications to finite difference methods. *J. Comput. Phys.* **40**, 263–293.

Thomas, J., van Leer, B., & Walters, R. W. 1985 Implicit flux split scheme for the Euler equations. Paper AIAA 85-1680.

Toro, E. F., Spruce, M., & Speares, W. 1994 Restoration of the contact surface in the HLL–Riemann solver. *Shock Waves J.* **4**, 25–34.

van Leer, B. 1982 *Flux-Vector Splitting for the Euler Equations*. NASA Langley Research Center: ICASE Report 82-30.

Zoltak J., & Drikakis, D. 1998 Hybrid upwind methods for the simulation of unsteady shock-wave diffraction over a cylinder. *Comput. Meth. Appl. Mech. Eng.* **162**, 165–185.

10 Incompressible Wall-Bounded Flows

Christer Fureby, Mattias Liefvendahl, Urban Svennberg, Leif Persson, and Tobias Persson

10.1 Introduction

Almost all flows of practical interest are turbulent, and thus the simulation of turbulent flow and its diversity of flow characteristics remains one of the most challenging areas in the field of classical physics. In many situations the fluid can be considered incompressible; that is, its density is virtually constant in the frame of reference, moving locally with the fluid, but density gradients may be passively convected with the flow. Examples of such flows of engineering importance are as follows: external flows, such as those around cars, ships, buildings, chimneys, masts, and suspension bridges; and internal flows, such as those in intake manifolds, cooling and ventilation systems, combustion engines, and applications from the areas of biomedicine, the process industry, the food industry, and so on. In contrast to free flows (ideally considered as homogeneous and isotropic), wall-bounded flows are characterized by much less universal properties than free flows and are thus even more challenging to study. The main reason for this is that, as the Reynolds number increases, and the thickness of the viscous sublayer decreases, the number of grid points required to resolve the near-wall flow increases.

The two basic ways of computing turbulent flows have traditionally been direct numerical simulation (DNS) and Reynolds-averaged Navier–Stokes (RANS) modeling. In the former the time-dependent Navier–Stokes equations (NSE) are solved numerically, essentially without approximations. In the latter, only time scales longer than those of the turbulent motion are computed, and the effect of the turbulent velocity fluctuations is modeled with a turbulence model. More recently, the time-scale separation requirement in RANS is often relaxed allowing the mean flow to evolve according to its natural instabilities. In such unsteady RANS (URANS), the resolved flow is usually taken to be unsteady, and a model is applied to properly account for the turbulence and its effects. Early on, it was realized that DNS was too expensive for industrial use, being also limited to low Reynolds numbers, while RANS was too dependent on the characteristics of particular flows to be used as a method of general applicability. Large eddy simulations (LES) was then developed as an intermediate approximation between RANS and DNS; the general idea was that the large, nonuniversal scales of the flow were to be computed explicitly, as in DNS, while the small scales were modeled (see Chapter 3). The hope was that the small scales, which are removed from the flow inhomogeneities and boundary conditions by several steps of the turbulent cascade, would be universal and isotropic enough for a simple model to be able to represent its effects in most situations. More recently, attempts have been made to combine RANS with LES in order to exploit the best of both methods in a complementary manner. Such hybrid RANS–LES methods have been developed that restrict the use of LES to flow regions where it is needed while RANS is used elsewhere.

In this chapter we discuss the computation of wall-bounded incompressible flows by use of classical and implicit LES models. Our overall aim is to examine the predictive capabilities of classical and implicit LES models, and to compare explicit or implicit subgrid-scale (SGS) models. We proceed by examining a series of flow problems, ranging from canonical cases to building block flows. We carry out comparisons between classical and implicit LES predictions with reference to experimental or DNS data. In some cases, we also include comparison to state-of-the-art RANS results. The need for such systematic comparison and validation studies is clear from the ongoing research activities (e.g., AFRL 2004; AGARD 2001). The numerical method used in this chapter are described in Chapter 4a.

10.1.1 Overview of CFD models for incompressible flows

In RANS models (Launder and Spalding 1972; Pope 2000), we obtain equations for the statistical average of the flow by averaging the NSE over homogeneous directions, time, or an ensemble of flows. Since most flows involve inhomogeneous flow, time averaging is most appropriate, resulting in

$$\nabla \cdot (\langle \mathbf{v} \rangle \otimes \langle \mathbf{v} \rangle) = -\nabla \langle p \rangle + \nabla \cdot (\langle \mathbf{S} \rangle - \mathbf{R}), \quad \nabla \cdot \langle \mathbf{v} \rangle = 0, \qquad (10.1)$$

where $\mathbf{v}$ is the velocity, p is the pressure, $\mathbf{S} = 2\nu \mathbf{D}$ is the viscous stress tensor, $\mathbf{D} = \frac{1}{2}(\nabla \mathbf{v} + \nabla \mathbf{v}^T)$ is the rate-of-strain tensor and ν is the molecular viscosity, $\langle \ \rangle$ is the time average, $\mathbf{v} = \langle \mathbf{v} \rangle + \mathbf{v}'$, and $\mathbf{R} = \langle \mathbf{v}' \otimes \mathbf{v}' \rangle$ is the Reynolds stress tensor. Note that URANS modeling (Ferziger 1983) is a more recent alternative for incorporating the large-scale flow that is excluded in RANS. To close the RANS model (10.1), we use a statistical turbulence model. This is a combination of algebraic and transport equations relating the turbulence parameters to the flow variables. To this end, $\mathbf{R}$ is often modeled by use of Boussinesq's hypothesis, $\mathbf{R} = 2\nu_t \langle \mathbf{D} \rangle_D$, where $\nu_t = c_\mu k^2 / \varepsilon$ is the turbulent viscosity, c_μ is a model coefficient, k is the turbulent kinetic energy, and ε is its dissipation rate. A well-known model is the $k - \varepsilon$ model (Jones and Launder 1972),

$$\nabla \cdot (k \langle \mathbf{v} \rangle) = P_k + \nabla \cdot [(\nu + \nu_t / \sigma_k) \nabla k] - \varepsilon,$$

$$\nabla \cdot (\varepsilon \langle \mathbf{v} \rangle) = P_\varepsilon + \nabla \cdot [(\nu + \nu_t / \sigma_\varepsilon) \nabla \varepsilon] - R, \qquad (10.2)$$

where $P_k = \mathbf{R} \cdot \langle \mathbf{D} \rangle = 2\nu_t \langle \mathbf{D} \rangle^2$ and $P_\varepsilon = c_1 \langle \mathbf{D} \rangle \varepsilon$ are the production terms and $R = c_2 \varepsilon^2 / (k + \sqrt{\nu \varepsilon})$ is the destruction term (Launder and Sharma 1974). The model coefficients take the values $\sigma_k = 1.00$, $\sigma_\varepsilon = 1.30$, $c_\mu = 0.09$, $c_1 = 1.44$, and $c_2 = 1.92$. There are more advanced models, such as the Reynolds stress equation models (Launder, Reece, and Rodi 1975), for example,

$$\nabla \cdot (\mathbf{R} \otimes \langle \mathbf{v} \rangle) = \mathbf{P}_D + \nabla \cdot (\nu_k \mathbf{R} \nabla \mathbf{R}) + c_1 \tfrac{\varepsilon}{k} \mathbf{R} - \tfrac{2}{3} c_2 \varepsilon \mathbf{I}, \tag{10.3}$$

where $\mathbf{P} = \mathbf{R} \langle \mathbf{D} \rangle^T + \langle \mathbf{D} \rangle \mathbf{R}$ and $\nu_t = c_\mu k^2 / \varepsilon$ are needed in order to incorporate, for example, curvature effects that are important in complex geometry flows. The model coefficients are $c_\mu = 0.22$, $c_1 = 0.60$, and $c_2 = 1.80$, respectively. For the near-wall treatment, either wall models (Wilcox 1998) or low-Reynolds-number versions of the RANS models can be used; see Wilcox (1998).

In explicit LES, as described in Chapter 3, the dependent variables and the NSE are low-pass filtered to eliminate the small-scale flow features, appearing instead as the SGS stress terms, whereas in implicit LES, the discretization filters out the small-scale flow features, leaving the leading-order truncation error to act as a built-in (implicit) SGS model. The two LES models can best be described and compared by use of the modified equations analysis (MEA) as discussed in Chapter 4a, Chapter 5, and references therein. Following the MEA in Chapter 4a, we find that the explicit and implicit LES models can be expressed in the same general functional form, that is,

$$\nabla \cdot \mathbf{v} = m^\rho,$$

$$\partial_t (\mathbf{v}) + \nabla \cdot (\mathbf{v} \otimes \mathbf{v}) = -\nabla p + \nabla \cdot (\mathbf{S} - \mathbf{B}) + \tau + \mathbf{m}^v, \tag{10.4}$$

where $\mathbf{B}$ is the explicit or implicit SGS stress tensor, $\tau = \nabla \cdot [(\tfrac{1}{6} \nu \nabla^3 \mathbf{v} - \tfrac{1}{8} \nabla^2 \mathbf{v})(\mathbf{d} \otimes \mathbf{d}) + \cdots]$ is the leading-order truncation error (resulting from a second-order-accurate finite-volume scheme), and $m^\rho = [G*, \nabla] \mathbf{v}$ and $\mathbf{m}^v = [G*, \nabla](\mathbf{v} \otimes \mathbf{v} + p\mathbf{I} - \mathbf{S})$ is the commutation error (resulting from interchanging filtering and derivative operations). The commutation error terms are exclusive to conventional LES and do not occur in implicit LES. In explicit LES, the SGS stress tensor must be modeled to close (10.4). Many such models are available (Sagaut 2005, see also Chapter 3). Here we use the following.

- We use SGS viscosity models, in which $\mathbf{B} = -2\nu_k \mathbf{D}$, where we model the SGS viscosity, ν_k, by using any of the following submodels:
  - the one-equation eddy viscosity model (OEEVM; Kim and Menon, 1999; Schumann 1975), in which $\nu_k = c_k \Delta \sqrt{k}$, where $\partial_t(k) + \nabla \cdot (k\mathbf{v}) = -\mathbf{B} \cdot \mathbf{D} + \nabla \cdot (\mu_k \nabla k) + \varepsilon$, in which $\varepsilon = c_\varepsilon k^{3/2} / \Delta$ is the SGS dissipation. The model coefficients (c_k and c_ε) are typically evaluated under the assumption of an infinite inertial subrange.
  - the Smagorinsky model (Smagorinsky 1963), in which $\nu_k = c_D \Delta^2 \|\mathbf{D}\|$. The model coefficient (c_D) is evaluated either under the assumption of an infinite inertial subrange (denoted by SMG), or by use of a dynamic approach (denoted by DSMG; Germano et al. 1994).

- the mixed model MM; (Bardina, Ferziger, and Reynolds 1980), which results from combining a scale-similarity and a SGS viscosity model to result in $\mathbf{B} = (\overline{\mathbf{v} \otimes \mathbf{v}} - \bar{\mathbf{v}} \otimes \bar{\mathbf{v}}) - 2\nu_k \mathbf{D}$. Here, ν_k is obtained from the OEEVM. The MM is not only dissipative but redistributes energy between the smallest resolved scales similar to that of the exact SGS stresses.
- the differential stress equation model (DSM; Deardorff 1973, Fureby et al. 1997), which is based on solving $\partial_t(\mathbf{B}) + \text{div}(\mathbf{B} \otimes \mathbf{v}) = -(\mathbf{L}\mathbf{B}^T + \mathbf{B}\mathbf{L}^T) + \nabla \cdot (\nu_k \nabla \mathbf{B}) - c_1 \varepsilon \mathbf{B}_D + \frac{2}{5}k\mathbf{D}_D - \frac{2}{3}\varepsilon \mathbf{I}$, where $\varepsilon = c_\varepsilon k^{3/2}/\Delta$ is the SGS dissipation.

As for the implicit LES model (ILES) used here, the hybridization of a low-order and high-order flux function for the convective terms results in a built-in (or implicit) SGS stress tensor of the form $\mathbf{B} = \mathbf{C}(\nabla \mathbf{v})^T + (\nabla \mathbf{v})\mathbf{C}^T + \chi^2(\nabla \mathbf{v})\mathbf{d} \otimes (\nabla \mathbf{v})\mathbf{d}$, where $\mathbf{C} = \chi(\mathbf{v} \otimes \mathbf{d})$ and $\chi = \frac{1}{2}(1 - \Psi)(\beta^- - \beta^+)$; Ψ is the flux limiter. The built-in SGS stress tensor can be split into $\mathbf{B}^{(1)} = \mathbf{C}(\nabla \mathbf{v})^T + (\nabla \mathbf{v})\mathbf{C}^T$ and $\mathbf{B}^{(2)} = \chi^2(\nabla \mathbf{v})\mathbf{d} \otimes (\nabla \mathbf{v})\mathbf{d}$, in which the former is a tensor-valued SGS viscosity model and the latter is of a form similar to the Clark model (Clark, Ferziger, and Reynolds 1979).

Prescribing and enforcing wall-boundary conditions presents practical challenges, because we must deal with viscous boundary layers that typically cannot be resolved by practical computational grids. As Reynolds number increases and the thickness of the viscous sublayer decreases, the number of grid points required to resolve the near-wall flow increases. Bagget, Jiménez, and Kravchenko (1997) estimated the number of grid points required for a wall-resolved LES scales as $O(\text{Re}_\tau^2)$, where Re_τ is the friction velocity Reynolds number. Unless the grid is sufficiently fine, the anisotropy of the flow will cause anisotropy of the SGS flow, necessitating SGS models capable of handling simultaneous flow and grid anisotropy. Following Fureby, Alin, et al. (2004), we have that an appropriate grid is such that $\Delta x^+ < 200$, $\Delta y^+ < 2$, and $\Delta z^+ < 30$, where the superscript plus sign denotes nondimensionalization by the viscous length scale, ν/u_τ, and the friction velocity, $u_\tau = \tau_w^{1/2}$, where $\tau_w = \nu(\partial v/\partial y)|_w$ is the wall-shear stress. High Reynolds number flows in complex geometries are too expensive to compute with LES, unless particular techniques are invoked to alleviate the severe resolution requirements near the wall. In principle, we can achieve this by following one of these approaches:

1. Modify a traditional SGS model to accommodate integration all the way to the wall by adding a van Driest type of damping function, $f = 1 - e^{-u_\tau y/A\nu}$ (henceforth denoted by adding "+ VD" to the baseline SGS model name).
2. Use methods similar to detached eddy simulation (DES; Spalart et al. 1997).
3. Use SGS wall models (Wikström et al. 2004).
4. Use SGS simulation models, which may be based on multiscale methods (Fureby, Persson, and Svanstedt 2002; Kemenov and Menon 2002).

In order to make LES of wall-bounded flows computationally feasible and physically justifitable, we pursue the use of SGS wall models here. Since the near wall-flow is dominated by the streaks, which are almost as numerous and universal as the small eddies in free turbulence, they are also amenable to modeling in the filtering context. Hence, a good candidate to start with is the law of the wall, since it is a quite robust feature of most boundary layers, although we expect an erosion of its domain of validity in strongly stimulated flows such as by a pressure gradient. The equations governing for the wall layer can be approximated by

$$\partial_y \left[\nu(\partial_y \bar{v}_i) - B_{iy} \right] = f_i; \quad f_i = \partial_i \bar{p} + \partial_t \bar{v}_i + \partial_j (\bar{v}_i \bar{v}_j) \tag{10.5}$$

(Pope 2000). Assuming that $f_i = \partial_i \bar{p}$, we can integrate (10.5) analytically to give the law of the wall,

$$\bar{v}^+ = y^+ + \frac{\nu(y^+)^2}{2u_\tau^3}\partial_i \bar{p} \quad \text{if} \quad y^+ + \frac{\nu y^+}{u_\tau^3}\partial_i \bar{p} \left(\frac{y^+}{2} - \frac{1}{\kappa} \right) - \frac{1}{\kappa} \ln |y^+| - B < 0,$$

$$\frac{1}{\kappa} \ln |y^+| + \frac{\nu y^+}{\kappa u_\tau}\partial_i \bar{p} + B \quad \text{if} \quad y^+ + \frac{\nu y^+}{u_\tau^3}\partial_i \bar{p} \left(\frac{y^+}{2} - \frac{1}{\kappa} \right) - \frac{1}{\kappa} \ln |y^+| - B > 0,$$
$$\tag{10.6}$$

where $\kappa \approx 0.41$ is the von Karmán constant and $B \approx 5.2$. Note that if $\partial_i \bar{p} = 0$, the classical law of the wall is recovered. For the full case, (10.5) has to be solved numerically. This approach has been used by Wang (1997), in which (10.5) is solved on an embedded near-wall grid by use of a mixing-length model. Following Fureby, Alin et al. (2004), we use (10.6) to modify the SGS model by adding a SGS wall viscosity, ν_{BC}, to ν on the wall so that the effective viscosity, $\nu + \nu_{BC}$, becomes

$$\nu + \nu_{BC} = \tau_w/(\partial v_y/\partial x)_P = u_\tau x_{y,P}/v_{y,P}^+, \tag{10.7}$$

where the subscript P denotes that the quantity is to be evaluated at the first grid point away from the wall. This model can, in principle, be combined with any other SGS model; in the notation, we add "+ WM" to the baseline SGS model name.

10.1.2 Summary of numerical algorithms

For complex geometries, the finite-volume (FV) method is the best choice. Here the computational domain, D, is partitioned into nonoverlapping cells, Ω_P. The cell average of f over the Pth cell is $f_P = \frac{1}{\delta V} \int_\Omega f dV$ so that Gauss' theorem can be used to derive the semidiscretized LES equations. By integrating these over time, using, for example, a multistep method (Hirsch 1999), we have

$$\frac{\beta_i \Delta t}{\delta V_P} \sum_f \left[F_f^C \right]^{n+i} = 0, \tag{10.8a}$$

$$\sum_{i=0}^{m} \left(\alpha_i (\bar{v})_P^{n+i} + \frac{\beta_i \Delta t}{\delta V_P} \sum_f \left[\mathbf{F}_f^{C,v} + \mathbf{F}_f^{D,v} + \mathbf{F}_f^{B,v} \right]^{n+i} \right) = -\beta_i (\nabla \bar{p})_P^{n+i} \Delta t,$$
$$\tag{10.8b}$$

where m, α_i, and β_i are numerical parameters, and $F_f^{C,p}$, $\mathbf{F}_f^{C,v}$, $\mathbf{F}_f^{D,v}$, and $\mathbf{F}_f^{B,v}$ are the convective, viscous, and SGS fluxes. To complete the FV discretization, the fluxes (at cell face f) have to be reconstructed from variables at adjacent cells. This requires interpolation for the convective fluxes and difference approximations for the inner derivatives of the other fluxes. The interpolation used for the convective fluxes is $\bar{\mathbf{v}}_f = \bar{\mathbf{v}}_f^H - (1 - \Psi)[\bar{\mathbf{v}}_f^H - \bar{\mathbf{v}}_f^L]$, where H and L denote high- and low-order schemes, respectively, and Ψ is the flux limiter. Besides using $\Psi = 1$ and the flux-corrected transport (FCT) limiter of Boris and Book (1973), we use the limiters discussed in Chapter 4. We decouple the equations in (10.8) by combining the continuity and momentum equations. We accomplish this by inserting the face interpolate, $F_f^{C,\rho}$, as derived from (10.8b), into (10.8a), thus resulting in a Poisson equation for $\bar{p}$. The pressure–velocity coupling is handled with a PISO procedure based on a modified Rhie–Chow interpolation for cell-centered data storage. The equations are finally solved sequentially with iteration over the coupling terms with a CFL number of about 0.3.

10.2 Fully developed turbulent channel flows

The first test case is the fully developed turbulent channel flow. The channel is confined between two smooth parallel plates $2h$ apart, where h is the channel half-width. The flow is driven by a fixed mass flow in the streamwise ($\mathbf{e}_x$) direction. We use no-slip conditions in the cross-stream ($\mathbf{e}_y$) direction, and we use periodic conditions in the spanwise ($\mathbf{e}_z$) direction. For initial conditions we use a parabolic $\bar{\mathbf{v}}$ distribution. After reaching a statistically steady state, we continued the runs for a sufficient number of time units (h/u_τ) to collect the first- and second-order statistical moments for comparison with DNS and experimental data. For $Re_\tau = 395$ and 590, DNS data with $256 \times 193 \times 192$ and $384 \times 257 \times 384$ grid points are available from Moser, Kim, and Mansour (1999), and for $Re_\tau = 1800$, experimental data are available from Wei and Willmarth (1989). The physical channel size is $6h \times 2h \times 3h$ in the streamwise, cross-stream, and spanwise directions, respectively. All LES and ILES with the exception of the two-level simulation (TLS) model (Kemenov and Menon 2002) use 60^3 grids with uniform spacing in the stream and spanwise directions, whereas geometrical progression is used in the wall-normal direction to cluster the mesh toward the walls. In terms of viscous wall units, the spatial resolution varies between $(\Delta x^+, \Delta y^+, \Delta z^+) = (40, 0.5, 20)$ and $(180, 2, 90)$, respectively. The TLS model uses a $32 \times 40 \times 32$ grid with $(\Delta x^+, \Delta y^+, \Delta z^+)$ between $(78, 13, 39)$ and $(116, 19, 60)$, along with $237 \times 257 \times 257$ embedded one-dimensional grids in each of the three coordinate directions. Previous comparisons of channel flow ILES and LES were reported in Fureby & Grinstein (2002) and Grinstein & Fureby (2002).

Figure 10.1 shows the main flow features of the channel flow in terms of vortex lines, contours of $\langle \bar{v}_x \rangle$, and isosurfaces of the second invariant of the velocity gradient, $Q = \frac{1}{2}(\|\mathbf{W}\|^2 - \|\mathbf{D}\|^2)$. The flow near walls is affected by shear, kinematic blocking, by fluctuating pressure reflections, and by moving internal shear layers produced by the large-scale structures. Elongated streamwise vortices are formed, which mix high-momentum fluid in the outer part of the boundary layer with low-momentum fluid from the near-wall region. These processes are believed to be responsible for the generation of turbulent shear stresses and the subsequent modification of the mean flowfield. The

Figure 10.1. Fully developed turbulent channel flow: Perspective of a fully developed turbulent channel flow at $Re_\tau = 1800$ from ILES, together with the wall model (ILES + WM) using the FCT algorithm. (See color inset, Plate 9).

spanwise resolution is found more important than the streamwise resolution for the prediction of coherent structure dynamics. The wall-normal resolution is important for the accurate prediction of τ_w, which, in turn, is important for estimating the drag.

In Figure 10.2 we compare LES and ILES data of the time-averaged streamwise velocity, $\langle \bar{v}_x \rangle$ (integrated over x and z) with DNS and experimental data. At $Re_\tau = 395$ all LES' examined show very agreement with the DNS data across the entire channel. When the flow is well resolved the details of the subgrid model are of little importance to the resolved flow, since most of the energy and structures are resolved on the grid. For $Re_\tau = 595$ the time-averaged velocity profiles from the LES still show good agreement with the DNS data, but some deviations are starting to show up in the buffer region and in the core. The DES model gives the least accurate result,

Figure 10.2. Fully developed turbulent channel flow: Time-averaged streamwise velocity profiles, $\langle v_x \rangle^+$, at $Re_\tau = 395, 590,$ and 1800. The results for $Re_\tau = 590$ and 1800 are shifted 10 and 20 units, respectively, in the vertical direction.

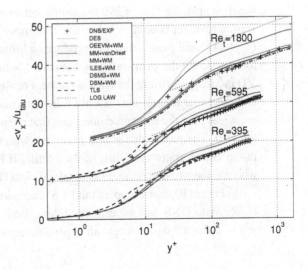

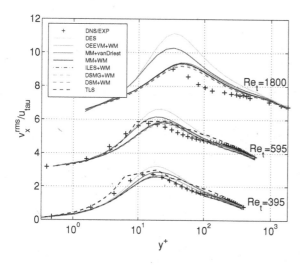

Figure 10.3. Fully developed turbulent channel flow: rms–velocity fluctuation profiles (v_{rms}) at $Re_\tau = 395$, 590, and 1800. Note that the results for $Re_\tau = 590$ and 1800 are shifted 3 and 6 units in the vertical direction, respectively.

suggesting that the near-wall treatment offered by DES is not sufficient unless the near-wall resolution is sufficiently fine, i.e. that $\Delta_y^+ < 1$. The MM + VD model also starts to show some deviations as compared to DNS data, whereas all models using the wall-model give good results. For $Re_\tau = 1800$, most LES, with the exception of MM + VD and DES, still show reasonable agreement with the experimental data with the ILES + WM model performing the best. The log-law, $\langle \bar{v}_x \rangle^+ = \kappa^{-1} \ln(y^+) + B$, is generally well predicted with $B \approx 5.2$ and $\kappa \approx 0.41$. Whereas the predictions of the skin friction coefficient, C_f, on the other hand are unsatisfactory. For $Re_\tau = 395$ the models underpredict C_f by about 1%, whereas at $Re_\tau = 1800$, C_f is typically underpredicted by about 5 to 10%.

In Figure 10.3 we compare the different LES and ILES predictions of the resolvable axial rms-velocity fluctuation, $\bar{v}_x'^+ = \bar{v}_x'/u_\tau$, where $\bar{v}_x' = \sqrt{\langle (\bar{v}_x - \langle \bar{v}_x \rangle)^2 \rangle}$, with DNS and experimental results. For $Re_\tau = 395$ all LES models show good agreement with the DNS data across the channel. All LES models, with the exception of the TLS model, overpredict the peak value with a few percent. The TLS model, on the other hand, shows a somewhat more edgy profile than the other models, probably caused by the coarse baseline grid. For $Re_\tau = 590$ the scatter between the LES model predictions increases. Best agreement is obtained with the TLS model, which correctly predicts the location ($y^+ \approx 15$) and peak value, but produces a broader profile. All other LES models tend to overpredict the peak value by between 5% and 15%, with the DES and the MM + VD model performing the least accurate. For $Re_\tau = 1800$ the agreement between LES and the experimental data is only fair. The overall trend is that the predicted profiles are wider than the measured ones, and that the predicted profiles are shifted away from the wall. The peak value is best predicted with the DSM + WM model, which results in the largest transverse shift of the profile. ILES + WM provides reasonable results whereas the worst agreement is obtained with DES and MM + VD.

In Figure 10.4 we compare our LES predictions of the resolvable shear stress, $\bar{R}_{xy} = \langle \bar{v}_x' \bar{v}_y' \rangle$ with DNS and experimental data. Both R_{xy} and $\bar{R}_{xy}$ profiles are bounded by $u_\tau^2 y/h$, which thus constitute an upper/lower bound for R_{xy}. Hence, as the Re_τ-number

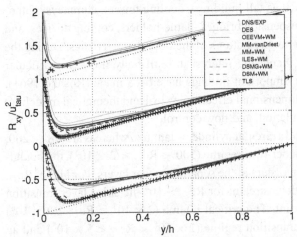

Figure 10.4. Fully developed turbulent channel flow: Reynolds shear stress profiles (R_{xy}) at Re$_\tau$ = 395 and 590, respectively. Note that the data for Re$_\tau$ = 590 and 1800 are shifted 1 and 2 units in the vertical direction, respectively.

increases, the resolved fraction (i.e., $\bar{R}_{xy}$) decreases, which therefore emphasizes the fact that the demands on the subgrid model increase with increasing Re-number given a fixed resolution. For Re$_\tau$ = 395 good agreement with the DNS data is found for all models. DES and ILES + WM show somewhat lower values of $\bar{R}_{xy}$ than the other models. For Re$_\tau$ = 590 the scatter between the $\bar{R}_{xy}$ profiles is larger, but the shapes are generally well reproduced. For Re$_\tau$ = 1800 this trend is continued, with $\bar{R}_{xy}$ covering a lower fraction of R_{xy} than for Re$_\tau$ = 590 and 395.

As demonstrated, the most significant challenge and the pacing item for LES is improved methodologies for modeling near-wall turbulence at high Reynolds numbers. Since we cannot afford to resolve the details of the near-wall flow, we are forced to model the small-scale flow features that are important in the near-wall region. In the channel flow simulations, these features are present on some level – either resolved or partially resolved – depending on the Reynolds number and the grid. In the more complicated cases these processes are partially resolved – at best – and the SGS model is of more importance to the outcome of the LES computation.

10.3 Flow around a circular cylinder at Re$_D$ = 3900 and Re = 140,000

The flow around bluff bodies is often quite complicated, especially if the geometry of the body is such that the separation lines are not fixed. Such flows typically involve boundary-layer separation, flow-regime transition, transition to turbulence, vortex shedding, and coherent structures. The wake consists of curved shear layers enclosing a region of high complexity that is characterized by vortices, but also including the entrainment of irrotational flow into the wake from the surroundings. If the body is symmetric, the wake usually exhibits self-induced periodicity from vortices being shed from alternate sides of the body, generating fluctuating forces on the body. Many industrial and environmental structures such as tall buildings, cooling towers, chimneys, antennae, periscopes, masts, cables, suspension bridges, flame-holders, cooling devices, and flow-metering devices require better predictions of the flow characteristics. During the past decade, there has been an increasing interest in bluff-body flows, in particular for flows past circular cylinders. Important findings have been made (Roshko 1961), especially with regard to three-dimensional effects, physical and theoretical modeling, flow instabilities, numerical simulation, and flow control.

At $Re_D < 150$, the flow around a circular cylinder is laminar, whereas at $Re_D \approx 200$, transition starts to occur. In the subcritical regime ($300 < Re_D < 2 \times 10^5$), the boundary layer is laminar and transition occurs in the separated shear layers. For higher Re_D, the flow becomes turbulent far downstream; for $Re_D > 10^4$ the shear layer transition occurs close to the separation point. The critical regime ($2 \times 10^5 < Re_D < 3 \times 10^6$) can be subdivided into a lower-transition regime ($2 \times 10^5 < Re_D < 5 \times 10^5$) and an upper-transition regime ($5 \times 10^5 < Re_D < 3 \times 10^6$). In the lower-transition regime, C_D drops from 1.2 to 0.3 as a result of an increase in p at $Re_D \approx 3.6 \times 10^5$, and the separation point moves from the front to the downstream side of the cylinder, causing the width of the wake to decrease to about $1D$. In the upper-transition regime, C_D increases from 0.3 to 0.7 as a result of a monotone decrease in C_P from -0.2 to -0.5. With increasing Re_D the separation point moves forward, but it remains on the downstream side and the near wake widens, but stays smaller than $1D$. In the postcritical regime, $Re_D > 3.5 \times 10^6$, the boundary layer becomes turbulent before separation, and the separation–reattachment bubble almost disappears, whereas C_D remains constant at $C_D \approx 0.7$.

Most of the literature dealing with experiments of flow past cylinders documents phenomena in the subcritical Reynolds-number regime (e.g., Cantwell and Coles 1983, Cardell 1993; Laurenco and Shih 1993; Ong and Wallace 1996). At high Re_D the amount of experimental data is more limited, but C_P has been measured on the cylinder surface at $Re_D = 6.7 \times 10^5$ and at 1.2×10^6 (Warshauer and Leene 1971). DNS has been performed at $Re_D = 3900$ by Ma, Karamanos, and Karniadakis (2000), and by Tremblay, Manhart, and Friedrich (2000), whereas Kravchenko and Moin (2000) studied the same case with LES by using the SMG model. They found that although the time-averaged velocity statistics were rather insensitive to the numerics, the predicted spectral energy content could be improved by use of high-order central schemes. Using second-order FV methods and the SMG model, Breuer (1998) and Fröhlich et al. (1998) found good agreement with data (Lourenco & Shih, 1993) with the exception of a shorter recirculation bubble. Breuer (1998) performed wall-resolved LES, using the SMG model, at $Re_D = 1.4 \times 10^5$, and Catalano et al. (2003) carried out wall-modeled LES, using the DSMG model, combined with an embedded one-dimensional SGS wall model at $Re_D = 5 \times 10^5$, 1×10^6, and 2×10^6, respectively. Travin et al. (1999) also recently carried out DES of the flow past a circular cylinder at $Re_D = 2 \times 10^6$.

Here, we examine the flow past a circular cylinder at subcritical Reynolds numbers of $Re_D = 3900$ and 1.4×10^5, using two LES models (OEEVM + WM and MM + WM), ILES using the FCT limiter and the wall model, two RANS models ($k - \varepsilon$; Launder and

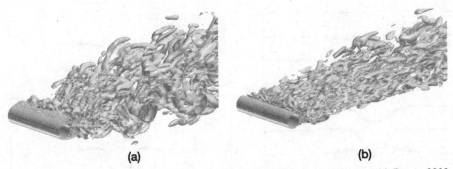

(a)   (b)

Figure 10.5. Flow past a circular cylinder: Perspective views of the flow at (a) Re$_D$ = 3900 and (b) Re$_D$ = 1.4×10^5 in terms of isosurfaces of the second invariant of the velocity gradient tensor, Q.

Sharma 1974), and the Reynolds stress equation model (RSM; Launder et al, 1975) as well as DNS, and experimental data. In both cases, the computational domain is of rectangular form with a spanwise extent of $1.5\pi D$, in order to facilitate the comparisons with the DNS (Lourenco & Shih, 1993; Ong & Wallace, 1996; Cantwell & Coles 1983). The cylinder is located 10D downstream of the inflow plane and 20D upstream of the outflow plane, and the vertical extent of the domain is 20D. The angular position at which separation occurs is defined clockwise from the upstream side of the cylinder. A block-structured H-grid has been used with an inner O-grid of radius 3D. The height of the first cell at the cylinder surface is $y^+ \approx 10$ for both Reynolds numbers, resulting in grids with about 1.00×10^6 and 1.30×10^6 cells, respectively, with 45 cells in the spanwise direction. At the inlet, $\bar{\mathbf{v}} = v_0\mathbf{n}$ and $(\nabla \bar{p} \cdot \mathbf{n}) = 0$; at the outlet, $\bar{p} = p_0$ and $(\nabla\bar{\mathbf{v}})\mathbf{n} = \mathbf{0}$ at the sides, periodic conditions are used; at the top and bottom sides, slip conditions are used; and at the cylinder wall, no-slip conditions are used. The simulations are initiated with quiescent conditions and after the flow reached a fully developed state, statistics were sampled under $200D/v_0$.

Figures 10.5(a) and 10.5(b) show perspective views of the Re$_D$ = 3900 and 1.4×10^5 flows, in terms of isosurfaces of the second invariant of the velocity gradient Q identifying vortical regions. The influence of the Reynolds number on the wake is evident from Figure 10.5: at Re$_D$ = 3900 the wake is fairly wide, but at Re$_D$ = 1.4×10^5 the width of the wake is much narrower, since the separation moves downstream with increasing Re$_D$. The coherent structures in the Re$_D$ = 1.4×10^5 case appear to be not as well organized as in the von Karman street at Re$_D$ = 3900. Independent of Re$_D$, the wake consists of curved shear layers enclosing a vortical region with vortices shed from alternate sides of the cylinder. The three-dimensional vortex structures and the shear layers are clearly visible, although the complexity of the structures obscures the view. In both cases, the vortical structures consist mainly of spanwise quasi-two-dimensional vortex rollers, or $\bar{\omega}_3$-vortices, and secondary longitudinal structures, $\bar{\omega}_{12}$-vortices. The $\bar{\omega}_3$-vortices are formed by Kelvin–Helmholz instabilities in the shear layers originating along the separation line on the cylinder. These $\bar{\omega}_3$-vortices undergo helical pairing as they are advected downstream whilst experiencing vortex stretching caused by the three-dimensional flow features. Once the $\bar{\omega}_3$-vortices have completed their initial

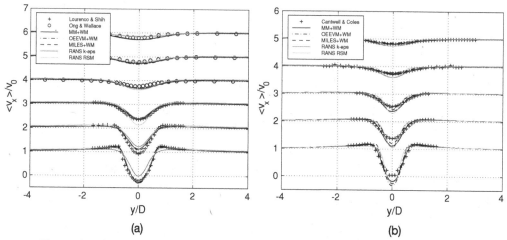

Figure 10.6. Flow past a circular cylinder: Time-averaged velocity at $x/D = 1.06, 1.54, 2.02, 4.00, 7.00$, and 10.0 for (a) $Re_D = 3900$ and at $x/D = 1.00, 1.50, 2.00, 4.00$, and 7.00 for (b) $Re_D = 1.4 \times 10^5$.

roll-up, a secondary instability generates longitudinal $\bar{\omega}_{12}$-vortices. These counterrotating vortices form in the braid region between adjacent $\bar{\omega}_3$-vortices and consist of finger-like structures entwining and distorting the primary $\bar{\omega}_3$-vortices, while exchanging momentum and energy.

In Figure 10.6 we show the time-averaged streamwise mean velocity normalized by the free-stream velocity $\langle \bar{v}_x \rangle$ for the $Re_D = 3900$ and $Re_D = 1.4 \times 10^5$ cases at different cross sections. We made comparisons with data (Lourenco and Shih 1993; Ong and Wallace 1996) for $Re_D = 3900$ and with the data (Cantwell and Coles 1993) for $Re_D = 1.4 \times 10^5$. We performed the $k - \varepsilon$ and RSM calculations with FLUENT, using the reference values for all model coefficients and a second-order-accurate upwind-biased convection scheme together with central differencing for the viscous fluxes. The grid we used for the RANS calculations consists of a slice through the three-dimensional grid used for the LES calculations, and the inflow–outflow boundary conditions are equivalent to those used in LES, whereas we used no-slip conditions together with a conventional wall model along the cylinder walls. At $x/D = 1.06$ and 2.02, we found very good agreement for all LES models, whereas at $x/D = 1.54$, LES underpredicts the velocity by about 5%. Further downstream, at $x/D = 4.00, 7.00$, and 10.00, we again obtained good agreement for all LES models. Virtually no difference can be observed among the OEEVM + WM, ILES + WM, and MM + WM models, suggesting that at this resolution the influence of the SGS model is small. Comparing $\langle \bar{v}_x \rangle / v_0$ along the centerline ($y = 0$; not shown) reveals that the calculated recirculation bubble is somewhat shorter than the measured one and with slightly higher peak reverse velocity. RANS show less accurate profiles, with the RSM being closer to the LES, DNS, and the data. For $Re_D = 1.4 \times 10^5$ we find larger differences among the computed profiles, and between these and the data. This is caused by the coarser resolution, forcing the SGS models to act in a wider range of scales. In particular, at $x/D = 1.00$, we find that all LES overpredicts the peak reverse velocity, and that the OEEVM + WM produces wider profiles than the other LES models. At $x/D = 1.50$ and 2.00, we find that all

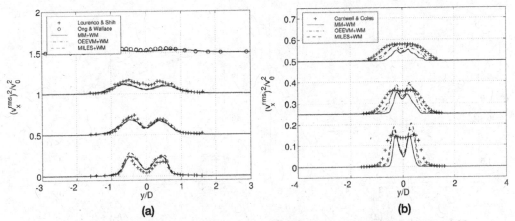

Figure 10.7. Flow past a circular cylinder. Time-averaged rms–velocity fluctuations at $x/D = 1.06$, 1.54, 2.02, and 4.00 for (a) $Re_D = 3900$ and at $x/D = 1.00$, 1.50, and 2.00 for (b) $Re_D = 1.4 \times 10^5$.

LES profiles are slightly narrower than the experimental data, and that OEEVM + WM underpredicts the velocity at the centerline, whereas both ILES + WM and MM + WM result in good agreement of the centerline velocity. Further downstream, at $x/D = 4.00$ and 7.00, we again obtained good agreement among the predicted profiles and between these and the experimental data. Both RANS models show less accurate profiles, with the RSM being closer to the LES and the experimental data.

The streamwise rms–velocity fluctuations v_x^{rms}, where $v_x^{rms} = \sqrt{(\bar{v}_x - \langle \bar{v}_x \rangle)^2}$, are presented in Figures 10.7(a) and 10.7(b) for $Re_D = 3900$ and 1.4×10^5, respectively. For $Re_D = 3900$, good agreement between LES and data is observed. As for the time-averaged streamwise velocity, small differences in streamwise rms–velocity fluctuations are observed between individual LES models, since most of the energy-containing flow structures are resolved on the grid. The differences observed between the LES and the data are most likely caused by insufficient experimental statistics, manifested by the wiggly and asymmetric v_x^{rms} profiles. At $Re_D = 1.4 \times 10^5$, we can find significant differences among the LES predictions, and between these and data. In contrast to the measured profiles, the LES profiles show the same bimodal character as experimentally and computationally observed for the $Re_D = 3900$ case. This bimodal character is expected since it represents the velocity fluctuations in the shear layers. The data of Cantwell and Coles (1983), however, hardly show any of these bimodal profiles, whereas this is clearly the case for all LES. We obtain the lowest resolved fluctuation levels by using the MM + WM, followed in turn by the ILES + WM and the OEEVM + WM, resulting in the highest fluctuation levels.

Figure 10.8 shows the static pressure coefficient and the energy spectra (along the wake centerline) for the $Re_D = 3900$ case. From Figure 10.8(a) it is clear that all LES models perform in a superior manner to the RANS models in the cylinder boundary, giving confidence in the wall modeling and the LES approach as such. From Figure 10.8(b), showing time-averaged energy spectra, it is evident that the difference between the LES models is small, and furthermore that all LES models predict a $|\mathbf{k}|^{-5/3}$ inertial subrange region consistent with the Kolmogorov theory.

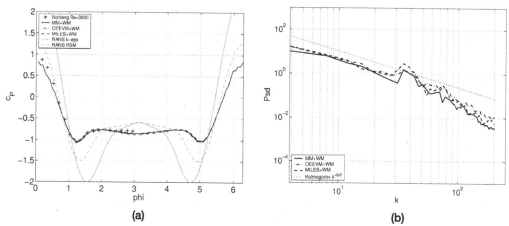

Figure 10.8. Flow past a circular cylinder: (a) Static pressure coefficient and (b) energy spectra along the wake for $Re_D = 3900$.

10.4 Symmetry breaking in a sudden expansion type of flow

The study of turbulent combustion has a long history and has to a large extent been driven by the development of heat engine technologies and airbreathing engines. The challenge of improving the efficiency and reliability of various types of engines has provided a forceful incentive in this field. LES has now replaced RANS modeling as the primary tool to study combustion, which also allows for the investigation of unsteady phenomena such as combustion instabilities (Poinsot and Veynante 2001; Fureby, Parmhed, and Grinstein 2004). To provide better validation data, The Centre National de la Recherche Scientifique developed the ORACLES (one rig for accurate comparison with LES) rig (Besson et al. 2000; Ngyen and Bruel 2003). The rig consists of (i) two mixing chambers, (ii) a 3-m-long approach channel, (iii) a 2-m-long combustion chamber with optical access, and (iv) an exhaust section. The inlet section is divided into two channels by a 10.0-mm-thick splitter plate that ends in a sharp tip with an opening angle of 14°. The tip is located 70.4 mm upstream of the dump plane to avoid flames anchoring at its trailing edge. The height of the approach channels is 30.4 mm, and the height of the dump planes is 29.9 mm, whereas the width of the system is 151 mm.

Here we focus on an inert case characterized by $Re = 23,000$, which is based on h and v_0 and the viscosity of air. The computational configuration is shown in Figure 10.9 and consists of an 80-cm-long combustion chamber attached to a 50-cm approach channel. The width of the domain is 5 cm, having periodic boundary conditions at the front and back. At the inlet, Dirichlet conditions are used for all variables but the pressure, for which zero Neumann conditions are applied. At the outlet, zero Neumann conditions are used for all variables but the pressure for which wave-transmissive conditions are used. Nonuniform stretched grids of 6×10^5 and 1.2×10^6 grid points, with 50 and 75 cells in the spanwise direction, are used together with no-slip conditions, for which the first grid point off the wall is at $y^+ \approx 8$, together with the wall model (Fureby, Alin et al. 2004).

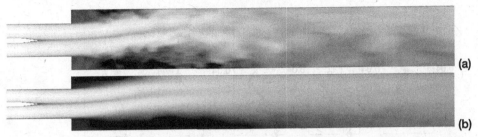

Figure 10.9. Plane sudden expansion: Contours of (a) axial velocity and (b) time-averaged axial velocity at the center plane of the ORACLES rig.

Although turbulent flow through a plane sudden expansion is relevant in many important applications, including combustion chambers, heat exchangers, and mixing devices, the number of studies reported is limited. A key parameter is the area-expansion ratio, $A_R = 1 + 2h/H$, where h is the step height and H is the height of the approach channel. Abbott and Kline (1962) were among the first to investigate systematically the influence of A_R on flow through a plane sudden expansion. They observed that for $A_R > 1.5$ the flow became asymmetric, with two unequal recirculation zones, whereas for $A_R < 1.5$ the flow became symmetric. This observation has been further confirmed in a number of other studies (Cherdron, Durst, and Whitelaw 1978; Durst, Melling, and Whitelaw 1974; Fearn, Mullin and Cliffe 1990; Restivo and Whitelaw 1978; and Smyth 1979) for different values of A_R and for different Reynolds numbers (see also Chapter 9). Furthermore, de Zilwa, Khezzar, and Whitelaw (2000) carried out Laser Doppler Anemometry measurements for $A_R = 2.86$ at Re = 26,500 and complementary RANS calculations using the $k - \varepsilon$ model. Escudier, Oliveira, and Poole (2002) recently conducted an experimental investigation for $A_R = 4$, further confirming the asymmetric flow feature, providing high-quality velocity, rms–velocity fluctuation, and shear stress data at Re = 55,500.

Figure 10.9 shows the axial velocity and the time-averaged axial velocity at the centerplane in the ORACLES configuration from an ILES calculation, using FCT. The most striking feature is the asymmetric mean flow with two unequal reattachment lengths on the top and bottom walls, respectively. For this configuration, $A_R = 1.84$, which is in the region ($A_R > 1.5$) where an asymmetric flow is to be expected. This is further corroborated by data from Besson et al. (2000) and Ngyen and Bruel (2003), also shown in Figure 10.10. Calculations using the OEEVM + WM and MM + WM SGS models show virtually identical results, and this behavior appears mainly to be caused by the coherent structure dynamics of the large-scale flow. All LES were started from quiescent conditions, and statistical data were sampled during eight flow-through times after which the flow was fully developed.

Figure 10.10 shows a comparison of the time-averaged axial velocity $\langle v_x \rangle$ and axial rms–velocity fluctuations v_x^{rms}, with $v_x^{rms} = \sqrt{\langle (v_x - \langle v_x \rangle)^2 \rangle}$, at five cross sections ($x/h = 0, 2, 4.4, 7.6$, and 10) downstream of the dump plane for the ORACLES rig.

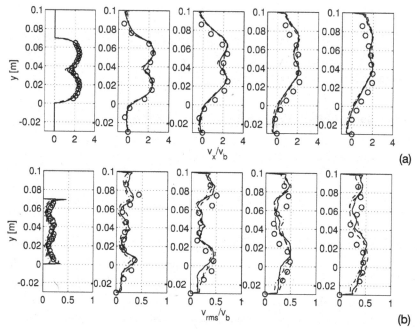

Figure 10.10. Plane sudden expansion: Statistical comparison of (a) axial velocity and (b) axial rms–velocity fluctuations at five cross sections in the center plane of the ORACLES rig. Solid line, ILES + WM; dashed line, MM + WM; dash-dotted line, OEEVM + WM; circles, experimental data.

Predictions are shown for the ILES + WM, MM + WM, and OEEVM + WM models, and are compared with the experimental data of Ngyen and Bruel (2003). For $\langle v_x \rangle$ reasonable agreement is found at all locations between the experiments and the LES and only small differences are observed separating the different LES model predictions. The asymmetry of the time-averaged flow is noticed in both the measured and predicted profiles, suggesting that the LES model correctly captures the spatial asymmetric evolution of the flow, further implying that the dynamics of the underlying coherent structures is reasonably well predicted. Concerning v_x^{rms}, larger differences between predicted and measured profiles can be observed, and also among the different LES model predictions. The asymmetry of the flow leads to the situation in which the shear layers have distinct peak values for v_x^{rms} and these peaks are located at different cross sections following the dividing streamlines.

In order to study the effect of the area-expansion ratio, and to see whether LES could correctly capture the underlying physics, we modified the computational domain used for the ORACLES configuration to alter the area-expansion ratio from $A_R = 1.84$ to $A_R = 1.45$ by reducing the step height from 29.9 mm to 16.0 mm. We carried out two more simulations with this modified geometry. We started the first from quiescent conditions and the second from the fully developed ORACLES solution. Both simulations converged rather quickly to the same type of flowfield, with virtually indistinguishable statistics. Figure 10.11 compares ILES predictions from the ORACLES configuration and the modified configuration. Clearly, the flow becomes symmetric

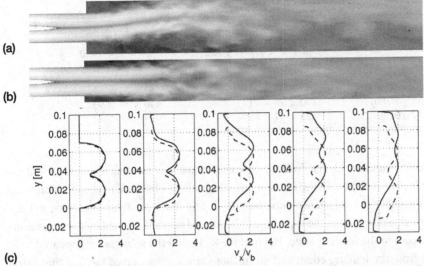

(a)

(b)

(c)

Figure 10.11. Comparison of numerical predictions of the axial velocity at the centerplane for different area-expansion ratios: (a) Axial velocity for $A_R = 1.84$. (b) Axial velocity for $A_R = 1.45$. (c) Comparisons of the time-averaged axial velocity at the centerplane. Solid lines denote results from the ORACLES rig with $A_R = 1.84$ and dashed lines denote results at $A_R = 1.45$.

when A_R drops from 1.84 to 1.45, which is consistent with all available measurement data. This indicates that the LES method is capable of resolving the underlying physical mechanisms, which currently is not very well understood. However, it seems that the shorter recirculation length is more controlled by the pressure inside the upper recirculation region, and the Coanda effect, which competes with the spreading of the shear layer and deflects the flow toward the wall. The longer reattachment length appears less controlled by the pressure, with lower values, and more by SGS diffusion that is considerably higher in the lower recirculation region.

10.5 Flow around a surface-mounted cube

The flow around a surface-mounted cube in a rectilinear channel is another example of a symmetric bluff body, showing many of the phenomena already alluded to in Section 10.4, including boundary layer separation, vortex shedding, and coherent structure dynamics. This particular flow problem was extensively investigated in 1995 when it appeared as a benchmark case for LES in a workshop (Rodi, et al. 1997) focusing on LES validation and comparison. The results presented here are aimed at validating LES models for atmospheric dispersion modeling, as a prototype for flow around a building, and therefore, we use rather coarse grids.

Early experimental results for a surface-mounted cube in a channel are presented by Castro and Robins (1997), who considered two cases. The first involves a uniform (except for a boundary layer with a thickness much smaller than the cube height) laminar approach flow, and the second involves a sheared turbulent approach flow (an emulated

Table 10.1. *Flow around a surface-mounted cube*

Grid	Small domain $-3 < x < 7, 0 < y < 2,$ $-3.5 < z < 3.5$	Large domain $-20 < x < 10, 0 < y < 2,$ $-6 < z < 6$
coarse, $\Delta y_w \approx 0.020h$	$83 \times 31 \times 57$	$134 \times 31 \times 75$
medium, $\Delta y_w \approx 0.015h$	$100 \times 38 \times 71$	$176 \times 38 \times 95$
fine, $\Delta y_w \approx 0.010h$	$125 \times 47 \times 88$	

atmospheric boundary layer). In both cases there were separating shear layers at the leading edges of the cube. In the laminar case the authors observed that, with increasing Reynolds number (based on the cube height, h, and the free-stream velocity, v_0), the shear layer separating from the top of the cube moved upward, caused by the transition's moving closer to the leading edge. Beyond Re $\approx$ 30,000 the shear layer appeared to be turbulent from the leading edge, and no further variations occurred beyond that value. Hunt et al. (1978) found that, in contrast to the laminar case, in the turbulent case the separating shear layers reattach to the body surface, probably intermittently, and that the size of the wake is significantly reduced. The flow around a surface-mounted cube at Re = 40,000 was investigated by Martinuzzi and Tropea (1993), using oil-film visualization, nonintrusive measurement techniques, and topological considerations. The data from this study formed the backbone for the LES workshop (Rodi et al. 1997).

Here, we present comparisons between different LES and ILS models and compare them with the data in Rodi et al. (1997). The large computational domain spans the true experimental domain, whereas we use a second smaller domain to enhance the resolution, cf. Table 10.1. The bulk velocity is $v_0 = 24$ m/s, the side length of the cube is $h = 0.025$ m, the height of the channel is $2h$, and the Reynolds-number, based on h and v_0 and the viscosity of air, is Re = 40000. The grids are clustered in the wall-normal direction at the roof, bottom, and cube walls, and also weakly stretched toward the inlet, outlet, and the sides of the domain. We use the same inlet boundary condition as Krajnovic and Davidsson (2002), namely a specified constant velocity profile (the time-averaged experimental profile from Martinuzzi and Tropea 1993), essentially corresponding to the mean profile of a fully developed turbulent channel flow. Shah and Ferziger (1997), in contrast, supply the inflow with velocity and pressure data from a separate channel flow simulation to mimic large-scale flow structures and the boundary layer dynamics. We apply periodic boundary conditions on the sides. At the outlet, we use Dirichlet conditions for the pressure and Neumann conditions for the velocity. On the upper and lower walls and on the cube, we use no-slip conditions. After we fully developed the flow, we started the statistical sampling, and we continued it for $300h/v_0$, corresponding to 10 flow-through times.

Streamwise vortices are generated within the shear layer and affect the flow near the cube, reorganizing the recirculation region and influencing the downstream recovery region. The main vortex structures consist of the horseshoe vortex, the arch-shaped wake

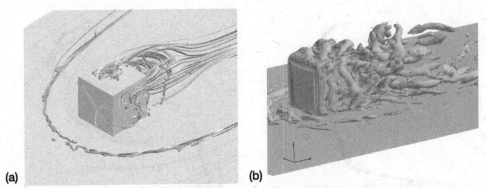

(a) **(b)**

Figure 10.12. Flow around a surface-mounted cube: (a) Streamlines of the time-average velocity, showing the vortex systems, and (b) isosurface of the second invariant of the velocity gradient.

vortex, and lateral vortices on the top and lateral sides of the cube; see Figure 10.12. The horseshoe vortex in the region upstream of the cube is extending over the whole width of the cube and is deflected downstream. Vertical vortices behind the cube entrain the surrounding fluid and convect it along the plane of symmetry. We find an exchange of flow between the separation regions, confirming the results of Hunt et al. (1978) that there is no evidence of closed separation bubbles. Clearly, as is pointed out by Larousee, Martinuzzi, and Tropea (2000), the flow is unsteady. The features of the time-averaged flow need not always be present in the instantaneous flowfield. For example, the arch vortex behind the cube is an effect of a quasi-periodic vortex shedding from the upstream vertical corners that resembles a von Karman vortex street. In addition, the locus of the horseshoe vortex varies considerably over time (Shah & Ferziger, 1997), as confirmed in the LES and ILES presented here. Martinuzzi and Tropea (1993) found that the flow oscillates between two states, characterized by different vortex locations, and thus the instantaneous velocity probability density function is bimodal. This is confirmed by Shah and Ferziger (1997), who also observed that the choice of SGS model affects the length of the recirculation zone considerably. The aforementioned features are indeed found in the LES calculations presented here.

From the present and previous LES' we find that the fluid, approaching the cube along the channel, develops a turbulent boundary layer along the floor. About one length upstream of the cube, the flow starts to separate from the floor of the channel and a horseshoe vortex system is formed in the recirculation region. This horseshoe vortex system actually consists of several vortices, oscillating back and forth, streamwise as well as spanwise, interacting to form a single horseshoe vortex when averaged in time. The horseshoe vortex is deflected downstream along the sides of the cube. The flow separates again at the leading edges of the cube, where the velocity and velocity gradients are large and turbulent kinetic energy and vorticity are generated. A reattachment line is occasionally found on the roof of the cube, very close to the downstream edge, where the flow separates again. Lateral vortices on the sides and roof are generated and shed quasi-periodically downstream, resembling a von Karman

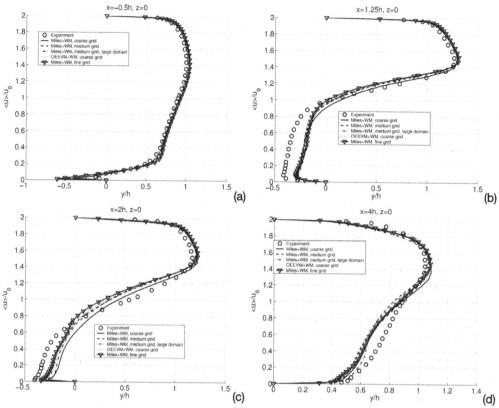

Figure 10.13. Flow around a surface-mounted cube. Statistical comparison of the first-order statistical moment, that is, the time-averaged axial velocity component at (a) $x/h = -0.5$, $z/h = 0.0$; (b) $x/h = 1.25$, $z/h = 0.0$; (c) $h = 2.00$, $z/h = 0.0$; and (d) $x/h = 4.00$, $z/h = 0.0$.

vortex street. Occasionally, two vortices are found on the roof and sides of the cube, but in the time-averaged results only one vortex remains. Furthermore, we find an exchange of fluid between the side and wake regions, in both directions as, occasionally, the vortex structures in the wake extend upstream to the side region. The surrounding fluid is entrained by vertical vortices in the wake behind the cube and transported along the plane of symmetry. The features of the mean flow need not always be present in the instantaneous flowfield; for example, the arch-shaped vortex observed in the wake behind the cube is a direct effect of the quasi-periodic vortex shedding.

In general, we find that the time-averaged streamline agrees reasonably well with the experiment of Martinuzzi and Tropea (1993), except for the detailed location of the horseshoe vortex in front of the cube. In addition, Krajnovic and Davidson (2002) fail to correctly predict the experimentally observed location of the time-averaged horseshoe vortex, which in the experiments is found closer to the front side of the cube than in the present and previous LES calculations. Reasons for this may be that the approach boundary layer is insufficiently resolved or that the wall model cannot handle this flow situation, or that the approach flow contains coherent structures that enhance the separation and roll-up in front of the cube, or that the boundary layer dynamics is of significant importance to the formation of the horseshoe vortex system. Precise comparison with data is found in Figure 10.13, along the four lines $x/h = -0.5$, $z/h = 0.0$ (including the horseshoe vortex in front of the cube); $x/h = 1.25$, $z/h = 0.0$ (including the wake vortex); $x/h = 2.0$, $z/h = 0.0$; and $x/h = 4.0$, $z/h = 0.0$ (capturing the spatial evolution of the wake). Virtually no differences in the velocity statistics are found between the LES model (using the OEEVM+WM SGS model) and the ILES model (using FCT and the wall model) for these grids and domains, except for the spanwise velocity at the horseshoe vortex downstream from the cube. This coherent structure contains low energy compared with the wake vortices; thus more influence of the SGS model might be expected. Compared with the laboratory measurement data, the main difference seems to be that the time-averaged velocity is underpredicted in the recirculation region behind the cube, and that the wake recirculation region is larger than experimentally observed. Further downstream the recovery of the velocity profile is slower in the ILES–LES as compared with the measured profiles, indicating lower levels of large-scale flow. These results are in close agreement with those of Krajnovic and Davidson (2002), suggesting that LES is self-consistent. As expected, the differences in the second-order statistical moments (rms–velocity fluctuations; not shown) are larger than the differences in the first-order statistical moments.

We conclude that, although most flow features observed in the experiments are represented by the LES models, the quantitative agreement is not sufficiently good. It appears that the influence of the explicit or implicit SGS model is comparatively small, and this modeling is not likely to affect the observed differences between predictions and measurements. Alternatively, we propose that these defects are more related to inappropriate supergrid modeling, discussed in Section 10.1.5, involving primarily an unsteady approach flow with embedded vortical structures that mainly affect the dynamic flow pattern in the approach boundary layer.

10.6 Flow around the KRISO KVLCC2 tanker hull

The flow around a ship hull or a submarine is extremely complex. The large size of cargo, cruise, tankers, and military ships in combination with the low viscosity of water makes the Reynolds number very high, despite the low speed of these ships compared with that of aircraft and cars. The high Reynolds number places the laminar to turbulent transition strip early at the bow. This usually makes it superfluous for one to model transition explicitly, except for submarine hulls, where transition is crucial for the placement of sonars and other sensors. Instead, the whole flowfield can be considered to be turbulent. The flow around the first two-thirds of the hull is usually not difficult to predict. Potential flow methods in combination with two-dimensional boundary layer equations usually give sufficient accuracy (Larsson 1978, 1980). The difficult part, however, is the stern, at which the thick boundary layer transforms into a wake; see Figure 10.14. Here the flow from the sides and the bottom of the ship meet at the curved surface between the side and bottom of the hull (the so-called bilge). This often results in a three-dimensional vortex separation over a curved surface, resulting in a pair of

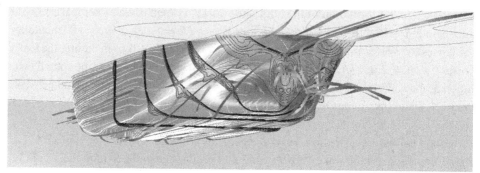

Figure 10.14. Ship flow: Perspective view from the stern of the flow past the KVLCC model, showing the time-averaged boundary layer profiles, surface streamlines, and streamlines released at the bow. (See color inset, Plate 10).

vortices – the bilge vortices. These are usually strong and propagate into the wake passing through the propeller. This causes distortion of the flow into the propeller and may cause vibration, noise, and cavitation, and may generally affect the efficiency of the propeller. The bilge vortex pair is found to be present in the time-averaged flowfield only, while the instantaneous flowfield consists of a large number of unsteady large-scale vortex structures (Pattenden, Turnock, and Bressloff 2004). This makes accurate predictions with RANS virtually impossible.

The time-averaged flow is important for the thrust and torque of the propeller; the time-dependent (unsteady) flow is important for vibration and cavitation. The efficiency of the propulsion unit is of interest for all types of ships. Vibration and noise are important issues for cruise ships because of the requirement for comfort, and for naval ships because of signature requirements. Since cavitation is a source of noise and vibrations, it is also of interest for cruise and naval ships, and indeed for cargo ships and tankers since cavitation erodes and damages propellers and rudders. Thus for the design of the hull and propeller, it is important to predict the bilge vortices and other details of the flow. Already this phenomenon generates a complex flow but there is also interest to add the water surface, propulsion, complex sea-states with incoming waves, maneuvering and time-dependent forces from control surfaces and incoming waves, resulting force-induced movements, and signatures. Furthermore, all these phenomena appear at full scale, which increases the Reynolds number from the range of 10^6–10^7 at model scale to 10^7–10^9.

Here we present results from LES and ILES computations over such a double model, the KVLCC2 (Van et al. 1998a, 1998b) as shown in Figure 10.15, for which experimental data are available, in order to test the predictive capabilities of LES and ILES and to

Figure 10.15. Ship flow: Side view of the KVLCC double model (the bow is to the left and the stern is to the right).

Table 10.2. *KVLCC2 simulations*

Grid/model	Quarter model	Half model	Full model
coarse	0.35×10^6	0.70×10^6	1.40×10^6
medium	0.70×10^6		
fine	1.40×10^6		
xfine	3.20×10^6		

learn more about the flow around a typical large ship hull. The KVLCC2 is a relatively modern tanker hull, with a bulbous bow and one propeller. This hull was chosen as one of the test cases for the Gothenburg 2000 Workshop (Larsson, Stern, and Bertram 2002). This was the third workshop held in Gothenburg in a series of workshops on numerical ship hydrodynamics (Kodama 1994; Larsson 1980; Larsson, Patel, and Dyne 1991). Numerous results with RANS codes were produced for this workshop, covering the entire range of RANS models – from two-equation models to the RSM.

For the purpose of this study, we used two ordinary LES models, the OEEVM+WM and MM+WM, and one ILES, using the Gamma scheme of Jasak, Weller, and Gosman (1999), together with five grids and three computational configurations (Table 10.2). We construct the different computational models by mirroring a half model in the vertical and horizontal planes. Since the hull is geometrically symmetric, we also expect the time-averaged flow to be symmetric. It is common to reduce the computational cost by use of geometrical symmetry. This is a correct approach for RANS, which is designed for capturing the mean flow only. However, for LES the time-averaged results are still expected to be symmetric but the instantaneous flow may not be symmetric. Enforcing symmetry in LES may affect the time-averaged flow. On the basis of the LES predictions, interactions across the symmetry planes have been observed, although not significantly affecting the time-averaged flow. A drawback with using the full models and double geometrical models is that they are much more expensive than the quarter model. This is, however, a price that must be paid when the flow is not known beforehand, as in most real flow problems.

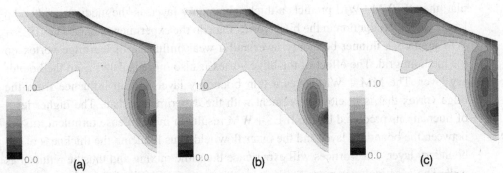

Figure 10.16. Ship flow: Axial velocity contours in the propeller plane on the fine grid for the (a) OEEVM, (b) ILES model, and (c) MM, respectively.

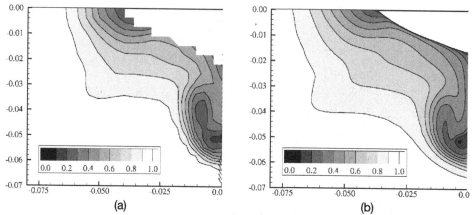

Figure 10.17. Ship flow: (a) Experimental axial velocity contours and secondary velocity vectors in the propeller plane, reproduced from Kim (2000). (b) Axial velocity contours and secondary velocity vectors in the propeller plane for the RSM on a finer grid, from Svennberg (2001).

One of the most important features of the flowfield is the axial velocity distribution in the propeller plane, illustrated in Figure 10.16 in terms of axial velocity contours, $\langle \bar{v}_x \rangle / v_\infty$. These contours show the effect of the large-scale vortex separation, the bilge vortices. The bilge vortices are located symmetrically, and are counterrotating one on each side of the hull. Fluid with high axial velocity is pushed toward the hull centerplane just below the stern and then downward along the hull centerplane. The fluid with lower axial velocity, which is supposed to be here, is pushed outward and upward from the propeller hub. This redistribution generates a hook shape in the velocity contours in the wake. The hook shape is typical for wide hulls generating strong bilge vortices and can be used to design hulls with a reasonably even velocity distribution in the propeller disk. The axial velocity and the secondary velocity vectors in the propeller plane on the fine grid are presented in Figure 10.16, whereas the experimental data are presented in Figure 10.17(a) and the state-of-the-art RSM RANS results of Svennberg (2001) are presented in Figure 10.17(b).

Comparison of Figures 10.16 and 10.17 shows that the LES predicts a flowfield in closer agreement with the experimental data than does the RANS model. In particular, the OEEVM+WM predicts a thicker boundary layer on the medium grid and a larger velocity reduction in the bilge vortex than do the experimental data. The ILES + WM predicts a thinner boundary layer and a weak influence of the bilge vortex on the medium grid. The effect of the bilge vortex is also more isolated from the boundary layer. The MM + WM gives a thin boundary layer and an influence from the bilge vortex that is in close agreement with the experimental data. The higher level of fluctuations predicted by the ILES + WM results in more intense turbulent mixing between the boundary layer and the outer flowfield, thus reducing the thickness of the boundary layer. The vortices will experience the same mixing and thus be reduced as well. The MM includes more SGS physics than the OEEVM + WM, which may explain the improved results. The stronger fluctuations predicted by the ILES + WM call for higher temporal resolution than the OEEVM + WM and MM + WM. At the same time, the ILES+WM appears more sensitive to grid skewness and warp-age than the conventional OEEVM + WM and MM+WM. The influence of the bilge vortex is reduced for both the OEEVM and the ILES model by the finer grid as compared with the medium grid. This reduction gives an improved result for the OEEVM + WM. The wake structure is improved for the ILES + WM, while it still gives a low-velocity reduction in the hook shape.

10.7 Concluding remarks

When LES and ILES are used to study canonical and "building-block" flows, for which the methods were originally invented, they generally produce excellent results. The dependency of the results on the SGS model is weak – almost all SGS models produce identical resolved flows. For wall-bounded flows, however, the situation becomes more complicated as the scales become smaller and smaller as the wall is approached. For low Reynolds numbers the principal structures can be resolved (wall-resolved LES), and the results are generally in good agreement with experiments and DNS. For higher Reynolds numbers the demands on the SGS model increase, and it either has to be modified in the near-wall region or we need separate wall models (wall-modeled LES). For these cases we can no longer afford to resolve the key features of the near-wall flow, although the trend is moving toward adaptive grids and local grid refinement. This is an important research area if we have the ambition to use LES for engineering flows, which usually are associated both with high Reynolds numbers and complex geometries. In order to address such flows, new modeling strategies such as ILES and multiscale methods may prove to be very useful, and they may improve the results. The numerical methods used in LES must recognize the task of an LES calculation, and the particular requirements put forward by the explicit or implicit SGS turbulence model – being accurate to at least second order in space and time.

Acknowledgment

We gratefully acknowledge financial support for the collection and compilation of the data from the Swedish Defence Research Agency (FOI).

REFERENCES

Abbott, D. E., & Kline, S. J. 1962 Experimental investigation of subsonic turbulent flow over single and double backward facing steps. *J. Basic Eng.* **84**, 317–325.

AGARD. *A Selection of Test Cases for the Validation of Large Eddy Simulations of Turbulent Flows.* Washington, DC: Advisory Report No 345 2001.

AFRL. 2004 *Research Interest of the Air Force Office of Scientific Research and Broad Agency Announcement.* Air Force Office of Scientific Research: AFRL. 2004–1.

Bagget, J. S., Jiménez, J., & Kravchenko, A. G. 1997 Resolution requirements in large eddy simulations of shear flows, in *CTR Annual Research Briefs 1997*. Stanford, CA: NASA Ames/Stanford University, 51–66.

Bardina, J., Ferziger, J. H., & Reynolds, W. C. 1980 Improved Subgrid Scale Models for Large Eddy Simulations. *AIAA Paper 80-1357*.

Besson, M., Bruel, P., Champion, J. L., & Deshaies, B. 2000 Experimental analysis of combusting flows developing over a plane-symmetric expansion. *J. Thermophys. Heat Transf.* **14**, 59–67.

Boris, J. P., & Book, D. L. 1973 Flux corrected transport I. SHASTA, a fluid transport algorithm that works. *J. Comput. Phys*. **11**, 38–69.

Breuer, M. 1998 Large eddy simulation of the subcritical flow past a circular cylinder: Numerical and modeling aspects. *Int. J. Numer. Meth. Fluids* **28**, 1281–1302.

Cantwell, B., & Coles, D. 1983 An experimental study of entrainment and transport in the turbulent near wake of a circular cylinder. *J. Fluid Mech.* **136**, 321–374.

Cardell, G. S. 1993 *Flow Past a Circular Cylinder with a Permeable Splitter Plate*. PhD. dissertation, Graduate Aeronautical Laboratory, California Institute of Technology.

Castro, I. P., & Robins, A. G. 1997 The flow around a surface-mounted cube in uniform and turbulent streams. *J. Fluid Mech.* **79**, 307–335.

Catalano, P., Wang, M., Iaccarino, G., & Moin, P. 2003 Numerical simulation of the flow around a circular cylinder at high Reynolds numbers. *Heat Fluid Flow* **24**, 463–469.

Cherdron, W., Durst, F., & Whitelaw, J. H. 1978 Asymmetric flows and instabilities in symmetric ducts with sudden expansions. *J. Fluid Mech*. **84**, 13–31.

Clark, R. A., Ferziger, J. H., & Reynolds, W. C. 1979 Evaluation of sub-grid scale models using an accurately simulated turbulent flow. *J. Fluid Mech.* **91**, 1–16.

De Zilwa, S. R. N., Khezzar, L., & Whitelaw, J. H. 2000 Flows through plane sudden expansions. *Int. J. Numer. Meth. Fluids* **32**, 313–329.

Deardorff, J. W. 1973 The use of subgrid transport equations in a three-dimensional model of atmospherical turbulence. *ASME J. Fluids Eng. Trans.* **95**, 429–438.

Durst, F., Melling, A., & Whitelaw, J. H. 1974 Low Reynolds number flow over a plane sudden expansion. *J. Fluid Mech.* **64**, 111–128.

Escudier, M. P., Oliveira, P. J., & Poole, R. J. 2002 Turbulent flow through a plane sudden expansion of modest aspect ratio. *Phys. Fluids* **14**, 3641–3654.

Fearn, R. M., Mullin, T., & Cliffe, K. A. 1990 Nonlinear flow phenomena in a symmetric sudden expansion. *J. Fluid Mech.* **211**, 595–608.

Ferziger, J. H. 1983 Higher level simulations of turbulent flow, in *Computational Methods for Turbulent, Transonic, and Viscous Flows*, ed. J.-A. Essers. New York: Hemisphere.

FLUENT reference manual, v. **6**.

Fröhlich, J., Rodi, W., Kessler, Ph., Parpais, S., Bertoglio, J. P., & Laurence, D. 1998 Large eddy simulation of flow around circular cylinders on structured and unstructured grids, in *Notes on Numerical Fluid Mechanics*. Wiesbaden: Vieweg-Verlag, 319–324.

Fureby, C. and Grinstein, F. F. 2002 Large eddy simulation of high Reynolds number free and wall bounded flows, *J. Comput. Phys*. **181**, 68–97.

Fureby, C., Alin, N., Wikström, N., Menon, S., Persson, L., & Svanstedt, N. 2004 On large eddy simulations of high Re-number wall bounded flows. *AIAA. J.* **42**, 457–468.

Fureby, C., Parmhed, O., & Grinstein, F. F. 2004 LES of combustion, Special Technology Sessions (STS) at ECCOMAS, July 24–28, Jyväskylä, Finland, Invited.

Fureby, C., Persson, L., & Svanstedt, N. 2002 On homogenisation based methods for large eddy simulation. *J. Fluids Eng.* **124**, 892–903.

Fureby, C., Tabor, G., Weller, H., & Gosman, D. 1997 On differential sub grid scale stress models in large eddy simulations. *Phys. Fluids* **9**, 3578–3580.

Germano, M., Piomelli, U., Moin, P., & Cabot, W. H. 1994 A Dynamic sub grid scale eddy viscosity model. *Phys. Fluids A* **3**, 1760–1765.

Grinstein, F. F. and Fureby, C. 2002 Recent progress on MILES for high Reynolds number flows. *J. Fluids Eng.* **124**, 848–886.

Hirsch, C. 1999 *Numerical Computation of Internal and External Flows*. New York: Wiley.

Hunt, J. C. R., Abell, C. J., Peterka, J. A., & Woo, H. 1978 Kinematical studies of the flows around surface-mounted obstacles; applying topology to flow visualization. *J. Fluid Mech.* **86**, 179–200.

Jasak, H., Weller, H. G., & Gosman, A. D. 1999 High resolution NVD differencing scheme for arbitrarily unstructured meshes. *Int. J. Numer. Meth. Fluids* **31**, 431–437.

Jones, W. P., & Launder, B. E. 1972 The prediction of laminarization with a two-equation model of turbulence. *Int. J. Heat Mass Transfer* **15**, 301–309.

Kemenov, K., & Menon, S. 2002 TLS: A New Two Level Simulation Methodology for high-Reynolds LES. *Paper AIAA No. 2002-0287.*

Kim, J., & Moin, P. 1979 Large eddy simulation of turbulent channel flow, presented at the AGARD Symposium on Turbulent Boundary Conditions, The Hauge, Netherlands.

Kim, S. E. 2000 Reynolds stress transport modeling of turbulent shear flow past a modern VLCC hull form, presented at the Workshop on Numerical Ship Hydrodynamics, September 14–16, Göteborg. Sweden.

Kim, W.-W., & Menon, S. 1999 A new incompressible solver for large-eddy simulations. *Int. J. Numer. Fluid Mech.* **31**, 983–1017.

Kodama. Y. 1994 CFD workshop Tokyo 1994, in *Proceedings, Ship Research Institute.* Tokyo, Japan.

Krajnovic, S., & Davidsson, L. 2002 Large-eddy simulation of the flow around a bluff body. *AIAA. J.* **40**, 927–936.

Kravchenko, A., & Moin, P. 2000 Numerical studies of flow around a circular cylinder at $Re_D = 3900$. *Phys. Fluids* **12**, 403–417.

Larousse, A, Martinuzzi, R., & Tropea, C. 2000 Flow around surface-mounted, three-dimensional obstacles, in *Proceedings of the 8th Symposium on Turbulent Shear Flows.* Munich, Germany.

Larsson, L., Patel, V. C., & Dyne, G. 1991 *1990 SSPA-CTH-IIHR Workshop on Ship Viscous Flow.* Göteborg, Sweden: Flowtech International AB, Research Report No. 2.

Larsson, L. 1978 *Proceedings of the International Symposium on Ship Viscous Resistance.* Göteborg, Sweden: SSPA.

Larsson, L. 1980 *SSPA-ITTC Workshop on Ship Boundary Layers.* Göteborg, Sweden: SSPA Report No. 90.

Larsson, L., Stern, F., & Bertram, V. 2002 *A Workshop on Numerical Ship Hydrodynamics.* Chalmers University of Technology, Göteborg, Sweden: Report. CHA/NAV/R-02/0073.

Launder, B. E., & Sharma, B. I. 1974 Application of the energy dissipation model of turbulence to the calculation of flow near a spinning disc. *Lett. Heat Mass Transf.* **1**, 131–138.

Launder, B. E., Reece, G. J., & Rodi W. 1975 Progress in the development of a Reynolds stress turbulence closure. *J. Fluid Mech.* **68**, 537–566.

Launder, D. B., & Spalding, D. B. 1972 *Mathematical Models of Turbulence.* London: Academic Press.

Lourenco, L. M., & Shih, C. 1993 *Characteristics of the Plane Turbulent Near Wake of a Circular Cylinder: A Particle Image Velocimetry Study.*

Ma, X., Karamanos, G.-S., & Karniadakis, G. E. 2000 Dynamics and low-dimensionality of a turbulent near wake. *J. Fluid Mech.* **410**, 29–65.

Martinuzzi, R., & Tropea, C. 1993 The flow around surface-mounted, prismatic obstacles placed in a fully developed channel flow. *J. Fluids Eng.* **115**, 85–92.

Moser, R. D., Kim, J., & Mansour, N. N. 1999 Direct numerical simulation of turbulent channel flow at Re = 590. *Phys. Fluids* **11**, 943.

Nguyen, P. D., & Bruel, P. 2003 Turbulent reacting flow in a DumCombustor: Experimental determination of the influence of an inlet equivalence ratio difference on the contribution of the coherent and stochastic motions to the velocity field dynamics. Paper AIAA 2003-0958.

Ong, J., & Wallace, L. 1996 The velocity field of the turbulent very near wake of a circular cylinder. *Exp. Fluids* **20**, 441–453.

Pattenden, R. J., Turnock, S. R., & Bressloff, N. W. 2004 The use of detached eddy simulation in ship hydrodynamics, in *Proceedings of the 25th Symposium on Naval Hydrodynamics.* Vol. **2**, 91–103.

Poinsot, T., & Veynante, D. 2001 *Theoretical and Numerical Combustion.* Ann Arbor: Edwards.

Pope, S. B. 2000 *Turbulent Flows.* New York: Cambridge University Press.

Restivo, A., & Whitelaw, J. H. 1978 Turbulence characteristics of the flow downstream of a symmetric plane sudden expansion. *J. Fluids Eng. Trans. ASME* **100**, 308–310.

Rodi, W., Ferziger, J. H., Breuer, M., & Pourquiré, M. 1997 Status of large eddy simulation: Results of a workshop. *ASME J. Fluids. Eng. Trans.* **119**, 248–262.

Roshko, A. 1961 Experiments on the flow past a circular cylinder at very high Reynolds numbers. *J. Fluid Mech.* **10**, 345–360.

Sagaut, P. 2005 *Large Eddy Simulation for Incompressible Flows.* 3rd edition, Heidelberg: Springer-Verlag.

Schumann, U. 1975 Subgrid scale model for finite difference simulation of turbulent flows in plane channels and annuli. *J. Comput Phys.* **18**, 376–404.

Shah, K. B., & Ferziger, J. H. 1997 A fluid mechanichians view of wind engineering: Large eddy simulation of flow past a cubic obstacle. *J. Wind Eng. Ind. Aerodyn.* **67–68**, 211–224.

Smagorinsky, J. 1963 General circulation experiments with the primitive equations. I. The basic experiment. *Mon. Weather Rev.* **91**, 99–165.

Smyth, R. 1979 Turbulent flow over a plane sudden expansion. *J. Fluids Eng.* **101**, 348–353.

Spalart, P. R., Jou, W.-H., Strelets, M., & Allmaras, S. R. 1997 Comments on the feasibility of LES for wings, and on a hybrid RANS/LES approach, in *Advances in DNS/LES*, ed. C. Liu. & Z. Liu. Columbus, OH: Greyden Press.

Svennberg, U. 2001 *On Turbulence Modeling for Bilge Vortices: A Test of Eight Models for Three Cases.* PhD. dissertation, Chalmers University of Technology, Göteborg, Sweden.

Szymocha, K. 1984 An experimental analysis of the turbulent flow downstream of a plane sudden expansion. *Arch. Mech.* **36**, 705–713.

Travin, A., Shur, M., Strelets, M., & Spalart, P. 1999 Detached eddy simulation past a circular cylinder. *Flow Turb. Combust.* **63**, 293–313.

Tremblay, F., Manhart, M., & Friedrich, R. 2000 DNS of flow around a circular cylinder at a sub-critical Reynolds number with Cartesian grids, in *Proceedings of the 8th European Turb. Conf., EUROMECH*, 659–662.

Van, S.H., Kim, W. J., Yim, D. H., Kim, G. T., Lee, C. J., & Eom, J. Y. 1998a Flow measurement around a 300K VLCC model, in *Proceedings of the Annual Spring Meeting, SNAK*, Ulsan, 185.

Van, S. H., Kim, W. J., Yim, G. T., Kim, D. H., & Lee, C. J. 1998 Experimental investigation of the flow characteristics around practical hull forms, in *Proceedings of the 3rd Osaka Colloquium on Advanced CFD Applications to ShiFlow and Hull Form Design.*

Wang, M. 1997 Progress in large eddy simulation of trailing edge turbulence and aeroacoustics, in *CTR Annual Research Briefs.* Stanford, CA: NASA Ames/Stanford University, CTR 37.

Warschauer, K. A., & Leene, J. A. 1971 Experiments on mean and fluctuating pressures of circular cylinders at cross flow at very high Reynolds numbers, in *Proc. Int. Conf. On Wind Effects on Buildings and Structures.* 305–311.

Wei T., & Willmarth W. W. 1989 Reynolds number effects on the structure of a turbulent channel flow. *J. Fluid Mech.* **204**, 57–95.

Wikström, N., Svennberg, U., Alin, N., & Fureby, C. 2004 LES of the flow past an inclined prolate spheroid. *J. Turb.* **5**, 29–47.

Wilcox, D. C. 1998 *Turbulence Modeling for CFD.* DCW Industries.

11 Compressible Turbulent Shear Flows

Christer Fureby, Doyle D. Knight, and Marco Kupiainen

11.1 Methodology

11.1.1 Governing equations

The governing equations for compressible turbulent shear flows are the spatially filtered, Favre-averaged compressible Navier–Stokes equations. The spatial filtering removes the small-scale (subgrid-scale) components, while the three-dimensional, time-dependent large-scale (resolved-scale) motion is retained. For a function f, its filtered form $\bar{f}$ and its Favre-filtered form $\tilde{f}$ are

$$\bar{f} = \frac{1}{V} \int_V G f \, dV, \qquad \tilde{f} = \frac{\overline{\rho f}}{\bar{\rho}}, \qquad (11.1)$$

respectively, where G is the filter function and ρ is the density. Favre filtering and spatial filtering of the Navier–Stokes equations yield

$$\frac{\partial \bar{\rho}}{\partial t} + \frac{\partial \bar{\rho} \tilde{u}_j}{\partial x_j} = 0, \qquad (11.2)$$

$$\frac{\partial \bar{\rho} \tilde{u}_i}{\partial t} + \frac{\partial \bar{\rho} \tilde{u}_i \tilde{u}_j}{\partial x_j} = -\frac{\partial \bar{p}}{\partial x_i} + \frac{\partial \mathcal{T}_{ij}}{\partial x_j}, \qquad (11.3)$$

$$\frac{\partial \bar{\rho} \tilde{e}}{\partial t} + \frac{\partial}{\partial x_j}(\bar{\rho} \tilde{e} + \bar{p}) \tilde{u}_j = \frac{\partial}{\partial x_j}(\mathcal{Q}_j + \mathcal{T}_{ij} \tilde{u}_i), \qquad (11.4)$$

$$\bar{p} = \bar{\rho} R \tilde{T}, \qquad (11.5)$$

where u_i is the velocity in the ith coordinate direction, p is the static pressure, T is the static temperature, and the Einstein summation notation is used (i.e., the appearance of a repeated index in a term implies summation over all values of the index). Additionally,

$$T_{ij} = \tau_{ij} + \bar{\sigma}_{ij}, \tag{11.6}$$

$$\tau_{ij} = -\bar{\rho}(\widetilde{u_i u_j} - \tilde{u}_i \tilde{u}_j), \tag{11.7}$$

$$\bar{\sigma}_{ij} = \mu(\tilde{T})\left(-\frac{2}{3}\frac{\partial \tilde{u}_k}{\partial \tilde{x}_k}\delta_{ij} + \frac{\partial \tilde{u}_i}{\partial x_j} + \frac{\partial \tilde{u}_j}{\partial x_i}\right), \tag{11.8}$$

$$\bar{\rho}\tilde{k} = \frac{1}{2}(\overline{\rho u_i u_i} - \bar{\rho}\tilde{u}_i\tilde{u}_i) = -\frac{1}{2}\tau_{ii}, \tag{11.9}$$

$$\mathcal{Q}_j = Q_j + \bar{q}_j, \tag{11.10}$$

$$Q_j = -\bar{\rho}c_p(\widetilde{Tu_j} - \tilde{T}\tilde{u}_j), \tag{11.11}$$

$$\bar{q}_j = k(\tilde{T})\frac{\partial \tilde{T}}{\partial x_j}, \tag{11.12}$$

$$\bar{\rho}\tilde{e} = \bar{\rho}c_v\tilde{T} + \frac{1}{2}\bar{\rho}\tilde{u}_i\tilde{u}_i + \bar{\rho}\tilde{k}. \tag{11.13}$$

The particular form of this energy equation, proposed by Knight et al. (1998), was found by Martin, Piomelli, and Candler (1999) to provide an accurate model of the subgrid-scale turbulent diffusion in decaying compressible isotropic turbulence.

Closure of the aforementioned system of equations requires specification of a model for the subgrid-scale stress, τ_{ij}, and heat flux, Q_j. Two basic categories of closure have been proposed, namely, (1) explicit subgrid-scale models and (2) the monotone integrated large eddy simulation (LES) method, known as MILES, the subclass of implicit LES models using mononicity-preserving high-resolution algorithms. We present examples of each category in the following sections.

11.1.2 Explicit subgrid-scale models

Historically, the earliest subgrid-scale models were based on explicit mathematical models for τ_{ij} and Q_j. In this chapter, we present results for several explicit subgrid-scale models. On the basis of experience and the results of a previous study (Fureby Nilsson, and Andersson 1999), we use functional modeling (Fureby 2001; Sagaut 2005, also see Chapter 3) to relate only the action of the subgrid scales on the resolved scales. More precisely, we conjecture that the energy transfer from the resolved to the subgrid scales is analogous to Brownian motion superimposed on the motion of the large-scale resolved motion, so that the models for τ_{ij} and Q_j are

$$\tau_{ij} = 2\mu_t\left(\tilde{S}_{ij} - \frac{1}{3}\tilde{S}_{kk}\delta_{ij}\right) - \frac{2}{2}\bar{\rho}\tilde{k}, \tag{11.14}$$

$$Q_j = \kappa_t c_p \frac{\partial \tilde{T}}{\partial x_j}, \tag{11.15}$$

where $\tilde{S}_{ij}$ is the rate-of-strain tensor,

$$\tilde{S}_{ij} = \frac{1}{2}\left(\frac{\partial \tilde{u}_i}{\partial x_j} + \frac{\partial \tilde{u}_j}{\partial x_i}\right). \tag{11.16}$$

Variables $\tilde{k}$, μ_t, and κ_t are the subgrid kinetic energy, eddy viscosity, and eddy diffusivity, respectively. To this end we may use the Smagorinsky (SMG) model (Smagorinsky 1963),

$$\tilde{k} = C_I \Delta^2 \tilde{S}_{kl}\tilde{S}_{kl}, \tag{11.17}$$

$$\mu_t = C_D \bar{\rho} \Delta^2 \sqrt{\tilde{S}_{kl}\tilde{S}_{kl}}, \tag{11.18}$$

and

$$\kappa_t = \frac{\mu_t}{\mathrm{Pr}_t'}, \tag{11.19}$$

where $\mathrm{Pr}_t' \approx 0.7$ is the LES turbulent Prandtl number, or the one-equation eddy viscosity model (OEEVM; Schumann 1975),

$$\mu_t = C_k \bar{\rho} \Delta \sqrt{\tilde{k}}, \tag{11.20}$$

with κ_t given by (11.19). We model the equation for $\tilde{k}$ after the exact subgrid turbulence energy equation, and we approximate it as

$$\frac{\partial \bar{\rho}\tilde{k}}{\partial t} + \frac{\partial \bar{\rho}\tilde{k}\tilde{u}_j}{\partial x_j} = \tau_{ij}\frac{\partial \tilde{u}_i}{\partial x_j} - \bar{\rho}\varepsilon + \frac{\partial}{\partial x_j}\left[(\mu_t + \bar{\mu})\frac{\partial \tilde{k}}{\partial x_j}\right], \tag{11.21}$$

where we model the subgrid dissipation, ε, in (11.21) by using simple dimensional arguments as

$$\varepsilon = C_\varepsilon \frac{\tilde{k}^{3/2}}{\Delta}. \tag{11.22}$$

Note that we obtain the SMG model from OEEVM under the assumption that production equals dissipation. We obtain the model coefficients (C_I, C_D, C_k, and C_ε) from the assumption of isotropy and a $k^{-5/3}$ inertial subrange behavior or by using a dynamic evaluation of the coefficients (Germano et al. 1991), which is called the dynamic Smagorinsky model (DSMG). We can approximate the solution to these equations with a finite-volume method using a second-order central differencing scheme in space yielding a formally second-order-accurate method.

11.1.3 Monotone integrated large eddy simulation

The MILES approach assumes that the subgrid-scale model is implicit in the numerical algorithm for the inviscid fluxes and therefore formally sets $\tau_{ij} = 0$ and $Q_j = 0$ in (11.3) and (11.4). This is a bold approach, and despite the significant evidence of its success as described in this book, it is still controversial. The approach is based on the concept that the principal function of the subgrid-scale model is to provide the correct transfer of energy from the resolved scales to the subgrid (unresolved) scales (Boris 1990; Boris et al. 1992; Ferziger 1997; Fureby 2001; Fureby and Grinstein 2002; Grinstein and DeVore 1996; see also Chapter 1). That is, the principal function is to accurately emulate the turbulence energy cascade.

The principal task is the selection of the numerical algorithm for the inviscid fluxes. This is by no means a simple and foolproof process, since a complete statement of the requirements for the inviscid flux algorithm does not exist, nor does a formal mathematical proof of the efficacy of the MILES approach overall. This should not discourage research using the MILES approach, however, as virtually all progress in modeling of turbulent flows has been based upon a combination of conjecture, intuition, and known mathematical constraints (e.g., dimensional consistency, Galilean invariance). Fureby and Grinstein (1999) discuss an intuitive set of properties for the inviscid flux algorithm. The first property is *local monotonicity preserving* (LMP). That is, the numerical algorithm does not permit the formation of spurious (i.e., nonphysical) oscillations whose length scale is of the order of the local grid spacing, Δ. Such oscillations are nearly always unphysical and therefore should not appear in the computation. The second property is *total variation diminishing* (TVD), which was originally proposed by Harten (1983; also see Laney 1998). In the context of one-dimensional flow, the total variation (TV) of a function $f(x_i, t)$ defined at a discrete set of points in space $x_i, i = 1, \ldots, m$ and time $t^n, n = 1, \ldots$ is

$$\mathrm{TV}[f(x, t^n)] = \sum_{i=2}^{i=m} |f(x_i, t^n) - f(x_{i-1}, t^n)|. \qquad (11.23)$$

The TVD condition is

$$\mathrm{TV}[f(x, t^{n+1})] \leq \mathrm{TV}[f(x, t^n)]. \qquad (11.24)$$

The extension to multidimensional flows is discussed in Fureby and Grinstein (1999). The appearance of spurious oscillations tends to increase the TV, and therefore an inviscid flux algorithm that possesses the TVD property would necessarily avoid such unphysical behavior. Numerical algorithms possessing the TVD property are also LMP (at least, this can be shown in one dimension; see Laney 1998).

We may summarize the basic approach to selection of an algorithm for the inviscid fluxes of mass, momentum, and energy as follows. In general, we use nonlinear algorithms to construct the implicit subgrid-scale model in order to maintain formal second-order accuracy in smooth regions of the flow. Moreover, we require these schemes to provide a leading-order truncation error that vanishes as the grid size goes to zero, so that the method remains consistent with the Navier–Stokes equations. To this end we introduce a flux limiter, $\Gamma = \Gamma[\Psi(x_i, t)]$, to combine a second-order flux function, $\Psi_f^{(H)}$, that is well behaved in smooth flow regions with a first-order dispersion-free convective flux function, $\Psi_f^{(L)}$, being well behaved near sharp gradients, so that the total convective flux function for mass, momentum, or energy becomes

$$\Psi_f = \Psi_f^{(H)} - (1 - \Gamma)\left[\Psi_f^{(H)} - \Psi_f^{(L)}\right]. \tag{11.25}$$

Here, we obtain $\Psi_f^{(H)}$ from linear interpolation and $\Psi_f^{(L)}$ from an upwind-biased piece-wise constant approximation. The flux limiters we use in this study are an ENO-like method (Yan, Knight, and Zheltovodov 2002) for LES of shock-wave turbulent boundary layer interactions, and flux corrected transport (FCT)-type methods (Boris and Book 1973) for supersonic base flows.

The modified equations provide the most suitable platform for comparing MILES and LES. Following Fureby and Grinstein (2002), Grinstein and Fureby (2002) and Chapter 4a, we see that the implicit (or built-in) models for subgrid-scale stress τ and heat flux $\mathbf{Q}$ are

$$\tau = \bar{\rho}[\mathbf{C}(\nabla\mathbf{u})^T + (\nabla\mathbf{u})\mathbf{C}^T + \chi^2(\nabla\mathbf{u})\mathbf{d} \otimes (\nabla\mathbf{u})\mathbf{d}], \tag{11.26}$$

$$\mathbf{Q} = \bar{\rho}[\mathbf{C}(\nabla e) + \chi^2(\nabla e \cdot \mathbf{d})(\nabla\mathbf{u})\mathbf{d}], \tag{11.27}$$

where $\mathbf{d}$ is the distance between two adjacent cell centers, $\mathbf{C} = \chi(\mathbf{u} \otimes \mathbf{d})$, and $\chi = \frac{1}{2}(1 - \Gamma)(\beta^- - \beta^+)$, with

$$\beta^{\pm} = \frac{1}{2}\frac{\mathbf{u}_f \cdot d\mathbf{A}_f \pm |\mathbf{u}_f \cdot d\mathbf{A}_f|}{|\mathbf{u}_f \cdot d\mathbf{A}_f|}. \tag{11.28}$$

We can split the built-in subgrid stress tensor into $\tau = \tau_1 + \tau_2$, where

$$\tau_1 = \bar{\rho}\left[\mathbf{C}(\nabla\mathbf{u})^T + (\nabla\mathbf{u})\mathbf{C}^T\right], \tag{11.29}$$

$$\tau_2 = \bar{\rho}\chi^2(\nabla\mathbf{u})\mathbf{d} \otimes (\nabla\mathbf{u})\mathbf{d}. \tag{11.30}$$

One can regard (11.29) as a generalized eddy viscosity term in which $\mathbf{C}$ is a tensor-valued eddy viscosity whereas (11.30) is of a form similar to the scale-similarity part in a mixed model. In Borue and Orszag (1998), it is shown that terms of type (11.30) improve the correlations between the exact and the modeled subgrid stress tensor.

11.2 Supersonic flat-plate boundary layer

11.2.1 Introduction

The supersonic turbulent boundary on a flat plate in the absence of a pressure gradient is the simplest bounded turbulent shear flow and is generally well understood. A detailed description of the mean flow and turbulence properties is presented in Smits and Dussauge (1996). Here we present a brief summary of the salient features.

We may describe the mean streamwise velocity profile in terms of three regions. Consider a Cartesian coordinate system $(x_1, x_2, x_3) = (x, y, z)$ with x, y, and z corresponding to streamwise, wall normal, and spanwise directions, respectively, and corresponding velocities $(u_1, u_2, u_3) = (u, v, w)$. In the immediate vicinity of the boundary, the unsteady streamwise momentum equation reduces to

$$0 = \frac{\partial \mathcal{T}_{xy}}{\partial y},$$

$$(11.31)$$

where $\mathcal{T}_{xy}$ is the total unsteady shear stress (subgrid scale plus molecular),

$$\mathcal{T}_{xy} = \tau_{xy} + \bar{\sigma}_{xy}$$

$$= -\bar{\rho}(\widetilde{uv} - \tilde{u}\tilde{v}) + \bar{\mu}\frac{\partial \tilde{u}}{\partial y}.$$

$$(11.32)$$

A time average of the streamwise momentum equation yields

$$0 = \frac{\partial \overline{\mathcal{T}}_{xy}}{\partial y},$$

$$(11.33)$$

where the overbar implies a time average

$$\overline{f} = \frac{1}{t_f - t_i} \int_{t_i}^{t_f} f \, dt,$$

$$(11.34)$$

where t_i and t_f are the initial and final times. Since $\tilde{u} = \mathcal{O}(y)$ as $y \to 0$, the molecular shear stress dominates and therefore

$$U = \frac{\tau_w y}{\bar{\mu}_w} + \mathcal{O}(y^2),$$

$$(11.35)$$

where U is the time-averaged streamwise velocity and τ_w and $\bar{\mu}_w$ are the mean wall shear stress and the molecular viscosity evaluated at the wall. The shear stress varies with streamwise position along the wall, and the molecular viscosity may also. Thus, the mean streamwise velocity profile is linear in the immediate neighborhood of the boundary. This region is known as the *viscous sublayer*. Its extent is measured in terms of the *inner coordinate*, y^+, defined as

$$y^+ = \frac{y u_\tau}{\nu_w},$$

$$(11.36)$$

where ν_w is the kinematic molecular viscosity evaluated at the wall and u_τ is the *friction velocity*, defined as

$$u_\tau = \sqrt{\frac{\tau_w}{\rho_w}},$$

$$(11.37)$$

and is a function of streamwise position. Typically, the viscous sublayer corresponds to approximately $y^+ \leq 10$.

At larger distances from the boundary, turbulent shear stress dominates the molecular shear stress, and the convection of streamwise momentum is still negligible as a result of the intensive turbulent mixing. Integrating Eq. (11.33) from the boundary, we have

$$-\overline{\bar{\rho}(\widetilde{uv} - \tilde{u}\tilde{v})} = \tau_w.$$

$$(11.38)$$

Applying Prandtl's mixing length theory, we have

$$-\overline{\rho(\widetilde{uv} - \tilde{u}\tilde{v})} = \rho \ell^2 \frac{\partial U}{\partial y} \left| \frac{\partial U}{\partial y} \right|, \tag{11.39}$$

where ρ is the time-averaged density. We assume the mixing length, ℓ, to be

$$\ell = \kappa y, \tag{11.40}$$

where $\kappa = 0.4$ is von Karman's constant. Since the static pressure is constant across the boundary layer,

$$\frac{\rho_w}{\rho} = \frac{T}{T_w}, \tag{11.41}$$

where T is the time-averaged temperature. A similar analysis of the energy equation (Smiths and Dussauge 1996) yields the Crocco relation,

$$\frac{T}{T_w} = 1 + B\frac{U}{U_\infty} - A^2 \left(\frac{U}{U_\infty}\right)^2, \tag{11.42}$$

where

$$A = \sqrt{\frac{(\gamma - 1)}{2} \Pr_t M_\infty^2 \frac{T_\infty}{T_w}}, \tag{11.43}$$

$$B = -\frac{\Pr_t q_w U_\infty}{c_p T_w \tau_w}, \tag{11.44}$$

where $\Pr_t$ is the mean flow turbulent Prandtl number. We can further show (White 1974) that

$$B = \frac{T_{\text{aw}}}{T_w} - 1, \tag{11.45}$$

where T_{aw} is the adiabatic wall temperature,

$$T_{\text{aw}} = T_\infty \left[1 + \frac{(\gamma - 1)}{2} \Pr_t M_\infty^2\right]. \tag{11.46}$$

Combining these equations yields the Van Driest–Fernholz and Finley (Fernholtz and Finaly 1980; Van Driest 1951; White 1974) velocity profile,

$$U_{\text{VD}} = \frac{u_\tau}{\kappa} \log \left(\frac{y u_\tau}{v_w}\right) + C u_\tau, \tag{11.47}$$

where U_{VD} is the Van Driest transformed velocity:

$$U_{\text{VD}} = \frac{U_\infty}{A} \left\{ \sin^{-1} \left[\frac{\left(2A^2 \frac{U}{U_\infty} - B\right)}{\sqrt{B^2 + 4A^2}}\right] + \sin^{-1} \left[\frac{B}{\sqrt{B^2 + 4A^2}}\right] \right\}. \tag{11.48}$$

We may also write this as

$$U_{\text{VD}} = u_\tau f_1(y^+), \tag{11.49}$$

where $f_1(y^+)$ is a universal function (Smits and Dussauge 1996). This defines the *logarithmic region*. The extent of the logarithmic region depends on the local-flow Reynolds number, $\text{Re} = \rho_\infty U_\infty \theta / \mu_\infty$, where θ is the compressible momentum thickness. A typical range may be $30 \leq y^+ \leq 10^3$ to 10^4.

Coles (1956) proposed a scaling law for the outermost portion of an incompressible flat-plate turbulent boundary layer that was later extended to compressible flows by Sun and Childs (1973):

$$U_{VD} = \frac{u_\tau}{\kappa} \log \left(\frac{y u_\tau}{\nu_w} \right) + C u_\tau + \frac{\Pi u_\tau}{\kappa} w \left(\frac{y}{\delta} \right). \tag{11.50}$$

Here δ is the local boundary layer thickness and $w(y/\delta) = 2 \sin^2(\pi y / 2\delta)$. For high Reynolds numbers, $\Pi = 0.55$. Applying Eq. (11.50) at the edge of the boundary layer $y = \delta$ and substracting Eq. (11.50) yields

$$U_{VD}^\infty - U_{VD} = u_\tau \left\{ -\frac{1}{\kappa} \log \left(\frac{y}{\delta} \right) + \frac{\Pi}{\kappa} \left[1 - w \left(\frac{y}{\delta} \right) \right] \right\}. \tag{11.51}$$

We may rewrite this as

$$U_{VD}^\infty - U_{VD} = u_\tau f_2(\eta), \tag{11.52}$$

where $\eta = y/\delta$ and $f_2(\eta)$ is a universal function (Smits and Dussauge 1996). This defines the *outer* or *wake region*. It extends from the logarithmic region to the edge of the boundary layer.

The streamwise velocity fluctuations in a compressible flat-plate turbulent boundary layer at supersonic and lower hypersonic speeds appear to scale in accordance with

$$\frac{\rho \overline{u'^2}}{\tau_w} = f_3 \left(\frac{y}{\delta} \right), \tag{11.53}$$

where $\overline{u'^2}$ is the time average of the square of the conventional fluctuating streamwise velocity, u'; $f_3(y/\delta)$ is a universal function for a flat plate, zero-pressure-gradient boundary layer; and Eq. (11.53) holds in the logarithmic and outer regions of the boundary layer (Suits and Dussauge 1996).

11.2.2 Boundary conditions

LES of supersonic turbulent shear flows requires the specification of appropriate boundary conditions. Consider the LES of a supersonic turbulent boundary on a flat plate in the absence of an external pressure gradient. We choose the computational domain to be a hexaedron (Figure 11.1). There are five types of boundary conditions, namely, inflow, outflow, solid (lower) wall, upper surface, and lateral surfaces. In this section, we describe each type of boundary condition.

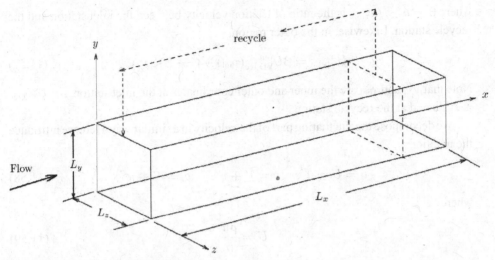

Figure 11.1. Computational domain.

11.2.2.1 Inflow boundary conditions

There are two possible choices for the inflow boundary condition. In the first choice, we assume that the inflow boundary is upstream of the region of transition from laminar to turbulent flow. The inflow boundary condition is therefore a steady laminar boundary layer profile (see, e.g., White 1974) or possibly uniform inflow corresponding to a location upstream of the leading edge of the flat plate. If the focus of the computation is the turbulent boundary layer, the laminar and transitional regions may not be of particular interest.

In the second choice, we simulate a turbulent inflow corresponding to a position downstream of the transition region. Urbin and Knight (2001) developed a compressible extension of the method of Lund, Wu, and Squres (1998) for self-generation of an unsteady turbulent inflow. The simulation generates its own inflow conditions through a sequence of operations where the flowfield at a downstream station is rescaled and reintroduced at the inflow boundary (Figure 11.1). The idea is to decompose each flowfield component into a mean and fluctuating part, and then to apply the appropriate scaling law to each one separately. In the inner region, Eq. (11.49) is

$$U_{\mathrm{VD}}^{\mathrm{inn}} = u_\tau(x) f_1(y^+).$$ (11.54)

In the outer region, Eq. (11.52) is

$$U_{\mathrm{VD}}^\infty - U_{\mathrm{VD}}^{\mathrm{out}} = u_\tau(x) f_2(\eta).$$ (11.55)

Consider a location (the "recycle" location) downstream of the inflow boundary (Figure 11.1). The Van Driest transformed velocity in the inner region, $U_{\mathrm{VD,rec}}^{\mathrm{inn}}$, at the downstream station and the corresponding velocity, $U_{\mathrm{VD,inl}}^{\mathrm{inn}}$, at the inflow ("inlet") are therefore related according to

$$U_{\mathrm{VD,inl}}^{\mathrm{inn}} = \beta \, U_{\mathrm{VD,rec}}^{\mathrm{inn}}(y_{\mathrm{inl}}^+),$$ (11.56)

where $\beta = u_{\tau, \text{inl}}/u_{\tau, \text{rec}}$ is the ratio of friction velocity between the inlet station and the recycle station. Likewise, in the outer region,

$$U_{\text{VD,inl}}^{\text{out}} = \beta U_{\text{VD,rec}}^{\text{out}}(\eta_{\text{inl}}) + (1 - \beta)U^{\infty}. \tag{11.57}$$

Note that y_{inl}^{+} and η_{inl} are the inner and outer coordinates at the inlet station, but $U_{\text{VD,rec}}$ is evaluated at the recycle station.

We decompose the fluctuating part of the velocity in a similar way. Here we introduce the notation

$$u = U + u'', \quad v = V + v'', \quad w = W + w'', \tag{11.58}$$

where

$$U = \frac{\overline{\rho u}}{\bar{\rho}}, \tag{11.59}$$

where we use the long time average. A similar definition holds for V, W, and T.

The velocity fluctuation reads in the inner domain and in the outer domain:

$$u''^{\text{inn}}_{\text{inl}} = \beta u''_{\text{rec}}(y_{\text{inl}}^{+}, z, t), \tag{11.60}$$

$$u''^{\text{out}}_{\text{inl}} = \beta u''_{\text{rec}}(\eta_{\text{inl}}, z, t). \tag{11.61}$$

We assume the scaling for the wall-normal velocity between the recycle and the inlet stations to be

$$V^{\text{inn}}_{\text{inl}} = V_{\text{rec}}(y_{\text{inl}}^{+}), \tag{11.62}$$

$$V^{\text{out}}_{\text{inl}} = V_{\text{rec}}(\eta_{\text{inl}}). \tag{11.63}$$

We use this relatively simple scaling as a convenient approximation to avoid the computation of the derivatives, $\partial u_{\tau}/\partial x$ and $\partial \delta/\partial x$, needed to be consistent with the streamwise velocity scaling. Such approximation appears to be sufficient for a zero-pressure-gradient boundary layer (Arad and Wolfshtein 1999). No scaling of the spanwise mean velocity component is required since it is zero. We assume the v'' and w'' velocity fluctuations scaling to be the same as for u''.

We scale the mean temperature to account for compressibility effects as well. As the streamwise pressure gradient is negligible compared with the wall-normal temperature gradient, we assume the scaling to be

$$T^{\text{inn}}_{\text{inl}} = T_{\text{rec}}(y_{\text{inl}}^{+}), \tag{11.64}$$

$$T^{\text{out}}_{\text{inl}} = T_{\text{rec}}(\eta_{\text{inl}}). \tag{11.65}$$

In a supersonic turbulent boundary layer, the pressure fluctuations are negligible compared with the temperature fluctuations (Bradshaw 1977), and thus we assume

$$T''^{\text{inn}}_{\text{inl}} = T''_{\text{rec}}(y_{\text{inl}}^{+}, z, t), \tag{11.66}$$

$$T''^{\text{out}}_{\text{inl}} = T''_{\text{rec}}(\eta_{\text{inl}}, z, t). \tag{11.67}$$

We obtain the profiles for the entire boundary layer by forming a weighted average of the inner and outer profile, as proposed by Lund et al. (1998):

$$u_{\text{inl}} = \left(U^{\text{inn}}_{\text{inl}} + u''^{\text{inn}}_{\text{inl}} \right) [1 - W(\eta_{\text{inl}})] + \left(U^{\text{out}}_{\text{inl}} + u''^{\text{out}}_{\text{inl}} \right) W(\eta_{\text{inl}}). \tag{11.68}$$

We define the weighting function, W, as

$$W(\eta) = \frac{1}{2} \left\{ 1 + [\tanh(4)]^{-1} \tanh \left[\frac{4(\eta - B)}{(1 - 2B)\eta + B} \right] \right\}, \tag{11.69}$$

using $B = 0.2$ to provide a smooth transition at $y/\delta = 0.2$ (Lund et al. 1998).

The last step of the rescaling process consists of determining u_τ and δ at the inlet. Urbin and Knight (2001) obtained the best results by imposing ratios $u_{\tau,\text{inl}}/u_{\tau,\text{rec}}$ and $\delta_{\text{inl}}/\delta_{\text{rec}}$ according to the combined law of the wall and wake (Smits and Dussauge 1996) and the classical empirical correlation,

$$\frac{\delta_{\text{rec}}}{\delta_{\text{inl}}} = \left[1 + \frac{(x_{\text{rec}} - x_{\text{inl}})}{\delta_{\text{inl}}} 0.37^{\frac{1}{1-1/5}} \text{Re}_{\delta_{\text{inl}}}^{-\frac{1/5}{1-1/5}} \right]^{1-1/5}, \tag{11.70}$$

where $\text{Re}_{\delta_{\text{inl}}}$ is the Reynolds number based on the incoming boundary layer thickness.

11.2.2.2 Outflow boundary conditions

For a supersonic turbulent boundary layer, the instantaneous flow normal to the outflow boundary is nearly everywhere supersonic, except in a very thin region near the wall where subsonic flow exists. Therefore, any error associated with an approximate outflow boundary condition will not propagate upstream to a significant extent. Consequently, the typical outflow boundary condition is zero-gradient extrapolation,

$$\frac{\partial \tilde{u}_i}{\partial x} = 0, \tag{11.71}$$

$$\frac{\partial \tilde{T}}{\partial x} = 0, \tag{11.72}$$

$$\frac{\partial \bar{p}}{\partial x} = 0, \tag{11.73}$$

where x is direction normal to the outflow boundary.

However, Urbin and Knight (2001) observed that the occurrence of an instantaneous inflow at the downstream boundary (caused by the time-dependent nature of the flowfield) was numerically destabilizing, and consequently they applied an ad hoc instantaneous correction to force the flow velocity normal to the outflow boundary to be directed outward at the free-stream condition. This correction was rarely needed, and consequently it had negligible effect on the computed flowfield. A detailed comparison of the computed and experimental results for mean flow and turbulence statistics showed excellent agreement (Urbin and Knight 2001).

11.2.2.3 Solid-wall boundary conditions

At the solid wall the velocity is zero,

$$\tilde{u}_i = 0, \tag{11.74}$$

and either wall temperature T_w is fixed,

$$\tilde{T} = T_w, \tag{11.75}$$

or heat flux q_w is specified,

$$-k\frac{\partial \tilde{T}}{\partial y} = q_w, \tag{11.76}$$

where k is the thermal conductivity and y is the normal distance to the wall. We obtain an additional boundary condition on the pressure from the projection of the momentum equation onto the direction normal to the wall. In the absence of transpiration, this boundary condition becomes (to a first approximation)

$$\frac{\partial \bar{p}}{\partial y} = 0. \tag{11.77}$$

11.2.2.4 Upper surface boundary conditions

The upper surface boundary condition should allow time-dependent acoustic disturbances (generated by the turbulent eddies in the boundary layer) to exit the computational domain. We typically achieve this by using an approximate Riemann boundary condition at the upper surface, which in effect assumes that the disturbance is propagating normal to the outer boundary. We therefore determine the instantaneous fluxes of mass, momentum, and energy at the outer boundary by solving the one-dimensional Riemann problem (see Toro 1997).

11.2.2.5 Lateral surface boundary conditions

We apply a periodic boundary condition on the lateral boundaries. Of course, this boundary condition is not precisely correct, as it imposes a direct correlation between flow variables over the spanwise length, L_z, of the computational domain. Nevertheless, it is commonly applied in both LES and direct numerical simulation (DNS) studies, and may be justified provided that the computed results for the mean flowfield and turbulence statistics are rather insensitive to the value selected for L_z, and that experimental data indicate that the selected value of L_z is large compared with the experimental spanwise correlation length for the turbulence fluctuations. Kline et al. (1967; also see Cantwell 1981) observed that the experimental spanwise wall streak spacing in an incompressible turbulent boundary layer λ_z is approximately $100\,\nu/u_\tau$, where ν is the kinematic viscosity and $u_\tau = \sqrt{\tau_w/\rho}$, where τ_w is the local wall shear stress. Thus, assuming that the viscous sublayer dynamics for a compressible turbulent boundary layer have a

Table 11.1. *Flat-plate boundary layer*

Case no.	M_∞	Re_δ	T_w/T_{aw}	Δx^+	Δy^+	Δz^+
1	2.88	2.0×10^4	1.0	20	1.8	7.0
2	2.88	2.0×10^4	1.1	18	1.6	6.4
3	2.88	8.2×10^4	1.0	17	1.8	17
4	4.0	2.0×10^4	1.0	12	1.8	4.0
5	4.0	2.0×10^4	1.1	11	1.6	3.4

similar scale to the incompressible case, we see that spanwise length L_z must be large compared with λ_z (where ν and ρ are evaluated at the wall) in order that the sublayer turbulence dynamics be properly simulated. Additionally, the spanwise length L_z must be large compared with the correlation length of the large eddies in the outer portion of the boundary layer, which implies $L_z \gg \delta$ where δ is the boundary layer thickness.

11.2.3 Results

Yan, Knight, and Zheltovodov (2002a, 2002b, 2002c) performed a series of LES studies using MILES for a supersonic flat-plate boundary layer at adiabatic and isothermal wall conditions. Details of the flow conditions and grid resolution in the vicinity of the boundary are presented in Table 11.1.

Computed mean streamwise velocity profiles for the five cases are shown in Figures 11.2 (Cases 1 and 2), 11.3 (Cases 1 and 3), and 11.4 (Cases 4 and 5). The ordinate is the Van Driest transformed velocity (11.48), and the abscissa is the dimensionless inner

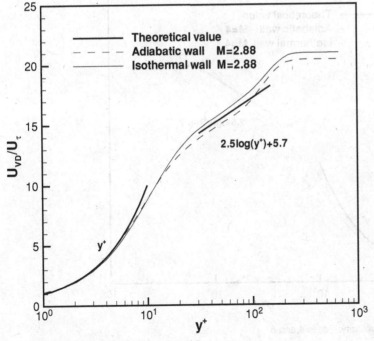

Figure 11.2. Mean velocity: Cases 1 and 2.

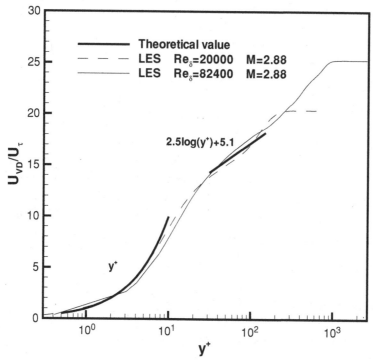

Figure 11.3. Mean velocity: Cases 1 and 3.

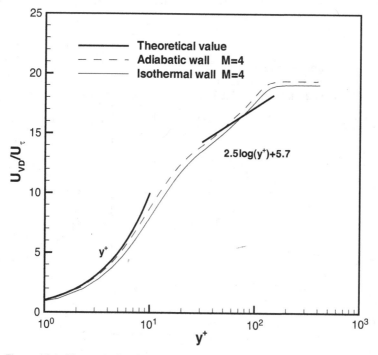

Figure 11.4. Mean velocity: Cases 4 and 5.

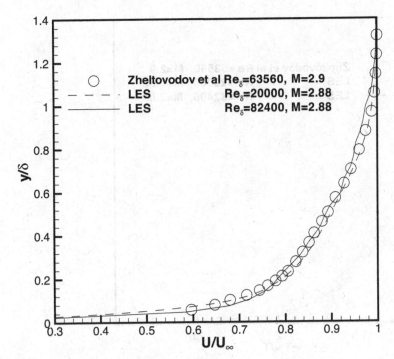

Figure 11.5. Mean velocity: Cases 1 and 3.

coordinate, $y^+ = y u_\tau / \nu_w$. The theoretical profile is the combined law of the wall and wake,

$$U_{\mathrm{VD}} = u_\tau \left[\frac{1}{\kappa} \ln y^+ + C + \frac{2\Pi}{\kappa} \sin^2 \left(\frac{\pi}{2} \frac{y}{\delta} \right) \right], \qquad (11.78)$$

where $\kappa = 0.4$ is von Karman's constant, $C = 5.1$, and the wake parameter $\Pi = 0.12$ at $\mathrm{Re}_\delta = 2 \times 10^4$ and 0.50 at $\mathrm{Re}_\delta = 8.2 \times 10^4$. Good agreement between the computed and theoretical profiles is observed. In Figure 11.5, the computed profiles for Cases 1 and 3 ($\mathrm{Re}_\delta = 2 \times 10^4$ and 8.2×10^4) are compared closely with the experimental data of Zheltovodov and Yakovlev (1986) and Zheltovodov et al. (1990). The computed skin-friction coefficient is within 10% of the experimental correlation for all cases (Yan et al. 2002a, 2002b, 2002c).

The computed mean temperatures for Cases 1 and 3 are shown in Figure 11.6, together with the experimental data of Zheltovodov and Yakovlev (1986) and Zheltovodov et al. (1990). We can see close agreement. The computed adiabatic wall temperatures for Cases 1, 3, and 4 are within 3% of the experimental correlation (11.46), and the computed heat transfer coefficient agrees with the experimental correlation within 12% or less for Cases 2 and 5 (Yan et al. 2002a, 2002b, 2002c).

The computed streamwise Reynolds stress, $\overline{\rho u'' u''}$, is displayed in Figures 11.7 (Cases 1 and 2), 11.8 (Cases 4 and 5), and 11.9 (Cases 1 and 3). The ordinate is the streamwise Reynolds stress nondimensionalized by the local wall shear stress, τ_w, and the abscissa is the vertical distance, y, from the wall normalized by local boundary layer

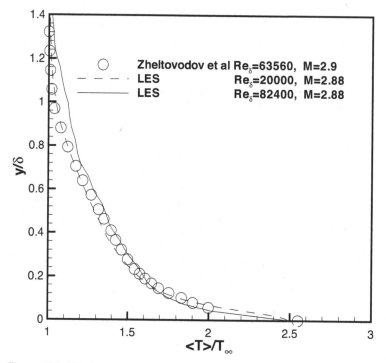

Figure 11.6. Mean temperature: Cases 1 and 3.

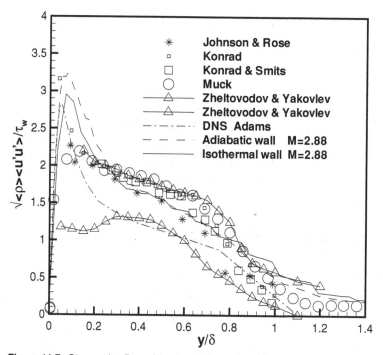

Figure 11.7. Streamwise Reynolds stress: Cases 1 and 2.

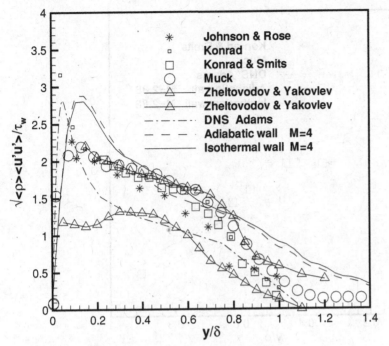

Figure 11.8. Streamwise Reynolds stress: Cases 4 and 5.

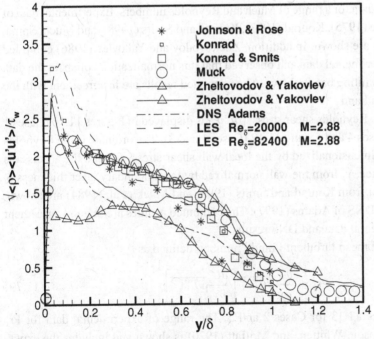

Figure 11.9. Streamwise Reynolds stress: Cases 1 and 3.

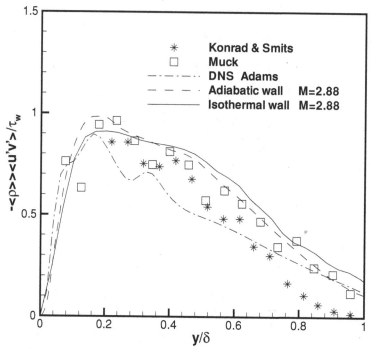

Figure 11.10. Reynolds shear stress: Cases 1 and 2.

thickness δ. Smits and Dussauge (1996) note that this normalization tends to collapse the experimental data for a range of Mach and Reynolds numbers. Experimental data of Johnson and Rose (1975), Konrad (1993), Konrad and Smits (1998), and Muck, Spina, and Smits (1984) are shown. In addition, Zheltovodov and Yakovlev (1986) compiled a large set of experimental data and observed that this normalization collapses the data to lie within the limiting bands shown. The computed results are in agreement with the experimental data band.

The computed Reynolds shear stress, $\overline{\rho u'' v''}$, is displayed in Figures 11.10 (Cases 1 and 2), 11.11 (Cases 4 and 5), and 11.12 (Cases 1 and 3). The ordinate is the Reynolds shear stress nondimensionalized by the local wall shear stress, τ_w, and the abscissa is the vertical distance, y, from the wall normalized by local boundary layer thickness δ. Experimental data from Konrad and Smits (1998) and Muck et al. (1984) are shown, together with the DNS of Adams (1997). The computed profiles are in good agreement with the experimental data and DNS results.

The computed mean turbulent Prandtl number, defined as

$$\mathrm{Pr}_t = \frac{\partial \tilde{T}}{\partial y} \overline{\rho u'' v''} \left[\frac{\partial U}{\partial y} \overline{\rho T'' v''} \right]^{-1}, \tag{11.79}$$

is shown in Figure 11.13 for Cases 2 and 5. The range of experimental data for Pr_t compiled by Simpson, Whitten, and Moffatt (1970) is shown and includes the experimental data of Meier and Rotta (1971) at Mach 1.75 to 4.5 and Horstman and Owen (1972) at Mach 7.2. The computed turbulent Prandtl number falls within the range of the experimental data.

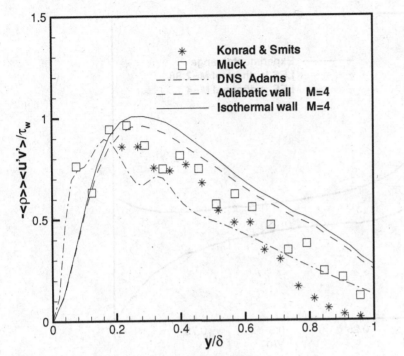

Figure 11.11. Reynolds shear stress: Cases 4 and 5.

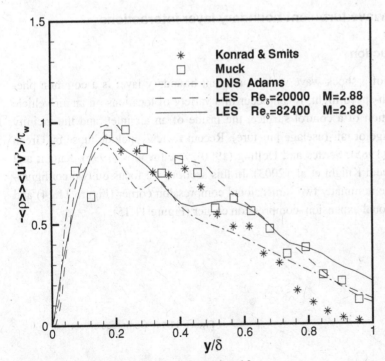

Figure 11.12. Reynolds shear stress: Cases 1 and 3.

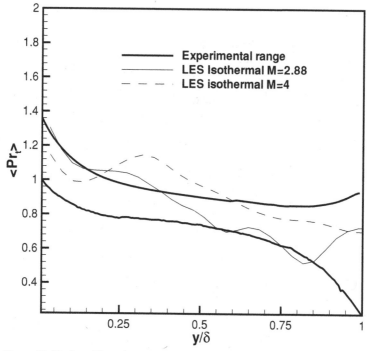

Figure 11.13. Prandtl number.

11.3 Shock-wave turbulent boundary layer interactions

11.3.1 Introduction

The interaction of a shock wave with a turbulent boundary layer is a common phenomenon in high-speed flight. It can occur in a variety of locations on an air vehicle (e.g., the deflection of a control surface, the inside of an air inlet, and the vicinity of a wing-fuselage or tail-fuselage juncture). Recent reviews include those by Green (1970), Delery (1985), Settles and Dolling (1990), Zheltovodov (1996), Knight and Degrez (1998), and Knight et al. (2003). In this section, we focus on two configurations, namely the nominally two-dimensional compression corner (Figure 11.14) and the two-dimensional expansion–compression corner (Figure 11.15).

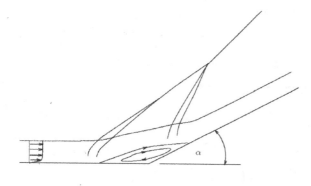

Figure 11.14. Compression corner.

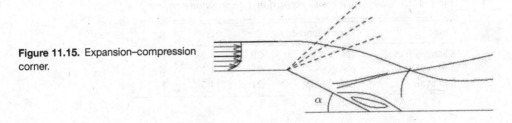

Figure 11.15. Expansion–compression corner.

In the two-dimensional compression corner (Figure 11.14), an equilibrium supersonic turbulent boundary layer is deflected by a compression corner of angle α. The deflection generates a shock system. For a sufficiently large pressure rise, the boundary layer separates, a plateau pressure is observed, and a λ-shock is formed. Zheltovodov and Schülein (1988) and Zheltovodov, Schülein, and Horstman (1993) correlated the separation length with the plateau pressure and Reynolds number. Settles, Perkins, and Bogdonoff (1981) developed a correlation for the upstream influence length (i.e., the distance between the initial pressure rise and the corner) at Mach 3 as a function of corner angle α. Zheltovodov (1990) defined the turbulence structure through a series of extensive experiments. As illustrated in Figure 11.16, they include amplification of the turbulence by shock waves in the boundary layer (1) and external flow (2), suppression of turbulence by the expansion fan (3), formation of a new layer in the near-wall portion of the reattaching flow (4), generation of Görtler vortices caused by the curvature of the mean streamlines (5), and reverse transition, that is, turbulent to laminar (6), in the separation region that is due to the favorable pressure gradient in the direction of the reverse flow and lower effective local Reynolds number associated with the reverse flow. Unsteadiness of the shock system and separation region has been discussed in detail by Dolling (1998, 2001), Smits and Dussauge (1996), and Adams (2000).

In the two-dimensional expansion–compression corner (Figure 11.15), an equilibrium supersonic turbulent boundary layer is accelerated through an expansion angle α and subsequently recompresses through a compression angle of the same value. The expansion causes several changes to the boundary layer. First, the boundary layer thickness

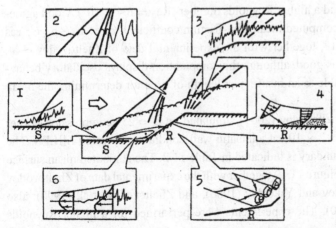

Figure 11.16. Turbulence structure in a two-dimensional compression corner.

Table 11.2. *Compression and expansion–compression corners*

Case	Type of corner	Δx^+	Δy^+	Δz^+
Kannepalli et al.	comp.	20	1	7
Yan et al.	comp.	21	1.7	7
Yan et al.	exp.–comp.	21	1.7	7

increases more rapidly than it does on a flat plate for the same distance (Dawson, Samimy, and Arnette 1994). Second, the turbulence intensity decreases rapidly as a result of the stabilizing effects of bulk dilatation, convex curvature, and favorable pressure gradient. For example, Dussauge and Gaviglio (1981) observed a 50% reduction in the streamwise turbulence intensity. Third, the boundary layer may indeed relaminarize if the expansion is sufficiently strong. Narasimha and Sreenivasan (1973) proposed a model for relaminarization based upon the local pressure gradient and wall shear stress.

The shock wave–boundary layer interaction at the second corner was studied in detail by Zheltovodov et al. (1993). The boundary layer immediately upstream of the recompression has typically not reached equilibrium, and therefore the interaction results in different detailed turbulence behavior than in the nominal two-dimensional compression corner just described. For example, the reduced turbulence levels in the incoming turbulent boundary layer at the second corner result in a lower surface heat transfer compared with the two-dimensional compression corner. For sufficiently large pressure rise at recompression, boundary layer separation occurs.

11.3.2 Compression corner

Kannepalli, Aunajatesan, and Dash (2002) simulated an $\alpha = 24°$ compression corner at Mach 3 and $Re_\delta = 2 \times 10^4$ at adiabatic wall conditions by using MILES. The grid resolution at the boundary is indicated in Table 11.2. The computed mean surface pressure is displayed in Figure 11.17 together with the experimental data of Zheltovodov, Schülein, and Yakovlev (1983), Zheltovodov and Yakovlev (1986), and Zheltovodov et al. (1990); also see Borisov et al. (2000). The experiments were performed for a slightly different ramp angle ($\alpha = 25°$) and a higher Reynolds number ($Re_\delta = 6.4 \times 10^4$). Good agreement is observed. The computed mean skin-friction coefficient, $c_f = \tau_w / \frac{1}{2}\rho_\infty U_\infty^2$, is displayed in Figure 11.18, together with the experimental data of Zheltovodov et al. Overall, the agreement is good, although the computation displays oscillatory behavior that may be attributable to an insufficient length of time for determining the mean values.

Yan, Knight, and Zheltovodov (2001) simulated an $\alpha = 25°$ compression corner at Mach 2.88 and $Re_\delta = 2 \times 10^4$ at adiabatic wall conditions by using MILES. The grid resolution at the boundary is indicated in Table 11.2. The computed mean surface pressure is displayed in Figure 11.19 together with the experimental data of Zheltovodov et al. (1983), Zheltovodov and Yakovlev (1986), and Zheltovodov et al. (1990); also see Borisov et al. (2000). The experiments were performed for a higher Reynolds

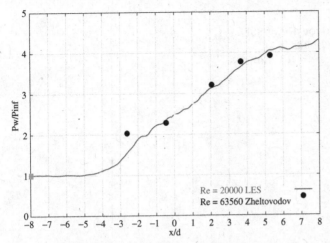

Figure 11.17. Surface pressure (Kannepalli et al.). See color plates at the middle of this book.

number ($Re_\delta = 6.4 \times 10^4$). The agreement is poor. The computed mean skin-friction coefficient is shown in Figure 11.20, together with the experimental data of Zheltovodov et al. Overall, the agreement is reasonable.

11.3.3 Expansion–compression corner

Knight, Yan, and Zheltovodov (2001) simulated an $\alpha = 25°$ expansion–compression corner at Mach 2.88 and $Re_\delta = 2 \times 10^4$ at adiabatic wall conditions by using MILES. The grid resolution at the boundary is indicated in Table 11.2. The computed mean surface pressure is displayed in Figure 11.21, together with the experimental data of Zheltovodov et al. (1990), Zheltovodov and Schülein (1988). Zheltovodov et al. (1993), and Zheltovodov et al. (1987); see additional references cited in

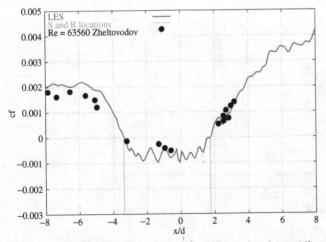

Figure 11.18. Skin friction (Kannepalli et al.). See color plates at the middle of this book.

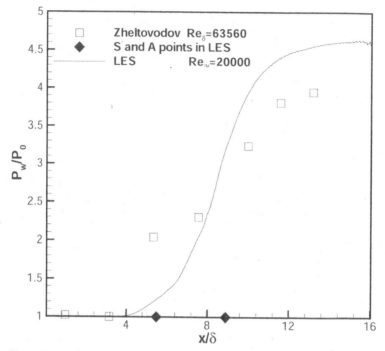

Figure 11.19. Surface pressure (Yan et al.).

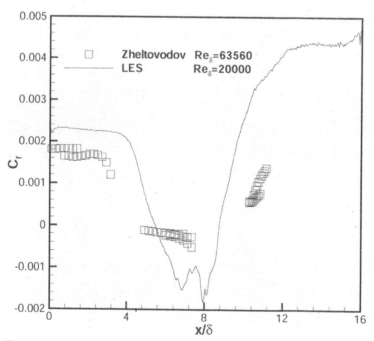

Figure 11.20. Skin friction (Yan et al.).

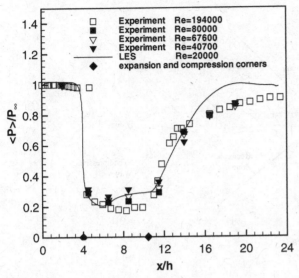

Figure 11.21. Surface pressure (Yan et al.).

Knight et al. (2003). The experiments were performed for higher Reynolds numbers ($Re_\delta = 4.1 \times 10^4$ to 1.94×10^5). The agreement between the computed profile (at $Re_\delta = 2 \times 10^4$) and experimental data at the closest Reynolds number ($Re_\delta = 4.1 \times 10^4$) is good. In particular, the computed upstream influence point (i.e., the location at which the pressure begins to rise, following the expansion) is in close agreement with the experiment. The computed mean skin-friction coefficient is shown in Figure 11.22, together with the experimental data of Zheltovodov et al. at $Re_\delta = 8 \times 10^4$ and 1.94×10^5. Overall, the agreement is reasonable.

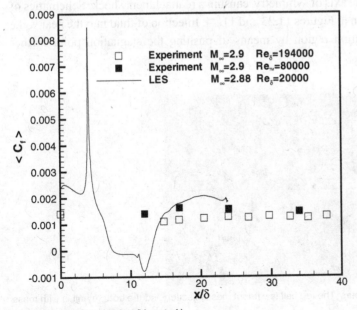

Figure 11.22. Skin friction (Yan et al.).

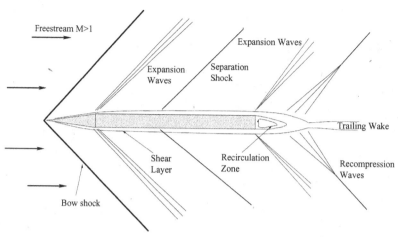

Figure 11.23. Flowfield characteristics.

11.4 Supersonic base flows

Flows around missiles, rockets, and projectiles often experience base-flow separation. This means that the pressure behind the base of the projectile is considerably lower than the free-stream pressure, causing base drag that often constitutes a large portion of the total drag (Rollstin 1987). A recirculation zone is formed just behind the blunt base. The size of this region determines the flow turning angle that is coming off the base, and hence the strength of the expansion waves. A smaller recirculation region causes the flow to turn more sharply, leading to a stronger expansion wave and lower pressures behind the base. Small separated regions therefore cause lower base pressures and larger base drag than do large regions. As the shear layer reattaches, the flow is forced to turn along the axis of symmetry, causing a reattachment shock. Schematics of the flowfield are shown in Figures 11.23 and 11.24. Injection of fluid into the near wake expands the recirculation region by means of pushing the stagnation point further downstream.

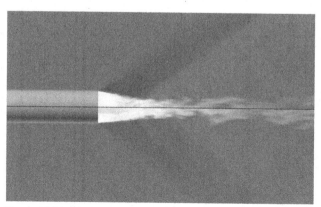

Figure 11.24. Mach contours: The top half is without mass injection and the bottom half is with mass injection.

Attempts to solve after-body flows with Reynolds-averaged Navier–Stokes (RANS) methods have been ongoing since the late 1970s. Putnam and Bissinger (1985) summarized these early attempts and concluded that the current methods were unable to predict these flows accurately. Recent RANS studies, such as those by Peace (1991), Kaurinkoski (2000), Sahu (1994), and Chuang and Chieng (1996), present improved results, as a result of improved mesh handling, turbulence modeling, and numerical methods. Based on the unsatisfactory results of RANS, other approaches such as detached eddy simulation (DES) or LES should be considered. As a compromise between LES and RANS, Forsythe, Hoffmann, and Squires (2002) performed DES on the supersonic axisymmetric base flow. Baurle et al. (2001) used a hybrid RANS–LES approach to simulate axisymmetric base flow. A separate RANS simulation was used to generate the upstream fully developed turbulent flow just prior to the base and then MILES was performed in the base region. The agreement with experiments was quite good. LES of supersonic axisymmetric base flow has been performed, for example, by Fureby et al. (1999) and Fureby and Kupiainen (2003a), in which agreement with experimental data is good when the grid resolution is sufficient.

11.4.1 Computational configuration

The computational configuration that we adopt is based on the experimental work by Dutton et al. (Bourdon et al. 1998; Herrin and Dutton 1994b; Mathur and Dutton 1996), carried out in the blowdown-type wind tunnel at the Gas Dynamic Laboratory at the University of Illinois at Urbana-Champaign. In the experiments, high-pressure air enters a stagnation chamber and passes through a flow-conditioning module. The air is expanded to a nominal design Mach number of approximately $M_\infty \approx 2.46$ in the test section by use of a converging diverging nozzle with a Reynolds number of $Re = 2.86 \times 10^6$, which is based on the free-stream velocity and the after-body diameter. The air exits through a conical diffuser and an exhaust duct to the atmosphere. A more detailed description of the design of the wind tunnel is given in Herrin (1993). The free-stream velocity, pressure, and temperature are

$$U_\infty = 576 \text{ m/s}, \quad p_\infty = 515 \text{ kPa}, \quad T_\infty = 294 \text{ K},$$

respectively, for the zero-bleed cases and

$$U_\infty = 576 \text{ m/s}, \quad p_\infty = 471 \text{ kPa}, \quad T_\infty = 300 \text{ K},$$

respectively, for the cases with base bleed or mass injection. The cylindrical after body utilized in the experiments has a radius of $r_b = 31.75$ mm, and for the base-bleed cases it is equipped with a bleed orifice with radius $r_j = 12.70$ mm, so that the area of the base bleed jet is 16% of the total base area.

The computational configuration and the boundary conditions were designed to resemble, as closely as possible, the laboratory setup, and they consist of a circular cylinder with an outer radius of $6r_b$ and extending $8r_b$ upstream of the base plane and

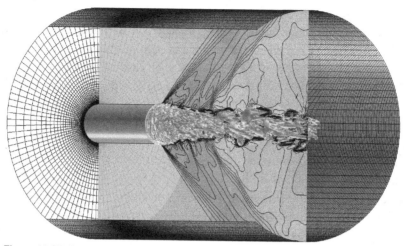

Figure 11.25. Perspective view.

$12r_b$ downstream of the base plane. Details of the grid are shown in Figure 11.25. The grid was refined in the shear layer regions.

We impose supersonic inflow conditions at the upstream end of the simulated test section, and we impose subsonic inflow conditions for the base-bleed jet orifice when applicable. We impose open boundary conditions, using supersonic outflow conditions (Poinsot and Lele 1992), at the downstream end of the simulated test section, and in the cross-stream direction. Furthermore, we impose no-slip adiabatic boundary conditions on the after body and the base plane. The flowfield is initially set to the free-stream conditions everywhere and is then advanced in time until the initial transients have disappeared, whereafter the statistical sampling is activated. We monitor the time evolution of the flow in terms of time histories of the velocity, velocity–rms fluctuations, and pressure to determine when the flow is fully developed. We perform the statistical sampling over an additional time sequence of about $5T$, where $T = U_\infty/20r_b$ is the typical flow-through time. Simulations with different subgrid models, or bleed rates, are restarted from previous simulations and therefore require shorter time for initial transients to disappear. We used three block-structured grids, with 700,000, 1,400,000 and 2,100,000 cells, respectively, in the present study to investigate the effects of spatial resolution. The average wall-normal distance, Δy^+, is between 5 and 20. We expand the mesh radially from the cylinder-shaped after body by using geometrical progression. In the streamwise direction we use geometrical progression to expand the grid, but here we also take into consideration the expansion of the shear layer by using a finer mesh resolution in the appropriate direction. We obtain the inflow velocity profile (at $x/r_b = -8.0$) by running a separate boundary layer code at the prevailing conditions.

11.4.2 Flow results

The base flows are strongly three dimensional and unsteady, having the additional effects of weakening the corner expansion and the downstream recompression. At the base corners, the expansion fans can be recognized by the increase in Mach number

Figure 11.26. Vorticity: The top half is without mass injection and the bottom half is with mass injection.

followed by the gradual recompression in the free-stream flow associated with a decrease in Mach number. At low bleed rates, near the base, the flow is generally reversed virtually out to the corner; however, at higher bleed rates, the flow only becomes reversed in the annular region around the bleed jet and downstream of the bleed jet, depending on the bleed rate. The free annular shear layer is turbulent and viscous, forming at the base corner and developing downstream to reattachment, whereupon it forms the far wake. The pressure gradient and the shear stress gradients in the shear layer are reduced as the shear layer matures when turbulent mixing develops, entraining ambient fluid from the free stream into the wake. Base bleed seems to narrow the annular shear layer and move it further from the line of symmetry and seems to increase the mixing near the corners, leading to reduction of shear stresses, but by directing most of the injectant into the recirculation region the downstream mixing seems not to be fully developed.

The vorticity distribution is extremely complicated and strongly dependent on the mass injection parameter, $I = (\dot{m}/\rho_\infty U_\infty A_b)$, where $\dot{m}$ is the mass flow rate, ρ_∞ is the ambient density, and A_b is the base area. For $I = 0$, the vorticity is dominated by the azimuthal and axial vortices produced in the annular shear layer downstream of the base corner and in the region of mean reattachment. The axial vortices are unstable and exhibit a shedding-like behavior. For $I \neq 0$, the annular Kelvin–Helmholtz instability, along the jet border, yields additional high-intensity vortex structures mainly consisting of axisymmetric rings enclosing a region of intense vorticity. The axisymmetric vortex rings are short lived and are therefore only present in the near-wake region and may occasionally alternate with helicoidal vortices. In the region of reattachment, large coherent structures, having a characteristic size of about $r_b/10$ (considerably larger than the mesh resolution), are found in simulations and experiments (Bouston et al. 1998). For $I \neq 0$, the developments of the flow in the reverse and redevelopment regions are different, as may be seen in Figures 11.26 and 11.27, which display contours of vorticity and density with (top) and without (bottom) mass injection. In the shear layer the bleed jet dominates the flowfield close to the mean reattachment, although the differences are less apparent; further downstream, in the developing wake, differences emerge as a result of the different flow, governed by the interactions of the free shear layers, the bleed jet, and the expansion and recompression waves.

11.4.3 Statistical results

One of our objectives is to examine the influence of grid resolution and turbulence models upon the time-averaged flow quantities. We do this by performing numerical

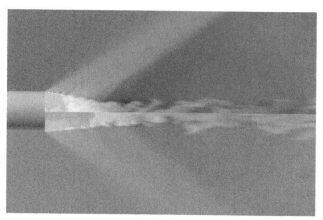

Figure 11.27. Density: The top half is without mass injection and the bottom half is with mass injection.

simulations of the no-bleed ($I = 0$) case with three increasingly fine grids and with the OEEVM, DSMG, and MILES models. For each run we start the statistical sampling (performed over five flow-through times) after the intermediate transients have declined (after about two-flow through times). The experimental data (Bourden et al. 1998; Herrin and Dutton 1994a, 1994b; Mathur and Dutton 1996) are generally considered very reliable, extensive, and well documented and thus lend themselves to detailed comparison in order to validate the computational models, although there is no measurement data available for the Strouhal number, $St = r_b f / U_\infty$, where f is the frequency. The primary objective of this investigation is not to examine and document the *detailed* behavior of different subgrid models in supersonic base flow, but rather to demonstrate the applicability of LES–MILES and to examine the behavior of supersonic base flows, with and without mass injection. Therefore we limit this study to a comparison of the first- and second-order statistical moments of the flowfield in the near wake.

In Figure 11.28, results of grid refinements are shown for different numerical simulations of the no-bleed case. We determine the location of the peak reverse velocity, its value, the rear stagnation point (defined by the "closure" of the recirculation region), and the forward stagnation point (the point where the bleed jet and the reverse flow meet) from the mean velocity, $\langle \tilde{u} \rangle$. The effects of jet bleed seem to be harder to capture, as seen in Figure 11.29, where MILES underpredicts the location of the peak reverse velocity by $\approx \frac{x}{2D_0}$. The magnitude of the reverse velocity is somewhat overpredicted by approximately 15%. The local peak velocity behind the base bleed jet is also underpredicted by approximately 16%, but the approximations become better as the grid is refined.

By comparing global parameters such as the length of the recirculation bubble, we observed that the effect of using different subgrid models is comparatively small. For example, Figures 11.30 and 11.31 indicate that the computed mean streamwise velocity at the centerline is virtually the same for all subgrid models tested. In the free stream ($r/R > 1$) we can find no differences between the various models, whereas

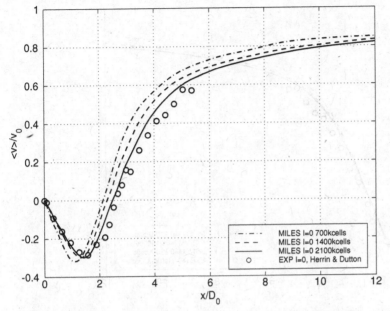

Figure 11.28. Effect of grid refinement on mean streamwise velocity ($I = 0$).

within the wake we observe minor differences. This behavior is also observed in Brandt (2004), where it is shown that the numerical error dominates the subgrid-scale terms in magnitude when a numerical second-order central scheme is used. In particular, we note that all models overpredict the width of the recirculation bubble, with MILES

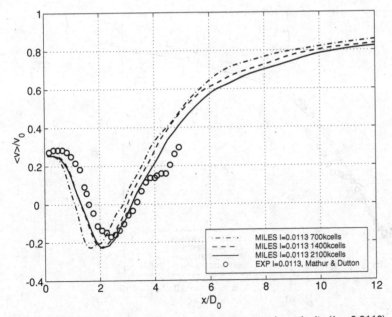

Figure 11.29. Effect of grid refinement on mean streamwise velocity ($I = 0.0113$).

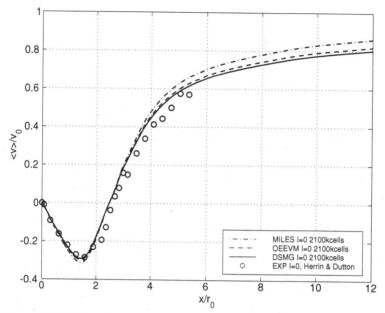

Figure 11.30. Effect of different subgrid models on mean streamwise velocity ($I = 0$).

giving the least satisfactory agreement with results. For the case $I = 0$, OEEVM gives the best agreement with experimental data, when we are comparing the base pressure level, whereas MILES produces the least satisfactory results.

The general influence of the bleed rate is shown in Figures 11.32 to 11.37, where contour plots of the mean axial velocity component, $\langle \tilde{u}_x \rangle$, and mean radial velocity

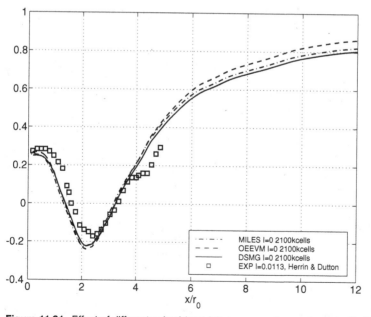

Figure 11.31. Effect of different subgrid models on mean streamwise velocity ($I = 0.0113$).

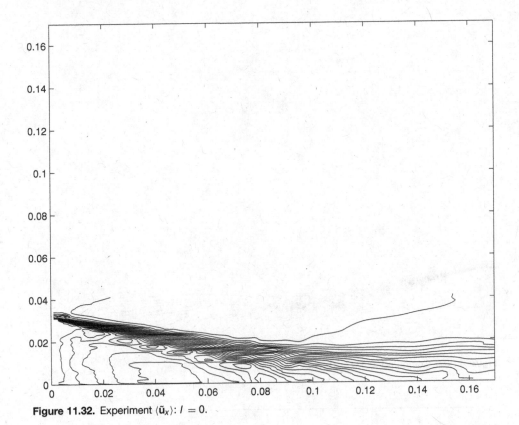

Figure 11.32. Experiment $\langle \tilde{\mathbf{u}}_x \rangle$: $I = 0$.

component, $\langle \tilde{\mathbf{u}}_r \rangle$ at the centerplane for $I = 0$ and 0.0113 are presented. The initial portion of the free shear layer is characterized by very high gradients of $\langle \tilde{\mathbf{u}}_x \rangle$, and it is in good qualitative agreement with the experimental data (Herrin and Dutton 1994b; Mathur and Dutton 1996), for $I = 0$. The spreading of the axial velocity gradients is indicative of the development of the shear layer prior to reattachment, and also of the further wake evolution. Furthermore, the distribution of $\langle \tilde{\mathbf{u}}_x \rangle$ in the free shear layer downstream of the base appears to be separated in an inner region and an outer region, of which the former appears to diverge more rapidly toward the centerline.

The inner region finally overtakes the outer region and consumes most of the total shear layer width. These findings are in good qualitative accordance with the experimental data (Herrin and Dutton 1994b; Mathur and Dutton 1996). The shape of the mean-velocity defect, and the associated recovery of $\langle \tilde{\mathbf{u}}_x \rangle$ downstream of reattachment, is dependent on I; along the centerline, the flow reaccelerates to sonic speed at $x/r_b \approx 4.8$ and 5.8, for $I = 0$ and 0.0113, respectively. Note that $\langle \tilde{\mathbf{u}}_r \rangle$ is smaller than $\langle \tilde{\mathbf{u}}_x \rangle$, emphasizing the anisotropy in the near wake. The high radial velocity gradient at the base corner marks the turning of the flow through the expansion fan, whereas the radial velocity gradient further downstream, occurring at the recompression region, marks the start of the trailing wake. As the outer flow approaches the axis of symmetry, the magnitude of the radial component of $\langle \tilde{\mathbf{u}} \rangle$ continues to increase as a result of the axisymmetric effect, to a peak value of about $0.24\mathbf{u}_\infty$ for $I = 0$ at approximately $2r_0$

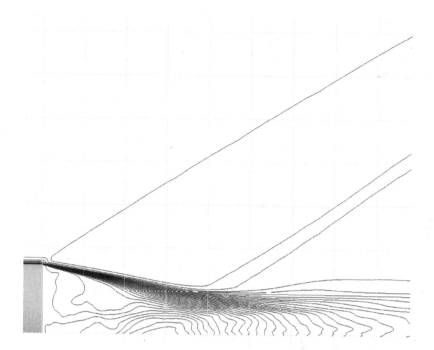

Figure 11.33. Computed $\langle \bar{u}_x \rangle$: $I = 0$.

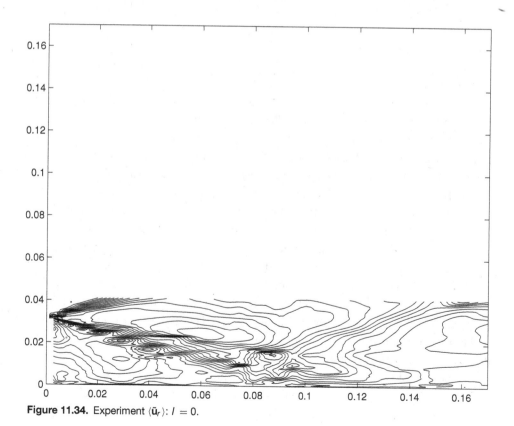

Figure 11.34. Experiment $\langle \bar{u}_r \rangle$: $I = 0$.

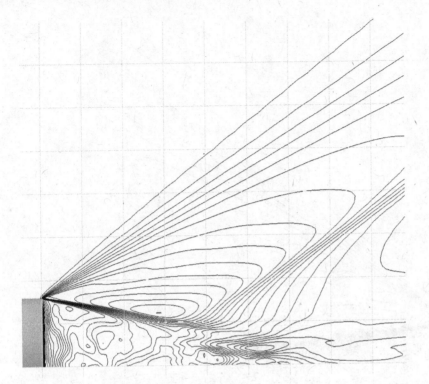

Figure 11.35. Computed $\langle \tilde{u}_r \rangle$: $l = 0$.

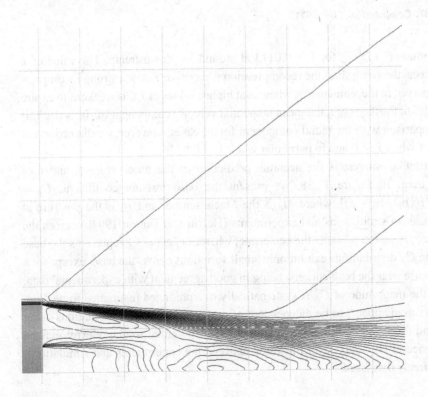

Figure 11.36. Computed $\langle \tilde{u}_x \rangle$: $l = 0.0113$.

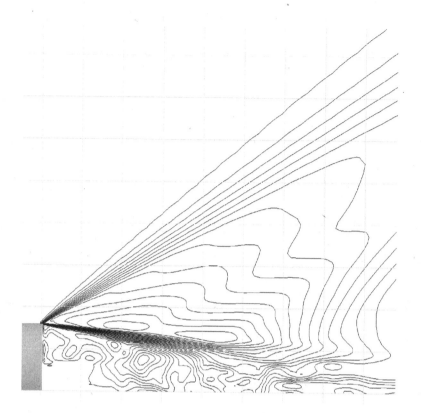

Figure 11.37. Computed $\langle \tilde{u}_r \rangle$: $I = 0.0113$.

downstream, and $0.29\mathbf{u}_\infty$ for $I = 0.0113$ at around $3r_0$ downstream. The value of I clearly affects the strength of the recompression waves: For $I = 0$ a strong recompression is observed in the simulations, whereas at higher values of I this weakens to expire at $I = 0.0226$. Furthermore, the mean tangential velocity component of $\langle \tilde{\mathbf{u}} \rangle$ is negligible in comparison with the radial component for all cases; however, we did observe it to be larger when $I \neq 0$ and in particular when $I = 0.0226$.

Of particular interest is the accurate prediction of the base pressure, and thus the base drag. In Figure 11.38, we present the base-pressure coefficient, $C_p = 2(\gamma M_o^2)^{-1}[(\langle p_b \rangle / p_o - 1)]$, where M_o is the Mach number and p_o is the pressure at the nozzle, from simulations and experiments (Herrin and Dutton 1994b) across the base. Here $r/r_b = 0$ represents the centerline, whereas $r/r_b = \pm 1$ represents the base corner. The C_p distributions exhibit only small variations across the base, except for a slight increase near the base corners, being in good agreement with experimental data. However, the magnitude of C_p is systematically overpredicted (estimated to be 5%). This observation appears to be independent of the resolution, since simulations using the fine grid for $I = 0$ do not improve predictions. A potential source of error may be the difference in boundary layer thickness between the experiments and simulations, as subsequently discussed.

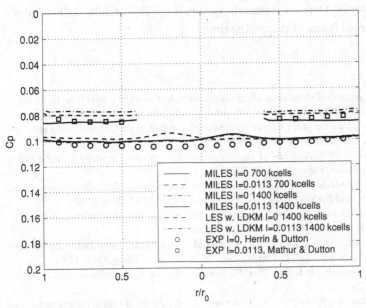

Figure 11.38. Base-pressure coefficient versus r/r_o.

11.5 Conclusions

In this chapter we discuss the use of MILES for compressible turbulent shear flows. In MILES the effects of subgrid physics on the resolved scales are incorporated in the functional reconstruction of the convective fluxes by use of high-resolution nonlinear methods. An analysis based on the modified equations shows that MILES based on a particular class of flux-limiting schemes provides an implicitly implemented anisotropic subgrid model that is dependent on the specifics of the particular numerical scheme, that is, on the flux limiter, on the choice of low- and high-order schemes, and on the gridding. Here we exemplify the versatility and accuracy of MILES methods as applied to supersonic flat-plate boundary layers, shock-wave turbulent boundary layer interactions, and supersonic base flows. We make a comparison with DNS, other conventional LES methods, and experimental data.

Acknowledgments

D. Knight gratefully acknowledges the financial support of the U.S. Air Force Office of Scientific Research under the program management of Dr. John Schmisseur, and the assistance of Dr. Hong Yan, who performed the supersonic turbulent boundary layer and shock wave–turbulent boundary layer computations described herein. M. Kupianinen and C. Fureby gratefully acknowledge the financial support of SAAB Bofors Dynamics (with J. Ekeroot serving as contract monitor). They thank C. Dutton for providing data and helpful information from his experimental studies and J. Forsythe for valuable discussions. The Swedish Defence Research Agency (FOI) is gratefully acknowledged for providing numerous hours of computing time.

REFERENCES

Adams, N. 1997 Direct numerical simulation of supersonic turbulent boundary layer flow, in *Advances in DNS/LES*. Columbus, OH: Greyden Press, 29–40.

Adams, N. 2000 Direct simulation of the turbulent boundary layer along a compression ramp at $M = 3$ and $Re_\theta = 1685$. *J. Fluid Mech.* **420**, 47–83.

Arad, E., & Wolfshtein, M. 1999 Analysis of equilibrium adverse pressure gradient boundary layer using large eddy simulation. Paper AIAA 99-0420.

Baurle, R., Tam, C.-J., Edwards, J., & Hassan, H. 2001 An assessment of boundary treatment and algorithm issues on hybrid RANS/LES solution strategies. Paper AIAA 2001-2562.

Boris, J., & Book, D. 1973 Flux corrected transport. I. SHASTA, a fluid transport algorithm that works. *J. Comput. Phys.* **11**, 38–69.

Boris, J. 1990 On large eddy simulations using subgrid turbulence models, in *Whither Turbulence? Turbulence at the Crossroads*, ed. J. L. Lumley. New York: Springer-Verlag, 344–353.

Boris, J., Grinstein, F., Oran, E., & Kolbe, R. 1992 New insights into large eddy simulation. *Fluid Dyn. Res.*, **10**, 199–228.

Borisov, A., Zheltovodov, A., Maksimov, A., Fedorova, N., & Shpak, S. 2000 Experimental and numerical study of supersonic turbulent separated flows in the neighborhood of two-dimensional obstacles. *Fluid Dyn.* **34**, 181–189.

Borue, V., & Orszag, S. 1998 Local energy flux and subgrid-scale statistics in three dimensional turbulence. *J. Fluid Mech.* **366**, 1–31.

Bourdon, C., Dutton, J. C., Smith, K., & Mathur, T. 1998 Planar visualisations of large scale turbulent structures in axisymmetric supersonic base flows. Paper AIAA 98-0624.

Bradshaw, P. 1977 Compressible turbulent shear layers. *Annu. Rev. Fluid Mech.* **9**, 33–54.

Brandt, T. 2004 Discretization errors in LES using second-order central-difference scheme, *ECCOMAS 2004*, ed. P. Neittaanmäki, T. Rossi, K. Majava, O. Pironneau, W. Rodi, & P. Le Quere. University of Jyvaskyia, Finland, July 24–28 2004.

Cantwell, B. 1981 Organized motions in turbulent flow. *Annu. Rev. Fluid Mech.* **13**, 457–515.

Coles, D. 1956 The law of the wake in the turbulent boundary layer. *J. Fluid Mech.* **1**, 191–226.

Chuang, C., & Chieng, C. 1996 Supersonic base flow computations by higher order turbulence models. *J. Spacecr. Rockets* **33**, 374–380.

Delery, J. 1985 Shock wave turbulent boundary layer interaction and its control. *Progr. Aerospace Sci.* **22**, 209–280.

Dolling, D. 1998 High-speed turbulent separated flows: consistency of mathematical models and flow physics. *AIAA J.* **36**, 725–732.

Dolling, D. 2001 Fifty years of shock wave boundary layer interaction research: What next? *AIAA J.* **39**, 1517–1531.

Dawson, J., Samimy, M., & Arnette, S. 1994 Effects of expansions on supersonic boundary layer: Surface pressure measurements. *AIAA J.* **32**, 2169–2177.

Dussauge, J., & Gaviglio, J. 1981 Bulk dilatation effects on Reynolds stress in the rapid expansion of a turbulent boundary layer at supersonic speed, in *Proceedings of the Symposium on Turbulent Shear Flows*. New York: Springer Verlag **2**, 33–38.

Fernholtz, H., & Finley, P. 1980 *A Critical Commentary on Mean Flow Data for Two-Dimensional Compressible Turbulent Boundary Layers*. Washington, DC: AGARD Technical Report 253.

Ferziger, J. 1997 Large eddy simulation: An introduction and perspective, in *New Tools in Turbulence Modeling*, ed. O. Métais & J. Ferziger. New York: Springer, 29–47.

Forsythe, J., Hoffmann, K., & Squires, K. 2002 Detached-eddy simulation with compressibility corrections applied to a supersonic axisymmetric base flow. Paper AIAA 02-0586.

Fureby, C., Nilsson, Y., & Andersson, K. 1999 Large eddy simulation of supersonic base flow. Paper AIAA 99-0537.

Fureby, C. 2001 Towards large eddy simulation of complex flows, in *Direct and Large Eddy Simulation IV*. Dordrecht: Kluwer.

Fureby, C., & Grinstein, F. 2002 Large eddy simulation of high Reynolds-number free and wall bounded flows. *J. Comput. Phys.* **181**, 68–97.

Fureby, C., & Grinstein, F. 1999 Monotonically integrated large eddy simulation of free shear flows. *AIAA J.* **37**, 544–556.

Fureby, C., & Kupiainen, M. 2003 Large eddy simulation of supersonic axisymmetric baseflow, in *Proceedings of Turbulent Shear Flow Phenomena III*, ed. N. Kasagi, J. Eaton, R. Friedrichs, J. Humphrey, M. Leschziner, & T. Miyauchi.

Germano, M., Piomelli, U., Moin, P., & Cabot, W. 1991 A dynamic subgrid-scale eddy viscosity model. *Phys. Fluids A* **3**, 1760–1765.

Green, J. 1970 Interactions between shock waves and turbulent boundary layers. *Progr. Aerospace Sci.* **11**, 235–341.

Grinstein, F., & DeVore, C. 1996 Dynamics of coherent structures and transition to turbulence in free square jets. *Phys. Fluids* **8**, 1237–1251.

Grinstein, F., & Fureby, C. 2002 Recent progress on MILES for high Reynolds number flows. *J. Fluids Eng.* **124**, 848–861.

Harten, A. 1983 High resolution schemes for hyperbolic conservation laws. *J. Comput. Phys.* **49**, 357–393.

Herrin, J. 1993 *An Experimental Investigation of Supersonic Axisymmetric Base Flow Including the Effects of Afterbody Boattailing*. Urbana, IL: University of Illinois at Urbana-Champaign.

Herrin, J., & Dutton, J. C. 1994a Supersonic near-wake afterbody boattailing effects on axisymmetric bodies. *J. Spacecr. Rockets* **31**, 1021–1028.

Herrin, J., & Dutton, J. C. 1994b Supersonic base flow experiments in the near wake of a cylindrical afterbody. *AIAA J.* **32**, 77–83.

Horstman, C., & Owen, F. 1972 Turbulent properties of a compressible boundary layer. *AIAA J.* **10**, 1418–1424.

Johnson, D., & Rose, W. 1975 Laser velocimeter and hot wire anemometer comparison in a supersonic boundary layer. *AIAA J.* **13**, 512–515.

Kannepalli, C., Arunajatesan, S., & Dash, S. 2002 RANS/LES methodology for supersonic transverse jet interactions with approach flow. AIAA Paper 2002-1139.

Kaurinkoski, P. 2000 *Simulation of the Flow Past a Long-Range Arillery Projectile*. Helsinki: Helsinki University of Technology.

Kline, S., Reynolds, W., Schraub, F., & Runstadler, P. 1967 The structure of turbulent boundary layers. *J. Fluid Mech.* **30**, 741–773.

Knight, D., & Degrez, G. 1998 *Shock Wave Boundary Layer Interactions in High Mach Number Flows – A Critical Survey of Current CFD Prediction Methodologies*. Washington, DC: AGARD AR-319.

Knight, D., Yan, H., Panaras, A., & Zheltovodov, A. 2003 Advances in CFD prediction of shock wave turbulent boudnary layer interactions. *Progr. Aerospace Sci* **39**, 121–184.

Knight, D., Yan, H., & Zheltovodov, A. 2001 Large eddy simulation of supersonic turbulent flows in expansion-compression corner, in *DNS/LES Progress and Challenges*. Columbus, OH: Greyden Press, 183–194.

Knight, D., Zhou, G., Okong'o, N., & Shukla, V. 1998 Compressible large eddy simulation using unstructured grids. Paper AIAA 98-0535.

Konrad, W. 1993 *Three Dimensional Supersonic Boundary Layer Generated by Isentropic Compression*. PhD dissertation, Princeton University.

Konrad, W., & Smits, A. 1998 Turbulence measurements in a three dimensional boundary layer in supersonic flow. *J. Fluid Mech.* **372**, 1–23.

Laney, C. 1998 *Computational Gas Dynamics*. Cambridge: Cambridge University Press.

Lund, T., Wu, X., & Squires, K. 1998 Generation of turbulent inflow data for spatially-developing boundary layer simulations. *J. Comput. Phys.* **140**, 233–258.

Mathur, T., & Dutton, J. C. 1996 Base-bleed experiments with a cylindrical afterbody in supersonic flow. *J. Spacecr. Rockets* **33**, 30–37.

Martin, M., Piomelli, U., & Candler, G. 1999 A priori tests of SGS models in compressible turbulence, presented at the Third ASME/JSME Joint Fluids Engineering Conference, July 10–23, 1999, San Francisco, CA.

Meier, H., & Rotta, J. 1971 Temperature distributions in supersonic turbulent boundary layer. *AIAA J.* **9**, 2149–2156.

Muck, K., Spina, E., & Smits, L. 1984 *Compilation of Turbulence Data for an 8 Degree Compression Corner at Mach 2.9*. Princeton University, Princeton, NJ: Report MAE-1642.

Narasimha, R., & Sreenivasan, K. 1973 Relaminarization in high accelerated turbulent boundary layers. *J. Fluid Mech.* **61**, 417–447.

Peace, A. 1991 Turbulent flow predictions for afterbody/nozzle geometries including base effects. *J. Propul. Power* **7**, 396–403.

Poinsot, T., & Lele, S. 1992 Boundary conditions for direct simulation of compressible viscous reacting flows. *J. Comput. Phys.* **101**, 104–129.

Putnam, L., & Bissinger, N. 1985 Results of AGARD assessment of prediction capabilities for nozzle afterbody flows. AIAA Paper 85-1464.

Rollstin, L. 1987 Measurement of inflight base pressure on an artillery-fired projectile. Paper AIAA 87-2427.

Sagaut, P. 1998 *Large Eddy Simulation for Incompressible Flows*. 2005 New York: Springer. 3rd. edition.

Sahu, J. 1994 Numerical computations of supersonic base flow with special emphasis on turbulence modeling. *AIAA J.* **32**, 1547.

Schumann, U. 1975 Subgrid scale model for finite difference simulation of turbulent flows in plane channels and annuli. *J. Comput. Phys.* **18**, 376.

Settles, G., & Dolling, D. 1990 Swept shock boundary layer interactions – tutorial and update. Paper AIAA 90-0375.

Settles, G., Perkins, J., & Bogdonoff, S. 1981 Upstream influence of 2D and 3D shock/turbulent boundary layer interactions at compression corners, Paper AIAA 81-0334.

Simpson, R., Whitten, D., & Moffatt, R. 1970 An experimental study of the turbulent Prandtl number of air with injection and suction. *Int. J. Heat Transfer* **13**, 125–143.

Smagorinsky, J. 1963 General circulation experiments with the primitive equations, I. The basic experiment. *Mon. Weather Rev.* **91**, 99–164.

Smits, A., & Dussauge, J.-P. 1996 *Turbulent Shear Layers in Supersonic Flow*. Woodbury, NY: American Institute of Physics.

Sun, C., & Childs, M. 1973 A modified wall-wake velocity profile for turbulent compressible boundary layers. *J. Aircraft* **10**, 381–383.

Toro, E. 1997 *Riemann Solvers and Numerical Methods for Fluid Dynamics*. New York: Springer.

Urbin, G., & Knight, D. 2001 Large eddy simulation of a supersonic boundary layer using an unstructured grid. *AIAA J.* **39**, 1288–1295.

Van Driest, E. 1951 Turbulent boundary layer in compressible fluids. *J. Aeronaut. Sci.* **18**, 145–160.

Yan, H., Knight, D., & Zheltovodov, A. 2001 Large eddy simulation of supersonic compression corner using ENO scheme, in *DNS/LES Progress and Challenges*. Columbus, OH: Greyden Press, 381–388.

Yan, H., Knight, D., & Zheltovodov, A. 2002a Large eddy simulation of supersonic flat plate boundary layer part I. Paper AIAA 2002-0132.

Yan, H., Knight, D., & Zheltovodov, A. 2002b Large-eddy simulation of supersonic flat-plate boundary layers using the monotically integrated large-eddy simulation (MILES) technique. *J. Fluids Eng.* **124**, 868–875.

Yan, H., Knight, D., & Zheltovodov, A. 2002c Large eddy simulation of supersonic flat plate boundary layer part II. Paper AIAA 2002-4286.

White, F. 1974 *Viscous Fluid Flow*. New York: McGraw-Hill.

Zheltovodov, A. 1990 Peculiarities of development and modeling possibilities of supersonic turbulent separated flows, in *Separated Flows and Jets*, ed. V. Kozlov, & A. Dovgal, Berlin: Springer, 225–236.

Zheltovodov, A. 1996 Shock wave turbulent boundary layer interactions – fundamental studies and applications. Paper AIAA 96-1977.

Zheltovodov, A., & Schülein, E. 1988 The peculiarities of turbulent separation development in disturbed boundary layers [in Russian]. *Model Mekh* **2**, 53–58.

Zheltovodov, A., Schülein, E., & Horstman, C. 1993 Development of separation in the region where a shock interacts with a turbulent boundary layer perturbed by rarefaction waves. *J. Appl. Mech. Tech. Phys.* **34**, 345–354.

Zheltovodov, A., Schülein, E., & Yakovlev, V. 1983 *Development of Turbulent Boundary Layer under Conditions of Mixed Interaction with Shock and Expansion Waves* [in Russian]. Novosibirsk: USSR Academy of Sciences (ITAM preprint 28–43).

Zheltovodov, A., Trofimov, V., Schülein, E., & Yakovlev, V. 1990 *An Experimental Documentation of Supersonic Turbulent Flows in the Vicinity of Forward- and Backward-Facing Ramps.* USSR Academy of Sciences, Novosibirsk: ITAM Report 2030.

Zheltovodov, A., & Yakovlev, V. 1986 *Stages of Development, Gas Dynamic Structures and Turbulence Characteristics of Turbulent Compressible Separated Flows in the Vicinity of 2-D Obstacles.* Novosibirsk: USSR Academy of Sciences (ITAM preprint 27–86).

Zheltovodov, A., Zaulichniy, E., Trofimov, V., & Yakovlev, V. 1987 *The Study of Heat Transfer and Turbulence in Compressible Separated Flows* [in Russian]. Novosibirsk: USSR Academy of Sciences (ITAM preprint 22–87).

12 Turbulent Flow Simulations Using Vorticity Confinement

John Steinhoff, Nicholas Lynn, Wenren Yonghu, Meng Fan, Lesong Wang, and Bill Dietz

12.1 Introduction

In Chapter 4e, a computational method based on vorticity confinement (VC) is described that has been designed to capture thin vortical regions in high-Reynolds-number incompressible flows. The principal objective of the method is to capture the *essential* features of these small-scale vortical structures and model them with a very efficient difference method *directly* on an Eulerian computational grid. Essentially, the small scales are modeled as *nonlinear solitary waves* that "live" on the lattice indefinitely. The method allows convecting structures to be modeled over as few as two grid cells with no numerical spreading as they convect indefinitely over long distances, with no special logic required for merging or reconnection. It can be used to provide very efficient models of attached and separating boundary layers, vortex sheets, and filaments. Further, the method easily allows boundaries with no-slip conditions to be treated as "immersed" surfaces in uniform, nonconforming grids, with no requirements for complex logic involving "cut" cells.

There are close analogies between VC and well-known shock- and contact-discontinuity-capturing methodologies. These were discussed in Chapter 4e to explain the basic ideas behind VC, since it is somewhat different than conventional computational fluid dynamics (CFD) methods. Some of the possibilities that VC offers toward the very efficient computation of turbulent flows, which can be considered to be in the implicit large eddy simulation (ILES) spirit, were explored. These stem from the ability of VC to act as a negative dissipation at scales just above a grid cell, but that saturates and does not lead to divergence. This feature allows the following:

- It allows for an approximate cancellation of numerical diffusion, so that more complex, high-order–low-dissipation schemes can be avoided. Small-scale vortical structures at the grid-cell level can then be captured, resulting in very efficient use of the available degrees of freedom on the grid.
- It allows for an approximate treatment of backscatter. This involves the addition of (modeled) subgrid kinetic energy to the flow in a natural way, without requiring

stochastic forcing, and so restores some of the instabilities that are removed by the (implicit) filtering.

The important point is that VC, even in its simplest zeroth-order form with *constant* coefficients, on a coarse grid, can capture most of the main features of high-Reynolds-number flows. This is mainly because VC involves, in addition to a positive eddy-type viscosity, a *negative* one that does not diverge but automatically saturates. This allows one to use a much simpler turbulence modeling approach. Further, arguments (presented in Chapter 4e) show that just such a negative viscosity should be required; even with *no* numerical dissipation issues, to accurately simulate a filtered field, in certain regions of the flow one should add a term to the Euler equations that acts like such a negative dissipation with saturation.

In Chapter 4e two VC formulations were described: VC1 and VC2, which involve first- and second-order derivatives, respectively. VC1 converges more rapidly. The main difference is that VC1 closely conserves momentum whereas VC2 explicitly conserves it, making the latter more accurate for computations of vortex trajectories over long distances. In this chapter, we present some sample results from recent computations for both formulations.

We feel that, for practical turbulent flows, computed surface pressures or mean velocities are not sufficient to describe the flow: The study of the flowfield should also involve the examination of vortical scales that are small enough to allow reasonable physical resolution. For this reason, and to get an understanding of the basic features of VC applied to a range of increasingly complex flow configurations, we use visualizations of the vorticity field to demonstrate that correct qualitative properties of the flow are captured, while also demonstrating that hard quantitative comparisons can be established on the basis of detailed plots of available spectral, pressure, and velocity reference data.

12.2 Results

12.2.1 Forced turbulence

We performed computations for three-dimensional randomly stirred turbulence by using the VC1 formulation. We used a coarse, uniform $64 \times 64 \times 64$ Cartesian grid with periodic boundary conditions. The CFL number for the computation was 0.2.

We added forcing every time step for the first 10,000 time steps:

$$\hat{u}_x = A\,[\partial_y\phi_3 - \partial_z\phi_2],$$

$$\hat{u}_y = A\,[\partial_z\phi_1 - \partial_x\phi_3],$$

$$\hat{u}_z = A\,[\partial_x\phi_2 - \partial_y\phi_1],$$

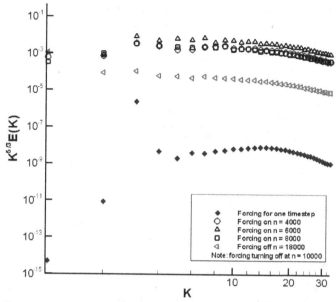

Figure 12.1. Compensated energy spectrum for simulated forced turbulence.

where A is a constant and ϕ_1, ϕ_2, and ϕ_3 are potential functions with the following expressions:

$$\phi_1 = \alpha_0 \sin\left[2\pi x/L_x + \alpha_1\right] \sin\left[2\pi y/L_y + \alpha_2\right] \sin\left[2\pi z/L_z + \alpha_3\right],$$

$$\phi_2 = \beta_0 \sin\left[2\pi x/L_x + \beta_1\right] \sin\left[2\pi y/L_y + \beta_2\right] \sin\left[2\pi z/L_z + \beta_3\right],$$

$$\phi_3 = \gamma_0 \sin\left[2\pi x/L_x + \gamma_1\right] \sin\left[2\pi y/L_y + \gamma_2\right] \sin\left[2\pi z/L_z + \gamma_3\right].$$

Here, α_0, α_1, α_2, α_3, β_0, β_1, β_2, and β_3 are random numbers generated every time step; x, y, and z are coordinates; and L_x, L_y, and L_z are the overall lengths of the computational grid in each direction, as denoted by the subscripts.

The spectrum of the forcing, which is limited to small wave numbers, is proportional to the spectrum shown in Figure 12.1 after one time step. The results presented include the compensated energy spectrum at several times and instantaneous isosurfaces of the vorticity field at one selected time. Figure 12.1 shows the compensated energy spectrum of the fully developed turbulence, and the spectrum after the forcing is turned off for a sufficiently long period, suggesting that the turbulence decays in self-similar fashion. The compensated (multiplied by $K^{5/3}$) energy spectrum, which we plotted using a log scale in both the wave-number and energy (K and E, respectively) axis, appears consistently flat, in accordance with Kolmogorov's theory. For reference, the spectrum of the forcing function after one single time step is also plotted in the figure. Figure 12.2 shows vorticity isosurfaces for the fully developed turbulence field at Time Step 8000.

These results suggest that the backscatter in our model effectively contributes to a "flattening" of the longer-wavelength modes. This can be thought of as a zeroth-order approach, with a confinement coefficient (ε) that is constant throughout the field.

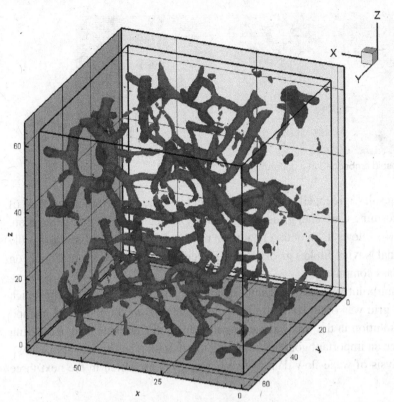

Figure 12.2. Vorticity isosurfaces at Time Step 8000 for simulated forced turbulence.

As in other LES methods, which model their coefficients as functions of the flow variables, the effects of a dependence of our coefficient on the flow variables should be studied. It is significant, however, that even with *no* functional dependence, the VC results are very good. This is true of all of the results presented in this chapter.

12.2.2 Ellipsoid

Flow over a blunt body is of considerable interest in CFD, as these flows are characterized by large-scale separation that is difficult and costly to simulate by use of conventional CFD methods. We used the VC1 formulation to compute the flow over a 6:1 ellipsoid for several angles of attack. Figure 12.3 depicts the body embedded in a coarse uniform Cartesian grid ($188 \times 70 \times 100$). The length of the ellipsoid was 120 cells and the diameter was 20 cells.

The configuration was also run with a conventional incompressible finite-volume (FV) Navier–Stokes flow solver (with a $k - \varepsilon$ turbulence model) using body-fitted structured grids (Tsai and Whitney 1999). Comparisons between VC results and the experiment for angles of attack of 20° and 25° are reproduced in Figure 12.4. At these high angles of attack, the flow is characterized by large-scale separation and the development of steady vortical structures on the lee side of the configuration. In both

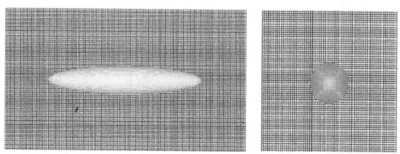

Figure 12.3. Ellipsoid embedded in a uniform grid.

cases, the VC results agree well with the experiment and the FV computation. Use of non-body-conforming Cartesian grids eliminates any difficulties with grid generation, and the use of VC allows us to obtain results on much coarser meshes than is possible with conventional Navier–Stokes grids. The VC calculations required about 5 hours on a Pentium II-class computer.

An important point is that these comparisons involve surface pressure, for which the body-fitted grid was optimized. Unlike the conventional method, the VC method did not lose resolution in the wake, since it used a uniform Cartesian grid. Resolving the wake is often an important goal in computing these flows (such as for submarines). A specific analysis of wake-flow data from VC studies is presented in the next three applications.

12.2.3 Flow over a circular cylinder

We computed the flow around a circular cylinder by using both the VC1 and VC2 formulations. In both cases, Re = 3900. We used a coarse, uniform $181 \times 121 \times 61$ Cartesian grid with an immersed boundary for the cylinder that was only 15 cells in diameter. The results presented here are from the VC2 study, although the VC1 formulation shows no significant difference (results for the VC1 formulation can be found in Fan et al. 2002). Both VC formulations compared very well with each other and with the experiment.

We subsequently present the results for the VC2 formulation: Vorticity magnitude isosurfaces are shown in Figure 12.5, where the isosurface magnitude has a value of one-fourth of the maximum. Plots corresponding to computed average streamwise velocity along lines behind the cylinder are shown in Figure 12.6, and rms streamwise velocity fluctuations are presented in Figure of 12.7. The lines where the measurements were taken are shown in Figure 12.8. We can see good agreement with the experiment. The pressure distribution on the cylinder surface also compares very well with experimental data, as can see in Figure 12.9.

The important point here is that by setting only one parameter, ε, which was constant throughout the field, we find that the computed results agreed closely with experimental data (Norberg 1987) for all six curves plotted in Figures 12.6 and 12.7. Additional

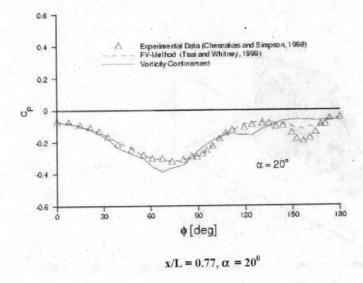

x/L = 0.77, α = 20⁰

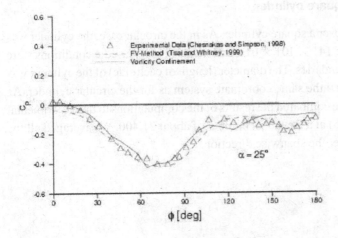

x/L = 0.77, α = 25⁰

Figure 12.4. Comparison of experimental and computed pressure coefficients for flow over an ellipsoid.

comparisons with experiment at different Reynolds number will be required to calibrate the Reynolds number dependence of this parameter.

It must be emphasized that the instabilities and chaotic behavior that result when ε is increased are only from three-dimensional effects, as in physical turbulence, and are not due to numerical instabilities: Extensive studies have been done over a much wider range of ε values than that studied here for flows in two dimensions, where no instabilities were expected. These only showed nonchaotic flow. These studies involved vortices shedding from a two-dimensional cylinder with pairing. Other studies involved isolated, shed wing-tip vortices in three dimensions.

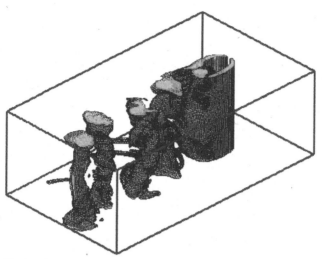

Figure 12.5. Vorticity isosurfaces for simulated flow over a cylinder with VC ($\mu = 0.15$, $\varepsilon = 0.325$).

12.2.4 Flow over a square cylinder

We also calculated flow over a square cylinder. As in the circular case, the cylinder was "immersed" in a uniform $141 \times 101 \times 61$ Cartesian grid and periodic conditions were imposed at the lateral boundaries. The diameter (length of each side) of the cylinder was also 15 grid cells. We used the same coordinate system as for the circular cylinder. As shown in Figure 12.10, we compared the results of the computations to the experimental results of Lyn et al. (1995) at a Reynolds number of about 21,400. We averaged all the computational results over the spanwise direction.

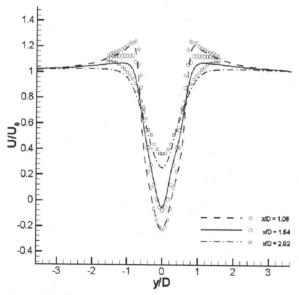

Figure 12.6. Mean streamwise velocity profiles for the VC simulations in Fig. 12.5. (Symbols correspond to experimental data by Norberg 1987.)

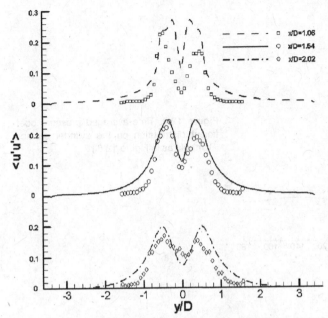

Figure 12.7. Streamwise Reynolds stress profiles. (Symbols as in Figure 12.6.)

As in the circular case, we held the diffusion coefficient, μ, and the confinement coefficient, ε, constant throughout the field. We adjusted the confinement coefficient, ε, to impose different levels of confinement so as to simulate the effects of different Reynolds numbers. Figure 12.11 depicts the comparison with experimental data of the time-averaged streamwise velocity along a streamwise line extending downstream from the middle of the leeward face of the cylinder. Results for one value of the confinement coefficient are plotted. Figure 12.12 shows the time-averaged velocity along a line normal to the cylinder axis and the mean stream at $x = 1$. Symbols represent the experimental data. The numerical results can be seen to agree well with the experimental data. Comparisons of the computed streamwise rms velocity fluctuations with the experimental results also show good agreement in Figure 12.13.

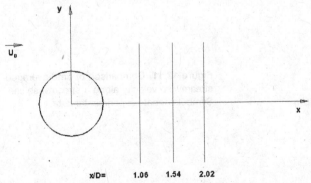

Figure 12.8. Measurement locations for a circular cylinder case study.

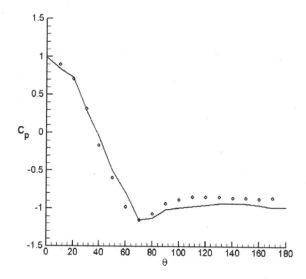

Figure 12.9. Time-averaged pressure coefficient distribution on the cylinder surface. (Symbols as in Figure 12.6.)

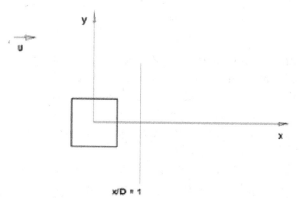

Figure 12.10. Measurement locations for the square cylinder case study.

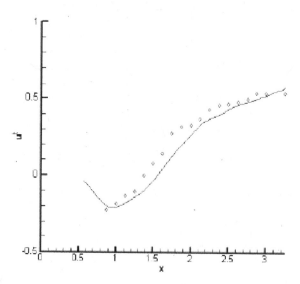

Figure 12.11. Comparison of time-averaged streamwise velocity along a streamwise line. Symbols denote experimental data.

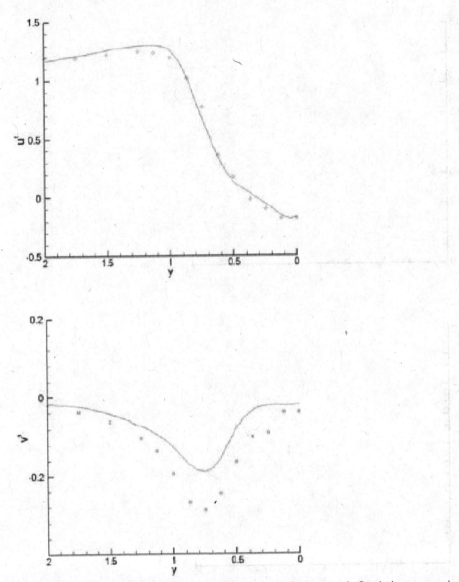

Figure 12.12. Comparison of time-averaged velocity profiles at $x = 1$. Symbols are experimental data.

12.2.5 Flow over a disk

We performed computations on a $185 \times 101 \times 101$ uniform Cartesian grid for an impulsively started disk moving along its axis. The disk width and diameter were 1 and 30 grid cells, respectively. The CFL number was set to 0.25, with $\mu = 0.15$ and $\varepsilon = 0.25$. The entire computation took about 52 hours on a single 933-MHz Pentium 3 processor for a nondimensional time of over 90.

The computed wake was visualized by vorticity isosurfaces of roughly 25% of maximum. These are shown for six nondimensional times, $T = 2.5, 5, 10, 20, 40$, and

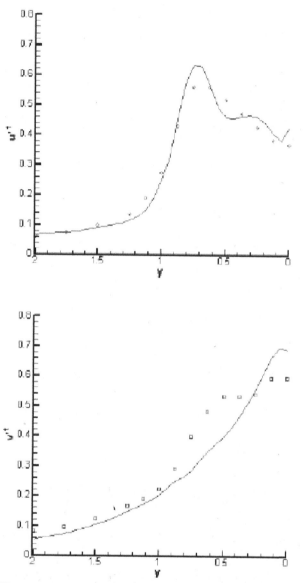

Figure 12.13. Comparison of rms velocity fluctuation profiles at $x = 1$. Symbols are experimental data.

80, in Figure 12.14. It can see that the flow resulted in instabilities and streamwise vorticity. This is very close to the visualizations from wind-tunnel experiments shown in Figures 2 and 3(a) of Berger, Scholz, and Schumm (1990). For comparison, vorticity isosurfaces from a computation using a conventional difference scheme are shown in Figure 2 of Johari and Stein (2000), where large unphysical effects of numerical diffusion are evident. These unphysical effects also occur in our code when we turn off confinement. The drag-coefficient history for the entire computational time is shown in Figure 12.15. This is in agreement with other computational results using more expensive finite-element methods (Figures 4 and 20 of Johari and Stein 2002).

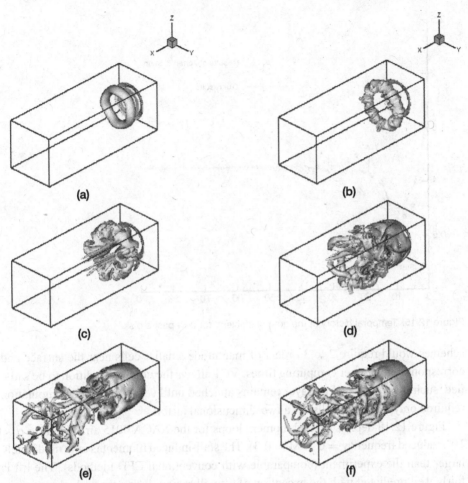

Figure 12.14. Vorticity isosurfaces for flow over a disk ($\mu = 0.15$, $\varepsilon = 0.3$): (a), (b), (c), (d), (e), and (f) are for $T = 2.5, 5, 10, 20, 40$, and 80, respectively. Normalized isosurface levels $= \pm 0.25$.

The computations on the same grid were also done for the flow over a disk that is undergoing unsteady motion. Here $\mu = 0.15$ and $\varepsilon = 0.35$, which are temporal and spatial constants. The velocity and drag-coefficient time histories are shown in Figure 12.16. This is in very close agreement to available experimental data and computational results (Figure 8 of Higuchi, Balligand, and Strickland 1996).

12.2.6 Dynamic stall – NACA 0015

The "TURNS" code was modified to include VC. This code is compressible, but VC was implemented in a similar way as described in Chapter 4e (see Dietz et al. 2004 for more details). However, the results presented here are for a Mach number of 0.3, so that compressibility effects are small. The computational grid-cell sizes are close to those in typical inviscid computations, even though the results are for a high-Reynolds-number viscous case (see Figure 12.17). Traditional Reynolds-averaged Navier–Stokes (RANS)

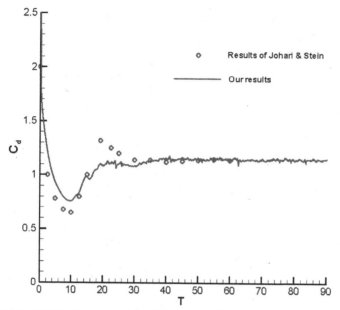

Figure 12.15. Temporal history of the drag coefficient, for flow past a disk.

schemes would require 2 to 3 orders of magnitude smaller cells near the surface and correspondingly longer computing times. VC1 allows the no-slip condition to be satisfied, such that the boundary layer remains attached until separation. The computation required only 6500 gridcells for a two-dimensional airfoil.

Figure 12.18 depicts lift and moment loops for the NACA 0015 airfoil ($11° < \alpha < 19°$, reduced frequency $= 0.1$, M $= 0.3$). The stall-induced moment is only moderately larger than the experiment (comparable with conventional CFD methods). The lift is fairly well predicted with the exception of a small region during the early downswing. The predicted moments for this case are fairly close to the data, with the exceptions of somewhat underpredicting the moment peak and the occurrence of another peak in the early downswing. The computed early downswing is marked (for this case) by the occurrence of a second vortex eruption, which is not apparent in the data.

A higher reduced frequency is shown in Figure 12.19 (for $13° < \alpha < 21°$, reduced frequency $= 0.13$, M $= 0.3$). In this case, it appears that the actual flow is always at least partially separated (to judge by the lift difference seen on the upswing). However, the moment comparison is not unreasonable. For this case, the moment peak is somewhat overpredicted. The accuracy of the comparisons seen in Figures 12.18 and 12.19 is close to what is required for an engineering model – especially considering the computation time, which is of the order of a minute on a PC for a single cycle.

12.2.7 Flow over a Comanche helicopter fuselage

We computed the flow about a realistic helicopter body (Comanche). We included rotating shanks in the computation because they could have a significant effect on

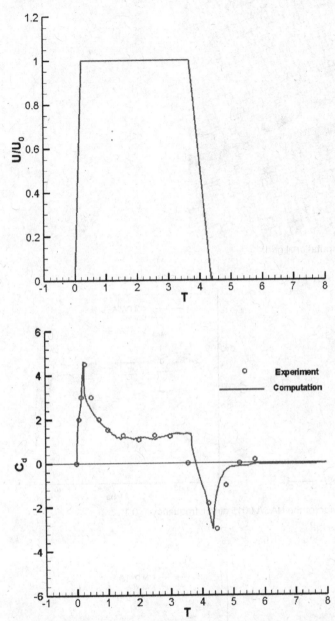

Figure 12.16. Time histories: (a) Velocity and (b) drag (deceleration to rest: $A_p = 3$, $D_p = 1$, and $Re_D = 12{,}400$).

the flow behind the pylon. However, the resolution of the shank geometry should be consistent with that of the vortices that they shed, which are spread over only about four grid cells. Accordingly, we used a simple analytic representation for the shanks (as opposed to the main body, which was accurately represented by surface points). We then used the shank and body definitions to compute the geometry – defining level set function for the flow computation (see Chapter 4e).

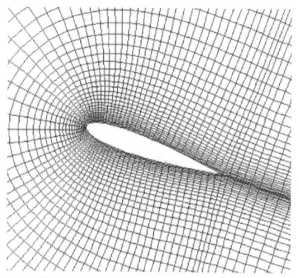

Figure 12.17. NACA 0015 computational grid.

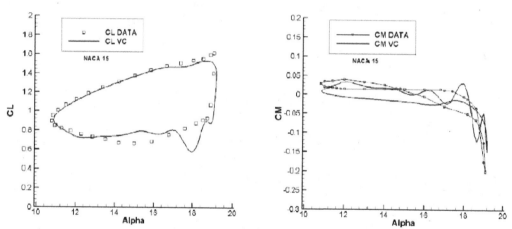

Figure 12.18. Hysteresis behavior for the NACA 0015 airfoil: frequency = 0.1.

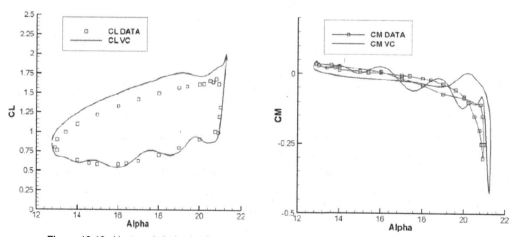

Figure 12.19. Hysteresis behavior for the NACA0015 airfoil: frequency = 0.13.

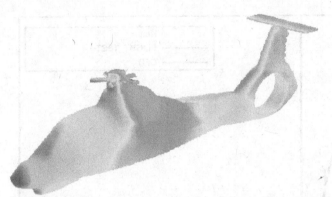

Figure 12.20. Interpolated surface pressure on the surface of the Comanche fuselage with rotating shanks.

We used a simple uniform Cartesian grid in the computation that had $288 \times 64 \times 128$ cells in the streamwise, horizontal, and vertical directions, respectively. Figure 12.20 shows the pressure distribution on the body surface at one time. We can see in Figure 12.21 that strong concentrated vorticity is shed from the pylon. This will cause strong pressure fluctuations on the tail, which is seen in the experiment. In fact, we computed a five-per-revolution oscillating pressure that corresponded closely to flight test data. Although we computed only three revolutions and more are required to get better statistics, good agreement in the comparison of power spectra can be seen in Figure 12.22. We performed this computation on a PC (Intel Pentium II, 266 MHz, 256 MB RAM), and it required 6 hours per revolution. These Comanche results were initially shown in Steinhoff (1998). The flight-test power spectra in Figure 12.22 are also shown in Duque and Meadowcraft (1999). (The power-spectrum results, based on our VC computation, were computed by Ted Meadowcraft of Boeing, Philadelphia.)

12.2.8 Flow over a missile

We have performed computations around an unpowered missile. We implemented VC as a subroutine in the "OVERFLOW-2" code. Since this computation involved supersonic flow, we used a different version of VC2 (Dietz 2004). VC is important because it can capture the *essential* features of the separating boundary layer – maintaining its

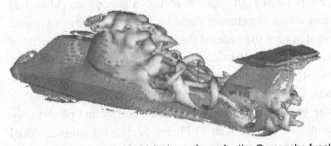

Figure 12.21. Computed vorticity isosurfaces for the Comanche fuselage with rotating shanks.

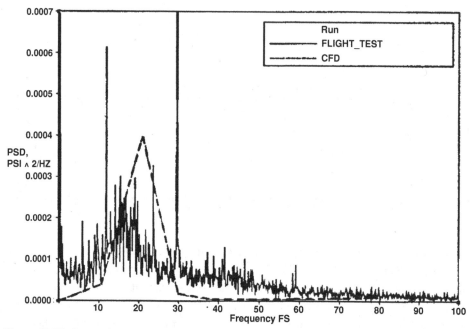

Figure 12.22. Comparison between computation and flight test of the power spectrum of pressure fluctuations at the tail.

small thickness and correct circulation. Further, VC allows coarse inviscid-sized grids (yet still maintains the no-slip condition at the surface) and can convect the separated vortices downsteam in a computationally efficient manner. The only other comparably accurate alternative is the much more expensive RANS approach, since conventional, economical inviscid approaches lead to inaccurate boundary layers and expansion fans.

Normal force and pitching moment versus angle of attack are shown in Figures 12.23 and 12.24 for a Mach number of 1.2. For a nonsymmetric (three-fin) configuration, it is expected that the force and moment curves will be nonsymmetric, as the proximity of the side fins to vortices shed from the body depends on the angle of attack. Very good agreement is found between the VC computation and experiment. The physical asymmetries apparent in the experiment are replicated in the computation for both normal force and moment. Results at higher Mach numbers are shown in Dietz (2004).

Figure 12.25 depicts the missile at high angle of attack in supersonic (M = 1.2) flow. Missiles at high angle of attack are characterized by two main types of vortical structure: a vortex pair generated near the nose of the missile, and vortical structures generated on lifting surfaces such as fins. Physically, these structures may interact and merge while convecting downstream and have a profound impact on any downstream body. To demonstrate the need for VC, we ran cases with and without confinement for comparison. Without VC, the vortices dissipate almost immediately and persist only slightly downstream of the missile tail, as seen in Figure 12.25(a). Figure 12.25(b) depicts the isosurfaces when confinement is applied. With VC, the vortices persist

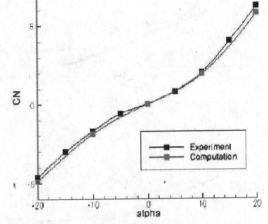

Figure 12.23. Normal force for a missile at M = 1.2.

almost indefinitely, and are free to interact with each other and any downstream object in a physically consistent manner. The capability to model the wake vortices can be critical for multiple configurations. Without confinement, the resolution of the wake vortices would require dense grids and a large number of grid points, with attendant large computational resources. This would not allow a sufficient number of cases to be run in a reasonable time. We can see that VC gives us a way of capturing these structures without resorting to fine meshes.

12.2.9 Other relevant studies

The cases presented above demonstrate the uses of VC for simulating selected complex fluid phenomena. Out of the large number of practical applications in the past several

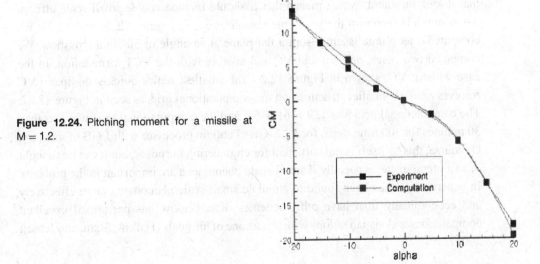

Figure 12.24. Pitching moment for a missile at M = 1.2.

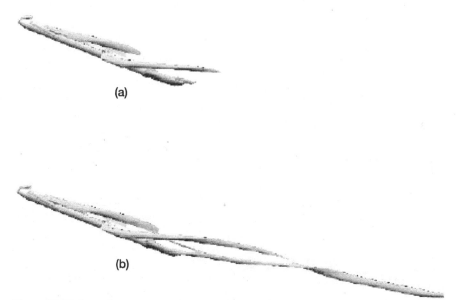

(a)

(b)

Figure 12.25. Isosurfaces of vorticity magnitude for a missile wake (M = 1.2): (a) without and (b) with confinement. See color plates at the middle of this book.

years (e.g., http://www.flowanalysis.com), two additional results should be mentioned here because they demonstrate additional uses of VC, namely, vortex propagation and interaction with an airfoil (blade vortex interaction), and simulations and visualizations of turbulent flow for special effects. The ability of VC to economically simulate the propagation of concentrated vortices has made possible the recent parametric studies of two-dimensional blade vortex interaction (Morvant 2004; Morvant et al. 2003). The latter studies also utilized a compressible version of VC, and demonstrated excellent agreement between computations and experiments.

For special effects, the important aspect of a turbulent simulation is, of course, that it *look* turbulent, which means that it should include *visible* small-scale effects. An example is shown in the jet-plume visualizations in Figures 12.26 and 12.27. We computed a jet plume issuing from a flat plane at an angle of 30° in a crossflow. We computed two cases, one without VC and another with the VC1 formulation. In the case without VC, shown in Figure 12.26, the smallest scales quickly dissipate. VC resolves *realistic* smaller structures on the computational grid, as seen in Figure 12.27. The computational grid was 129 × 65 × 65, and the computational time was less than 30 minutes for 300 time steps, for a 1.8-GHz Pentium processor with 1 GB of memory. Of course, this by itself is not sufficient for engineering purposes, but it can be thought of as a prerequisite, especially if small-scale phenomena are important in the problem. In general, VC1 has been found to simulate small-scale phenomena more effectively and economically than have other schemes. Ron Fedkiw has performed excellent computations and visualizations with this as one of his goals (Fedkiw, Stam, and Jensen 2001).

Figure 12.26. Jet-plume simulation without VC.

12.3 Conclusions

The VC method has been presented here (and in Chapter 4e) in more comprehensive detail than has previously been available. Although the basic ideas are somewhat different than those in conventional CFD, there is some commonality with a number of well-known (shock-capturing) computational methods proposed as basis for ILES elsewhere in this volume, although VC was developed independently. Extensive use of analogies with shock-capturing methods have been made to explain the basic motivations of VC.

The main goal of VC is to efficiently compute complex high-Reynolds-number flows, including blunt bodies with extensive separation and shed vortex filaments that convect over long distances. Almost all of the vortical regions in these flows are turbulent. This means that, for any feasible computation, they must be modeled. The remaining aspects of the flows are largely irrotational and largely defined once the vortical distributions are. Further, these vortical regions are often very thin.

For these reasons, the basic approach of VC is to efficiently model these vortical regions. The most efficient way to do this appears to be to develop model equations *directly* on the computational grid, rather than to first develop model partial differential equations and then attempt to accurately discretize them in these very thin regions.

These goals are easily achieved in the large number of flows where the essential features of the main flow are not sensitive to the internal structure of thin vortical regions. Then, VC can easily be used to capture these regions over only a couple of

Figure 12.27. Jet-plume simulation with VC.

grid cells and propagate them, essentially as nonlinear solitary waves that "live" on the computational lattice. Flows with these feature that are treatable with the present state of VC include blunt bodies with separation from edges and other well-defined locations. These configurations include complex geometries that can be easily immersed in uniform Cartesian grids using VC. These flows also include vortex filaments that be convected, with no numerical spreading, even over arbitrarily long times, and that can merge automatically with no requirement for special logic. Flows that involve separation from smooth surfaces, and that depend on the turbulent state of the boundary layer, require more detailed modeling, including parametric calibration. This is an area of current investigation. However, initial studies show that even these cases can be simulated by the adjustment of one parameter, which is constant throughout the field.

In contrast, a large amount of effort has been expended over a number of years by a large number of workers to develop and calibrate turbulent models based on partial differential equations for conventional eddy-viscosity-based CFD schemes, such as RANS and LES. These schemes can be quite complex and can require very fine grids. The important point is that VC, even in its simplest zeroth-order form, with constant coefficients, on a coarse grid, can capture most of the main features of high-Reynolds-number flows. This is mainly because VC involves, in addition to a positive eddy-type viscosity, a *negative* one that does not diverge but automatically saturates. This allows a much simpler turbulence modeling approach. Further, arguments (presented in Chapter 4e) show that just such a negative viscosity should be required: Even with *no* numerical dissipation issues, to accurately simulate a filtered field, in certain regions of the flow one should add a term to the Euler equations that acts like such a negative dissipation with saturation.

Preliminary results, some of which were demonstrated here, suggest that very large computer savings can be achieved, even with the simplest form of VC.

REFERENCES

Berger, E., Scholz, D., & Schumm, M. 1990 Coherent vortex structures in the wake of a sphere and a circular disk at rest and under forced vibrations. *J. Fluids Struct.* **4**, 231–257.

Dietz, W., 2004 Application of vorticity confinement to compressible flow. Paper AIAA 2004-0718.

Dietz, W., Wenren, Y., Wang, L., Chen, X., & Steinhoff, J. 2004 *Scalable Aerodynamics and Coupled Comprehensive Module for the Prediction of Rotorcraft Maneuver Loads*. Washington, DC: SBIR Final Report, RDECOM TR 04-D-45.

Duque, E., & Meadowcraft, E. 1999 Numerical evaluation of main rotor pylon flowfields on the RAH-66 Comanche helicopter, presented at the AHS 55th Annual Forum and Technology Display, May 25–27, Montreal, Canada.

Fan, M., Wenren, Y., Dietz, W., Xiao, M., & Steinhoff, J. 2002 Computing blunt body flows on coarse grids using vorticity confinement. *J. Fluids Eng.* **124**, 876–885.

Fedkiw, R., Stam, J., & Jensen, H. W. 2001 Visual simulation of smoke. *Proceedings of SIGGRAPH 2001*, 23–30.

Higuchi, H., Balligand, H., & Strickland, J. H. 1996 Numerical and experimental investigations of the flow over a disk undergoing unsteady motion. *J. Fluids Struct.* **10**, 705–719.

Johari, H., & Stein, K. 2000 Temporal evolution of the wake of an impulsively started disk, presented at the ASME Fluids Engineering Summer Conference, June, Boston, MA, FEDSM 2000-11170.

Johari, H., & Stein, K. 2002 Near wake of an impulsively started disk. *Phys. Fluids* **14**, 3459–3474.

Lyn, D. A., Einav, S., Rodi, W., & Park, J. H. 1995 A laser-doppler velocimetry study of ensemble-averaged characteristics of the turbulent near wake of a square cylinder. *J. Fluid Mech.* **304**, 285–319.

Norberg, C. 1987 *Effects of Reynolds Number and Low-intensity Freestream Turbulence on the Flow Around a Circular Cylinder*. Gothenburg, Sweden: Chalmer University of Technology, publication 87 (2).

Morvant, R. 2004 *The Investigation of Blade–Vortex Interaction Noise Using Computational Fluid Dynamics*. PhD Dissertation, University of Glasgow, United Kingdom.

Morvant, R., Badcock, K., Barakos, G., & Richards, B. 2003 Aerofoil–vortex interaction simulation using the compressible vorticity confinement method, presented at the 29th European Rotorcraft Forum, September 16–18, Friedrichshafen, Germany, paper 30.

Steinhoff, J. 1998 in *Proceedings of the 3rd ARO Workshop on Vorticity Confinement and Related Methods*. Tullahoma, TN: University of Tennessee Space Institute (UTSI preprint).

Tsai, C. Y., & Whitney, A. K. 1999 Numerical study of three-dimensional flow separation for a 6:1 ellipsoid. Paper AIAA 99-0172.

13 Rayleigh–Taylor and Richtmyer–Meshkov Mixing

David L. Youngs

13.1 Introduction

Rayleigh–Taylor (RT) instability (see Sharp 1984) occurs when the interface between two fluids of different density is subjected to a normal pressure gradient with a direction such that the pressure is higher in the less dense fluid. The related Richtmyer–Meshkov (RM) process (see Holmes et al. 1999) occurs when a shock wave passes through a perturbed interface. These instabilities are currently of concern for researchers involved in inertial confinement fusion (ICF). RT and RM instabilities can degrade the performance of ICF capsules, where high-density shells are decelerated by lower-density thermonuclear fuel. In these applications and in many RT or RM laboratory experiments, the Reynolds number is very high. Turbulent mixing will then occur. Direct numerical simulation (DNS) is feasible at a moderate Reynolds number. However, for most experimental situations, the calculation of the evolution of turbulent mixing requires some form of large eddy simulation (LES).

The flows of interest here involve shocks and density discontinuities. It is then highly desirable to use monotonic or total variation diminishing (TVD) numerical methods, either for calculating the mean flow or the development of instabilities. Hence, for three-dimensional turbulent flows, monotone-integrated LES (MILES) is very strongly favored.

My purpose in this chapter is to show that a particular form of MILES gives good results for RT and RM mixing. I consider the mixing of miscible fluids, and I assume the Reynolds number to be high enough for the effect of the Schmidt number to be unimportant. I investigate the validity of the MILES results by a comparison with experiment, a comparison with alternative techniques, and by mesh-resolution studies.

Section 13.2 gives an overview of previous research on the numerical simulation of RT and RM instabilities, from early single-mode simulations to the more recent three-dimensional turbulent mixing simulations. Numerical simulation of self-similar RT mixing is described in Section 13.3, and results for a shock-tube (RM) mixing experiment are given in Section 13.4. MILES is feasible for some relatively complex experimental situations but not for the full range of engineering applications. In the concluding section, I argue that MILES is now able to make an essential contribution to the construction and validation of the engineering models that can be applied to complex applications.

I obtained the results shown here by using the TURMOIL3D Lagrangian remap hydrocode described in Chapter 4c, which calculates the mixing of two compressible fluids. The numerical method solves the Euler equations plus an advection equation for the mass fraction of one of the fluids. The explicit calculation for each time step is divided into two parts, a Lagrangian phase and an advection or remap phase. The Lagrangian phase is nondissipative (unless shocks are present); it uses von Neumann artificial viscosity rather than a Godunov technique. The remap phase uses a third-order van Leer monotonic advection method. The monotonicity constraints in the van Leer method, which are a major benefit for modeling flows with discontinuities, provide dissipation at high wave numbers (hence the method is an example of MILES). I argue that this implicit dissipation makes the use of an explicit subgrid-scale (SGS) model unnecessary.

13.2 Overview of previous simulations

There have been a large number of papers on the numerical simulation of RT and RM instability, ranging from two-dimensional single-mode simulations to three-dimensional turbulent mixing simulations. A comprehensive review is beyond the scope of this chapter. Instead, I give a summary of some of the key developments. The first two-dimensional single-mode RT simulations were performed at the Los Alamos National Laboratory in the 1960s (see Daly 1967). Single-mode simulations have since given researchers considerable insight into the dynamics of the large-scale structures that occur in the mixing process. The early work on RT simulations is reviewed by Sharp (1984).

The simplest situation in which RT instability occurs consists of fluid with density ρ_1 resting initially above fluid with density $\rho_2 < \rho_1$ in a gravitational field g. I described two-dimensional multimode incompressible simulations for this case, triggered by short-wavelength random perturbations, elsewhere (Youngs 1984). These simulations showed an increase in the dominant length scale as mixing evolved, in proportion to gt^2, and suggested that loss of memory of the initial conditions should occur (as previously reported for various turbulent shear flows). The depth to which the mixing zone penetrated the denser fluid was given by

$$h_1 = \alpha \frac{\rho_1 - \rho_2}{\rho_1 + \rho_2} gt^2 = \alpha A g t^2, \tag{13.1}$$

where α was approximately 0.04. This value of α is somewhat smaller than typical observed values, $\alpha \sim 0.06$ (see Dimonte et al. 2004). The calculations I described in earlier research (Youngs 1984) did not use an interface-tracking technique; I used the monotonic advection method of van Leer to transport fluid volume fractions. Glimm et al. (1990) performed two-dimensional multimode simulations by using an accurate front-tracking technique. After a few generations of bubble merger, α is ~ 0.055 to

0.065 (close to the observed values). However, toward the end of the simulations, when a fine-scale structure developed, α dropped to about 0.04.

Three-dimensional simulations of RT instability began to appear around 1990. Dahlburg and Gardner (1990) showed single-mode simulations for ablatively driven RT instability. Tryggvason and Unverdi (1990) used an interface-tracking technique for single-mode viscous RT instability. Results for single-mode growth in spherical implosions were described by Sakagami and Nishihara (1990) and Town and Bell (1991). Three-dimensional multimode simulations for spherical implosions were described by Marinak et al. (1998).

For the truly turbulent stage of three-dimensional RT mixing, the main emphasis has been on self-similar mixing, which is the situation described by Eq. (13.1). In an earlier study (Youngs 1991) I showed simulations using the TURMOIL3D code with mesh sizes up to $160 \times 128 \times 128$. Other simulations at a somewhat higher resolution (Youngs 1994), using short-wavelength initial perturbation, gave self-similar growth – see Eq. (13.1) – but with $\alpha \sim 0.03$, which is significantly smaller than the observed values. On the other hand, Glimm et al. (2001) and Oron et al. (2001), using interface-tracking techniques (at lower resolution), gave higher values of α, closer to the experimental values. Dimonte et al. (2004) compared results for seven different MILES codes in order to investigate the discrepancies in the values of α. That paper also contains a review of previous three-dimensional RT simulations. DNS for RT mixing at moderate Reynolds numbers is described by Cook and Dimotakis (2001) and Cook and Zhou (2002). These results are of particular interest because they provide an independent check on the MILES results. A three-dimensional MILES simulation of the effect of initial conditions on the value of α is described by Ramaprabhu, Dimonte, and Andrews (2005). This provides a likely explanation of the higher observed values of α. I discuss the controversy over the value of α in the next section.

Examples of two-dimensional single-mode RM simulations are discussed by Holmes et al. (1999). Comparisons of two-dimensional multimode simulations with shock-tube RM experiments are shown by Mügler and Gauthier (2000) and Baltrusaitis et al. (1996). There is further research showing two-dimensional multimode simulations, such as that by Mikaelian (2005). Three-dimensional single-mode RM simulations are described by, for example, Cloutman and Wehner (1992) and Li and Zhang (1997). However, compared with RT mixing, there have been fewer three-dimensional RM turbulent mixing simulations. Very high resolution, up to $2048 \times 2048 \times 1920$, piecewise parabolic method simulations (see Chapter 4b) of two-scale RM instability were described by Cohen et al. (2002). For self-similar RM mixing caused by a single shock, the mixing zone with should vary as t^p where p is a fractional power. Estimates of p for three-dimensional simulations are given by me (Youngs 1994) for an Atwood number of 0.5 and by Oron et al. (2001) for a range of Atwood numbers. However, as pointed out by Inogamov (1999), using simple theoretical arguments, one sees that index p should be very dependent on the initial perturbation spectrum. There is scope for much further use of three-dimensional simulation to understand RM self-similar mixing.

Three-dimensional simulations for more complex RM experiments are shown in Holder et al. (2003). I discuss further simulations of such experiments in Section 13.4 in order to illustrate the full capability of the MILES technique.

13.3 Self-similar Rayleigh–Taylor mixing

13.3.1 Experimental results

Dimonte et al. (2004) gives a good review of the experimental results for RT mixing. Values of the growth rate coefficient, α, in Eq. (13.1) are usually in the 0.05–0.07 range. Many of the experiments, such as those by Read (1984) and Dimonte and Schneider (2000), used immiscible fluids. High acceleration, g, was used to reduce the effects of surface tension. It was possible to measure the growth rate of the mixing zone width for a wide range of density ratios, $\rho_1/\rho_2 \sim 2$–3 (liquid/liquid), ~ 10–20 (liquid/compressed gas), and ~ 1000 (liquid/gas), but further diagnostics were difficult to use. Much more detailed information on the internal structure of the turbulent zone is available from experiments using miscible fluids at low Atwood numbers. The static tank experiments of Dalziel, Linden, and Youngs (1999), which used fresh water/salt water, provided estimates of the degree of molecular mixing and indicated the existence of a $k^{-5/3}$ spectrum for concentration fluctuations at high wave numbers. The water-tunnel experiments of Ramaprabhu and Andrews (2004), which used hot water/cold water, provided data on both density and velocity fluctuations.

13.3.2 Ideal initial conditions

In my earlier simulations (Youngs 1991, 1994), the initial perturbations consisted of a random combination of short-wavelength modes, and my aim was to calculate the growth of the dominant length scale by mode coupling where complete loss of memory of initial conditions should occur. However, the simulations gave $\alpha \sim 0.03$, which is significantly smaller than the experimental values. On the other hand, Glimm et al. (2001) and Oron et al. (2001), using interface-tracking techniques (at somewhat lower resolution), obtained higher values of α with similar initial conditions, in agreement with experiments. In order to throw light on these discrepancies, Dimonte et al. (2004) showed results for a comparative study using seven different computational fluid dynamics codes (including TURMOIL3D). All gave a low value of $\alpha \sim 0.025 \pm 0.003$. It should be noted that the participating codes were all forms of MILES (it seems that the test case with the initial discontinuity favored the use of this approach). Some results were given by Dimonte et al. (2004) using an interface-reconstruction technique, rather than assuming SGS mixing as in TURMOIL3D. It has sometimes been argued that this is more appropriate for the experiments, most of which used immiscible liquids. However, the effect on α was small and did nothing to improve agreement with experiment.

The main purpose of this section is to show that MILES does indeed give good results for the RT mixing problem and that the discrepancies with experiment are almost certainly due to the assumption made about the initial conditions rather than excessive numerical diffusion or any other flaw in the MILES technique. I repeated the test problem, similar to the one I used in earlier research (Youngs 1994), with the mesh resolution increased by a factor of $\sim$3. The number of meshes I used is $720 \times 600 \times 600$ and the computational region is $-0.65 < x < 0.8, 0 < y, z < 1.0$. I used a uniform mesh (690 zones) in the interval $-0.5 < x < 0.65$. I added extra coarse meshes at $x < -0.5$ and $x > 0.65$ to reduce the effect of the boundaries on the growth of the mixing zone. I used periodic boundary conditions in the y and z directions. I gave the initial distribution of Fluid 1 by

$$m_1 = \begin{cases} 1, & x < \zeta_R, \\ 0, & x > \zeta_R, \end{cases}$$

where ζ_R is the random interface perturbation:

$$\zeta_R = S \sum_{m,n>0} a_{mn} \cos(mk_0 y) \cos(nk_0 z) + b_{mn} \cos(mk_0 y) \sin(nk_0 z)$$

$$+ c_{mn} \sin(mk_0 y) \cos(nk_0 z) + d_{mn} \sin(mk_0 y) \sin(nk_0 z), \qquad (13.2)$$

where $k_0 = 2\pi/L$ and $L =$ box width $= 1.0$.

I set coefficients a_{mn}, b_{mn}, c_{mn}, and d_{mn} to Gaussian random variables if

$$k_{min} < k = k_0\sqrt{m^2 + n^2} < k_{max}.$$

Otherwise, I set the coefficients to zero. The wave-number range (k_{min}, k_{max}) corresponded to wavelengths $\lambda = 4\Delta x$ to $8\ \Delta x$, and I chose the scaling factor S to give

$$(\zeta_R^2)^{1/2} = 0.02\ \lambda_{min}.$$

At the interface, the fluid densities were $\rho_1 = 3$ and $\rho_2 = 1$; Ag was equal to unity. Within each initial fluid, I used hydrostatic equilibrium with an adiabatic variation. I chose the initial pressure at the interface high enough to give a low-Mach-number flow ($M < 0.2$), and this gives a good approximation to the mixing of incompressible fluids.

The isosurfaces ($f_1 = 0.99$) in Figure 13.1 show the increase in scale length as time proceeds (f_1 and f_2 are the volume fractions for Fluids 1 and 2). For a mixture in pressure and temperature equilibrium (see Chapter 4c),

$$f_1 = \frac{(\gamma_1 - 1)c_{v1}m_1}{(\gamma_1 - 1)c_{v1}m_1 + (\gamma_2 - 1)c_{v2}m_2}. \qquad (13.3)$$

For the RT problem, the specific heats are chosen to give temperature equilibrium initially at the interface. Then, for incompressible mixing,

$$f_1 = \frac{\rho - \rho_2}{\rho_1 - \rho_2}.$$

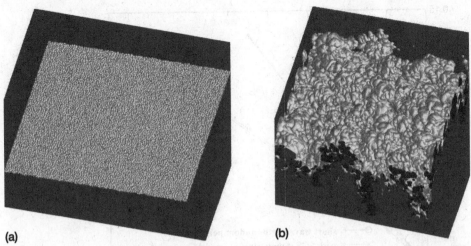

(a) **(b)**

Figure 13.1. RT mixing: Three-dimensional perspective views, showing (a) $t = 0.4$ and (b) $t = 4.0$. Isosurfaces are for $f_1 = 0.99$.

At $t = 0.4$ there are of the order of 100 bubbles across the width of the computation region. At the end of the calculation, $t = 4.0$, there are approximately three large-scale structures across the domain width. The large increase in scale size should give loss of memory of the initial conditions. In order to demonstrate the gt^2 growth of the mixing zone width, I use an integral measure of the width (which is insensitive to statistical fluctuations):

$$W = \int \overline{f}_1 \overline{f}_2 dx,$$

where $\overline{f}_1$ and $\overline{f}_2$ are values of the volume fractions averaged over yz planes. If $\overline{f}_1$ varies linearly with x throughout the mixing zone, then the bubble penetration should be $h_1 = 3W$. The edges of the mixing zone are slightly diffuse, and I use the approximation $h_1 = 3.3W$ here. Figure 13.2 shows a plot of W against Agt^2. Except at early times there is a very good linear correlation. The slope of the line is 0.00803 and this implies $\alpha = 3.3 \times 0.0803 = 0.027$. The results of this high-resolution calculation confirm the earlier results.

A further analysis of the results is needed to confirm that self-similar mixing has been established. Key integral properties of the mixing zone are examined that should approach constant values in the self-similar regime. These are as follows:

$$\frac{D}{P} = \frac{\text{KE dissipated}}{\text{Loss of potential energy}},$$

$$\theta = \int \overline{f}_1 \overline{f}_2 dx \Big/ \int \overline{f}_1 \overline{f}_2 dx,$$

where KE is the kinetic energy. I previously used the quantity θ (Youngs 1991) to give a measure of the "chemical reaction" between components in Fluids 1 and 2. It gives the

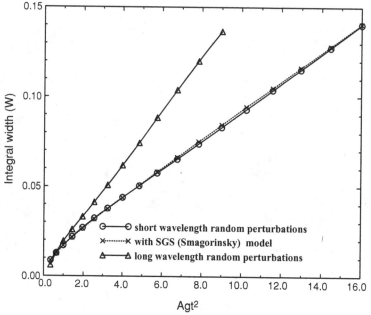

Figure 13.2. RT mixing: Integral mixing zone widths.

total reaction rate divided by the reaction rate if there is no concentration fluctuation in each horizontal plane. Figure 13.3 shows how these integral properties vary with time. Quantities D/P and θ approach approximately constant values, and this is a good indication that self-similar mixing is accurately modeled.

13.3.3 Use of SGS models

It may be argued that the monotonic advection method used in TURMOIL3D is essential for advection of ρ and m_1, where overshoots and undershoots will clearly give non-physical results. However, the need for monotonic momentum advection is less obvious. Therefore, I repeated the calculation with van Leer limiting removed from momentum advection and with a SGS model included instead (as described in Chapter 4c). I used a second-order-accurate method for momentum advection, which minimizes KE dissipation, and I used a simple Smagorinsky model, with constant $C_s = 0.17$ (about the middle of the range of values typically used in LES).

Before I discuss the RT results, I show a comparison for the Taylor–Green vortex problem described in Chapter 4a. Figure 13.4. shows a plot of the dissipation rate, $\varepsilon = \dot{D}$ per unit mass, as defined in Eq. (4c.8). A logarithmic scale is used to highlight the early-time behavior when the flow is smooth. The main criticism of the Smagorinsky model is that it is too dissipative in regions of smooth flow. It is interesting to note from Figure 13.4 that, for the 128^3 mesh simulations at early times, the MILES approach gives at least a factor 10 less dissipation than that using the Smagorinsky model. At

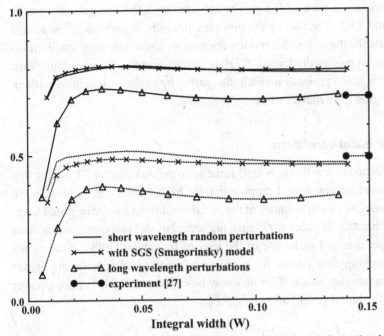

Figure 13.3. Plots of molecular mixing fraction, θ (solid lines), and dissipation fraction, D/P (dotted lines).

late times when the dissipation is high, the two methods give similar results. It is also of interest to note that, for the MILES simulations, increasing the resolution from 64^3 to 128^3 reduces the initial dissipation by a factor of 8, indicating third-order accuracy for this part of the flow. The results suggest that the MILES technique for momentum advection is equivalent to using an SGS model superior to the Smagorinsky model.

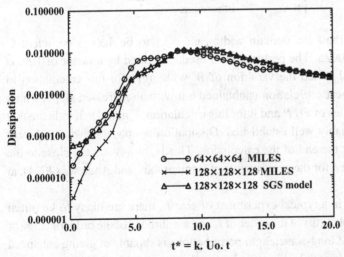

Figure 13.4. Variation of KE dissipation (per unit mass) with time for the Taylor–Green vortex problem.

For the RT problem, I have added results obtained with the Smagorinsky model to Figures 13.2 and 13.3. The plot of the mix region width, W, versus gt^2 is almost unchanged. Results for the molecular mixing fraction, θ, show only very small differences. The fraction of energy dissipated, D/P, is also similar for the two simulations: The use of the SGS model replaces implicit dissipation by explicit dissipation without significantly changing the behavior of the mixing zone.

13.3.4 Realistic initial conditions

The conclusion reached here is that $\alpha \sim 0.03$ is the appropriate value for RT mixing that is evolving by pure mode coupling. I propose that the higher observed values of α are explained by incomplete loss of memory of the initial conditions caused by initial long-wavelength perturbations. In order to obtain a numerical model that corresponds more closely to the experiments, I make use of the idea of Inogamov (1999), who argued that self-similar mixing also occurs for a multimode perturbation with amplitudes proportional to the wavelengths, λ. This idea may be expressed in the form of a power spectrum, $P(K)$, for initial perturbation height ζ_R:

$$\sigma_\lambda = \left\{ \int_{2\pi/\lambda}^{\infty} P(k)\,dk \right\}^{1/2} = \varepsilon\lambda \Rightarrow P(k) \sim 1/k^3,$$

where ε is a constant.

I have performed a third three-dimensional simulation with a random long-wavelength perturbation. The power spectrum for the long-wavelength perturbation is

$$P(k) = \begin{cases} C/k^3, & \text{if } \dfrac{2\pi}{\lambda_{\max}} < k < \dfrac{2\pi}{\lambda_{\min}}, \\ 0 & \text{otherwise.} \end{cases}$$

I took $\lambda_{\max}$ to be 0.5 (half the domain width) and $\lambda_{\min}$ to be $4\Delta x$. Coefficient C corresponds to $\varepsilon = 0.00025$. The perturbation I used, amplified by a factor of 100, is shown in Figure 13.5. I added the variation of W with Agt^2 for this calculation to Figure 13.2. A good linear correlation is obtained but with an increased slope, which implies $\alpha = 0.049$. Values of D/P and θ for this calculation, Figure 13.3, indicate that self-similarity is reasonably well established. Dissipation is now somewhat reduced; for example, $\theta \sim 0.7$ at the end of the calculation. This is, however, very close to the value of $\theta = 0.7$ reported for the experiments of Ramaprabhu and Andrews (2004), at low Atwood numbers.

I propose here that, in a typical experiment of size L, there are likely to be initial perturbations with wavelengths of the order of L and amplitudes of the order of 0.1% of L. This very low level of long-wavelength perturbations is capable of giving enhanced mixing compared with growth by pure mode coupling ($\alpha = 0.03$), and it explains the higher observed values of α.

Figure 13.5. RT mixing: Initial random long-wavelength perturbation × 100.

13.3.5 Conclusions

The fact that MILES simulations with the ideal initial perturbations give $\alpha \sim 0.03$, which is significantly smaller than the experimental values, has been a major cause of controversy over the past decade. Many researchers have suggested that the discrepancy is due to excessive diffusion in the MILES techniques. However, the following points support the reliability of the MILES results:

1. The high-resolution simulations discussed here, with ideal initial conditions, give very similar results to lower-resolution simulations (Dimonte et al. 2004; Youngs 1994). Results are insensitive to the mesh size.
2. The plots shown in Figures 13.2 and 13.3 also indicate insensitivity to mesh size; the key quantities, growth rate, θ, and D/P, are almost the same for an increase in mixing zone width of a factor of 6, that is, for an increase in mesh resolution of a factor of 6.
3. The molecular mixing fraction for ideal initial conditions, $\theta \sim 0.8$, is very similar to that obtained with DNS (see Cook and Dimotakis 2001 and Cook and Zhou 2002). This indicates that the dissipation of concentration fluctuations is not excessive.
4. For the simulation with long-wavelength initial perturbations, which gave a typical experimental growth rate, $\theta \sim 0.7$ and $D/P \sim 0.4$, close to the observed values of Ramaprabhu and Andrews (2004), $\theta \sim 0.7$ and $D/P \sim 0.5$. Again there is no evidence for excessive dissipation in the simulations.

Results for seven MILES computational fluid dynamics codes, all giving $\alpha \sim 0.03$ for ideal initial conditions, were discussed by Dimonte et al. (2004). This more detailed investigation provided further evidence for the validity of the MILES approach. Hence the conclusion reached here is that the presence of additional long-wavelength pertur-bations is the probable explanation of the higher experimental growth rates. (However, for some of experiments with immiscible fluids, surface tension inhibited fine-scale mixing, and this may also have had some effect on growth rates.) The same conclu-sions were reached in the paper by Ramaprabhu et al. (2005) in which both MILES simulations and a simple model indicated a logarithmic dependence of α on initial perturbation amplitudes. Growth purely by mode coupling is so slow that complete loss of memory of the initial conditions is unlikely to occur in experimental situations. Some researchers will probable disagree with this, but if this is confirmed, MILES simulations will have made a major contribution to our understanding of RT mixing.

13.4 Richtmyer–Meshkov mixing

13.4.1 Experimental results

Shock-tube experiments have been used to investigate RM mixing ever since the pioneering work of Meshkov and his co-workers; see Andronov et al. (1976). Mixing at planar interfaces has been investigated in a number of subsequent papers, such as those by Mügler and Gauthier (2000), Vetter and Sturtevant (1995), and Poggi, Thorembey, and Rodriguez (1998). For the experiments described by Poggi et al., velocity measurements were made. However, detailed quantitative diagnostics are generally lacking because of the transient nature of the experiments with relatively short (millisecond) times scales. Most RM experiments have used thin nitrocellulose membranes to separate the gases initially (as in Meshkov's original experiments), and this does lead to some uncertainty in the initial perturbations. The initial conditions are more clearly defined in the membrane-free gas curtain experiments of Rightley et al. (1999). Such experiments have provided very good two-dimensional computational fluid dynamics code validation tests (see Baltrusaitis et al. 1996), but because of the initial diffuse interface they are not highly turbulent and are less well suited to MILES validation.

Shock-tube experiments of many varieties have recently been carried out at Atomic Weapons Establishment (AWE) (see Holder et al. 2003), and they may be used for validating the MILES technique in relatively complex situations. The type of experiment I used is shown in Figure 13.6. The SF_6 layer is initially separated from the air regions by thin nitrocellulose membranes (~ 0.5 μm thick). In this case the membranes are supported on very fine wire grids. These ensure that the incident shock ruptures the membranes and allows mixing of the gases to begin. The incident shock ($M = 1.26$ in air) accelerates the SF_6 layer. Shock waves reflected from the end wall then decelerate the layer, and this results in a high degree of turbulent mixing. A two-dimensional feature (Figure 13.6 shows a chevron shape) may be added to one of the interfaces. This gives a flow which is two dimensional on average but with superimposed three-dimensional turbulence. The main purpose of these experiments is to provide results for the validation of two-dimensional Reynolds-averaged Navier–Stokes (RANS) models. A laser sheet diagnostic is used. The SF_6 layer is seeded with olive oil droplets and the image produced by Mie scattering is recorded on cine film. The success of the TURMOIL3D code for simulating such experiments is demonstrated first by comparison with observed results and then by investigating the insensitivity of the results to the mesh resolution.

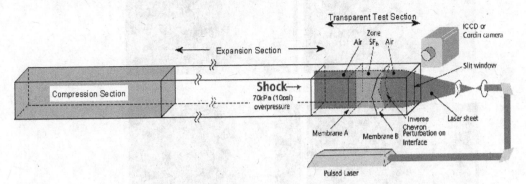

Figure 13.6. Layout for AWE shock-tube experiments.

13.4.2 Simulations

For simulation of the shock-tube experiments, I used the semi-Lagrangian option in TURMOIL3D. The x-direction (shock-propagation) mesh moves with the mean x-direction velocity. I achieve this by making a very simple modification to the remap phase, that is, a remap back to a displaced mesh rather than the original mesh. Moreover, at the left-hand boundary, I use a one-dimensional Lagrangian region. This calculates the behavior of the reflected shocks away from the region of turbulent mixing. For comparison with the chevron experiment, the initial three-dimensional zonings I used were ($x = 0$ corresponds to the left-hand interface)

	Coarse mesh	Fine mesh
-150 mm $< x < 350$ mm:	400 zones	1000 zones
$0 < y < 200$ mm:	320 zones	800 zones
$0 < z < 100$ mm:	160 zones	400 zones

The two air/SF$_6$ interfaces are randomly perturbed to represent the effect of membrane rupture. The perturbation used has a form similar to that used for the RT simulations. As I use reflective free-slip boundary conditions in the vertical direction, I use only cosine modes in this direction; that is, $c_{mn} = d_{mn} = 0$ in Eq. (13.2). The wavelength range is 5 to 50 mm and the standard deviation is

$$\left\langle \zeta_R^2 \right\rangle^{1/2} = 0.1 \text{ mm.}$$

Figure 13.7 shows a comparison of results from the fine-mesh three-dimensional numerical simulation with the observed laser-sheet images. The plane sections of SF$_6$ density show the same gross behavior as in the experiment. The formation of the central jet, its impact on the end wall, and subsequent spreading out are very well represented in the simulations. There is, however, much more detailed structure in the simulations. This is due to the effect of multiple Mie scattering in the observed images. We obtain a better comparison with experiment by taking the three-dimensional SF$_6$ distribution from the numerical simulation and using a Monte Carlo technique to calculate the image expected if scattering is taken into account. The simulated images are now in

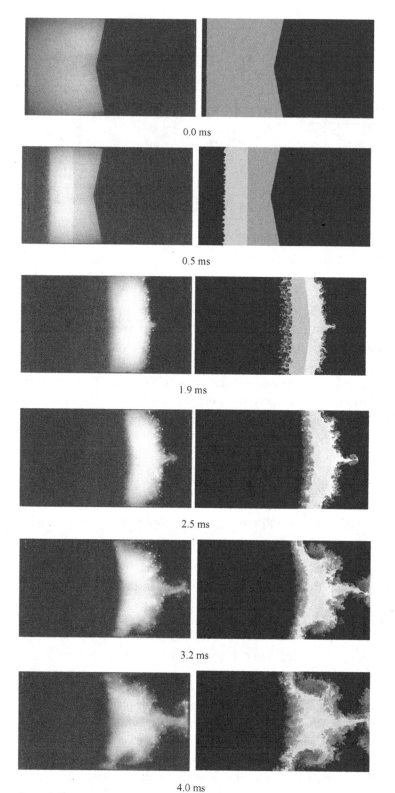

Figure 13.7. The chevron shock-tube experiment: Comparison of the experiment (laser-sheet image, on left) with simulation (plane section, on right).

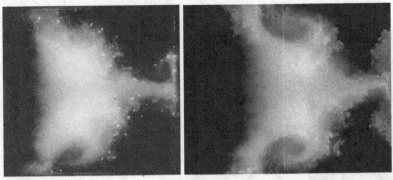

Figure 13.8. The chevron shock-tube experiment: Comparison of the experimental image at 4.0 ms with the image from the three-dimensional simulation allowing for multiple Mie scattering.

better agreement with the experiment; see Figure 13.8. Three-dimensional perspective views of the jet formation are shown in Figure 13.9.

Similar comparisons have been made with alternative choices for the interface shapes (see Holder et al. 2003). The TURMOIL3D simulations give consistently good qualitative agreement with the experiment, and this gives us considerable confidence in the numerical technique.

As a check on numerical accuracy, I repeated the chevron calculation shown here by using a reduced mesh resolution, $400 \times 320 \times 160$ zones. The initial perturbation remained the same. As the mesh resolution is increased, more random fine-scale structure appears in the simulations. Hence it does not make sense to investigate convergence simply by comparing two-dimensional sections, as shown in Figure 13.7. Instead, in order to examine the effect of mesh size, I compare appropriate averaged quantities. Figure 13.10 shows distributions of z-averaged quantities: $\overline{f}_1$ and $\sigma = \sqrt{\overline{(f_1 - \overline{f}_1)^2}}$, where f_1 is the SF_6 volume fraction, as defined by Eq. 13.3.

The distribution of $\overline{f}_1$ is affected little by the change in mesh size. Peak values of σ are somewhat lower for the finer resolution, but the effect is not large. Another quantity I examined is the total amount of turbulence KE, defined as

$$K_{\text{TOT}} = \int \overline{\rho} k dx dy,$$

where

$$k = \int \frac{1}{2}\rho[(u - \widetilde{u})^2 + (v - \widetilde{v})^2 + w^2]dz \bigg/ \int \rho dz$$

and $\widetilde{u}, \widetilde{v}$ are mass-weighted, z-averaged x and y velocities.

Figure 13.11 shows plots of K_{TOT} versus time for the two mesh sizes. For the finer mesh, turbulence KE production is initially somewhat higher and decays more rapidly at the end of the simulation. Both the production of turbulence that is due to the incident shock and the turbulence dissipation by the cascade to high wave numbers are better resolved in the higher-resolution simulation. The overall conclusion is that the effect of

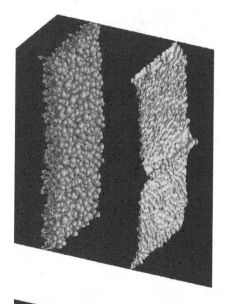

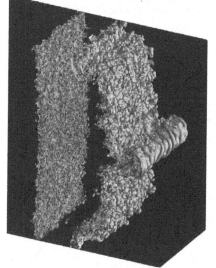

Figure 13.9. The chevron shock-tube experiment: Three-dimensional perspective views of the air/SF$_6$ boundaries, with $t = 1.9$ ms (top) and $t = 2.5$ ms (bottom).

mesh size is not very large. The finer-mesh simulation is probably not fully converged but should give adequate results.

Note that $\overline{f}_1$ and $\sigma = \sqrt{\overline{(f_1 - \overline{f}_1)^2}}$, as shown in Figure 13.10, are two of the key quantities needed for validation of two-dimensional RANS models. It is very difficult to obtain their distributions experimentally. The results shown here illustrate how MILES can, at present, make an essential contribution to the development of RANS models for complex flows.

I carried the shock-tube calculations on the 1920-processor IBM RS6000/SP computer at AWE. The lower-resolution simulation ran for 8 hours, using 32 processors; the higher-resolution simulation ran for 38 hours, using 320 processors.

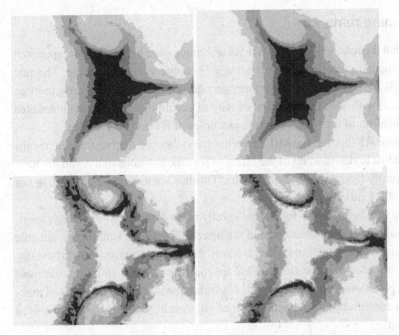

$400 \times 320 \times 160$ meshes $1000 \times 800 \times 400$ meshes

Figure 13.10. Three-dimensional simulation of the chevron shock-tube experiment: The effect of mesh resolution. Top row: SF_6 volume fraction, $\overline{f}_1$ (contour values 0.05, 0.3, 0.7, 0.95). Bottom row: $\sqrt{(f_1 - \overline{f}_1)^2}$ (contour values 0.05, 0.1, 0.2, 0.3).

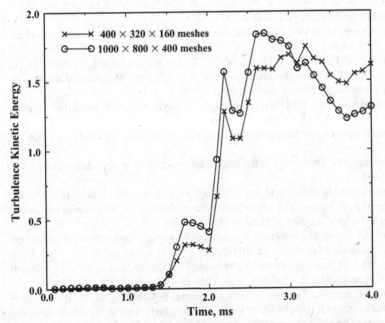

Figure 13.11. Three-dimensional simulation of the chevron shock-tube experiment: Total turbulence KE, K_{TOT}.

13.5 Concluding remarks

I do not claim that the relatively simple MILES technique used here is the best approach for such problems as a detailed understanding of homogeneous turbulence. The real advantage of MILES is the ability to perform simulations of turbulent mixing in more complex problems with shocks and initial density discontinuities. I have demonstrated this by the application of the method to two examples of RT and RM mixing.

For self-similar RT mixing, the MILES technique is thought to give accurate results and has provided an understanding of the effect of initial conditions that could not have been obtained solely from experimental results. This has led to a major advance in our understanding of RT mixing.

Results have also been shown for a significantly more complex RM mixing experiment. Good results have been obtained and the three-dimensional simulations are able to give distributions of key quantities that are very difficult to measure experimentally. On present-day supercomputers, the MILES approach is not able to model the most complex real applications. However, the MILES technique is currently capable of making an essential contribution to the validation of engineering models that can be used for such problems.

REFERENCES

Andronov, V. A., Bakhrakh, S. M., Meshkov, E. E., Mokhov, V. N., Nikiforov, V. V., Pevnitskii, A. V., & Tolshmyakov, A. I. 1976 Turbulent mixing at contact surface accelerated by shock waves. *Sov. Phys. JETP* **44**, 424–427.

Baltrusaitis, R. M., Gittings, M. L., Weaver, R. P., Benjamin, R. F., & Budzinski, J. M. 1996 Simulation of shock-generated instabilities. *Phys. Fluids* **8**, 2471–2483.

Cloutman, L. D., & Wehner, M. F. 1992 Numerical simulation of Richtmyer–Meshkov instabilities. *Phys. Fluids* **A4**, 1821–1830.

Cohen, R. H., Dannevik, W. P., Dimits, A. M., Eliason, D. E., Mirin, A. A., Zhou, Y., Porter, D. H., & Woodward, P. R. 2002 Three-dimensional simulation of a Richtmyer–Meshkov instability with a two-scale initial perturbation. *Phys. Fluids* **14**, 3692–3709.

Cook, A. W., & Dimotakis, P. E. 2001 Transition stages of Rayleigh–Taylor instability between miscible fluids. *J. Fluid Mech.* **443**, 69–99.

Cook, A. W., & Zhou, Y. 2002 Energy transfer in Rayleigh–Taylor instability. *Phys. Rev. E* **66**, 026312 (12 pages).

Dahlburg, J. P., & Gardner, J. H. 1990 Ablative Rayleigh–Taylor instability in three dimensions. *Phys. Rev. A* **41**, 5695–5698.

Daly, B. J. 1967 Numerical study of two-fluid Rayleigh–Taylor instability. *Phys. Fluids* **10**, 297–307.

Dalziel, S. B., Linden, P. F., & Youngs, D. L. 1999 Self-similarity and internal structure of turbulence induced by Raylegh–Taylor instability. *J. Fluid Mech.* **399**, 1–48.

Dimonte, G., & Schneider, M. 2000 Density ratio dependence of Rayleigh–Taylor mixing for sustained and impulsive accelerations. *Phys. Fluids* **12**, 304–321.

Dimonte, G., Youngs, D. L., Dimits, A., Weber, S., Marinak, M., Wunsch, S., Garasi, G., Robinson, A., Andrews, M. J., Ramaprabhu, P., Calder, A. C., Fryxell, B., Bello, J. L., Dursi, L., Macniece, P., Olson, K., Ricker, P., Rosner, R., Timmes, F., Tufo, H., Young, Y.-N., & Zingale, M. 2004 A comparative study of the turbulent Rayleigh–Taylor instability using high-resolution three-dimensional numerical simulations: The alpha-group collaboration. *Phys. Fluids* **16**, 1668–1693.

Glimm, J., Grove, J. W., Li, X. L., Oh, W., & Sharp, D. H. 2001 A critical analysis of Rayleigh–Taylor growth rates. *J. Comput. Phys.* **169**, 652–677.

Glimm, J., Li, X. L., Menikoff, R., Sharp, D. H., & Zhang, Q. 1990 A numerical study of bubble interactions in Rayleigh–Taylor instability for compressible fluids. *Phys. Fluids* **A2**, 2046–2054.

Holder, D. A., Smith, A. V., Barton, C. J., & Youngs, D. L. 2003 Shock-tube experiments on Richtmyer–Meshkov instability growth using an enlarged double-bump perturbation. *Laser Part. Beams* **21**, 411–418.

Holmes, R. L., Dimonte, G., Fryxell, B., Gittings, M. L., Grove, J. W., Schneider, M., Sharp, D. H., Velikovich, A. L., Weaver, R. P., & Zhang, Q. 1999 Richtmyer–Meshkov instability growth: Experiment, simulation and theory. *J. Fluid Mech.* **389**, 55–79.

Inogamov, N. A. 1999 The role of Rayleigh–Taylor and Richtmyer–Meshkov instabilities in astrophysics: An introduction. *Astrophys. Space Phys.* **10**, 1–335.

Li, X. L., & Zhang, Q. 1997 A comparative numerical study of the Richtmyer–Meshkov instability with nonlinear analysis in two and three dimensions. *Phys. Fluids* **9**, 3069–3077.

Marinak, M. M., Haan, S. W., Dittrich, T. R., Tipton, R. E., & Zimmerman, G. B. 1998 A comparison of three-dimensional multimode hydrodynamic instability growth on various National Ignition Facility capsule designs with HYDRA simulations. *Phys. Plasmas* **5**, 1125–1132.

Mikaelian, K. O. 2005 Richtmyer–Meshkov instability of arbitrary shapes. *Phys. Fluids* **17**, 034101 (13 pages).

Mügler, C., & Gauthier, S. 2000 Two-dimensional Navier–Stokes simulations of gaseous mixtures induced by Richtyer–Meshkov instability. *Phys. Fluids* **12**, 1783–1798.

Oron, D., Arazi, L., Kartoon, D., Rikanati, A., Alon, U., & Shvarts, D. 2001 Dimensionality dependence of the Rayleigh–Taylor and Richtmyer–Meshkov instability late-time scaling laws. *Phys. Plasmas* **8**, 2883–2889.

Poggi, F., Thorembey, M.-H., & Rodriguez, G. 1998 Velocity measurements in turbulent gaseous mixtures induced by Richtmyer–Meshkov instability. *Phys. Fluids* **10**, 2698–2700.

Ramaprabhu, P., & Andrews, M. J. 2004 Spectral measurements of Rayleigh–Taylor mixing at small Atwood number. *J. Fluid Mech.* **502**, 233–271.

Ramaprabhu, P., Dimonte, G., & Andrews, M. J. 2005 A numerical study of the influence of initial perturbations on the turbulent Rayleigh–Taylor instability. *J. Fluid Mech.* **536**, 285–319.

Read, K. I. 1984 Experimental investigation of turbulent mixing by Rayleigh–Taylor instability. *Physica* **12D**, 45–58.

Rightley, P. M., Vorobieff, P., Martin, R., & Benjamin, R. F. 1999 Experimental observations of the mixing transition in a shock-accelerated gas curtain. *Phys. Fluids* **11**, 186–200.

Sakagami, H., & Nishihara, K. 1990 Three-dimensional Rayleigh–Taylor instability of spherical systems. *Phys. Rev. Lett.* **65**, 432–435.

Sharp, D. H. 1984, An overview of Rayleigh–Taylor instability. *Physica* **12D**, 3–18.

Town, R. P. J., & Bell, A. R. 1991 Three-dimensional simulations of the implosion of inertial confinement fusion targets. *Phys. Rev. Lett.* **67**, 1863–1866.

Tryggvason, G., & Unverdi, S. O. 1990 Computations of three-dimensional Rayleigh–Taylor instability. *Phys. Fluids* **A2**, 656–659.

Vetter, M., & Sturtevant B. 1995 Experiments on the Richtmyer–Meshkov instability of an air–SF_6 interface. *Shock Waves* **4**, 247–252.

Youngs, D. L. 1984 Numerical simulation of turbulent mixing by Rayleigh–Taylor instability. *Physica* **12D**, 32–44.

Youngs, D. L. 1991 Three-dimensional numerical simulation of turbulent mixing by Rayleigh–Taylor instability. *Phys. Fluids* **A3**, 1312–1320.

Youngs, D. L. 1994 Numerical simulation of mixing by Rayleigh–Taylor and Richtmyer–Meshkov instabilities. *Laser Part. Beams* **12**, 725–750.

SECTION D

FRONTIER FLOWS

14 Studies in Geophysics

Piotr K. Smolarkiewicz and Len G. Margolin

14.1 Introduction

Prediction of the Earth's climate and weather is difficult in large part because of the ubiquity of turbulence in the atmosphere and oceans. Geophysical flows evince fluid motions ranging from dissipation scales as small as a fraction of a millimeter to planetary scales of thousands of kilometers. The span in time scales (from a fraction of a second to many years) is equally large. Turbulence in the atmosphere and the oceans is generated by heating and by boundary stresses – just as in engineering flows. However, geophysical flows are further complicated by planetary rotation and density–temperature stratification, which lead to phenomena not commonly found in engineering applications. In particular, rotating stratified fluids can support a variety of inertia-gravity and planetary waves. When the amplitude of such a wave becomes sufficiently large (i.e., comparable to the wavelength), the wave can break, generating a localized burst of turbulence. If one could see the phenomena that occur internally in geophysical flows at any scale, one would be reminded of familiar pictures of white water in a mountain stream or of breaking surf on a beach. The multiphase thermodynamics of atmosphere and oceans – due to ubiquity of water substance and salt, respectively – adds complexity of its own.

Because of the enormous range of scales, direct numerical simulation (DNS) of the Earth's weather and climate is far beyond the reach of current computational technology. Consequently, all numerical simulations truncate the range of resolved scales to one that is tractable on contemporary computational machines. However, retaining the physicality of simulation necessitates modeling the contribution of truncated scales to the resolved range. While a number of phenomenological models may be considered, due to the intermittent nature of geophysical turbulence, the preferred approach in the numerical prognosis of atmospheric and oceanic flows is large eddy simulation (LES); see Chapter 3 for an extended discussion. Conventionally, LES is understood as a numerical integration of coarse-grained (filtered) Navier–Stokes' equations, where all scales of motion larger than some multiple of grid interval ΔX are resolved explicitly, but the effects of finer scales are modeled on the basis of universal properties of fully

Figure 14.1. Geophysical turbulence. Scales of motion $\mathcal{O}(10^7)$, $\mathcal{O}(10^4)$, and $\mathcal{O}(10^{-2})$ m, from left to right, respectively. (See color inset, Plate 11).

developed turbulence. The goal of LES is to account for the effects of subgrid-scale (SGS) motions on those resolved on the grid.

The idea underlying LES, to represent the effects of the unresolved scales in terms of the resolved scales, is simple and intellectually appealing. Unfortunately, it is difficult to put into practice for a multiscale atmospheric or oceanic simulation. For the Navier–Stokes' equations, the formalism of decomposing flow variables into resolved and unresolved scales of motion by spatial filtering leads straightforwardly to altered equations, modified by the appearance of the divergence of the SGS stress tensor. In general, the elements of the SGS stress tensor contain products of various components, resolved and unresolved, of the flow variables. Such elements are not computable a priori for the obvious reason that the unresolved flow variables are not available. To proceed, one must model these elements in terms only of the resolved components of the flow variables. That one can do so accurately is the fundamental assumption of the LES approach, an assumption that is only justified by expediency and ultimately by comparison with observations. In meteorology, the most popular formal SGS model postulates an SGS stress tensor proportional to the rate of strain (of the resolved flow) by means of a local eddy-viscosity coefficient. The eddy coefficient itself is assumed to depend on the magnitude of the local strain rate. This model, first proposed by Smagorinsky, is a multidimensional generalization of the artificial viscosity developed by von Neumann to regularize shock waves (Smagorinsky 1993; see Chapter 2 for a discussion). The simplicity of this model, mainly the result of assuming isotropy and stationarity of the unresolved scales, has made this a widely used choice for LES. The need to incorporate additional physics has led to various elaborations of the original Smagorinsky model that address transient behavior, inhomogeneity, and anisotropy of the turbulence (Germano et al. 1991; Kosović 1997; Lilly 1967, 1992).

Universally valid SGS models do not exist as yet, and it is easy to provide examples clearly violating the assumptions behind the established LES approach (Smolarkiewicz and Prusa 2002b). Furthermore, the formal SGS models are not necessarily simple, and when combined with nonorthogonal time-dependent geometries (forming a base for mesh-refinement or boundary-fitting schemes), they become overly complicated both implementationally and computationally. In effect, numerical models used in the simulation of atmospheric–oceanic circulations often adopt simplifications that compromise the fidelity to principles of LES in the name of practicality. In this context, the implicit large eddy simulation (ILES) approach – by means of nonoscillatory finite-volume (NFV) schemes – becomes particularly useful. In the framework of a momentum–velocity representation of the governing equations and the formulation of the associated elliptic problem(s) at the discrete level (viz. exact projection), ILES facilitates turbulent-flow studies by obviating the task of evaluating the more cumbersome of the differential expressions such as vorticity, scalar Laplacian, and viscous stress. The latter is not meant to imply the liberation from ever having to programming higher-order differential expressions. First, we do require the methods suitable for ILES to be equally effective for DNS and LES, as each parameter regime is relevant to numerical study of geophysical circulations (we shall provide illustrative examples later). Second, the strain-rate tensor – a key element of explicit SGS models – is useful beyond standard LES applications for diagnosing SGS fluctuations and flow uncertainties, whereas accurate evaluation of vorticity, potential vorticity, and their budgets is important for analyzing complex vortical flows (cf. Rotunno, Grubišić, and Smolarkiewicz 1999), regardless of the adopted numerical approach.

Since finite-volume advection schemes yield truncation terms in the form of a flux divergence (see Chapter 5) the truncation fluxes of the momentum transport effectively define a SGS stress tensor. Because of the inherent nonlinearity and complexity of NFV solvers, deriving the leading higher-order truncation error is cumbersome, whereupon the form of an effective SGS stress tensor is, in general, unknown. Successful simulations of turbulent flows, while relying on only the dissipative properties of nonoscillatory advection schemes, have been reported in a variety of regimes and applications for over 10 years. Examples relevant to natural turbulent flows include the work of Margolin, Smolarkiewicz, and Sorbjan (1999), in which the authors demonstrate that a mesoscale atmospheric code based on the MPDATA advection scheme (see Chapter 4d, this volume, or, for a more detailed review, Smolarkiewicz and Margolin 1998) accurately reproduces – in close agreement with field and laboratory data and the existing benchmark computations – the dynamics of the convective planetary boundary layer; and the work of Smolarkiewicz and Prusa (2002a), in which this same numerical model is shown to capture nonstationary inhomogeneous anisotropic turbulence (with both upscale and downscale energy flow) induced by gravity-wave breaking in a deep stratified atmosphere. Complementarily, the ILES property of MPDATA has been quantified in simulations of the canonical, decaying-turbulence problems (Domaradzki, Xiao, and Smolarkiewicz 2003; Margolin, Smolarkiewicz, and Wyszogrodzki 2002; Smolarkiewicz and Prusa 2002a). Although the experimental (numerical) evidence in support of implicit turbulence modeling has been gathered for over a decade, the attempt to establish a theoretical rationale for implicit turbulence modeling has been made only recently. In particular, Margolin and Rider (2002) have derived a finite-scale (i.e., filtered) version of the pointwise Burgers' equation – appropriate for describing the dynamics of finite volumes of (Burgers') fluid. They compared this filtered equation to the MPDATA approximation of the pointwise equation and showed that MPDATA already accounts for the finite-scale effects. Since each computational cell is a finite volume, they rationalized that MPDATA more accurately represents the filtered equations of motion than the pointwise equations. This derivation has been extended to the two-dimensional Navier–Stokes equations in Chapter 2, this volume.

14.2 SGS properties of MPDATA

The iterative application of upwinding in MPDATA (see Chapter 4d) has important physical consequences. In particular, since the upwind scheme filters high frequencies on the grid, and each subsequent step reverses the dissipative error of the preceding step, MPDATA is reminiscent of generalized similarity models, in which an estimate of the full unfiltered Navier–Stokes velocity (which enters the SGS stress tensor) is obtained by an approximate inversion of the filtering operation (cf. Domaradzki and Adams 2002; Margolin and Rider 2002).

In general, nonoscillatory advection schemes are dissipative. In practice, this means that they tend to dissipate (rather than conserve) the quadratic integrals ("energy" or "entropy") of the transported variable (in contrast to centered-in-time-and-space, CTS, schemes like Arakawa-type methods; see Arakawa 1966). By no means does dissipativity imply poor accuracy – although statements in this spirit appear in the literature, blaming nonoscillatory methods for excessive implicit viscosity. In order to substantiate the foregoing assertion as well as to illustrate the SGS properties of NFV methods at work, here we highlight the results of three benchmark simulations of turbulent flows using the incompressible Boussinesq model based on the MPDATA advection scheme.[*]

Our first example – adopted from Domaradzki et al. (2003) – is an ILES of decaying, isotropic turbulence, with zero explicit viscosity (i.e., infinite Reynolds number). The calculations employed 64^3 mesh points, and the velocity field was initialized in spectral space with amplitudes consistent with the Kolmogorov $k^{-5/3}$ spectrum and random phases. The simulation was run with time step $\Delta t = 0.005$ until final time $t = 5$ (≈ 2.5 large-eddy turnover times). For such a flow, a good SGS model should preserve the $-5/3$ spectral slope during decay. The corresponding energy spectra are shown in Figure 14.2. The initial $-5/3$ spectral slope in the range of small wave numbers is preserved reasonably well, but for higher wave numbers the slope appears somewhat shallower.[†] A better way of assessing spectral slopes is by plotting a wave-number-dependent Kolmogorov function,

$$C_K(k) = \varepsilon^{-2/3} k^{5/3} E(k), \tag{14.1}$$

which is a constant for a perfect inertial range spectrum $E(k) \sim \varepsilon^{2/3} k^{-5/3}$. We compute the dissipation rate, ε, as a numerical decay rate, and the Kolmogorov function (14.1)

---

[*] All three examples exclusively use the second-order-accurate, flux-form, monotone (flux-corrected transport), gauge-transformation [i.e., linearized around a large background constant; section 3.2(4) in Smolarkiewicz and Margolin 1998] option of MPDATA.

[†] Clearly, the employed option of MPDATA is slightly under dissipative; see Domaradzki and Radhakrishnan (2005) for further discussion.

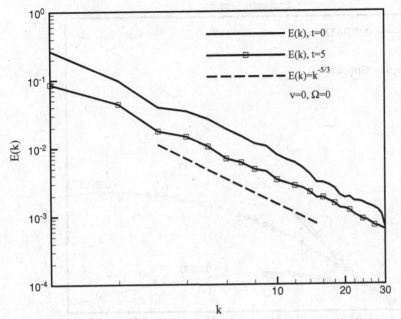

Figure 14.2. Kinetic energy spectra for flows with $\nu = 0.0$. Solid line, the initial spectrum; solid line with symbols, MPDATA; broken line, $k^{-5/3}$ line.

at the end of run is plotted in Figure 14.3. In the range of lower wave numbers, its value is about 1.5; at higher wave numbers, it increases to about 2.0. The latter value is often observed in DNS of forced isotropic turbulence with a marginally resolved inertial range (Kerr 1990). The former value is consistent with the generally accepted

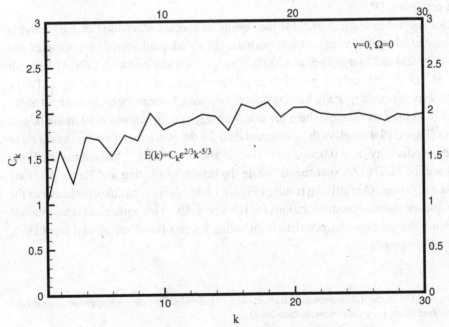

Figure 14.3. Kolmogorov function $C_K(k) = \varepsilon^{-2/3} k^{5/3} E(k)$ for a flow with $\nu = 0.0$.

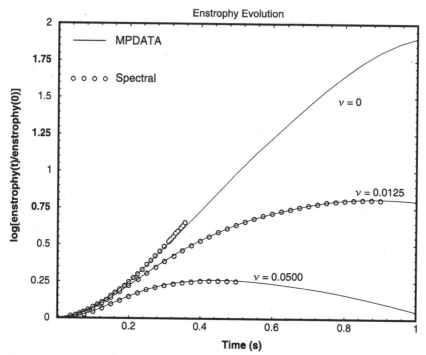

Figure 14.4. Direct and implicit modeling of isotropic, decaying turbulence.

experimental results (Monin and Yaglom 1981) and higher-resolution DNS (Bogucki, Domaradzki, and Yeung 1997). In Domaradzki et al. (2003), the authors concluded that, overall, the prediction for the inertial-range spectrum is of the same quality as that observed in LES performed with standard eddy-viscosity models as well as in high-resolution DNS.

Our second example highlights the results of another simulation of the decaying incompressible turbulence in a triply periodic cube – adopted from Smolarkiewicz and Prusa (2002a) and Margolin et al. (2002). The NFV simulations with MPDATA mirror the 256^3 DNS and inviscid pseudospectral simulations of Herring and Kerr (1993), which were intended to study the transient development of enstrophy. In contrast to the problem in the first example, here the initial energy is concentrated only in the largest eddies. Figure 14.4 displays the numerical data for the evolution of enstrophy for three values of viscosity, $\nu = 0.0500$, $\nu = 0.0125$, and $\nu = 0$ m^2 s^{-1} (as indicated). Solid lines are for MPDATA experiments, while the results of Herring and Kerr (1993) are marked as circles. One striking result in Figure 14.4 is the remarkable agreement of the NFV and the pseudospectral solutions for DNS ($\nu > 0$).[‡] This agreement is maintained uniformly for all flow characteristics, including spectra (see Herring and Kerr 1993, for other diagnostics).

[‡] For $\nu = 0.0125$, the Kolmogorov scale is about one grid interval and the energy dissipation is marginally resolved (Kerr, personal communication 2002).

Table 14.1. *Ratio of the left and right-hand sides of Eq. (14.2)*
as a function of time, verifying Kolmogorov's Four-Fifths Law

Time	1.00	1.25	1.50	1.75	2.00
$-15\langle u_x^3 \rangle \Delta X^2 / (4\dot{E})$	0.785	0.933	1.028	1.054	1.019

The $\nu = 0$ results expose the essential difference between the pseudospectral (point-wise) and MPDATA (finite-volume) approaches. Without viscous dissipation, enstrophy blowup occurred in the pseudospectral solution. The rapid growth of enstrophy was accompanied by an energy buildup at the highest wave numbers; the spectral calculations became unstable and were terminated after ~ 0.35 s (Herring and Kerr 1993). Up to this point, the spectral and MPDATA results agree closely. Beyond this point, MPDATA continues to produce a plausible solution. While it is not clear a priori whether the MPDATA simulation is physically realizable, the quantitative analyses attest the SGS properties of MPDATA. For example, in the computational study highlighted in Figure 14.4, Margolin et al. (2002) showed that $\nu = 0$ energy spectra at a given resolution closely match the results generated at other resolutions, and converge uniformly to an asymptotic spectrum, representing the continuum limit. They also demonstrated uniform convergence of $\nu > 0$ asymptotic spectra to this same inviscid limit. Here we employ the MPDATA simulations of the transient turbulence benchmark once again to point out another remarkable result consistent with observed properties of the developed turbulence.[§]

In the first example discussed, we focused attention on the $-5/3$ slope of the inertial-range energy spectra, consistent with the experimental Two-Thirds Law, proportionating the mean square velocity increment between two points to the 2/3 power of the distance (Chapter 5 in Frish 1995). In the present experiment the slope does not reach the asymptotic value $-5/3$ even at time $t = 2$ (Herring and Kerr 1993), long after the global energy dissipation reaches its maximum; see Figure 14.5. However, the results appear consistent already with Kolmogorov's Four-Fifths Law (Section 6.2 in Frish 1995). The latter states that in the Re $\nearrow \infty$ limit, the third-order (longitudinal) structure function of homogeneous isotropic turbulence, evaluated for small increments δ (compared with the integral scale), equals $-(4/5)\dot{E}\delta$, where $\dot{E}$ denotes the energy dissipation per unit mass. Keeping in mind that we have uniform spacing ΔX in all three coordinate directions, and that the turbulence is assumed to be isotropic, the Four-Fifths Law on a discrete grid becomes

$$3\langle u_x^3 \rangle = -\frac{4}{5}\frac{\dot{E}}{\Delta X^2}, \qquad (14.2)$$

where u_x denotes a finite-difference derivative and $\dot{E}$ can be inferred from Figure 14.5.

In Table 14.1, we show the ratio of the left- and right-hand sides of Eq. 14.2. At time $t = 1.0$, the ratio is less than unity, probably indicating that the flow is not yet

[§] The actual analyzed calculation employed 255^3 grid points.

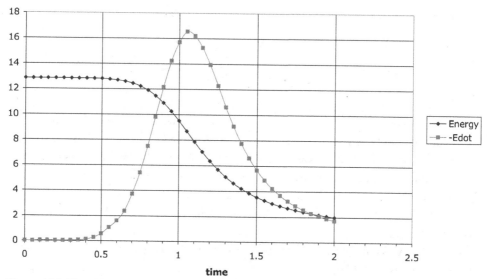

Figure 14.5. Time history of the total energy and energy dissipation rate (Energy & -Edot histories (*N* = 255)) in the MPDATA simulation of the Herring–Kerr decaying-turbulence problem.

sufficiently isotropic. At later times, the agreement is excellent. The Four-Fifths Law expresses the fact that the energy dissipation rate is independent of the viscosity and is controlled, both physically and numerically, by the large scales of the flow (section 7.2 in Frisch 1995) – consistent with the experimental Law of Finite Energy Dissipation (chapter 5 in Frish 1995).

Our last example highlights the results of explicit (LES) versus implicit turbulence modeling in the context of the convective planetary boundary layer (PBL) simulations – after Margolin et al. (1999). The three curves shown in Figure 14.6 represent mean profiles of the normalized resolved heat flux (horizontally averaged product of temperature perturbation and vertical velocity) from three different simulations: the short-dashed curve is from the LES benchmark simulations of Schmidt and Schumann (1989) using a CTS model; the long-dashed curve is from explicit LES simulations with MPDATA, and the solid curve is from implicit LES with MPDATA. Circles represent field and laboratory data superimposed by Schmidt and Schumann. The comparability of all the results with the data is excellent (for other characteristics of the flow, see Margolin et al. 1999). Without the implicit SGS-model result, one might be tempted to argue that the dissipativity of the employed NFV approach is simply negligibly small. However, the results reveal a more interesting story.

A full appreciation of the results in Figure 14.6 intertwines with understanding the mechanics of nonoscillatory schemes. NFV schemes are nonlinear (even for linear problems), as they employ coefficients that depend on the transported variables. In other words, these schemes are *self-adaptive* as they design themselves in the course of the simulation. Thus, in contrast to linear CTS, methods, different realizations of the same turbulent flow use different numerical approximations to the governing equations

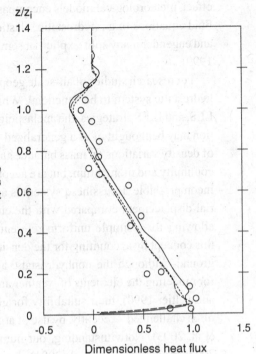

Figure 14.6. Turbulence modeling of a convective PBL: ILES (solid line) and LES (long dashes) versus Schmidt and Schumann LES benchmark results (short dashes) and data (circles).

of motion. When the explicit SGS model is included (LES) then the resolved flow is sufficiently smooth, and the entire machinery assuring nonoscillatory properties of the numerics is effectively turned off (there is no need to limit or adjust linear components of the scheme). In the absence of an explicit SGS model the nonoscillatory machinery adapts the numerics "smartly" so as to assure solutions that are apparently as smooth as those generated with physically motivated explicit SGS models. Thus, insofar as the dissipativity per se of the NFV methods is concerned, there is no simple unique quantification, since the resulting transport scheme can be effectively either nondissipative or dissipative, depending upon the presence or absence, respectively, of an explicit SGS model.

14.3 A multiscale geophysical fluid model

14.3.1 Motivation

Because of the enormous span of the spatial and temporal scales, and the wave phenomena important in geophysical fluids, explicit integrations of generic compressible equations are impractical for the majority of applications. In order to account for this broad range of scales, while deriving a numerical model still useful with existing computational resources, one has no choice but to invoke analytic or numerical approximations that allow for reasonably large time-step integrations of the governing equations. In effect, meteorological models encompass a variety of approximate (filtered) systems of fluid equations (e.g., hydrostatic, elastic, anelastic, Boussinesq; cf. Davis et al. 2003) and engender many split-explicit or semi-implicit methods for their integration (Durran 1999).

For research studies of all-scale geophysical fluids, we have found the anelastic non-hydrostatic system to be beneficial. While nonhydrostacy is a prerequisite of the DNS, LES, and ILES strategies, the anelasticity is merely an option. The anelastic approximation may be thought of as a generalized Boussinesq approximation in which the effects of density variations on mass balance and inertia are neglected in the equations of mass continuity and momentum, but are accounted for in the buoyancy forces. The classical, incompressible, Boussinesq system is applicable to shallow motions with small material displacement compared with the characteristic vertical scale of the fluid, thereby allowing for a simple uniform reference state. The anelastic approximation extends this concept by accounting for the density–temperature stratification of the static background. Although the nonhydrostatic anelastic equations were shown to be accurate for modeling the elements of weather and climate up to the synoptic scale (Polavarapu and Peltier 1990), their suitability for global weather and climate prediction has been often challenged – recently, by use of arguments of linear normal mode analysis (Davis et al. 2003). Notwithstanding, our numerical results (Grabowski and Smolarkiewicz 2002; Smolarkiewicz et al. 2001) document that the anelastic equations adequately represent a range of idealized planetary flows. This has important practical consequences. Inherent in the anelastic system are (i) the Boussinesq linearization of the pressure gradient forces and the mass fluxes in the momentum and mass continuity equations, respectively, and (ii) the anelasticity per se, equivalent to taking the limit of an infinite speed of sound. Working in concert, these two approximations greatly simplify the design of second-order-accurate, flexible, and computationally efficient (viz., implicit with respect to inertia-gravity waves) research models for a broad range of geophysical flows. This is especially important within the class of nonoscillatory forward-in-time (NFT) models, where two-time-level self-adaptive nonlinear numerics leads inevitably to difficult nonlinear elliptic problems for the implicit discretization of the fully compressible Euler equations. From the viewpoint of numerical engineering, the anelastic model is converted easily into either a compressible–incompressible Boussinesq, or an incompressible Euler system (Smolarkiewicz et al. 2001). The examples of the convective boundary layer and of decaying turbulence discussed earlier employ the incompressible Boussinesq option of our NFV model, named EULAG for its capability to solve the fluid equations in either an Eulerian (flux form) or a Lagrangian (advective form) framework.

14.3.2 Analytic formulation

The scope of this chapter justifies a brief symbolic description of the governing anelastic model; for more thorough mathematical expositions, refer to Prusa and Smolarkiewicz

(2003), Wedi and Smolarkiewicz (2004), Smolarkiewicz and Prusa (2005), and references therein. Our model can be applied to a broad class of problems ranging from simulating atmospheric–oceanic circulations to biomechanics and solar physics applications (Cotter, Smolarkiewicz, and Szczyrba 2002; Elliott and Smolarkiewicz 2002). To address such a broad class of flows in a variety of domains – with, optionally, Dirichlet, Neumann, or periodic boundaries in each direction – we formulate (and solve) the governing equations in transformed time-dependent curvilinear coordinates:

$$(\bar{t}, \bar{\mathbf{x}}) \equiv [t, \mathcal{F}(t, \mathbf{x})]. \tag{14.3}$$

The key assumptions are that (i) both physical and transformed domains are topologically either cuboidal, toroidal, or spheroidal; (ii) the coordinates $(t, \mathbf{x})$ of the physical domain are orthogonal and stationary; (iii) time flow is the same in both domains; and (iv) the transformed horizontal coordinates, $(\bar{x}, \bar{y})$, are independent of the vertical coordinate, z. Given the transformation in Eq. (14.3), the anelastic equations of Lipps and Hemler (1982) can be compactly written as follows:

$$\bar{\nabla} \cdot (\rho^* \bar{\mathbf{v}}^s) = 0, \tag{14.4}$$

$$\frac{d\mathbf{v}}{d\bar{t}} = -\widetilde{\mathbf{G}}(\bar{\nabla}\pi') - \mathbf{g}\frac{\theta'}{\theta_o} - \mathbf{f} \times \mathbf{v}' + \mathbf{M}' + \mathbf{D}, \tag{14.5}$$

$$\frac{d\theta'}{d\bar{t}} = -\bar{\mathbf{v}}^s \cdot \bar{\nabla}\theta_e + \mathcal{H}, \tag{14.6}$$

where, because of the coordinate transformation, the physical and geometrical aspects intertwine each other. Insofar as the physics is concerned, $\mathbf{v}$ denotes the *physical velocity* vector; θ, ρ, and π denote potential temperature, density, and a density-normalized pressure; $\mathbf{g}$ is the acceleration of gravity, and $\mathbf{f}$ the vector of the "Coriolis parameter"; $\mathbf{M}$ symbolizes the inertial forces of geospherical metric accelerations; whereas $\mathbf{D}$ and $\mathcal{H}$ symbolize viscous dissipation of momentum and diffusion of heat, respectively. Primes denote deviations from the geostrophically balanced ambient (alias, environmental) state $\mathbf{v}_e$, θ_e, and the subscript o refers to the basic state, that is, a horizontally homogeneous constant-stability hydrostatic reference state (cf. Section 2b in Clark and Farley 1984).

The geometry of the coordinates in (14.3) enters the governing equations as follows: In the mass continuity equation (14.4), $\rho^* \equiv \rho_o \overline{G}$, with $\overline{G}$ denoting the Jacobian of the transformation; in the momentum equation (14.5), $\widetilde{\mathbf{G}}$ symbolizes the renormalized Jacobi matrix of the transformation coefficients $\sim (\partial \bar{\mathbf{x}}/\partial \mathbf{x})$; $\bar{\nabla} \cdot \equiv \partial/\partial \bar{\mathbf{x}} \cdot$, and the total derivative is given by $d/d\bar{t} = \partial/\partial \bar{t} + \bar{\mathbf{v}}^* \cdot \bar{\nabla}$, where $\bar{\mathbf{v}}^* \equiv d\bar{\mathbf{x}}/d\bar{t} \equiv \dot{\bar{\mathbf{x}}}$ is the *contravariant velocity*. Appearing in the continuity equation, (14.4) and potential temperature equation, (14.6), is a weighted *solenoidal velocity*,

$$\bar{\mathbf{v}}^s \equiv \bar{\mathbf{v}}^* - \frac{\partial \bar{\mathbf{x}}}{\partial t}, \tag{14.7}$$

that readily follows – given $\rho_o = \rho_o(\mathbf{x})$, and the time-independent coordinate system in the physical space – from the generic (tensor-invariant) form of the anelastic continuity equation:

$$\overline{G}^{-1}\left[\frac{\partial \rho^*}{\partial \bar{t}} + \overline{\nabla} \cdot (\rho^* \overline{\mathbf{v}}^*)\right] \equiv 0. \tag{14.8}$$

Use of the solenoidal velocity facilitates the solution procedure because it preserves the incompressible character of the numerical equations. While numerous relationships can be derived that express any velocity (solenoidal, contravariant, or physical) in terms of the other, in either the transformed or physical coordinate system (Prusa and Smolarkiewicz 2003), a particularly useful transformation,

$$\overline{\mathbf{v}}^s = \tilde{\mathbf{G}}^T \mathbf{v}, \tag{14.9}$$

relates the solenoidal and physical velocities directly. For further details of the metric and transformation tensors as well as the formulation of viscous and dissipative terms in the governing equations, refer to Smolarkiewicz and Prusa (2005) and the references therein.

14.3.3 Numerical approximations

Given (14.8), each prognostic equation that forms the anelastic system (14.5) and (14.6) can be written in two equivalent forms, either as a Lagrangian evolution equation,

$$\frac{d\psi}{d\bar{t}} = R, \tag{14.10}$$

or Eulerian conservation law,

$$\frac{\partial \rho^* \psi}{\partial \bar{t}} + \overline{\nabla} \cdot (\rho^* \overline{\mathbf{v}}^* \psi) = \rho^* R. \tag{14.11}$$

Here ψ symbolizes components of $\mathbf{v}$ or θ', and R denotes the associated right-hand side.

We approximate either (14.11) or (14.10) to second-order accuracy in space and time by using the NFT approach; see Smolarkiewicz and Prusa (2002b) for a review. A particular NFT algorithm employed here can be formally written as

$$\psi_{\mathbf{i}}^{n+1} = \mathrm{LE}_{\mathbf{i}}(\tilde{\psi}) + 0.5\Delta t\, R_{\mathbf{i}}^{n+1} \equiv \widehat{\psi}_{\mathbf{i}} + 0.5\Delta t\, R_{\mathbf{i}}^{n+1}, \tag{14.12}$$

where $\psi_{\mathbf{i}}^{n+1}$ is the solution sought at the grid point $(\bar{t}^{n+1}, \overline{\mathbf{x}}_{\mathbf{i}})$, $\tilde{\psi} \equiv \psi^n + 0.5\Delta t\, R^n$, and LE denotes a two-time-level either advective semi-Lagrangian (Smolarkiewicz and Pudykiewicz 1992) or flux-form Eulerian (Smolarkiewicz and Margolin 1993) NFT transport operator (viz. advection scheme).‖

‖ The flux-form Eulerian transport operator, LE, invokes the multiplicative factor ρ^{*n}/ρ^{*n+1} to account for time variability of generalized density ρ^* that is due to coordinate dependence on time; see Smolarkiewicz and Prusa (2002b) for a discussion.

Equation (14.12) represents a system that is implicit with respect to all dependent variables in (14.5) and (14.6), because all principal forcing terms are assumed to be unknown at $n + 1$.[#] For the physical velocity vector $\mathbf{v}$, we can write it compactly as

$$\mathbf{v_i} = \widehat{\mathbf{v}}_i - 0.5\Delta t\big[\widetilde{\mathbf{G}}(\overline{\nabla}\pi')\big]_i + 0.5\Delta t\mathbf{R_i}(\mathbf{v}, \widehat{\theta})\,, \tag{14.13}$$

where

$$\mathbf{R_i}(\mathbf{v}, \widehat{\theta}) \equiv -[\mathbf{f} \times (\mathbf{v} - \mathbf{v}_e)]_i - g\frac{1}{\theta_o}\{\widehat{\theta}_i + 0.5\Delta t[(\widetilde{\mathbf{G}}^T\mathbf{v}) \cdot \overline{\nabla}\theta_e]_i\} \tag{14.14}$$

accounts for the implicit representation of the buoyancy by means of Eq. (14.6), and the superscript $n + 1$ has been dropped as there is no ambiguity. On grids unstaggered with respect to all prognostic variables (e.g., A and B Arakawa grids), (14.13) can be inverted algebraically to construct expressions for the solenoidal velocity components that are subsequently substituted into (14.4) to produce

$$\left(\frac{\Delta t}{\rho^*}\overline{\nabla} \cdot \rho^*\widetilde{\mathbf{G}}^T\big\{(\mathbf{I} - 0.5\Delta t\mathbf{R})^{-1}[\widehat{\mathbf{v}} - 0.5\Delta t\widetilde{\mathbf{G}}(\overline{\nabla}\pi')]\big\}\right)_i = 0, \tag{14.15}$$

that is, an elliptic equation for pressure,

$$\left\{\frac{\Delta t}{\rho^*}\overline{\nabla} \cdot \rho^*\widetilde{\mathbf{G}}^T\big[\widehat{\overline{\mathbf{v}}} - (\mathbf{I} - 0.5\Delta t\mathbf{R})^{-1}\widetilde{\mathbf{G}}(\overline{\nabla}\pi'')\big]\right\}_i = 0, \tag{14.16}$$

where $\widetilde{\mathbf{G}}^T[\widehat{\overline{\mathbf{v}}} - (\mathbf{I} - 0.5\Delta t\mathbf{R})]^{-1}\widetilde{\mathbf{G}}(\overline{\nabla}\pi'') \equiv \overline{\mathbf{v}}^s$ defined in (14.7); see Prusa and Smolarkiewicz (2003) for the complete development. Boundary conditions imposed on $\overline{\mathbf{v}}^s \cdot \mathbf{n}$, subject to integrability condition $\int_{\partial\Omega} \rho^*\overline{\mathbf{v}}^s \cdot \mathbf{n}d\sigma = 0$, imply the appropriate boundary conditions on π'' (Prusa and Smolarkiewicz 2003; Wedi and Smolarkiewicz 2004). The resulting boundary value problem is solved by using a preconditioned generalized conjugate residual, GCR(k), algorithm, a nonsymmetric Krylov subspace solver akin to the popular generalized minimum residual GMRES(k), scheme (Eisenstat, Elman, and Schultz 1983; Saad 1993). Given the updated pressure, and hence the updated solenoidal velocity, the updated physical and contravariant velocity components are constructed from the solenoidal velocities by using transformations (14.9) and (14.7), respectively.

14.4 Applications

The examples of the planetary boundary layer and of decaying turbulence discussed earlier address elementary small-scale atmospheric–oceanic dynamics posed on a

[#] Nonlinear terms in R^{n+1} (e.g., metric terms arising on the globe) may require outer iteration of the system of equations generated by (14.12) (Smolarkiewicz et al. 2001); when included, diabatic, viscous, and SGS forcings may be first-order accurate and explicit. For example, assume $SGS(\psi^{n+1}) = SGS(\psi^n) + \mathcal{O}(\Delta t)$ in R^{n+1}, thereby contributing to the right-hand side of the resulting elliptic problem; for extensions to moist processes, see Grabowski and Smolarkiewicz (2002).

Cartesian nonrotating domain. Here, we provide a cross section through a range of scales, problems, and numerical strategies (from ILES, through LES, to DNS) that is representative of an applied simulation of geophysical turbulence. All calculations discussed have been performed with the Eulerian (NFV) version of the massively parallel EULAG code.

14.4.1 ILES of idealized climate

The first application – the idealized climate problem of Held and Suarez (1994) – represents a thermally-forced baroclinic instability on the rotating sphere. In a sense, it bears striking resemblance to LES–ILES studies of convective boundary layers in which small differences in model setups can lead to totally different instantaneous flow realizations, and in which different model designs can lead to quite divergent integral flow characteristics. In other words, these simulated flows are both turbulent and stochastic.

Figure 14.7 illustrates the overall complexity of the flow, showing instantaneous vertical cross sections in the equatorial plane and surface plots of isentropes θ and isolines of zonal velocity u, after 3 years of simulated flow. The plates a′ and a show isentropes in the vertical equatorial plane and at the surface, respectively. The plates b′ and b display the zonal velocity contours with imposed flow vectors, respectively, in the equatorial plane and at the surface. Contour extrema and intervals are shown in the upper left corner of each plate (in plate a′ we used a variable contour to capture θ variability in the troposphere). Negative values are dashed. Maximum vector lengths are shown in the upper right corner of the two lower plates. The displayed results typify the response of an initially stagnant and uniformly stratified fluid to a diabatic forcing that mimics the long-term thermal and frictional forcing in the Earth's atmosphere. Figure 14.8 contrasts the complexity of the instantaneous flow in Figure 14.7 with the display of the resulting "climate," that is, zonally averaged 3-year means of zonal velocity and potential temperature. Figure 14.9 supplements Figure 14.8 with the equivalent displays of meridional and vertical velocities to complete the picture of the simulated climatic circulation.

The results of Figures. 14.7, 14.8, and 14.9 exemplify the role of implicit turbulence modeling of geophysical flows. Here, the globe is covered by a uniform spherical mesh with $nx \times ny = 64 \times 32$ grid intervals (no grid points at the poles); the $H = 32 \times 10^3$ m deep atmosphere is resolved with $nz = 40$ uniform grid intervals; and the time step of integration is $\Delta t = 900$ s; see Smolarkiewicz et al. (2001) for further details of the simulation performed. There is no explicit SGS model employed, and it would be unreasonable to expect that any standard turbulence model in the sense of LES could represent the actual turbulence of the Earth's atmosphere at such a low grid resolutions. Consider that, near the equator, the horizontal extent of the model grid boxes (circa 600×600 km^2) is comparable with the size of a small European country, where convection, terrain forcing, gravity wave breaking, and so on are responsible for generating energy of the SGS, but are not accounted for in standard turbulence models.

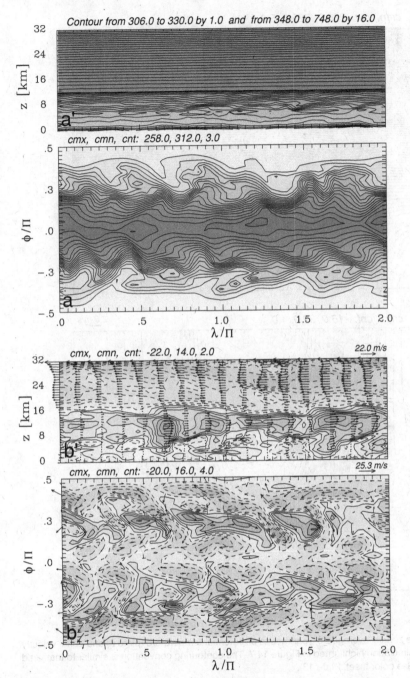

Figure 14.7. Instantaneous solutions of the idealized climate problem after 3 years of simulation. (See color inset, Plate 12).

14.4.2 LES of aeolian flows

An outstanding problem of interest to meteorology and environmental engineering is a rapid sediment transport on scales from micro to planetary. It covers a range of natural

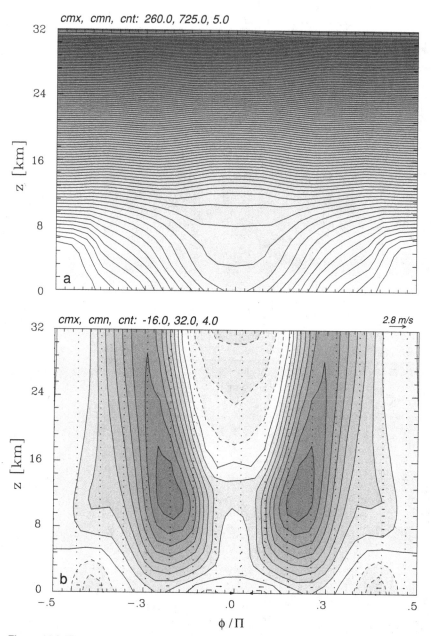

Figure 14.8. The zonally averaged 3-year means of potential temperature (plate a) and zonal velocity (plate b) for the simulation highlighted in Figure 14.7. The contouring convention is similar to that used in Figure 14.7. (See color inset, Plate 13).

phenomena under severe wind conditions, such as wind processes on beaches, flows past sand dunes, to dust storms. At present, the prediction of the sediment transport and coupled landform evolution is based on crude empirical formulae relating the transport to idealized wind profiles, surface slopes, and other ambient parameters, while assuming a smooth and negligibly slow variation of the controlling parameters (Andreotti,

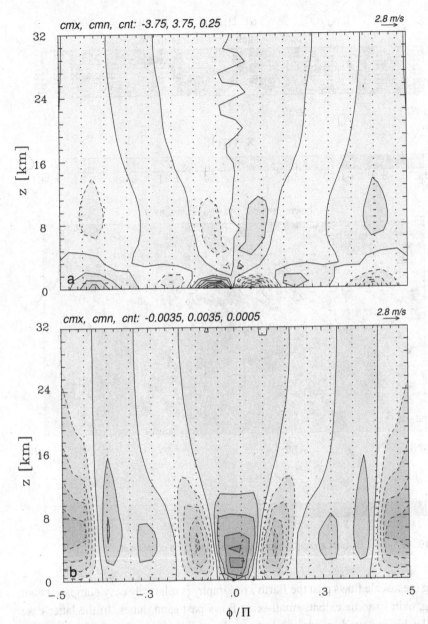

Figure 14.9. The zonally averaged 3-year means of meridional and vertical velocities (plates a and b, respectively) for the simulation highlighted in Figure 14.7. (See color inset, Plate 14).

Claudin, and Douady 2002; Ortiz, personal communication, 2005). Although simple, this approach yields inaccurate results wherever the underlying surface is irregular or rapidly evolving in time. A direct LES of the atmospheric boundary layer flow past the irregularities of the time-evolving terrain offers a viable alternative to crude phenomenological models.

Simulating geophysical (or engineering) flows becomes extremely challenging wherever boundary conditions cease being smooth either in time or space. For example,

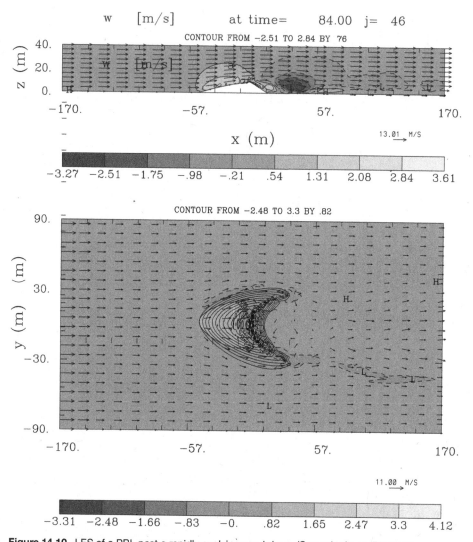

Figure 14.10. LES of a PBL past a rapidly evolving sand dune. (See color inset, Plate 15).

simulating mesoscale flows past the Earth's orography is relatively easy compared with simulating, akin to some extent, small-scale flows past sand dunes. In the latter case the irregular lower boundary evolves in time, depending on the flow itself, including boundary steepening and micro-avalanche dissipation, which act to preserve the critical slope (reminiscent of the steepening and overturning of atmospheric gravity waves). The geometric difficulty of the problem is enhanced by the ubiquity of the turbulence and separating surface PBLs, affecting the evolution of the dune–air boundary per se.

Figure 14.10 highlights LES (after Ortiz and Smolarkiewicz 2006) of a shear driven PBL over a sand dune, evolving in response to sand saltation (Bagnold 1941, 1956). The saltation rate depends essentially on surface boundary layer stress (Lettau and Lettau 1978). As the dune slope reaches the critical value ($\approx 32°$), grain avalanches prevent further steepening of a dune (Rajchenbach 2002). In the present simulation, the evolution of the dune and the flow aloft are fully coupled by means of time-dependent curvilinear coordinates adapting to the lower boundary (Prusa and Smolarkiewicz 2003; Wedi and Smolarkiewicz 2004), whose profile and time derivative are predicted from the saltation model. To accelerate the dune evolution, we have premultiplied the phenomenological constant controlling the saltation rate arbitrarily by 1440 so that 1 min of simulated time corresponds roughly to 1 day of dune evolution. The model domain $340 \times 180 \times 40$ m^3 was resolved with $\delta x = \delta y = 2$m, $\delta z = 1$ m grid intervals, and time step $\delta t = 0.05$ s was used. A 7.5-m-high conical pile of sand, centered at $(x, y) = (-57, 0)$, has been assumed for the initial condition. The result in Figure 14.10 is after 84 min; it evinces a characteristic barchan dune (Andreotti et al. 2002). Aloft, airflow vectors are superimposed on isolines of vertical velocity in the central vertical plane (upper panel) and at the lower surface (lower panel).

The result of this section exemplifies the importance and merit of explicit SGS models. In general, truncation terms of accurate advective transport schemes vanish in the absence of a flow, in particular, in a no-flow direction. Consequently, in applications where wall effects are essential, an explicit mechanism must be supplied for modeling transport of dependent variables normal to the boundary. Whether such a mechanism is built into the NFV transport scheme or provided separately, it constitutes an explicit SGS model. Here we employed a standard Smagorinsky-type turbulence model (Margolin et al. 1999). It is noteworthy that we attempted abbreviated approaches, but have found the explicit modeling of the near-boundary SGS turbulence a necessary prerequisite of a realistic response of the saltation model.

14.4.3 DNS of oceanic boundary–current separation

Intense currents form along the western boundaries of ocean basins as a result of the Earth's rotation, providing a mean flow of heat toward the poles. Examples include the Gulf Stream off the eastern shore of the United States, the Kuroshio Current off Japan, and the Agulhas Current that courses south along the southeastern coast of Africa.

The physical processes that control western boundary–current separation have proven to be complex, particularly in the case of the Gulf Stream, where no single physical process exerts dominant control but instead a number of processes contribute at first order (cf. Dengg, Beckmann, and Gerdes 1996 for a review). Much of what understanding we do have has been reached through the study of western boundary–current separation under simplifying assumptions. Laboratory and numerical simulation may be used together to gain insight, with the laboratory simulation providing a measure of truth against which to constrain the numerical simulation, lending confidence to the mathematical analysis of the simulation.

The projection of the Earth's rotation vector onto the local vertical varies with latitude, and it is this latitudinal dependence (commonly referred to in geophysics as the β effect) that is in fact responsible for western intensification of boundary currents (Stommel 1948). Here, β is defined as

$$\beta = \frac{1}{\mathcal{R}} \frac{\partial}{\partial \phi} 2\Omega \, \sin(\phi), \qquad (14.17)$$

where ϕ is the latitude, $\mathcal{R}$ the Earth's radius, and Ω the angular frequency of the Earth's rotation.

In the laboratory it is generally more feasible to vary the depth of a rotating tank of water rather than design an experimental apparatus for which β is nonzero. If friction is negligible, such that angular momentum is very nearly conserved, then the potential vorticity

$$PV = \frac{f + \zeta}{H} \qquad (14.18)$$

is conserved. Here, f is the projection of the rotation vector onto the local vertical, such that $f = f_0 + \beta y$, ζ is the relative vorticity and H is the depth of the fluid. If PV is conserved then a variation of H with y is equivalent to a variation of f with y, and so western boundary intensification and the process of separation from the boundary may be simulated in a rotating tank of water, with a depth that varies in the y direction, in the laboratory.

One such laboratory simulation of western boundary–current separation is that of Baines & Hughes (1996; hereafter BH96). Despite the enormous difference in scale between a rotating tank in the laboratory and an oceanic gyre, certain parameters, such as the ratio of advective to rotational time scales (the Rossby number; see Pedlosky 1986), are set to be within bounds that are applicable to the circulation of the ocean. In terms of numerical simulation, however, the small spatial scales place the simulation in the realm of DNS, in which the explicit dissipation is that of the molecular viscosity of water.

Although the flow is very nearly two dimensional in the experiment of BH96, we use an isotropic, uniform grid spacing in all three dimensions in our numerical simulation (except in the "western" half of the domain, in which a linear "continental slope" causes the vertical grid spacing to be reduced), allowing inclusion of the slight departures from vertical uniformity. The domain is 0.2 m in x, and 1.1 m in y, as in the tank of BH96. The rotational period is 4 s, and β is $0.091 \times f/H$, as in the right plate of Figure 14.11 (after Figure 3e of BH96). We use a maximum depth H of 0.21 m, but with a linear continental slope, of slope 0.5, which extends from the western boundary halfway across the domain.

Simulations at grid resolutions of 1, 0.5, and 0.25 cm show near-convergence of the solution at the 0.25 cm resolution, producing a simulation that compares well with that of BH96, as we show in Figure 14.11. Here, we indicate the stagnation point, where the vorticity along the western boundary changes sign, with the annotation "vort," and, upstream, the point where the normal derivative of the vorticity changes sign and the stream begins to detach from the boundary and to enter the interior, with the annotation "d_vx." Despite the underresolution of the frictional boundary layers along the sides, top, and bottom of the simulated rotating tank, a comparison of the numerical and laboratory results suggests that our modeling strategy is producing an acceptable DNS

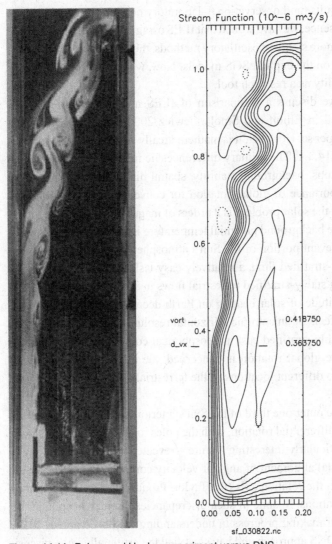

Figure 14.11. Baines and Hughes experiment versus DNS.

reproduction of the laboratory experiment: The separation occurs just slightly earlier in our numerical simulation, and compares well with the laboratory experiment in terms of the width of the current, distance between meanders, and the extent of lateral excursions. This degree of realism in the DNS simulation enables a model-based analysis of the physics of western boundary–current separation.

14.5 ILES as a research tool

So far, we have shown that NFV methods, which underlie the ILES approach to modeling geophysical turbulence, are adaptive and robust; that is, they provide quality results throughout a range of scales and problems. In particular, we have demonstrated that they are suitable for DNS and for explicit LES studies, and are capable of providing solutions comparable with LES results (and observational–laboratory data) in the absence of an explicit SGS model. In essence, we have argued that ILES using NFV methods is a viable alternative to LES using more standard oscillatory methods, the physical realizability of whose solutions depends on the explicit SGS models. Now, we depart from defending ILES and illustrate its utility as a research tool.

The next example, we discuss a comparison of ILES and DNS studies of solar convection (conducted in Elliott and Smolarkiewicz 2002), does not qualify as a geophysical turbulence per se. However, it is mathematically akin to the climate problem discussed in Section 14.4.1. The primary differences lie in the vertical–horizontal aspect ratios of the domains, background stability, spatial distribution of the energy sources, and relative importance of global rotation for convection organization. The geometric aspect ratio for the solar problem is 2 orders of magnitude larger than for the terrestrial climate, and the background potential temperature is nearly constant; thus it is easy to idealize the relevant portion of the Sun's atmosphere (so-called convection zone) as a deep neutrally-stratified fluid, a relatively easy task for numerical solution as compared with treating stably-stratified terrestrial flows in a thin spherical shell. On the other hand, the magnitude of solar heating on Earth decreases towards the poles, mitigating the adverse effects of anisotropic horizontal resolution (of grid-point models neutrally-stratified stably-stratified cast in geospherical coordinates) on simulated convection. Insofar as the global rotation is concerned, the differences between the two systems reduce just to different locations of the terrestrial and solar solutions in a common parameter space.

Deep convection in the outer one third of the Sun's interior leads to a characteristic pattern of time-averaged differential rotation, with the poles rotating about 25% slower than the equator. One particularly interesting feature – revealed by helioseismology in the late eighties – is a radial alignment of angular velocity contours, in contrast to the cylindrical alignment with the rotation axis (the Taylor–Proudman state) predicted by the early numerical simulations. In spite of such discrepancies, numerical simulations provide the best chance of making progress in understanding the observations. Many studies have adopted the DNS approach and have justified the artificially large viscosities and thermal diffusivities as means of modeling transport by unresolved eddies. LES techniques offer a superior alternative, but face the problem of defining a suitable turbulence closure. We avoid this problem by shifting the responsibility for truncating the turbulent cascade from an explicit turbulence closure to the numerical scheme itself by means of ILES. In Elliott and Smolarkiewicz (2002), the authors compared results of DNS simulations carried out with a spherical-harmonic anelastic code (Elliott, Miesch and Toomre 2000; Miesch et al. 2000) with DNS and ILES results they obtained by using the NFV option of anelastic model EULAG.

All three sets of results evinced similar patterns of vertical velocity in the convection zone, with banana-cell convective rolls and convection velocities of the order of a few hundred meters per second; see Figure 14.12. The DNS and the ILES solutions also produced very similar patterns of mean meridional circulation (not shown). The three differed, however, in predicting the pattern of the differential rotation. Inasmuch as the DNS with the spherical harmonic code gave a somewhat similar pattern to the DNS with

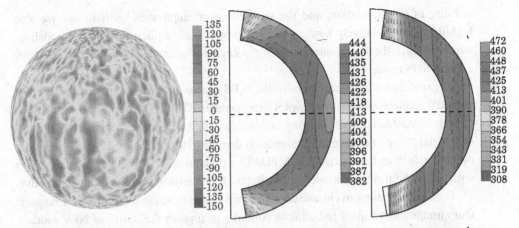

Figure 14.12. NFV simulations of solar convection. The left panel shows the vertical velocity (ms^{-1}) on a horizontal surface near the middle of the domain for the ILES run. Central and right panels show the time-averaged angular velocity (nHz) for, respectively, DNS and ILES runs (color inset, Plate 16).

the NFV code, the ILES with the NFV code was clearly different (Figure 14.12). Since the first two were carried out with almost identical equation sets and parameter values, one might have expected them to give similar results. The fact that there were differences at all indicates the sensitivity of the solutions to the details of the viscous stresses. On the other hand, the fact that the ILES simulation gave distinctly different results is intriguing and illustrates particularly well the strong effect of the artificial viscosity and diffusivity used in DNS simulations. Aside from the differences between the three simulations, each captured some elements of the observed results that the others did not; in other words, no one could clearly be qualified as the superior. There is merit in all three experiments, as comparing them provides hints as to the sensitivities of solar convection. Hence, understanding the origin and consequences of the discrepancies between the three simulations could illuminate the mechanism of the differential rotation. Since all the results failed to show uniformly good agreement with the observationally determined solar differential rotation, it is possible that some other aspect of the physics of the problem has been incorrectly modeled – perhaps the lack of magnetic fields, insufficient density contrast across the domain, or other factors. There is clearly much scope for further investigation of this fundamental problem.

14.6 Remarks

The overall accuracy of approximation is not the only quality measure of a numerical scheme. Modern numerical methods, based on NFV methods, embody physical constraints that enhance the realizability of simulations. Chief among these constraints is the second law of thermodynamics, which is closely related to the nonoscillatory property (see discussion in Chapter 2).

In this chapter we have highlighted a particular NFV model, EULAG, based on MPDATA NFV transport schemes. Among the chief properties of MPDATA are the preservation of sign of scalar quantities such as density, water content, etc.; the nonlinear stability of the simulation; and the suppression of unphysical oscillations. We also highlighted the relatively newly validated property of implicit turbulence modeling, which enables the representation of high-Reynolds-number flows without need for explicit SGS models.

To promote a wider acceptance of the ILES approach, we have provided several diverse calculations of turbulent flows. The agreement of MPDATA numerical simulations of high-Reynolds-number decaying turbulence with theoretical expectations and pseudospectral results (a community standard) demonstrates that this NFV scheme is not overly dissipative. Comparisons of MPDATA simulations of convective boundary layers with and without subgrid models show that the implicit turbulence property is adaptive.

Global simulations of climate contain broad ranges of scales of time as well as space. Our simulations of idealized climate continue to support the utility of NFV models. However, the Earth's climate system contains many additional processes beyond fluid dynamics, many of which cannot be resolved on computers of the present nor the foreseeable future. The construction of appropriate models for these processes, and their integration into NFV schemes, will continue to be a challenge for many years.

The ILES property of NFV schemes is not exclusive but complementary to LES and DNS strategies of numerical simulations. We showed that this same numerical model is suitable for ILES but performs equally well in explicit LES or DNS studies. This substantiates the adaptive character of NFV numerics.

Acknowledgments

We are grateful to Dr. Julian Elliott from Data Connection Ltd., Dr. Matthew Hecht from the Los Alamos National Laboratory, Professor Pablo Ortiz from the University of Granada, and Dr. Andrzej Wyszogrodzki from the National Center for Atmospheric Research for contributing input to Sections 14.5, 14.4.3, 14.4.2, and 14.2, respectively. We thank Professor Peter Baines from the University of Bristol for his comment and the permission to reproduce the left plate in Figure 14.11. In Figure 14.1 the left plate is the NASA Apollo 17 image; the photo by R. Symons in the central plate has been adopted from the web page of the Institute for Meteorology and Geophysics, University of Insbruck; and the original photograph of the cloud chamber experiment in the right plate was provided by Professor Szymon Malinowski from the Institute of Geophysics, Warsaw University. This work was supported in part by the Department of Energy Climate Change Prediction Program research initiative.

REFERENCES

Andreotti, B., Claudin, P., & Douady, S. 2002 Selection of dune shapes and velocities. Part 1: Dynamics of sand, wind and barchans. *Eur. Phys. J. B* **28**, 321–339.

Arakawa, A. 1966 Computational design for long-term numerical integration of the equations of fluid motions: Two-dimensional incompressible flow. *J. Comput. Phys.* **1**, 119–143.

Bagnold, R. 1941 *The Physics of Blown Sand and Desert Dunes*. London: Methuen.

Bagnold, R. 1956 The flow of cohesionless grains in fluids. *Phil. Trans. R. Soc. London Sec. A* **249**, 235–297.

Baines, P. G., & Hughes, R. L. 1996 Western boundary current separation: Inferences from a laboratory experiment. *J. Phys. Ocean.* **26**, 2576–2588.

Bogucki, D., Domaradzki, J. A., & Yeung, P. K. 1997 Direct numerical simulations of passive scalars with $Pr > 1$ advected by turbulent flow. *J. Fluid Mech.* **343**, 111–130.

Clark, T. L., & Farley, R. D. 1984 Severe downslope windstorm calculations in two and three spatial dimensions using anelastic interactive grid nesting: A possible mechanism for gustiness. *J. Atmos. Sci.* **41**, 329–350.

Cotter, C. S., Smolarkiewicz, P. K., & Szczyrba, I. N. 2002 A viscoelastic model for brain injuries. *Int. J. Numer. Meth. Fluids* **40**, 303–311.

Davis, T., Staniforth, A., Wood, N., & Thuburn, J. 2003 Validity of anelastic and other equation sets as inferred from normal-mode analysis. *Q. J. R. Meteor. Soc.* **129**, 2761–2775.

Dengg, J., Beckmann, A., & Gerdes, R. 1996 in *The Gulf Stream Separation Problem in the Warm Water Sphere of the North Atlantic Ocean,* 253–290., ed. W. Krauss. Berlin: Gebrüder Borntraeger.

Domaradzki, J. A., & Adams, N. A. 2002 Direct modeling of subgrid scales of turbulence in large eddy simulation. *J. Turb.* **3**, 1–19.

Domaradzki, J. A. & Radhakrishnan, S. 2005 Effective eddy viscosities in implicit modeling of decaying high Reynolds number turbulence with and without rotation. *Fluid Dyn. Res.* **36**, 385–406.

Domaradzki, J. A., Xiao, Z., & Smolarkiewicz, P. K. 2003 Effective eddy viscosities in implicit large eddy simulations of turbulent flows. *Phys. Fluids* **15**, 3890–3893.

Durran, D.R. 1999 *Numerical Methods for Wave Equations in Geophysical Fluid Dynamics.* New York: Springer-Verlag.

Elliott, J. R., Miesch, M. S., & Toomre, J. 2000 Turbulent solar convection and its coupling with rotation: The effects of Prandtl number and thermal boundary conditions on the resulting differential rotation. *Astrophys. J.* **533**, 546–556.

Elliott, J. R., & Smolarkiewicz, P.K. 2002 Eddy resolving simulations of turbulent solar convection. *Int. J. Numer. Meth. Fluids* **39**, 855–864.

Eisenstat, S. C., Elman, H. C., & Schultz M. H. 1983 Variational iterative methods for nonsymmetric systems of linear equations. *SIAM J. Numer. Anal.* **20**, 345–357.

Frish, U. 1995 *Turbulence.* New York: Cambridge University Press.

Germano, M., Piomelli, U., Moin, P., & Cabot, W. 1991 A dynamic subgrid-scale eddy-viscosity model. *Phys. Fluids A* **3**, 1760–1765.

Grabowski, W. W., & Smolarkiewicz, P. K. 2002 A multiscale anelastic model for meteorological research. *Mon. Weather Rev.* **130**, 939–956.

Held, I. M., & Suarez, M. J. 1994 A proposal for intercomparison of the dynamical cores of atmospheric general circulation models. *Bull. Am. Meteor. Soc.* **75**, 1825–1830.

Herring, J. R., & Kerr, R. M. 1993 Development of enstrophy and spectra in numerical turbulence. *Phys. Fluids* A **5**, 2792–2798.

Kerr, R. M. 1990 Velocity, scalar, and transfer spectra in numerical turbulence. *J. Fluid Mech.* **211**, 309–332.

Kosović, B. 1997 Subgrid-scale modeling for large-eddy simulation of high-Reynolds-number boundary layers. *J. Fluid Mech.* **336**, 151–182.

Lettau, K., & Lettau, H. 1978 Experimental and micro-meteorological field studies of dune migration, in *Exploring the Worlds Driest Climate,* ed. K. Lettau & H. Lettau, Madison: University of Wisconsin.

Lilly, D. K. 1967 The representation of small scale turbulence in a numerical experiment, in *Proceedings of the IBM Scientific Computing Symposium on Environmental Sciences.* White Plains, NY: IBM.

Lilly, D. K. 1992 A proposed modification of the Germano subgrid-scale closure method. *Phys. Fluids A* **4**, 633–635.

Lipps, F. B., & Hemler, R. S. 1982 A scale analysis of deep moist convection and some related numerical calculations. *J. Atmos. Sci.* **39**, 2192–2210.

Margolin, L. G., & Rider, W. J. 2002 A rationale for implicit turbulence modeling. *Int. J. Numer. Meth. Fluids* **39**, 799–819.

Margolin, L. G., Smolarkiewicz, P. K., & Sorbjan, Z. 1999 Large-eddy simulations of convective boundary layers using nonoscillatory differencing. *Physica D* **133**, 390–397.

Margolin, L. G., Smolarkiewicz, P. K., & Wyszogrodzki, A. A. 2002 Implicit turbulence modeling for high Reynolds number flows. *J. Fluids Eng.* **124**, 862–867.

Miesch, M. S., Elliott, J. R., Toomre, J., Clune, T., & Glatzmaier, G. A. 2000 Spherical shell convection with the ASH code. *Astrophys. J.* **532**, 593–615.

Monin, A. S., & Yaglom, A. M. 1981 *Statistical Fluid Mechanics: Mechanics of Turbulence, Vol II.* Cambridge, MA: MIT Press.

Ortiz, P., & Smolarkiewicz, P. K. 2006 Numerical simulation of sand dune evolution in severe winds. *Int. J. Numer. Meth. Fluids* **50**, 1229–1246.

Pedlosky, J. 1986 *Geophysical Fluid Dynamics.* New York: Springer-Verlag.

Polavarapu, S. M., & Peltier, W. R. 1990 The structure and nonlinear evolution of synoptic scale cyclones: Life cycle simulations with a cloud-scale model. *J. Atmos. Sci.* **47**, 2645–2673.

Prusa, J. M. & Smolarkiewicz, P. K. 2003 An all-scale anelastic model for geophysical flows: Dynamic grid deformation. *J. Comput. Phys.* **190**, 601–622.

Rajchenbach J. 2002 Dynamics of grain avalanches. *Phys. Rev. Lett.* **88**, 12, 014301-1 to 014301-4.

Rotunno, R., Grubišić, V., & Smolarkiewicz, P. K. 1999 Vorticity and potential vorticity in mountain wakes. *J. Atmos. Sci.* **56**, 2796–2810.

Saad, Y. 1993 A flexible inner-outer preconditioned GMRES algorithm. *SIAM J. Sci. Stat. Comput.* **14**, 461–469.

Schmidt, H., & Schumann, U. 1989 Coherent structure of the convective boundary layer derived from large-eddy simulation. *J. Fluid Mech.* **200**, 511–562.

Smagorinsky, J. 1993 Some historical remarks on the use of nonlinear viscosities, in *Large Eddy Simulation of Complex Engineering and Geophysical Flows*, ed. B. Galperin & S. A. Orszag. New York: Cambridge University Press, 3–36.

Smolarkiewicz, P. K., & Margolin, L. G. 1993 On forward-in-time differencing for fluids: Extension to a curvilinear framework. *Mon. Weather Rev.* **121**, 1847–1859.

Smolarkiewicz, P. K., & Margolin, L. G. 1998 MPDATA: A finite-difference solver for geophysical flows. *J. Comput. Phys.* **140**, 459–480.

Smolarkiewicz, P. K., Margolin, L. G., & Wyszogrodzki, A. A. 2001 A class of nonhydrostatic global models. *J. Atmos. Sci.* **58**, 349–364.

Smolarkiewicz, P. K., & Prusa, J. M. 2002a VLES modeling of geophysical fluids with nonoscillatory forward-in-time schemes. *Int. J. Numer. Meth. Fluids* **39**, 799–819.

Smolarkiewicz, P. K., & Prusa, J. M. 2002b Forward-in-time differencing for fluids: Simulation of geophysical turbulence, in *Turbulent Flow Computation*, ed. D. Drikakis & B. J. Guertz. Dordrecht: Kluwer, 279–312.

Smolarkiewicz P. K., & Prusa, J. M. 2005 Towards mesh adaptivity for geophysical turbulence. *Int. J. Numer. Meth. Fluids* **47**, 1369–1374.

Smolarkiewicz, P. K., & Pudykiewicz, J. A. 1992 A class of semi-Lagrangian approximations for fluids. *J. Atmos. Sci.* **49**, 2082–2096.

Stommel, H. 1948 The westward intensification of wind-driven ocean currents. *Trans. Am. Geophys. Union* **29**, 202–206.

Wedi, N. P., & Smolarkiewicz, P. K. 2004 Extending Gal-Chen and Somerville terrain-following coordinate transformation on time dependent curvilinear boundaries. *J. Comput. Phys.* **193**, 1–20.

15 Using PPM to Model Turbulent Stellar Convection

David H. Porter and Paul R. Woodward

15.1 Introduction

In this chapter we present the rationale behind, the validation of, and the results from our numerical models of the turbulent flow of gas expected in the convection zones of various types of stars. We review both local area models of convection expected near the surface of solar types of stars and global models of the convection zones of red giant stars. We made the local area models in slab geometry on Cartesian meshes. Even though the geometry of the full convection zone of a red giant star is basically spherical, we also made these models on Cartesian meshes. We made these two sets of models by using variants of the piecewise parabolic method (PPM) described in detail in Chapter 4b and applied to ideal turbulent flow in Chapter 7. Both the use of an Euler-based code such as the PPM to do first-principle studies of turbulent convection as well as the use of Cartesian meshes to do calculations of spherical regions motivated quite a few tests. Our focus in this chapter is to describe the effectiveness and limits of using the PPM to study turbulent stellar convection in a variety of geometries in terms of surprising yet verifiable results.

15.2 Rewards and challenges

Both simple and sophisticated one-dimensional models of stars, as well as observation, indicate that there are regions inside many types of stars that are unstable to convection. These regions are called *convection zones*. The convection zone in the nearby star we call "the Sun" spans roughly the outer third of its radius. The uppermost regions of the solar convection zone are seen on the surface as granulation. The convection zone in a red giant star covers nearly the entire star's volume: It extends from the surface, which might be at 1 astronomical unit (the Earth's orbital radius), all the way down to the star's core, which is roughly the size of the Earth. In these radial zones some fraction (usually large) of the star's luminosity is transported by, and drives, fluid motions. The enormous range of length scales between the size of a star and the scale where molecular dissipation is dominant implies that these convection zones run at very high Reynolds and Rayleigh numbers. The resulting flows are expected to be turbulent. The minimum Mach number required to convectively transport the energy flux associated with a star's luminosity can also be substantial. In the Sun, we observe Mach numbers approaching unity near the surface, and in red giants the lowest Mach numbers are expected to be found near midradius and are typically between Mach 0.1 and 0.5, and they increase both toward the interior and toward the surface. Hence, we expect strong compressibility as a result of both these transonic turbulent motions and the gravitational stratification in convections zones, which span many pressure scale heights.

The presence of turbulent and transonic convection zones introduces substantial questions for astronomers who are trying to understand the structure and evolution of stars. One question is this: How does the convection interact with the radial structure of the star? A temperature gradient in excess of adiabatic will drive convective flows. However, the convection itself will mix the gas and drive the radial profile to be adiabatic. In steady state, a self-consistent balance between convective mixing and superadiabatic driving, consistent with the background energy flux that is due to the star's luminosity, must be reached. The nature of this balance depends upon the character of the turbulence, which is usually guessed at by those doing one-dimensional models of stars. However, the result could affect the star's overall radial profile and size: Where the convective Mach numbers approach unity, the dynamic pressure may modify the star's radial structure. Another question is this: To what extent do motions in the convection zone penetrate and cause mixing in the surrounding regions? The extent of this mixing of the convection zone with the stable regions below could influence the mix of atoms observed at the surface of stars. In red giants, this mixing could greatly influence the flame zone, which one-dimensional models predict is just below the convection zone.

Given these questions, we find that three-dimensional numerical models of stellar convection, which directly follow a sufficient number of length scales, could provide valuable insight for those researchers trying to do stellar structure. Even models with highly idealized physics, with a simplified equation of state and simplified models of radiative transfer, could address the basic question of how turbulent convection behaves in, and affects, a stellar interior. This was our motivation in making our numerical models.

However, even with simplified physics, the numerical challenges were still great. Given the astronomically high Reynolds numbers, if all of the length scales, from the size of the convection zone to the molecular dissipation scale, were to be followed, then the resulting computational meshes would be far beyond the reach of computers for the foreseeable future. Clearly, some kind of truncation of length scales is needed, and the question of what is a sufficient range of length scales must be addressed. Further, the gravitationally stratified and transonic nature of stellar convection requires a numerical method that can handle compressibility and shocks. Large computational meshes would still be needed – enough to span many pressure scale heights and resolve even the smallest scale height with an adequate number of computational cells. Further, the number of time steps required for these high-resolution simulations would be very large, since the time step scales with the mesh size and length of time for the convective system to relax is hundreds of flow times. PPM, which is an Euler code designed to handle high-Mach-number compressible flows and is highly optimized for computational speed, seemed to be a natural fit.

15.3 Local area models of stellar convection

Our goal in this work was to understand the effect of three-dimensional (3D) turbulence on how thermally driven convection behaves inside of stars. The effects of physical properties, such as the presence of magnetic fields, variations of the thermal conductivity, ionization of the gas, rotation, curvature of the convection zone, and boundaries above and below are not only difficult to model and interpret, but also are expected to vary from one star to another. Hence, we found it appropriate to simplify the modeling and interpretation by simplifying our models to an ideal, polytropic, and inviscid gas with constant gravitational forcing and a constant thermal diffusivity. Let the dynamical variables be density, ρ, pressure, P, and velocity, $\vec{u}$. For a polytropic gas the temperature, T, and internal energy, ε, are related by a constant heat capacity at constant volume, $\varepsilon = c_V T$. We impose gravitational forcing by means of a constant acceleration that is due to gravity, g, pointed vertically downward (i.e., in the $-\hat{z}$ direction). The equation of state is $P = (\gamma - 1)\rho\varepsilon$, with a polytropic index of $\gamma = 5/3$. The total energy per unit mass is $E = u^2/2 + \varepsilon + gz$. We approximate radiative transfer of energy by a constant coefficient of heat conduction, κ. In contrast, the molecular diffusion of momentum in stars is very small, corresponding to extremely small Prandtl numbers. Hence we impose no explicit viscosity. The numerical viscosity of the method is the closest approximation to the much smaller actual viscosity in stellar convection that is achievable on a given computational mesh. With these definitions, we see that the fluid equations of motion for such a system are

$$\partial_t \rho + \vec{\nabla} \cdot (\rho\,\vec{u}) = 0,$$

$$\partial_t \vec{u} + \vec{u} \cdot \vec{\nabla}\vec{u} = -\frac{\vec{\nabla}P}{\rho} - g\hat{z},$$

$$\partial_t(\rho\,E) + \vec{\nabla} \cdot (\rho\,\vec{u}E) = -\vec{\nabla} \cdot (\vec{u}P - \kappa\vec{\nabla}T).$$

We numerically solve these equations of motion with PPM (Colella and Woodward 1984; Woodward 1986; Woodward and Colella 1984; see also Chapter 4b). It can be regarded as a method of numerically modeling the small-scale dissipation, just as large eddy simulations model the unresolved small scales with explicit transport coefficients. Numerical simulations performed with fully resolved Navier–Stokes equations and PPM produce similar results, in terms of both visualizations and velocity power spectra for the cases of two-dimensional (2D) and 3D homogeneous decaying isotropic turbulence (Porter, Pouquet, and Woodward 1992a; Sytine et al. 2000; see also Chapter 7). Results from PPM simulations of 3D decaying and driven turbulence also demonstrate convergence of velocity power spectra with increasing mesh resolution to a $k^{-5/3}$ law

(Porter, Pouquet, and Woodward 1991, 1992b, 1994; Porter, Woodward, and Pouquet 1998). Higher-order structure functions of velocity fields from these models exhibit anomalous intermittency scaling that is consistent with both laboratory experiment and direct numerical simulation (Porter et al. 1999; Porter, Pouquet, and Woodward 2002). To adapt PPM to do local area models of stellar convection, we include gravity as a body force (Woodward 1986), and we add an operator split, first-order, and finite-distance implementation of the thermal diffusivity onto each one-dimensional (1D) pass. The boundary conditions for the local area models are periodic in the two horizontal dimensions and free-slip, impenetrable walls at top and bottom. We impose a constant energy flux along the lower boundary, and we impose a spatially constant but time-dependent temperature along the top. We allow the temperature at the top to drift, in order to maintain the same average energy flux through the upper and lower boundary. In this way we can thermally relax these convective layers much more quickly, with only small changes to the overall vertical profile. Several of our publications (Porter et al. 1990; Porter and Woodward 1994, 2000; Porter, Woodward, and Mei 1991) contain detailed descriptions of these numerical models.

15.3.1 Results from local area models

These models are intended to approximate the gravitationally stratified outer layers of stars like the Sun. Convection in these systems efficiently mixes the gas, so the vertical stratification is very nearly adiabatic. We wish to model a layer of gas spanning several pressure scale heights. In the series of models presented here, we set the density to increase by a factor of 11 from top to bottom, corresponding to 4.5 pressure scale heights. We choose a modest horizontal to vertical aspect ratio (2:2:1 in the $X:Y:Z$ directions) for these models to maximize the mesh resolution in all three directions. Given the gas constants, geometry, and gravitational stratification, we parameterize these systems by the energy flux imposed along the lower boundary and the coefficient of thermal conductivity. The imposed energy flux, F_T, is the sum of the radiative energy flux, F_R, and the convective energy flux, F_C. Since the temperature gradient of the layer is nearly adiabatic, especially in the lower pressure scale heights, the radiative flux is nearly constant and equal to its adiabatic value: $F_R = -\kappa \, [dT/dz]_{\text{ad}}$. We must carry the remaining convective flux, $F_C = F_T - F_R$, by convection. A simple unit analysis relates the convective energy flux to an estimate for the convective Mach number, $M_C = (2F_C/\rho)^{1/3}/c$, where ρ is the density and c is the speed of sound at a given depth.

These local area models of stellar convection are listed in Table 15.1. We parameterize our models in terms of the ratio of the radiative to total energy flux (F_R/F_T, column 4 in Table 15.1) and the convective Mach number (M_c, column 5) at the base of the layer. We study two kinds of convective systems, with relatively high and low thermal conductivities. In the high-thermal-conductivity case, the radiative transport carries 80% of the energy flux. In the low-thermal-conductivity case, radiative transport carries only 8% of the imposed flux. In each case we perform a series of computations for increasing mesh resolutions (column 2 in Table 15.1). We run each simulation through

Table 15.1. *Local area models of stellar convection*

Run	Mesh	Time	F_R/F_T	M_c	κ
C21	$64 \times 64 \times 32$	1000			
C22	$128 \times 128 \times 64$	300	0.80	0.021	10^{-4}
C23	$256 \times 256 \times 128$	125			
C31	$64 \times 64 \times 32$	1000			
C32	$128 \times 128 \times 64$	200	0.08	0.035	10^{-5}
C33	$256 \times 256 \times 128$	200			
C34	$512 \times 512 \times 256$	46			

a span of time (column 3) long enough for the flow to reach statistical equilibrium in the sense that both the velocity power spectra and the convective fluxes are stationary. Units for this simulation are based on the depth of the layer, the average mass density, acceleration that is due to gravity, and the heat capacity at constant volume all being unity. In these units, the time for a downflow plume to cross the convection zone is about 10 for the C2x series of runs and about 5 for the C3x series of runs. In each series, we use mesh-refined final states of each simulation as the initial state of the next higher-resolution simulation.

Visualizations show the wealth of detail in these models. Figure 15.1 shows the vertical component of velocity, V_Z, from run C23. The viewpoint is slightly above the

Figure 15.1. Perspective volume rendering of the vertical component of velocity from slab convection run C23 (highest resolution of the high-thermal-diffusivity models) at a time after convective equilibrium is reached. The entire simulation volume is shown. However, the flow variable is rendered as opaque; hence, only a horizontal cut at the top and vertical cuts along two edges are visible from this view. The thick white lines indicate the bounding box of the simulation. Note the network of downflow lanes along the upper boundary, and turbulent velocity field in the interior.

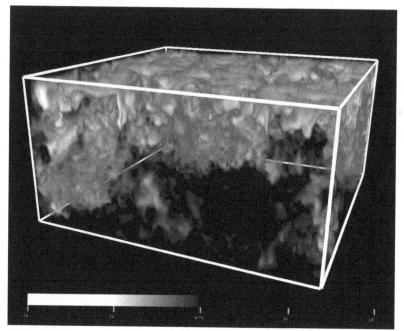

Figure 15.2. Same velocity field and view as shown in Figure 15.1, except small-amplitude velocities are rendered as transparent. The largest magnitude vertical velocities are organized into turbulent downflow lanes and plumes.

upper boundary of the simulation and at a distance that shows the entire simulation volume. We render all values of V_Z with 100% opacity here so that only the outer surfaces are visible. We see a cellular pattern of downflow plumes along the top boundary. We see disorganized upflows and downdrafts along the vertical edges. More of the 3D structure of the flow is revealed by only showing regions with strong vertical flow, where $|V_Z| > 0.5$, as shown in Figure 15.2. Here we see turbulent downflow plumes that span the depth of the layer. Updrafts, by contrast, are more disorganized. A vertical section of V_Z, which is selected to cut through a downflow plume, Figure 15.3, clearly shows that these plumes can span the entire vertical extent of the layer. This particular plume (at the right side of Figure 15.3) becomes increasingly turbulent with depth. The entropy in a different vertical slice is shown in Figure 15.4. A layer of relatively cool (i.e., low-entropy) gas can be seen along the upper boundary. This layer feeds cool downflow plumes. Again we see that the cool downflows are more concentrated and turbulent than the updrafts. Small-scale features are washed out along the upper boundary: The time scale for the relatively large and constant conductivity decreases with decreasing density, which decreases strongly with height near the upper boundary. The constant of conductivity is an order of magnitude smaller in the C3x series of runs. Temperature fluctuations in a horizontal slice near the upper boundary of run C34 (Figure 15.5) show a network of cool downflow lanes, dark gray in this figure, surrounding larger areas of warm updrafts, shown in light shades of gray. Figure 15.6 shows the temperature fluctuations one third of a pressure scale height lower at the same time. A

Figure 15.3. Same vertical velocity field as in Figures 15.1 and 15.2; here a thin vertical slice is shown. Note that the downflow plume near the right edge of the figure spans the vertical extent of the layer.

network of downflow lanes is still present; the finely spaced lanes above have merged into a courser network of downflow lanes and plumes.

In these models regions of downflow are strongly correlated with turbulence. Figure 15.7 shows a volume visualization of enstrophy, $\omega^2 = |\vec{\nabla} \times \vec{V}|^2$, in a horizontal section near the upper boundary of run C34 at the same time: Regions of large enstrophy match the down flows quite closely. Volume visualization of the enstrophy in a vertical section (Figure 15.8) shows regions of turbulence extending down from the upper boundary. Movies of the enstrophy as well as correlations with the vertical velocity show that these regions of turbulence are strongly correlated with downflows

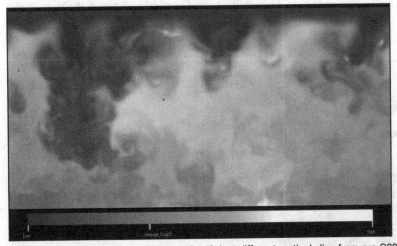

Figure 15.4. Entropy, evaluated as $\log(P/\rho^\gamma)$, in a different vertical slice from run C23. Note the thick thermal boundary layer along the top and relatively narrow and turbulent downflowing plume (dark gray column at the left side of this figure) compared with the warm updraft (light gray to the right of the plume).

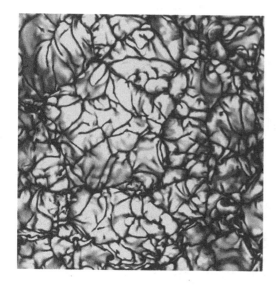

Figure 15.5. Temperature fluctuations near the top of run C34 (highest resolution of the low-thermal-diffusivity models) after convective equilibrium is reached. Note the network of cool downflowing lanes (very dark gray) that delineate convection cells with warm updrafts (light shades of gray) in their interiors.

(Porter and Woodward 2000). This correlation of turbulence – or at least shear – with downflow is present at all depths in both series of runs and can be understood in terms of the source terms for the vorticity (Porter and Woodward 1996, 2000). Near the upper boundary the baroclinic term, $2\vec{\omega} \cdot [\vec{\nabla} p \times \vec{\nabla}(1/\rho)]$, generates shear near the edges of the downflow lanes. Since the flow is subsonic, compression is nearly 100% correlated with vertical velocity so that downflows are always compressing and updrafts are always expanding, which leads to the situation in which the term that is due to compression $(-2\omega^2 \vec{\nabla} \cdot \vec{V})$ significantly increases or decreases the enstrophy, depending on the sign of V_Z. The stretching term $(S = 2\omega_i \omega_j \partial_i V_j)$ is the largest of the three terms and increases enstrophy everywhere.

However, S is strongest in downflows where the other two terms systematically enhance both enstrophy and shear. Vertically aligned vorticity is seen at the base of the

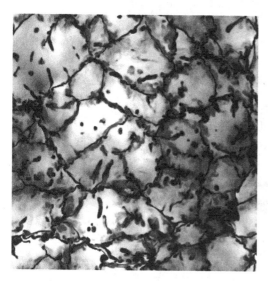

Figure 15.6. Temperature fluctuations from the same run and time as in Figure 15.5 in a horizontal slice one third of a pressure scale lower. The cool downflow lanes are more turbulent than in the slice above, while the updrafts show little or no small-scale temperature fluctuations. The typical size of the convection cells is seen to increase with depth.

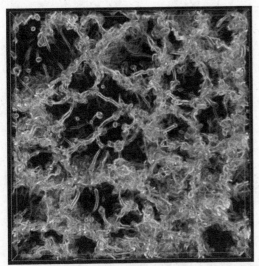

Figure 15.7. Magnitude of vector vorticity in a thick horizontal section near the top of run C34. The cellular pattern of intense vorticity seen here corresponds to the network of downflow lanes. Note the pairs of horizontally oriented vortex tubes and vortex rings, which correspond to the lowest ends of newly forming downflow lanes and plumes, respectively.

updrafts (bottom center of Figure 15.8) and at the top of the downdrafts (upper right of Figure 15.8). In both of these regions, the circulation of the large convection cell, which fills the simulation volume, produces stretching in the vertical direction. Conversely, at the top of updrafts, the convective circulation produces stretching primarily in the horizontal direction, resulting in horizontally aligned vortex tubes (see Figure 15.7).

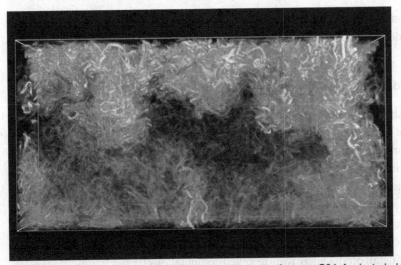

Figure 15.8. Magnitude of vorticity in a thick vertical section from run C34. Again, turbulent regions (i.e., where there are a concentration of intense vortex tubes) are seen to be associated with downflow lanes and plumes. Strong and vertically oriented vortex tubes are typically present at the upper boundary above downflow lanes (upper right) and along the lower boundary below updrafts (lower center). In both of these regions, the principal direction of strain induced by the large convection cell (which spans the entire volume) is in the vertical.

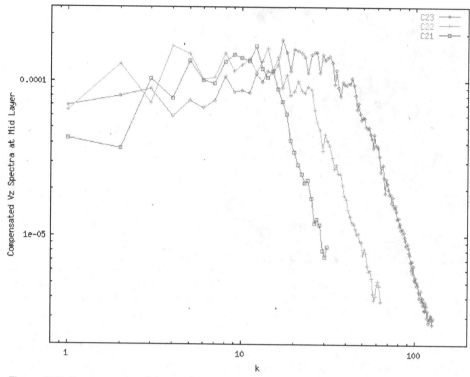

Figure 15.9. Power spectra of the vertical components of velocity from the series of high-thermal-diffusivity runs. These spectra are compensated for a $k^{-5/3}$ slope; hence, the predominantly horizontal trend for modes $k < 10$ are consistent with a Kolmogorov inertial range. Note that the amplitudes of these velocity spectra at low wave numbers converge with increasing mesh resolution.

These horizontal vortex tubes typically come in counterrotating pairs; each pair starts as the two sides of a small downflow lane.

15.3.2 Validations of local area models

If a computation produces unexpected results, how can you know that they are not simply numerical artifacts? Alternatively, if a computation produces known behavior, have you really learned anything? Similar to other methods of investigation (such as analytic theory, experiment, and observation), both value and credibility ultimately come from how results tie into other research in the field and how useful they are in general. New results from computation must be viewed as predictions, which might motivate new experiments or observations. Results, which have been seen before, may be viewed as confirmation if the new and old investigations used substantially different methods. The wealth of detail produced by well-resolved computations (in our largest cases $\sim 10^9$ degrees of freedom) of time-dependent fluid flow ($\sim 10^4$ time steps) can reveal unexpected features, or show how things work, even when some aspects of the flow, such as statistical reductions, are consistent with well-known behavior. Interpretation and utility to other investigators of results is enhanced by error analyses and descriptions

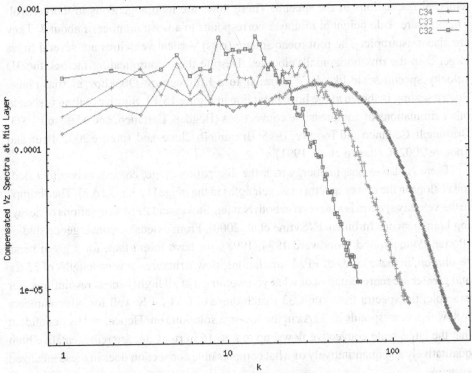

Figure 15.10. Power spectra of the vertical components of velocity from the three highest resolutions of low-thermal-diffusivity runs. Again, these spectra are compensated for a $k^{-5/3}$ trend (horizontal line). Spectra are seen to converge at low wave numbers with increasing mesh resolution. Kolmogorov-like ranges appear to be present for $k < 11$ in these models.

of where the results might apply. The richness of detail available from well-resolved numerical models facilitates traditional statistical reductions, which we use to quantify where our results are believable and relate our results to other work, both analytical and computational.

Velocity power spectra are a measure of how strongly the second-order moments of variations in a flow scale with separation. In these gravitationally stratified models of convection, there are systematic changes with height. To generate our velocity power spectra, we measure velocity variation for displacements in a horizontal plane, where we expect statistical uniformity. Figures 15.9 and 15.10 show Fourier power spectra of the vertical component of velocity fields for the three highest resolution models in each of the two series of runs. In both figures the spectra have been compensated for the trend that would be present if the full 3D velocity fields scaled like a Kolmogorov–Obukhov inertial range (Kolmogorov 1941; Obukhov 1941); this takes out most of the variation in these velocity spectra. We see that as the mesh resolution increases, the dissipation range, which is the band of wave numbers where velocity spectra rapidly roll off, scales with the Nyquist wave number. The velocity spectra at lower wave numbers appear to converge with increasing mesh resolution. In particular, the models with the largest two meshes in each series of runs agree at low wave numbers to within the temporal fluctuations in the spectra. These flows are inhomogeneous in the vertical: The pressure scale height at midlayer corresponds to a wave number of about 7. They are also anisotropic: The root mean square (rms) vertical velocities are several times larger than the rms horizontal velocities. Despite these complicating factors, the 3D velocity spectra scale like $k^{-5/3}$, similar to a Kolmogorov–Obukhov inertial range. Such a scaling is shown at the horizontal line in Figure 15.10. Similar scaling is seen in other simulations of compressible convections (Bogdan, Cattaneo, and Malagoli 1993; Brummell, Cattaneo and Toomre, 1995; Brummell, Clune, and Toomre 2002; Brun and Toomre 2002; Cattaneo et. al. 1991).

There is also excess in energy near the dissipation range, corresponding to a shallower slope in the power spectra in wavelengths in the range $[12\Delta x, 32\Delta x]$. The "bump" in the velocity spectra is also seen in both Navier–Stokes and PPM simulations of decaying homogeneous turbulence (Sytine et al. 2000). From extensive convergence studies (Porter, Pouquet, and Woodward 1994, 1998), we have found that, for a given mesh resolution, in these kinds of PPM simulations, flow structures at wavelengths of $32\Delta x$ and greater are representative of what you would get at all higher-mesh resolutions. For example, the spectra from run C33 match those of C34 fairly well for wave numbers < 8, which corresponds to $32\Delta x$ in the lower-resolution run. Hence, we feel confident that the large-scale convective flows, up to $k = 16$ in run C34, are representative both qualitatively and quantitatively of what compressible convection does in these idealized systems.

These velocity fluctuations represent the convective flow. At each height in the stratified layer, the amplitude of these velocity fluctuations is related to the local mass density and the energy flux that the flow must carry. A simple way to get the scale of these velocity fluctuations is based on relating the kinetic energy flux to the convective energy flux, which produces $\delta U \sim (2F_C/\rho)^{1/3}$. Figure 15.11 shows the vertical profiles of the rms Mach numbers from the two highest resolution runs from each series (points), along with the estimated Mach numbers based on this scaling for δU and the local speed of sound, $c = \sqrt{\gamma\, P/\rho}$. Since the total masses in both layers are identical, the total heat energies in both layers are nearly identical, both layers are very nearly adiabatic, the profiles of density and pressure are nearly the same in both runs, and the principal difference in the estimated Mach numbers $M \sim (2F_C/\rho)^{1/3}/(\gamma\, P/\rho)^{1/2}$ (lines in Figure 15.11) is due to the different convective energy fluxes. In general we should expect an order unity factor in the estimated Mach number. It just so happens that a factor of 1 works fairly well here. Hence, the amplitudes of the velocity fluctuations produced in these models are quite reasonable. Further, we see that the Mach number scales as we expected, both between the two runs and also with depth, away from the impenetrable walls at top and bottom.

The radial structure of stars is influenced by convection where it is present. Indeed, stellar astronomers are interested in exactly how convection, along with radiative transfer, ionization, and other physical effects, establishes vertical profiles of temperature and pressure in convection zones. The mixing caused by convection tends to drive the

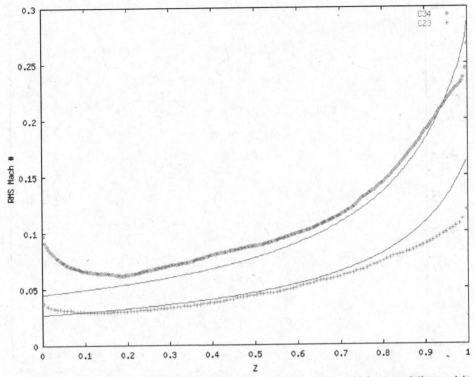

Figure 15.11. The rms Mach numbers as functions of height (Z) from the two highest-resolution models with high (C23) and low (C34) thermal diffusivities (points). Lines correspond to unit analysis estimates of the convective velocity based on local mass density convective energy flux for each model.

layer to be adiabatic. However, convection is driven by some part of the layer being superadiabatic, which for our simple polytropic models means

$$\frac{dT}{dz} < \left.\frac{dT}{dz}\right|_{ad} = -\frac{g}{\gamma\, C_V}.$$

In the absence of convection, typically radiative transfer carries all of the vertical energy flux in a star. If the corresponding radiative temperature gradient,

$$\left.\frac{dT}{dz}\right|_{rad} = -\frac{F_T}{\kappa};$$

is everywhere shallower than the adiabatic temperature gradient, that is, subadiabatic, $[-dT/dz]_{rad} < [-dT/dz]_{ad}$, then no convection is expected. However, if there is a region where a superadiabatic temperature gradient would be required to carry all of the energy flux by means of radiation, then convection is driven in that region. Hence, convective equilibrium is a balance between radiative transfer that generates superadiabatic gradients and convective mixing that flattens them to be very nearly adiabatic.

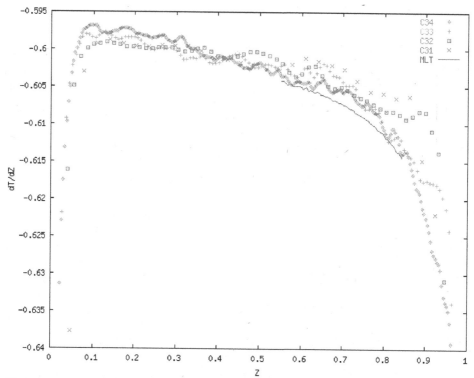

Figure 15.12. Average temperature gradients as functions of depth from the series of runs with low thermal diffusivity. These temperature gradients are seen to converge with increasing mesh resolution. The curve corresponds to a best-fit local mixing length model, which works for both series of runs away from the upper and lower boundaries. The adiabatic temperature gradient for a polytropic atmosphere is $-g/(\gamma C_V)$, which has a value of -0.6 in these models. Note the thermal boundary layers in terms of the superadiabatic temperature gradients at top ($Z = 1$) and bottom ($Z = 0$). The subadiabatic temperature gradient in the lower pressure scale height ($0.05 < Z < 0.3$) is due to the nonlocal effects produced by the coherent downflow plumes, which span the layer.

Figure 15.12 shows the vertical profiles of the temperature gradients from the four models where 92% of the energy flux is carried by convection. These temperature gradients are seen to converge with increasing mesh resolution. The lowest-resolution run (C31) agrees with all higher-resolution runs for $z \in [0.3, 0.6]$, the next higher-resolution run (C32) agrees with all higher-resolutions for $z \in [0.3, 0.75]$, and the two highest-resolution runs (C33 and C34) agree for $z \in [0.05, 0.9]$. Everywhere, there is a tendency for the higher-resolution runs to approach the highest resolution (C34) as mesh resolution is increased. The temperature gradients that these models converge to are sensible. Efficient convection drives the temperature gradients to be very close to the adiabatic value of $[-dT/dz]_{ad} = -0.6$.

There are superadiabatic thermal boundary layers along the impenetrable walls at $Z = 0$ and $Z = 1$. The extended superadiabatic region for $Z > 0.3$ is expected from analytic treatments, and it is also seen in other numerical models (Cattaneo et al. 1991; Chan and Sofia 1989, 1996; Kim et al. 1996; Singh and Chan 1993). Mixing length theory gives us a way to relate superadiabatic temperature gradients in convection zones to the vertical energy flux that the convection carries. The line corresponding to mixing length theory (MLT) in Figure 15.12 shows the expected temperature gradient for these models, based on local MLT, for efficient convection, with an effective ratio of mixing length to local pressure scale height of 2.68 (Porter and Woodward 2000). A detailed mixing length analysis shows that velocity and temperature fluctuations, as well as temperature gradients, can all be related with the same set of correlation coefficients for both series of runs (C2X and C3X) and over a range of depths that span several pressure scale heights (Porter and Woodward 2000). The slightly subadiabatic temperature gradient seen for $z \in [0.1, 0.3]$ has been seen in other simulations of convection (Cattaneo et. al. 1991; Chan and Sofia 1989), and can be attributed to nonlocal effects of convection (Chan and Sofia 1989; Porter, Woodward, and Jacobs 2000).

15.4 Convection in red giant stars

There are several interesting questions that simple hydrodynamic models of convection in spherical systems, such as in red giant stars, can address. Effects of geometry can be both unexpected and important. The local area models presented herein showed that convection cells tend to fill the simulation domain when the aspect ratio of the box is 2:1. The preferred aspect ratio of convection cells tends to be in the range of 3:1 to 5:1. Convection cells tend to be several times wider than they are high. Further, the largest convection cells span most, if not all, of the vertical extent of the domain, no matter how many pressure scale heights they go through. In red giants, convection zones fill the entire volume of the star except for a geometrically tiny region at the center; 1D models of these stars indicate that these convection zones span many pressure scale heights, and the lowest pressure scale heights span significant fractions of the stellar radius (Iben 1984; Kawaler 1998). Thus, curvature effects are inevitable even if convection cells span only one pressure scale height. Near the base of the convection zone, the local pressure scale heights are large compared with the local radius. This implies that there must be a very different geometry of convection in the lowest pressure scale heights of a red giant star. Either the size of the largest convection cells would have to become small (compared with a pressure scale height), or the aspect ratio of a convection cell would have to become very narrow, or the cell would fill the entire spherical volume at deeper radii. Any of these possibilities is radically different from what was seen in local area models of convection.

A second major hydrodynamics question raised by red giant stars is related to the high Mach numbers expected in these objects. Given the luminosity of these stars along with the mass densities and temperatures expected from 1D models (Iben 1984; Hansen and Kawaler 1994; Kawaler, private communication, 1998), simple MLT arguments, such as those given in the previous section, predict Mach numbers approaching unity in many red giant stars. For example, in a 3-solar-mass star that is well advanced into its asymptotic giant branch phase, minimum convective Mach numbers are expected to be about 0.1 near midradius. The convective Mach number is expected to increase, going to both greater and smaller radii. There should be prodigious production of acoustic waves in these turbulent and high-Mach-number flows. A fully developed turbulent spectrum of fluctuations, associated with the larger convection cells, could couple to the lowest-order modes of stellar pulsations; turbulent convection could drive stellar pulsations in red giants. These high-Mach-number flows would also generate shock waves, possibly influencing mass loss as the shocks penetrate the surface of these stars. The dynamic pressure associated with high Mach numbers could influence the overall radial structure of the star. Finally, high-Mach-number flows near the base of the convection cell could easily generate mixing with the convectively stable layer beneath, affecting both the abundance of heavier elements in the convection zone as well as the overall evolution of the star.

For the first series of numerical simulations, we defined the simplest models that would address these purely hydrodynamic issues. We chose to simplify the really complicated equation of state present in red giant stars to a polytropic γ-law gas. Similarly, we approximated radiative transfer with a diffusive conductivity in the interior coupled to a simple Stephan-Boltzmann-law cooling at the surface. In order to allow for both radial and nonradial pulsations, we let this cooling law act in the interior of the mesh: The thin region in which the cooling law has any substantial effect acts as a free surface for the convection zone. We also chose to only model the convection zone at radii that could be adequately resolved on the mesh by imposing a small impenetrable sphere at the center. With these approximations, we could efficiently produce flows that were in convective equilibrium and directly address the hydrodynamic issues. Then, given funding and interest, we could add the detailed physics into these models to see directly what difference each effect has. Since the exact strengths and forms of these physical effects vary from star to star, and can even be ambiguous as a result of unknowns such as the relative abundances of elements, we felt the idealized hydrodynamics was a reasonable starting point.

The question, then, turns to how to model such objects in a way that allows the hydrodynamics to do what ever it wants. We chose to use PPM on Cartesian meshes. PPM is well adapted to these kinds of very compressible and high-Mach-number flows. The advantages of using rectangular meshes to do these spherical systems have to be discussed. Alternative formulations of the problem, such as using spherical coordinates or spherical harmonics, would have significant problems with these models. Using spherical coordinates introduces coordinate singularities, strongly inhomogeneous and anisotropic numerical truncation terms, and very small grid sizes near the poles, which would drive the time step to be very small in any explicit method. Using a spherical harmonics method would avoid these problems, but such a moment code introduces a global solve that is expensive, has trouble with Euler discontinuities (such as shocks), and introduces unphysical global dependencies not present in high-Mach-number flows. Since a goal of these models is to directly assess both the nature and effects of fluid turbulence in these stars, directly following a very large number of degrees of freedom is required. Mesh resolutions for these spherical convection problems ranged from 128^3 to 1024^3 zones. Since high-order methods – like PPM – need only about 16 zones to represent virtually any geometry of flow, we felt that a sphere spanned by 100 or more zones across the diameter would be more than sufficient to accurately model the global characteristics. While the numerical diffusion is still anisotropic, being tied to the mesh directions, on a uniform mesh it is limited in amplitude and at any given wavelength decreases strongly with mesh resolution. In turbulent convection, where the velocity spectrum is fully developed (as we saw in the aforementioned slab models), at wavelengths larger than $8\Delta x$ the stresses that are due to numerical diffusion are negligible compared with the stresses that are due to the turbulence. Hence, we anticipated that our results would be unaffected by the mesh direction. This assumption is borne out in our models, as we discuss in the validation section.

Given that we can get away with using rectangular meshes, there are several strong reasons for using them. First, the numerical dissipation of the PPM is relatively simple, thoroughly tested, and characterized from our studies of decaying turbulence (Sytine et al. 2000) and slab convection (Porter and Woodward 1994; Porter et al. 1991). Further, since a large mesh, or more generally following a very large number of degrees of freedom, is required to directly follow turbulent flow, both speed and the capability to handle very large meshes are important factors. The 1D sweeps used in PPM are highly optimized for speed on modern cache-based architectures. We have also developed optimal ways of feeding the 1D PPM strip updated from moderately sized and logically regular 3D blocks of data. Further, we have developed a framework for decomposing really large meshes into moderately sized blocks, which can be efficiently read and updated by a parallel cluster of compute nodes (Woodward 1994; Woodward et al. 1995, 2001). These techniques, which respect cache, memory, disk, and network hierarchies of bandwidth and latency, have enabled us to efficiently compute very large problems on a variety of modern parallel architectures (Cohen et al. 2002; Woodward 1994; Woodward et. al. 1995, 2001).

15.5 Adapting PPM to do convection in spherical geometry

Given the goals and approximations just discussed, we needed to adapt PPM to model simple neutral fluid hydrodynamics in a spherical shell convection zone with a mechanically free outer boundary. The steps needed to implement this, on a Cartesian mesh, were fairly straightforward and are summarized as follows.

1. Impose a spherical gravitational body force.
2. Impose an impenetrable sphere at the base of the convection zone.
3. Approximate radiative energy transfer in terms of a thermal diffusivity in the interior, which turns out to be negligible in the stars we modeled.
4. Use Stephan-Boltzmann-law surface cooling in the interior of the mesh, which acts as a mechanically free boundary of the convection zone.
5. Impose a constant luminosity for the star by means of a heating law near the lower boundary of the convection zone.

We discuss each of these steps in turn.

We implemented the spherical gravitational field as a directionally split body force in each Cartesian 1D pass of PPM. Given the component of the body force in the longitudinal direction of the 1D strip in each zone of the strip being updated, we implemented the effects of the body force on the gas dynamics in the standard way for PPM (Woodward 1986). We derived the best estimate for this 1D component of the body force by following these steps: First, globally evaluate the total mass inside a uniformly spaced series of radii. Second, evaluate the gravitational potential and radial gravitational force corresponding to total mass within each radius. Third, fit, with a cubic polynomial, the gravitational potential in each radial interval based on the potential and radial force at the limits of the radial interval. Fourth, at the edge of each zone of the 1D strip evaluate the gravitational potential and longitudinal (in the direction of the 1D pass) gravitational force from the continuously differentiable 3D radial potential. Fifth, from the edge values of the gravitational potential and longitudinal component of the force, derive the cubic interpolated potential in each zone of the 1D strip being updated. Sixth, use this smooth description of the gravitational potential to evaluate the upwind-centered Riemann invariants as a part of the PPM update. Seventh, use the potential at the beginning and end of each 1D pass in the total energy equation.

While most of the simulations we did used a constant-in-time gravitational field based on the initial radial density profile, we also did some in which the radial gravitational potential was readjusted at each time step to reflect the current radial mass profile. The results were indistinguishable. The radial pulsations, which were present in both sets of models, produced only very small variation in the overall radial gravitational field. Hence, we could avoid the costly global sum at each time step.

We implemented the impenetrable sphere at the base of the convection zone in the Cartesian mesh by damping the velocity as a function of radius. If R_C is the radius of the impenetrable core, and Δx is the zone size, then at each time step the velocity is damped by a factor of $f_d = \min\{1, \max[0, (R^2 - R_C^2)/(2R_C \Delta x + \Delta x^2)]\}$, which ramps from 0 at $R = R_C$ to 1 for $R = R_C + \Delta x$ over one mesh spacing, and clips to values of 0 (no damping) and 1 (completely damped) outside this radial interval. While this extremely simple method of imposing an impenetrable sphere produces a stair-step representation, and also damps nonradial velocities, its effects decrease with increasing mesh resolution. We observed solutions to converge as we increased the mesh, as we subsequently discuss in the validation section.

In general, there are regions of stars where the diffusive approximation to radiative transport can be very effective. However, the radiative transport of energy in the bulk of the convection zone of a 3-solar-mass asymptotic giant branch (AGB) star, discussed here, is extremely small. Only at the edges of the convection zone, near the inner and outer radii, does the radiative transport become a nonnegligible fraction of the total energy flux. While an explicit thermal diffusivity was implemented in the code we ran, its amplitude was negligible in the region we modeled. Near the base of the convection zone in a real AGB star the radiative transport is substantial; however, these radii were inside the impenetrable sphere mentioned here. Radiative transport also becomes significant near the photosphere; however, the opacity transitions from being essentially opaque to optically thin over such a small distance (compared with our mesh) that it was a fair approximation to model the effects of radiative transport in and near the photosphere in terms of a cooling law.

Since the true base of the convection zone for an AGB star was not directly modeled, we drove these systems with a simple heating at the base of the convection zone we did model, which was just above the impenetrable sphere. The total heat energy injected in each time step corresponded to the constant luminosity of the star. The heating rate was a function of radius of the form

$$S_H(R) = S_0 f_d \, \max\left\{0, \left[(a^2 R_C^2 - R^2)\right]\right\}.$$

Here f_d is the velocity damping factor already discussed, R_C is the radius of the impenetrable core, $a R_C > R_C$ is the maximum radius at which any heat is injected, and S_0 is a normalization factor such that the total head injected is the imposed luminosity, L, of the star:

$$L = \int_{R_C}^{a R_C} S_H(R) \, 2\pi \, R^2 dR.$$

Finally, near the photosphere, a cooling law approximates the effect of radiative transfer on the energy equation. This cooling law is based on Stefan-Boltzmann surface cooling implemented in the volume of the 3D mesh. Averaging over the frequency dependence of the absorption and emission coefficients, and also assuming thermal equilibrium in a geometrically thin photosphere, we see that the integrated radiant luminosity per unit area on the outer surface of the photosphere is $\sigma \, T^4$. At an optical depth, τ, in the photosphere, the probability of a photon escaping is $e^{-\tau}$. Hence, we can be approximate the effective cooling rate per unit volume that is due to radiations escaping from the photosphere as

$$S_C = \sigma \, T^4 \, \nabla e^{-\tau}.$$

Note that the integral $\int |\nabla e^{-\tau}| \, ds$ over a path running through the photosphere along the gradient of τ is unity, so that volume-cooling rate S_C produces the desired cooling rate per unit area of the photosphere. Given that we are approximating the effects of radiative transfer, and not directly solving the equations for radiant intensity, we approximate the optical depth as a simple function of the mass density: $\tau = (\rho/\rho_o)^2$. Here ρ_o is a typical mass density at which the optical depth is unity.

Given this set of approximations, we can write the equations of motion as

$$\partial_t \rho + \vec{\nabla} \cdot (\rho \, \vec{u}) = 0,$$

$$\partial_t \vec{u} + \vec{u} \cdot \vec{\nabla} \vec{u} = -\frac{\vec{\nabla} P}{\rho} + \vec{g},$$

$$\partial_t (\rho \, E) + \vec{\nabla} \cdot (\rho \, \vec{u} E) = -\vec{\nabla} \cdot (\vec{u} P - \kappa \vec{\nabla} T) + S_H - S_C.$$

The heating and cooling terms, S_C and S_H, are as defined earlier. The energy per unit mass, $E = \frac{1}{2}\rho\, u^2 + \varepsilon + \phi$, includes the gravitational potential energy, $\phi = -M(R)/R$, where $M(R)$ is the total mass inside radius R. The vector acceleration that is due to gravity, $\vec{g} = -\vec{\nabla}\phi$, is implemented in the directionally split manner already discussed. As with the models in slab geometry, the gas modeled is ideal and polytropic.

From the point of view of doing a detailed model of any given red giant star, the approximations described here are severe. However, the detailed physics ignored here, such as effects from ionization, magnetic fields, and atomic abundances, leading to more realistic equations of state, opacities, and radiation pressure, are uncertain and vary from star to star. Since there were only guesses as to what any kind of convection would do, given the geometry and convective flux in a red giant star, we felt that these highly idealized models were a reasonable start. Putting all of the physical effects in at once would be expensive and complicate the interpretation of the results as well. We discuss the main results from these very simplified models next.

15.6 Description of convection in spherical geometry

We based the series of models discussed here on a 1D model of a 3-solar-mass AGB star provided by Kawaler (private communication, 1998). Kawaler's 1D models included fairly realistic forms for the gas state and opacity. The effective γ of the gas varied with radius. However, we found that, using a constant value of $\gamma = 1.6$, we could produce a fairly good fit to the radial profiles of density and temperature with our ideal gas models at several different points in the star's evolution. We performed several 3D simulations based on this star at several points along its AGB track. The principal dynamical change in such a star as it evolves in its AGB phase is that it becomes progressively more luminous, which results in the star's expanding. As it does so, mass density in the convection zone decreases, and the convective Mach number increases. Since the convective overturn time in such a star can be measured in months to years, while the time scale for such a star to change significantly in brightness is very much longer (millions of years), our 3D simulations only modeled the star at one epoch of its evolution. The highest-resolution simulations we performed, and the ones discussed here, correspond to this 3-solar-mass star when it is 4519 times brighter than the Sun. In such a star the minimum convective Mach number is about 0.1 and occurs at midradius. Such a star has an effective surface temperature of about 3600 K, a radius of 1.23×10^{11} m (or about 82% of the Earth's distance from the Sun), and a fundamental period (2 radial sound crossing times) of about 66.5 days.

We performed this series of simulations on computational meshes of 128^3, 256^3, 512^3, and 1024^3. Visualizations from the highest resolution run show a wealth of detail. Figure 15.13 shows temperature fluctuations in the full volume of the star. The opacity threshold is set to show only densities greater than or equal to the photosphere (where cooling occurs), so that the outer surface of the convection zone is seen here. Cool regions on the surface (dark shades of gray) are seen to form lanes, while relatively warm regions (light shades of gray) are seen in large regions between the cool lanes.

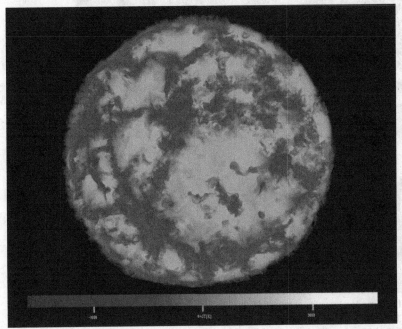

Figure 15.13. Fluctuations in the temperature relative to the mean temperature for the local pressure from a simulation of convection in spherical geometry. Here the entire simulation volume inside the model photosphere is shown. While regions with small temperature fluctuations are transparent, most of the volume near the photosphere has either relatively high (light gray) or low (aqua hues) temperature, so only temperature fluctuations near the modeled surface are seen here. Note the network of relatively cool regions. Like the models in slab geometry, these relatively cool lanes correspond to downflowing plumes, which delineate convection cells near the surface.

These lanes of cool temperature correspond to downflow lanes, similar to those seen from models of convection in slab geometry. Temperature fluctuations in a section of the volume, Figure 15.14, show that the cool lanes on the surface extend down into the interior as downflow plumes. The section in Figure 15.14 is centered on the star's core and oriented to reveal that cool fluctuations occur preferentially on one side of this model. In the left half of Figure 15.14, cool downflow lanes and plumes are seen to merge into one large relatively cool region, which extends all the way to the core. In the right half of Figure 15.14, relatively warm regions are seen near the core.

Radial velocity is shown in Figures 15.15 (full volume) and 15.16 (same section as in Figure 15.14) from the same viewpoint and orientation as used in Figures 15.13 and 15.14. Again, the opacity threshold is set to show only the region interior to the surface cooling. Like the temperature fluctuations, downflows are preferentially on the right half of these figures. The cool lanes in Figure 15.14 match, in detail, the downflow regions in Figure 15.16: Both fields simply show the cool downflow lanes and plumes, typical in thermally driven convection. A predominance of downflow on one side and upflow on the other suggests a dipolar circulation pattern. The angular component of velocity (Figure 15.17) in the same section confirms that there is a strong dipole component to the flow. The angular component of velocity shown here is relative to an axis along

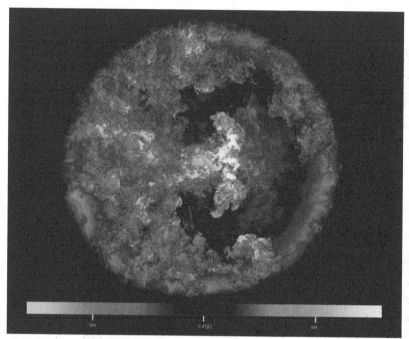

Figure 15.14. The same temperature fluctuation field as in Figure 15.13 in a thin section. This section passes through the center of the model and is aligned to cut through the principal downflow plume, which is seen as cool gas at on the left side of the figure. Animations show that the cellular pattern of downflow lanes near the surface (see Figure 15.13) are carried by the circulation to converge above the principal downflow plume. (See color inset, Plate 17).

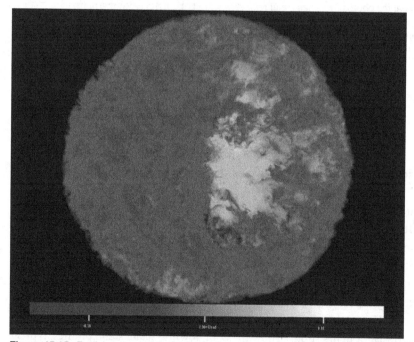

Figure 15.15. Radial component of velocity from a simulation of convection in spherical geometry. The entire volume inside the model photosphere is shown here. Regions with small radial velocity are transparent. However, large radial velocities nearly fill the volume, so only the surface layers are seen here. The orientation of this figure is the same as used in Figures 15.13 and 15.14. Hence, the predominantly negative radial velocities on the left-hand side of this figure are consistent with the cool downflow plume shown in Figure 15.14. Units of velocity, here and in Figures 15.16, 15.17, and 15.18, are based on scaling the radius and acoustic time of this model star to those of a 3-solar-mass, 4519-solar-luminosity red giant.

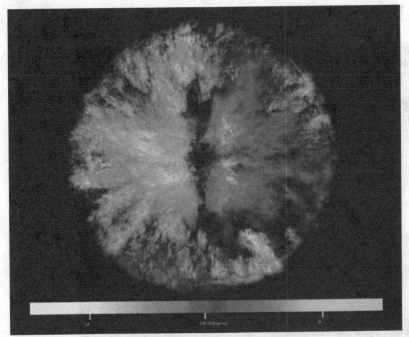

Figure 15.16. Radial velocity in the same section as shown in Figure 15.14. The predominantly negative and positive radial velocities on the left and right sides of this figure correspond to a dipolar flow. See color plates at the middle of this book (See color inset, Plate 18).

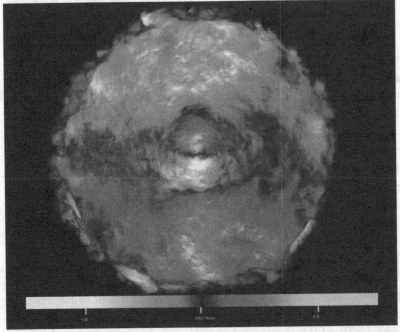

Figure 15.17. Angular component of velocity from a simulation of convection in spherical geometry. The view, orientation, and section shown here are the same as those used in Figures 15.14 and 15.16. Negative (dark shades of gray) and positive (light shades of gray) values of angular velocity correspond to counterclockwise and clockwise motion about the axis along the line of sight through the center of the model. The predominant pattern of angular motion shown here is consistent with the radial velocity shown in Figure 15.16, given that the largest convection cell is a dipolar flow filling the interior of the star. (See color inset, Plate 19).

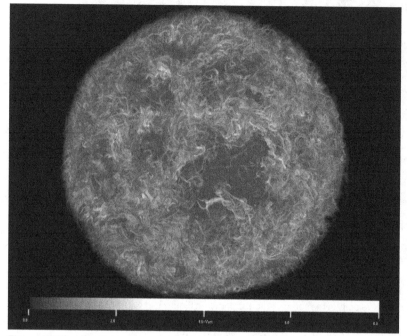

Figure 15.18. Magnitude of vector vorticity from a simulation of convection in spherical geometry. The volume inside the model photosphere is shown here. Only strong vorticity is shown here, which is organized in a myriad tangle of vortex tubes. The strongest vortex tubes (white) on the surface correspond to the cool downflow lanes most clearly seen in Figure 15.13.

the line of sight (from eye point to the center of the view) so that clockwise rotations around this axis are colored in light shades of gray and counterclockwise rotations are shown in dark shades of gray. In this view, flow near the core is seen to go from left to right, while at larger radii the flow re-circulates from right to left. Units of velocity in Figures 15.15 through 15.18 are based on scaling the radius and acoustic time of this model star to those of a 3-solar-mass red giant star when it is 4519 times as bright as the Sun. In the next section, we show that the rms radial velocities in these models are reasonable given the mass densities and luminosities of these stars. Hence, given that the dipole flow is dominant in such red giant stars, we can make a prediction about the velocity field that should be observable on a red giant star's surface. The angular component of velocity, shown in Figure 15.17, should lead to features at the surface systematically moving from one side of the star to the other with velocities of the order of 10 km/s for this 3-solar-mass, 4519-solar-luminosity red giant. 1D models (Iben 1984; Kawaler 1998; Schaller et al. 1992) have been fairly successful in matching the radial profiles of red giants as they evolve through their red giant and asymptotic red giant branch phases. These 1D models can be used to get estimates of the radius, mass density, and luminosity of red giants in general. Then, the local MLT scaling of radial energy flux and mass density to velocity amplitudes seen here (see validation sections for slab and spherical geometries) can be used to estimate the amplitude of the dipole flow, and predict the resulting velocity field on the surface.

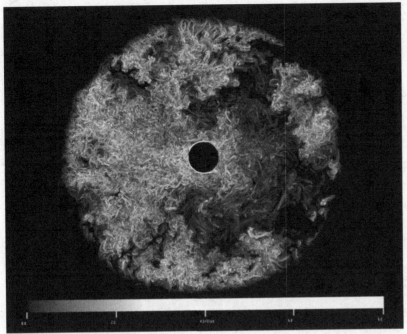

Figure 15.19. Magnitude of vector vorticity from a simulation of convection in spherical geometry. The view, orientation, and section shown here are the same as those used in Figures 15.14, 15.16, and 15.17. Just as with convection in slab geometry, the strongest turbulence (seen here as strong and tangled vortex tubes) invariably corresponds to downflows.

Figures 15.18 and 15.19 show the magnitude of vector vorticity in the full volume and in the same section as earlier, with the same view geometry as in Figures 15.13–15.17. Regions of strongest vorticity are organized in tangled clusters of vortex tubes, which is typical in fully developed 3D turbulence. The diameter of a vortex tube is tied to the dissipation scale, the size of which is several computational mesh widths for PPM. Typical vortex tube widths are seen to be small compared with structures in the flow, such as the downflow lanes and plumes, which delineate convection cells in each pressure scale height. Both near the surface as well as in the interior, the strongest turbulence is associated with downflow lanes and plumes. Evidently the same mechanisms of compression and vortex stretching that enhance the vorticity in downflows are at work here as were in the local area models of convection described earlier. Indeed, the strength of vorticity is seen to weaken in upflows; see, for example, the right half of Figure 15.19.

15.7 Validation of spherical models

Just as with the slab models, power spectra of the velocity field provide a measure of the amplitude of fluctuations as a function of scale. In Fourier analysis, modes of a specific scale or wavelength are characteristic functions of the Laplacian. The normal modes of

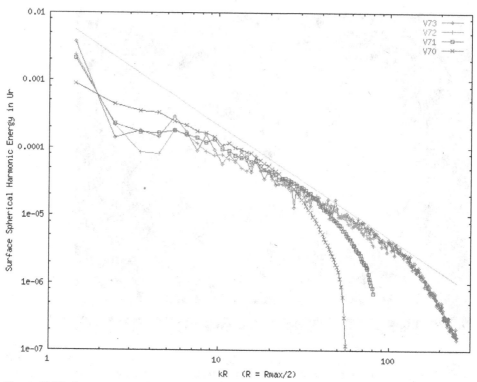

Figure 15.20. Power spectra of the radial component of velocity at midradius from a series of simulations of convection in spherical geometry at different mesh resolutions. The amplitude of the dipole flow (seen in Figures 15.14–15.18) is seen to converge with increasing mesh resolution. In addition, these spectra converge to be a short $k^{-5/3}$ range (light gray line) for $4 < kR < 30$.

the Laplacian in spherical geometry are spherical harmonics. Since we expect there to be systemic differences in the flow between different radii, it is appropriate to evaluate these modes on concentric spheres. Hence, we use surface spherical harmonics. Figure 15.20 shows the surface spherical harmonic power spectrum of the radial velocity at midradius from runs v70, v71, v72, and v73, which we ran on computational meshes ranging from 128^3 to 1024^3 in powers of 2. At a given radius R, wave number k is related to the surface spherical harmonic mode number, l, by means of the relation $kR = \sqrt{l(l+1)}$. The line in Figure 15.20 shows the power spectrum $E \sim k^{-5/3}$ associated with a Kolmogorov inertial range. These velocity spectra converge with increasing computational mesh resolution. Only the lowest resolution results differ qualitatively at low mode numbers. Velocity spectra in the higher three resolution models agree to within fluctuations up to mode $l = 40$. The highest two resolution modes (v72 and v73) agree up to mode $l = 80$. In the range of wave numbers $6 < kR < 30$, all of these spectra scale roughly as $k^{-5/3}$, similar to results from the local area models. The dipole mode dominates these velocity spectra at all four resolutions, especially in the higher-resolution models. Hence, the strong dipole mode, discussed earlier, is a robust and well-resolved feature of these models.

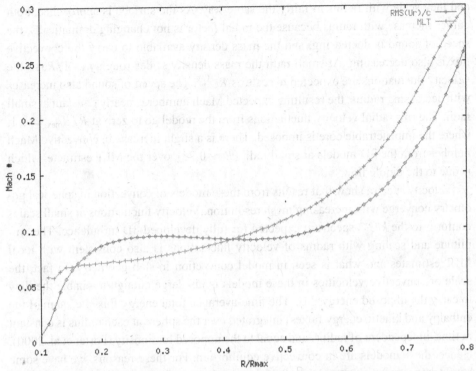

Figure 15.21. The rms Mach numbers of the radial component of velocity as a function of fractional radius (diamonds line), and an estimate based on the local radial convective energy flux and local mass density. Near the surface (i.e., model photosphere) where the convection cells are small compared with the radius, the geometry of the convection cells is similar to the slab convection models: Compare Figure 15.13 with Figures 15.5 and 15.6, and compare the outer radii in Figures 15.14 and 15.16 to Figures 15.3 and 15.4. Correspondingly, the rms Mach number is seen to scale in the same way, and with nearly the same unity coefficient, as the slab convection models (see Figure 15.11). At smaller fractional radii ($R/R_{max} < 0.5$), the rms velocities are large compared with the estimate because of the influence of the dipolar flow.

As with the slab convection models, the amplitude of velocity fluctuations can be estimated by a local unit piece analysis relating the local convective flux to the amplitude of the kinetic energy flux: $F = \frac{1}{2}\rho u^3$. In spherical geometry, the imposed and constant luminosity translates to a local energy flux, which decreases as the square of the radius: $F = L_o/(4\pi R^2)$. Hence, an estimate for the amplitude of velocity fluctuations is $\delta u = \sqrt[3]{L_o/(2\pi R^2 \rho)}$. Figure 15.21 shows a comparison of this estimate with the rms radial velocity fluctuations from the 3D model. Both are shown in terms of the Mach number as functions of radius. The estimate, which is consistent with a simplified local mixing length theory, is labeled MLT. Evaluating the rms radial velocity fluctuations as a function of radius and then dividing by the average speed of sound as a function of radius constructs the measure of convective Mach number shown here. This scaling, which worked well in the slab geometry models of convection, also appears to work fairly well here with the same order unity factor relating velocity fluctuations to convective flux. The MLT estimate and rms fluctuations (Figure 15.21) are of the same order, and they scale with radius in fairly the same way. As the surface is approached, both curves increase with radius because the radial factor is not changing dramatically, the speed of sound is decreasing, and the mass density available to carry the convective flux is also decreasing. At small radii the mass density scales roughly as $1/R$, so the velocity fluctuations are expected to scale as $R^{-1/3}$. The speed of sound also increases with decreasing radius; the resulting expected Mach number is nearly constant at small radii. The rms radial velocity fluctuations from the model go to zero at $R/R_{\max} = 0.1$, where the impenetrable core is imposed. There is a slight increase in convective Mach number from the 3D models at small radii ($R \sim 0.25$), over the MLT estimate, which is due to the dipole flow.

Velocity spectra show that results from these models of convection in spherical geometry converge with increasing mesh resolution. Velocity fluctuations at small scales conform to the $k^{-5/3}$ spectrum expected for fully developed 3D turbulence. The amplitude and scaling with radius of velocity fluctuations is also consistent with local MLT estimates and what is seen in model convection in slab geometry. In fact, the scale of convective velocities in these models is just large enough to enable the flow to carry the imposed energy flux. The time-averaged total energy flux (i.e., sum of the enthalpy and kinetic energy fluxes) integrated over the sphere at each radius is constant in time, independent of radius, and equal to the imposed luminosity (Porter et al. 2000). Hence, these models are in convective equilibrium. For these reasons, we have some confidence in our results, and the surprising dipolar flow seen here may well be a feature of convection in many red giant stars.

15.8 Conclusions and prospects

We have reviewed uses of PPM to study gravitationally stratified convection as it might occur in the interiors of stars. Our focus in this work was to understand the effects of 3D fluid turbulence on these systems. We simplified physical properties, such as radiative transfer and equation of state, as much as possible, and we neglected the effects of magnetic fields entirely. Since the ratio of dissipation to energy-containing scales in stellar convection zones is far too large to simulate all of the scales of motion directly, we used an Euler code (PPM) to simulate these systems. We studied both local area (slab geometry) and global area (spherical geometry) models of convection. We ran these simulations long enough for the spectra of fluctuations to reach a statistically steady state, and for convective equilibrium to be established. We found results to converge with increasing mesh resolution in all cases. We saw the amplitude of convective velocities to scale appropriately with mass density and convective energy flux. We found velocity spectra to scale with wave number as $k^{-5/3}$ in all of the higher-resolution runs. Hence, we have some confidence that results from these simulations are representative solutions to the idealized fluid problem posed. If flows in real stars were significantly different, then it would have to be the result of the physical approximation we made. Such a sensitivity of stellar convection on detailed physics would be of interest to astronomers, for then observation of gross convective features on the surface of stars could be used to help diagnose the complicated and uncertain microphysics going on within.

Results from local area models of convection showed that, at least for ideal neutral fluid convection, the large scales dominate the flow. We found local MLT to work fairly well for relating the amplitude of fluctuations and temperature gradients to the energy flux associated with the luminosity of the star. Our well-resolved models of slab convection confirmed mixing length analysis of earlier work (Chan and Sofia 1996; Kim et al. 1996), which used results from extremely low-resolution meshes. These low-resolution models worked as well as they did because the largest scales of the flow are energetically dominant, fairly independent of the smaller scales, and contribute the bulk of the kinetic energy and enthalpy fluxes. Well-resolved models of turbulence were required to establish this. Our slab convection models also showed that turbulence is enhanced in downflows and diminishes in upflows: This results in extremely turbulent downflowing lanes and plumes, which delineate nearly laminar updrafts.

The big result from our global models of convection in spherical geometry was the dominant dipolar flow. This was unexpected and should have observational consequences. Dipole flow in the interior leads to a circulation pattern on the surface, which goes from one side of the star (where the convection is welling up) to the other side (above the downflowing plume). Either Doppler shift or motion of surface features might show such a global circulation pattern. There are also consequences for the evolution of red giant stars. If the dipolar flow extends all the way down to the base of the convection zone, then it will produce a very strong wind past the core of the star, which might generate mixing with the convectively stable layers beneath. This could influence the flame zone, which is just below the base of the convection zone according to 1D models of these stars.

Natural extensions of this work would be to add the complicated physical effects, which we had neglected in the models of red giant stars. A more realistic equation of state could model ionization zones, which are known to drive convection very strongly. More realistic radiative transfer would be needed to accurately model the photosphere. Radiation pressure would be needed to model very luminous and large red giant stars. Magnetic fields might also be a major player in convection zones of red giants. Adding these physical properties, one at a time, would allow unambiguous interpretation of the effects of each.

Acknowledgments

This work was supported at the University of Minnesota by the Department of Energy, through Grants DE-FG02-87ER25035 and DE-FG02-94ER25207, contracts from Livermore and Los Alamos; by the National Science Foundation, through its Partnerships for Advanced Computational Infrastructure (PACI) program at National Center for Supercomputing Application (NCSA), through a research infrastructure grant, CDA-950297, and also through Grand Challenge Application Group Award ASC-9217394, through a subcontract from the University of Colorado; and from NASA, through Grand Challenge Team-Award NCCS-5-151, through a subcontract from the University of Chicago. We also acknowledge local support from the University of Minnesota's Minnesota Supercomputing Institute.

REFERENCES

Bogdan, T. J., Cattaneo, F., & Malagoli, A. 1993 On the generation of sound by turbulent convection. I – A numerical experiment. *Astrophys. J.* **407**, 316–329.

Brummell, N., Cattaneo, F., & Toomre, J. 1995 Turbulent dynamics in the solar convection zone. *Science* **269**, 1370–1379.

Brummell, N. H., Clune, T. L., & Toomre, J. 2002 Penetration and overshooting in turbulent compressible convection. *Astrophys. J.* **570**, 825–854.

Brun, A. S., & Toomre, J. 2002 Turbulent convection under the influence of rotation: Sustaining a strong differential rotation. *Astrophys. J.* **570**, 865–885.

Cattaneo, F., Brummell, N. H., Toomre, J., Malagoli, A., & Hurlburt, N. E. 1991 Turbulent compressible convection. *Astrophys. J.* **370**, 282–294.

Chan, K. L., & Sofia, S. 1989 Turbulent compressible convection in a deep atmosphere. IV – Results of three-dimensional computations. *Astrophys. J.* **336**, 1022–1040.

Chan, K. L., & Sofia, S. 1996 Turbulent compressible convection in a deep atmosphere. V – Higher order statistical moments for a deeper case. *Astrophys. J.* **466**, 372.

Cohen, R. H., Dannevik W. P., Dimits, A. M., Eliason, D. E., Mirin, A. A., Zhou, Y. K., Porter, D. H., & Woodward, P. R. 2002 Three-dimensional simulation of a Richtmyer–Meshkov instability with a two-scale initial perturbation. *Phys. Fluids* **14**, 3692–3709.

Colella, P., & Woodward, P. R. 1984 The piecewise-parabolic method (PPM) for gas dynamical simulations. *J. Comput. Phys.* **54**, 174–201.

Hansen, C. J., & Kawaler, S. D. 1994 *Stellar Interiors: Physical Principles, Structure, and Evolution.* New York: Springer-Verlag.

Iben, I. 1984 On the frequency of planetary nebula nuclei powered by helium burning and on the frequency of white dwarfs with hydrogen-deficient atmospheres. *Astrophys. J.* **227**, 333–354.

Kim, Y.-C., Fox, P. A., Demarque, P., & Sofia, S. 1996 Modeling convection in the outer layers of the sun: A comparison with predictions of the mixing-length approximation. *Astrophys. J.*, **461**, 499–506.

Kolmogorov, A. N. 1941 The local structure of turbulence in incompressible viscous fluid for very large Reynolds number. *Dokl. Akad. Nauk SSSR* **30**, 9–13 (reprinted in *Proc. R. Soc. Lond.* A **434**, 9–13, 1991).

Obukhov, A. M. 1941 Spectra energy distribution in a turbulent flow. *Izv. Akad. Nauk SSSR Ser. Geogr. Geofiz.* **5**, 453–466.

Porter, D. H., Pouquet, A., Sytine, I., & Woodward, P. R. 1999 Turbulence in compressible flows. *Physica A* **263**, 263–270.

Porter, D. H., Pouquet, A., & Woodward, P. R. 1991 Supersonic homogeneous turbulence, in *Proceedings, Large-Scale Structures in Hydrodynamics and Theoretical Physics, Vol. 392* in Lecture Notes in Physics, ed. J. D. Fournier & P. L. Sulem. New York: Springer-Verlag.

Porter, D. H., Pouquet, A., & Woodward, P. R. 1992a A Numerical study of supersonic homogeneous turbulence. *J. Theor. Comput. Fluid Dyn.* **4**, 13–49.

Porter, D. H., Pouquet, A., & Woodward, P. R. 1992b Three-dimensional supersonic homogeneous turbulence: A numerical study. *Phys. Rev. Lett.* **68**, 3156–3159.

Porter, D. H., Pouquet, A., & Woodward, P. R. 1994 Kolmogorov-like spectra in decaying three-dimensional supersonic flows, *Phys. Fluids A*, **6**, 2133–2142.

Porter, D. H., Pouquet, A., & Woodward, P. R. 2002 Measures of intermittency in driven supersonic flows. *Phys. Rev. E* **66**, 026301, 1–12.

Porter, D. H., & Woodward, P. R. 1994 High resolution simulations of compressible convection with the piecewise-parabolic method (PPM). *Astrophys. J. Suppl.* **93**, 309–349.

Porter, D. H., & Woodward, P. R. 1996 Numerical simulations of compressible convection, in *Les Houches School on "New Tools in Turbulence Modeling,"* ed. O. Metais & J. Ferziger. New York: Springer Verlag, 281–289.

Porter, D. H., & Woodward, P. R. 2000 3-D simulations of turbulent compressible convection. *Astrophys. J. Suppl.* **127**, 159–187.

Porter, D. H., Woodward, P. R., & Jacobs, M. L. 2000 Convection in slab and spheroidal geometries, in *Proceedings of the 14th International Florida Workshop in Nonlinear Astronomy and Physics: Astrophysical Turbulence and Convection*, Vol. 898 in the Annals of the New York Academy of Sciences, 1–20.

Porter, D. H., Woodward, P. R., & Mei Q. 1991 Simulation of compressible convection with the piecewise-parabolic method (PPM). *Video J. Eng. Res.* **1**, 1–24.

Porter, D. H., Woodward, P. R., & Pouquet, A. 1998 Inertial range structures in decaying turbulent flows. *Phys. Fluids* **10**, 237–245.

Porter, D. H., Woodward, P. R., Yang W., & Mei Q. 1990 Simultaion and visualization of compressible convection in 2- and 3-D, in *Nonlinear Astrophysical Fluid Dynamics*, Vol. 617 in the Annals of the New York Academy of Science, ed. R. Buchler & S. T. Gottesman. Maiden, MA: Blackwell, 234–258.

Schaller, G., Schaerer, D., Meynet, G., & Maeder, A. 1992 New grids of stellar models from 0.8 to 120 solar masses at $Z = 0.020$ and $Z = 0.001$ astronomy and astrophysics supplement series, **96**, 269–331.

Singh, H. P., & Chan, K. L. 1993 A study of the three-dimentional turbulent compressible convection in a deep atmosphere at various prandtl numbers. *Astron. Astrophys.* **279**, 107–118.

Sytine, I. V., Porter, D. H., Woodward, P. R., Hodson, S. H., & Winkler., K.-H. 2000 Convergence tests for piecewise parabolic method and Navier–Stokes solutions for homogeneous compressible turbulence. *J. Comput. Phys.* **158**, 225–238.

Woodward, P. R. 1986 Numerical methods for astrophysicists, in *Astrophysical Radiation Hydrodynamics*, ed. K.-H. Winkler & M. L. Norman. Dordrecht: Reidel, 245–326.

Woodward, P. R. 1994 Superfine grids for turbulent flows. *IEEE Comput. Sci. Eng.* **1**, 4–5.

Woodward, P. R., & Colella, P. 1984 The numerical simulation of two-dimensional fluid flow with strong shocks. *J. Comput. Phys.* **54**, 115–173.

Woodward, P. R., Porter, D. H., Edgar, B. K., Anderson, S. E., & Bassett, G. 1995 Parallel computation of turbulent fluid flow. *Comput. Appl. Math.* **14**, 97–105.

Woodward, P. R., Porter D. H., Sytine I., Anderson S. E., Mirin A. A., Curtis B. C., Cohen R. H., Dannevik W. P., Dimits A. M., Eliason D. E., Winkler K.-H., & Hodson S. W. 2001 Very high resolution simulations of compressible, turbulent flow, in *Computational Fluid Dynamics, Proceedings of the Fourth UNAM Supercomputing Conference*, ed. E. Ramos, G. Cisneros, R. Fernándes-Flores, & A. Santillán-González. Singapore: World Scientific, 3–15.

16 Complex Engineering Turbulent Flows

Niklas Alin, Magnus Berglund, Christer Fureby,
Eric Lillberg, and Urban Svennberg

16.1 Introduction

A grand challenge for computational fluid dynamics (CFD) is the modeling and simulation of the time evolution of the turbulent flow in and around different engineering applications. Examples of such applications include external flows around cars, trains, ships, buildings, and aircrafts; internal flows in buildings, electronic devices, mixers, food manufacturing equipment, engines, furnaces, and boilers; and supersonic flows around aircraft, missiles, and in aerospace engine applications such as scramjets and rocket motors. For such flows it is unlikely that we will ever have a really deterministic predictive framework based on CFD, because of the inherent difficulty in modeling and validating all the relevant physical subprocesses, and in acquiring all the necessary and relevant boundary condition information. On the other hand, these cases are representative of fundamental ones for which whole-domain scalable laboratory studies are extremely difficult, and for which it is crucial to develop predictability as well as establish effective approaches to the postprocessing of the simulation database.

The modeling challenge is to develop computational models that, although not explicitly incorporating all eddy scales of the flow, give accurate and reliable flowfield results for at least the large energy-containing scales of motion. In general terms this implies that the governing Navier–Stokes equations (NSE) must be truncated in such a way that the resulting energy spectra is consistent with the $|\mathbf{k}|^{-5/3}$ law of Kolmogorov, with a smooth transition at the high-wave-number cutoff end. Moreover, the computational models must be designed so as to minimize the contamination of the resolved part of the energy spectrum and to modify the dissipation rate in flow regions where viscous effects are more pronounced, such as the region close to walls. For flows of practical engineering interest the Reynolds number is usually high, placing the cutoff wave number in the lower part of the inertial subrange. This, together with the additional complication of walls, raises the demands on the modeling, and requires incorporating anisotropic features. Such complications usually occur in, or around, realistic geometries, which favors using unstructured and preferably also self-adaptive grids that can adjust locally to the flow.

The computational challenge is to solve the equations of the model as accurately as possible, which typically requires a large number of time steps and a large number of grid cells, in an efficient way. For flows in and around complex geometries, unstructured finite-volume methods are superior to finite-difference or finite-element methods since they are based on the integral formulation of the governing equations and are thus conservative by construction. However, for unstructured grids, efficient algorithms of higher spatial order than 2 are often arduous to develop and implement, yet unless we follow that route we cannot expect to be able to discriminate completely between the leading-order truncation error and the subgrid-scale (SGS) model. Alternately, we may instead design the overall algorithm, for example, guided by modified equation analysis approach (Chapters 4a and 5), to use the leading-order truncation error as an implicit or built-in SGS model. As seen in the examples in Chapters 10 and 11, this is appropriate as long as we are far from walls, but for wall-bounded flows we are in principle forced to model the near-wall flow. For some cases the flow is supersonic, for which explicit algorithms are ideally suited, and for some cases the flow will be essentially incompressible, for which semi-implicit algorithms are more appropriate. Other problems of interest in this context involve solving for the location of a free (water) surface, separating the flow of two fluids of different densities and viscosities. This requires specific algorithms that can handle steep density gradients, together with surface tension, breaking waves, and bubble formation and spray dynamics.

An additional challenge is to generate grids of sufficiently high quality (with low warpage, skewness, and stretching) to accurately resolve the flow features residing in the geometries of interest, including, for example, highly curved surfaces, small-scale geometrical details (such as struts and fins), and elements of relative motion. If all geometrical details cannot be properly represented on the grid, as a result of either lack of information or lack of computational resources to resolve these details properly, we must consider the effects of the necessary simplifications or approximations when we are evaluating the results. In some cases, submodels may be developed that can emulate the desired features, but this is not the general situation. In addition, to be of use in the design and evaluation of a set of concepts, the computations must be completed within a few days of building the grid representing the geometry and the flow domain.

We argue that, given the usually robust and accurate results of canonical and simple flows at moderate Reynolds numbers at marginal and higher resolutions (cf. Chapters 10 and 11), there are no obvious reasons why large eddy simulation (LES) in its conventional or implicit (ILES) versions would not be applicable to more complex high-Reynolds-number flows using similar spatial resolutions. Systematic experimentation is, however, needed to quantify the influence of the grid resolution and the requirements of the (explicit and implicit) SGS models when LES–ILES is applied to more complex flows. The cases we choose in this chapter are representative of such flows, although with somewhat simplified geometry and typically, but not always, in model scale to facilitate comparison with available experimental data. As will become evident from the results we present, LES–ILES offers much more information about the flow than Reynolds-averaged Navier–Stokes (RANS) approaches, and generally, with a few exceptions, also gives superior agreement with experimental data than do RANS approaches. The most significant problem, however, is the required specification of open inflow–outflow boundary conditions, for which LES–ILES requires much more information than do RANS approaches. This lack of inflow/outflow data is likely to hamper the usefulness of LES–ILES.

16.2 Submarine hydrodynamics

Submarine design, including modern propulsor concepts, is aimed at achieving, to the extent allowable within overall program constraints, a maximum in mobility and survivability and a minimum in detectability and vulnerability. The computational technologies required for such an undertaking are increasingly multidisciplinary as hydrodynamics becomes coupled with structural dynamics and hydroacoustics. The flow itself is complex and characterized by high Reynolds numbers, typically 10^9 in full scale and 10^7 in model scale, making CFD a formidable challenge. The flow is dominated by the laminar to turbulent transition at the bow, the evolving boundary layer, the wake and the horseshoe vortices created by the sail, and the complex flow around the stern. At the stern the rudders are attached to the hull, resulting in horseshoe and tip vortices together with an adverse pressure gradient caused by the tapering. The flow thus includes the unsteady-approach boundary layer and the slipstream and pressure effects from the propeller, causing vibrations and noise that are important to the operational performance of the submarine.

Present-day CFD models for hydrodynamics usually assume steady flow, or even potential flow. Turbulence is often included through steady-state RANS models (Launder and Spalding 1972) in which equations for the mean flow are solved together with a model for the statistical properties of the turbulence. More advanced simulation methods are required given the well-known limitations of RANS approaches, and the need to embody more complex flow physics than RANS models will ever handle when one is studying submarine hydrodynamics, such as flow noise, vortex–boundary layer interactions, and viscous wake dynamics. Potentially promising such improved approaches are unsteady RANS (URANS) models (Ferziger 1983), LES (Sagaut 2001), and detached eddy simulation (DES; Spalart et al. 1997). The direct computation of the large energy-containing eddies (being flow and geometry dependent) gives LES and DES more generality than RANS models, although at a higher cost. In the recent past, state-of-the-art RANS models of submarine hydrodynamics have been discussed (Van et al. 2003, Yang and Löhner 2003), whereas more advanced URANS approaches have been performed by Ramamurti, Sandberg, and Löhner (1994). Most recently, Alin, Svennberg, and Fureby (2003), Persson et al. (2004), and Alin et al. (2005) have started to analyze the usefulness of wall-modeled LES, for submarine hydrodynamics and for ship flows in general (Svennberg and Lillberg 2004). Concurrently, other groups, such

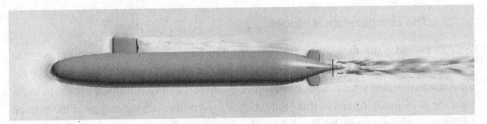

Figure 16.1. Submarine hydrodynamics: Schematic of the DARPA AFF-8 Suboff hull geometry (Huang et al. 1992) and the flowfield for a self-propelled model.

as Pattenden, Turnock, and Bressloff (2004), have started to investigate and explicitly test DES and hybrid RANS–LES approaches for general ships flows.

16.2.1 The DARPA AFF-8 Suboff case

Here we consider the DARPA AFF-8 Suboff configuration (Groves, Huang, and Chang 1989), which consists of hull, sail, and stern appendages. The geometry is defined analytically (Huang et al. 1992) and the model is shown in Figure 16.1. This is a 1:24 scale model, in which the hull has an overall length of $L = 4.36$ m and a diameter of $D = 0.51$ m. The sail is located on the hull at the top dead center, having its leading edge at $x = 0.92$ m and its trailing edge at $x = 1.29$ m. A cap of elliptical cross section is attached to the sail at a height of $h = 0.46$ m. The stern appendages consist of four NACA 0020 profiles, attached to the hull at $\varphi = 0°, 90°, 180°$, and $270°$, respectively, where φ is defined positive counterclockwise as viewed from astern. The appendage locations are defined with the trailing edge at $x = 4.00$ m. Experimental data are provided by the David Taylor Model Basin (DTMB) at a Reynolds number of $\mathrm{Re} = 12 \times 10^6$ based on the freestream velocity v_0, L, and ν (Huang et al. 1992). The total measurement uncertainty in the velocity data is estimated to 2.5% of the freestream velocity, v_0 (Huang et al. 1992).

The computational domain consists of the submarine model mounted in a cylinder having the same hydraulic diameter as the wind tunnel used in the DTMB experiments. The domain extends one hull length upstream of the model and two hull lengths downstream of the model, being $4L$ in overall length, and consists of about 6.3 million cells. The detailed geometry of the wind tunnel is not included, which may be a significant source of error. For the hull an C-O topology is used, whereas for the sail and stern appendages, C-O topologies are used and care is taken to ensure that the grid spacing and aspect ratios are appropriate for the SGS wall model (Fureby et al. 2004). About 10 cells are within the boundary layer on the parallel midsection of the hull, resulting in $y^+ \approx 15$. At the inlet, $\bar{\mathbf{v}} = v_0 \mathbf{n}$ and $(\nabla \bar{p} \cdot \mathbf{n}) = 0$; at the outlet, $\bar{p} = p_0$ and $(\nabla \bar{\mathbf{v}})\mathbf{n} = 0$; at the wind-tunnel walls, slip conditions are used, and on the hull, no-slip conditions are used. All LES are initiated with quiescent conditions and the unsteady flow evolves by itself. For one flow through time in the computational domain, 80,000 time steps are typically required.

16.2.2 The computational model

We can consider the flow around a submarine to be incompressible, and the governing equations are the incompressible NSE. In Chapters 3 and 4a, LES models for incompressible flows were discussed; therefore, here we only summarize them to highlight the SGS models used in this context. A finite-volume discretization of the filtered LES equations or the raw NSE results in the following modified equations (Fureby and Grinstein 2002; also see Chapter 4a):

$$\partial_t(\bar{\mathbf{v}}) + \nabla \cdot (\bar{\mathbf{v}} \otimes \bar{\mathbf{v}}) = -\nabla \bar{p} + \nabla \cdot (\bar{\mathbf{S}} - \mathbf{B}) + \tau + \mathbf{m}^v, \quad \nabla \cdot \bar{\mathbf{v}} = 0, \qquad (16.1)$$

where $\mathbf{v}$ is the velocity, p is the pressure, $\mathbf{m}$ is the commutation error, τ is the truncation error, and $\mathbf{B}$ is the (explicit or implicit) SGS stress tensor that is modeled by use of

- the SGS viscosity model (VM), $\mathbf{B} = -2\nu_k \bar{\mathbf{D}}$ (Sagaut 2001);
- the mixed model (MM), $\mathbf{B} = \overline{\bar{\mathbf{v}} \otimes \bar{\mathbf{v}}} - \bar{\bar{\mathbf{v}}} \otimes \bar{\bar{\mathbf{v}}} - 2\nu_k \bar{\mathbf{D}}$ (Bardina, Ferziger, and Reynolds 1980); or
- the implicit, or built-in (ILES) model, $\mathbf{B} = \mathbf{C}(\nabla \mathbf{v})^T + (\nabla \mathbf{v})\mathbf{C}^T + \chi^2(\nabla \mathbf{v})\mathbf{d} \otimes (\nabla \mathbf{v})\mathbf{d}$ (Boris et al. 1992).

Here $\mathbf{C} = \chi(\mathbf{v} \otimes \mathbf{d})$ and $\chi = \frac{1}{2}(1 - \Psi)(\beta^- - \beta^+)$, where Ψ is the flux limiter related to the flux reconstruction of the convective terms (see also Chapter 4a).

For both the VM and MM, we use the one-equation eddy viscosity model (OEEVM; Schumann 1975) to compute the SGS viscosity, ν_k. We evaluate the model coefficients under the assumptions of an infinite inertial sub-range, and that the grid cutoff lies within this inertial subrange. For the ILES model we use the second-order Flux Corrected Transport (FCT) scheme discussed in Chapter 4a. We combine the discretized momentum and continuity equations to form a Poisson equation for the pressure. We solve the scalar equations sequentially, with iteration over the coupling terms, and we handle the pressure-velocity coupling with a PISO-type procedure.

When simulating a self-propelled submarine hull, the simulation model must also be able to accurately represent the effects of the propeller. An affordable approach is to use an actuator-disc model in which a source term is added to the momentum equation in the region of the propeller that accelerates the flow, giving thrust, and a slipstream with rotation. In the model used here, we evaluate the source term by using a lifting-line technique (Li 1994) that employs a potential flow model and by using Goldstein's kappa theory (Tachmindji and Milam 1957) to account for the number of blades. The actuator disc reacts instantly to changes in the inflow conditions, but at the same time the resulting velocity is affected as the propeller effect is averaged over one revolution, and therefore blade frequency effects as well as tip and root vortices are not properly represented.

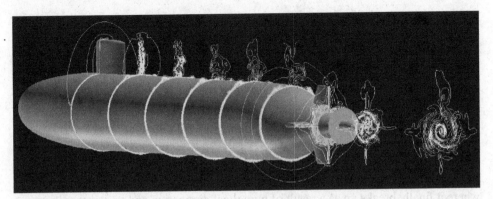

Figure 16.2. Submarine hydrodynamics: Axial velocity contours at nine cross sections for the self-propelled AFF-8 model (also note the actuator disc at the stern emulating the propeller).

16.2.3 Main flow features

The main flow features are evident from Figure 16.2, showing a perspective of the flow past the self-propelled AFF-8 hull. The flow is presented in terms of contours of the axial velocity component at nine cross sections. The flow past the forebody is dominated by the stagnation point region and the favorable pressure gradient over the foreshoulder. In the stagnation point region, the boundary layer is laminar and very thin, but it thickens when the favorable pressure gradient accelerates the flow past the foreshoulder. During the acceleration phase, the boundary layer becomes turbulent; that is, the velocity profile gradually approaches that of fully turbulent boundary layers, and hence pressure fluctuations, $p' = \bar{p} - \langle \bar{p} \rangle$, develop on the forebody. The computationally observed behavior is similar to that of transition (Landahl and Mollo-Christensen 1986), but the resolution is not sufficient to resolve this in appropriate detail. Accurate prediction of p' on the forebody will thus aid in positioning, adjusting, and optimizing bow and flank arrays, besides being useful in determining flow-induced noise and vibrations. To this end we notice that the spectral content of p' in this region is closely related to the geometrical shape of the forebody and to v_0.

The flow over the midbody section is dominated by the horseshoe-vortex pair, as evidenced by the velocity roll-up in the hull boundary layer of Figure 16.2 originating from the roll-up of the boundary layer just ahead of the sail, and by the sail wake. The separation line upstream of the sail oscillates around $x/L \approx 0.20$, and occasionally more than one vortex develops. The horseshoe vortex continues along the sail, forming additional longitudinal vortices by viscous interaction with the boundary layer. Toward the trailing edge of the sail, lateral vortices develop that interact with the unsteady-hull boundary layer and the horseshoe-vortex pair, forming a complex unsteady wake behind the sail. Since the horseshoe-vortex forces fluid outward away from the sail, next toward the hull, it may only be near the trailing edge of the sail that the joint effects of the hull curvature and the two vortex systems are adequate to deliver high-momentum flow back toward the sail, producing instabilities that create detachment and impingement. As the horseshoe vortex is convected further aft, it partly loses its coherence and flattens out toward the hull. During this process it interacts with the curved thickening hull boundary layer to create a very complex near-wall flowfield with embedded vortices approaching the tapered stern. Such a complex flow pattern on the sail side of the hull introduces transport of momentum across the hull, which affects the velocity and turbulent stress distribution in the hull boundary layer. The sail-tip vortex pair persists far downstream but does not interact with other vortices and passes well above the propulsor. The flow over the sail cap is dominated by the turning of the flow and the development of the sail-tip vortex pair at the junction between the sail and the sail cap. This sail-tip vortex pair is stable and maintains its coherence far astern, where it finally breaks up as a result of numerical dissipation and vortex stretching.

Toward the stern, the boundary layer thickens whereby viscous interactions become more important. Vortex sheets are formed along the upper and lower rudder surfaces, and counterrotating vortices are formed at the tips of the rudders, together with horseshoe-shaped vortices formed around each of the four rudders. These vortices are generally embedded in a thick boundary layer that is strongly affected by the tapering of the hull and the adverse pressure gradient in the stern. Such flow features are very hard to simulate, and the general opinion is that RANS approaches cannot include all the important flow features whereas LES approaches have potential of doing so if the resolution is sufficient and the SGS model is adequate. A thin intermittent recirculation region is observed just upstream of the leading edges of the rudders, resulting from a weak (almost) periodic separation of the boundary layer on the tapered stern. Redistribution of the Reynolds stresses are also found to take place at the stern as a result of the strong cross flow, partly caused by the unsteady sail wake and horseshoe-vortex. As can be seen in Figure 16.2, the presence of the propeller affects the flow over the stern. This is mainly caused by the pressure distribution caused by the suction induced by the propeller. Only in the immediate vicinity upstream of the propeller do we note the influence of the tangential and radial velocity distributions caused by the rotation of the propeller. Downstream of the propeller the flow is different and dominated by the swirl.

16.2.4 Statistical comparison

Next, we compare the results of the LES with experimental data (Huang et al. 1992) in order to quantify the accuracy of the LES model, and to examine the influence of grid resolution, SGS modeling, and propulsion. Figure 16.3 shows the distribution of the static pressure coefficient, C_P, along the meridian line of the hull for the towed and self-propelled AFF-8 case. For the towed hull, we include laboratory data along the upper meridian line of the hull, corrected for wind-tunnel effects (Huang et al. 1992), for comparison. We find excellent agreement along the upper meridian line of the hull for all the LES models, and we can detect virtually no difference between the MM + WM, OEEVM + WM, and ILES + WM models, where WM denotes the use of near-wall modeling. On the parallel midbody hull section, $C_P \approx -0.05$, which is in good agreement with channel-flow DNS and LES. Comparing the towed and self-propelled

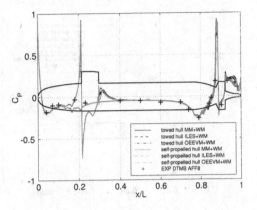

Figure 16.3. Submarine hydrodynamics: Comparison of predicted and measured static pressure coefficients, C_P, on the meridian line of the towed and self-propelled DARPA AFF-8 hull for the fine grid.

hulls, we find that the differences along the forebody and midbody sections are small. On the tapered part of the stern, however, we do find differences. These are caused by the propeller and show an increase in the axial pressure gradient, imposed by the propeller – causing suction over the stern.

Predicted and measured static pressure coefficients on the sail and on the upper rudder, at $\varphi = 0°$, are shown in Figure 16.4. In particular, Figure 16.4(a) presents C_P at 10% of the sail-tip chord length, and Figure 16.4(b) presents C_P at 50% of the stern appendage-tip chord length. By comparing the predicted C_P distribution with the experimental data, we find good agreement along both the sail and the upper stern appendage for all LES models, with essentially no difference between the explicit or implicit LES models. This is particularly interesting when we take into account the difference in boundary layer thickness over the sail and over the upper rudder. Furthermore, from Figure 16.4 we again find the influence of grid resolution to be marginal. For the lower rudder, at $\varphi = 180°$, we find lower C_P values toward the trailing edge, which is in agreement with the data. The observed difference in C_P between the upper and lower stern appendages, and its quantitative agreement with measurement data, is indicative of the predictive capabilities of LES.

Figure 16.5(a) presents a comparison of the time and azimuthally averaged streamwise velocity component, $\langle \bar{v}_1 \rangle_\varphi / v_0$, between LES and experimental data for the towed and self-propelled hulls, at $x/L = 0.978$. For comparison we have also included the experimental data from AFF-1, which is a bare-hull geometry. The azimuthal averaging basically removes all azimuthal velocity variations containing secondary velocity structures. The experiments (Huang et al. 1992) are carried out using a hot-film technique with an estimated uncertainty of 2.5% of v_0. We obtain good agreement between the LES results and the experimental data in the region where experimental data are available. Outside of this region, $\langle \bar{v}_1 \rangle_\varphi$ approaches v_0 asymptotically. The time and azimuthally averaged boundary layer is thick, because of both its natural development over the parallel midbody section of the hull and because of the adverse pressure gradient effects over the tapered part of the stern and the embedded vortices. The good agreement documented in Figure 16.5(a) does not, however, account for the azimuthal $\langle \bar{v}_1 \rangle$ distribution and the secondary velocity distribution, which we discuss later. In the self-propelled

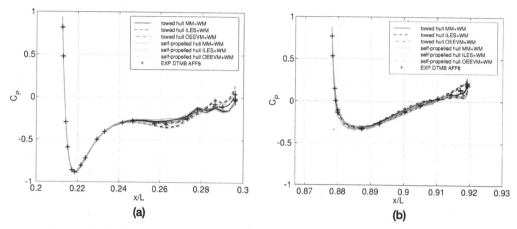

Figure 16.4. Submarine hydrodynamics: Comparison of predicted and measured static pressure coefficients, C_P, at 10% of sail-tip chord length (a) and at 50% of stern appendage-tip chord length (b) of the towed and self-propelled DARPA AFF-8 hull for the fine grid.

AFF-8 case, the slipstream effects of the propeller are clearly visible, revealing that the propeller entrains fluid mainly from the hull boundary layer, which is composed of the hull boundary layer and the appendage-generated vortex systems. This provides a boundary layer with high levels of fluctuations. Figure 16.5(b) presents a comparison of the streamwise rms–velocity fluctuations, $\langle \bar{v}'_1 \rangle_\varphi / v_0$, where $\bar{v}'_i = \sqrt{\langle (\bar{v}_i - \langle \bar{v}_i \rangle)^2 \rangle}$, between LES and experimental data for the towed and self-propelled hulls at $x/L = 0.978$, respectively. Again, the influence of the explicit or implicit SGS model is small. As compared with the towed hull, the self-propelled hull produces a bimodal rms–velocity fluctuation profile, with the outer peak associated with the radius of the propeller and the inner peak value associated with the peak in the mean velocity profile, associated with the boundary layer flow. This suggests that in order to limit the propeller-induced

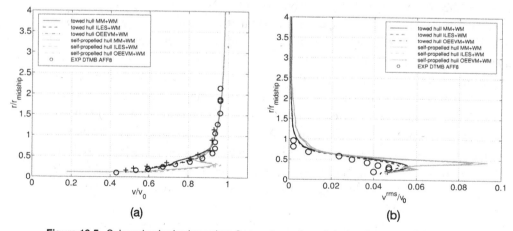

Figure 16.5. Submarine hydrodynamics: Comparison of predicted and measured time and azimutally averaged velocity, $\langle \bar{v}_1 \rangle_\varphi / v_0$, and velocity fluctuations, $\langle \bar{v}'_1 \rangle_\varphi / v_0$, where $\bar{v}'_i = \sqrt{\langle (\bar{v}_i - \langle \bar{v}_i \rangle)^2 \rangle}$, between ILES–LES and experimental data for the towed case, at $x/L = 0.978$ for the fine grid.

vibrations and noise, the propeller should be designed to take this rms–velocity fluctuation distribution into account. More specifically, this implies that the largest fluctuations occur in the propeller-tip region, where the blades are structurally weaker.

16.3 Flow in a multiswirl gas turbine combustor

The design of modern combustion chambers and air–fuel injectors for gas turbines relies heavily on RANS approaches, which predict the mean time- or ensemble-averaged part of the flow throughout the computational domain. In the continuous development of gas turbine air–fuel injectors and combustors, unexpected problems such as flashback, quenching, or combustion oscillations appear in many cases. Combustion instabilities are one of the most precarious phenomena, and the resulting oscillations are caused by the nonlinear coupling between the acoustics and the flame and may lead to lower combustion efficiency, higher emissions, or even the complete destruction of the combustor. Phenomena like this cannot be investigated by RANS approaches, since they cannot handle unsteady flows, or flows that involve other physical processes such as exothermicity, and therefore more detailed models and methods, such as DES or LES, must be used. Concerning the prediction of combustion instabilities, it is common to use acoustic analysis (in which the mean flow is provided by a RANS model) to predict the flow, frequency, and the growth rates of all modes (Eriksson et al. 2003). The weakest part of these models is the description of the flame response to acoustic perturbations but also the limited information supplied to the acoustic model by the RANS model. By replacing the RANS model by a LES calculation (using a variable-density formulation), we can make improved predictions. We can make even better predictions if we can carry out a fully compressible LES calculation (including the acoustics). The use of LES to predict turbulent combustion in simple configurations has been reported (e.g., Fureby 2000; Menon 2001; Pitsch and Duchamp De Lageneste 2002), but using LES to predict the reacting flow in a gas turbine, including air–fuel injectors, still remains to be demonstrated, although several successful attempts have recently been made (e.g., Kim, Menon, and Mongia 1999; Grinstein and Fureby 2004; Schoenfeld and Poinsot 1999).

In Kim et al. (1999) and Grinstein and Fureby (2004), LES has been used to predict the flame and the flow in a simplified version of the GEAE LM6000 combustor (Hura et al. 1998). Although successful in capturing the global features of the flow and the flame, the latter LES was limited by the use of unrealistically simple inflow conditions, in which air–fuel mixing-injectors were effectively replaced with constant-velocity profiles derived from experimental data. Compared with experimental data further downstream in the combustor, these LES calculations predict either a too short or too long flame – a problem believed to be caused by the constant-velocity-profile inflow condition. Grinstein et al. (2002) further investigated this issue, using the GEAE LM6000 combustor and a round combustor equipped with a mixer–fuel injector from GEAE and Goodrich Aerospace located at the combustor dome. They tested different constant inflow conditions, based on experimental data, RANS modeling, and theoretical results,

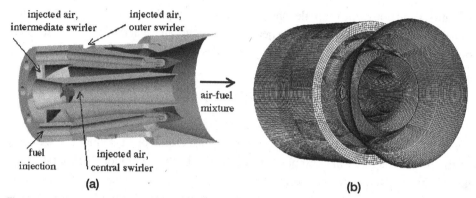

Figure 16.6. Flow in a multiswirl gas turbine combustor: (a) Schematic of the multiswirl fuel injector (TARS) and (b) perspective view of the computational model including the external grid.

in the framework of nonreacting LES, and they found that the flow in the combustor is very sensitive to the inflow boundary conditions. The natural extension of this study has been to perform LES of the flow within a whole integrated mixer–fuel injector combustor system (Fureby et al. 2006). This, however, is difficult because of the complex geometry of typical mixer–fuel injectors, involving geometrical details on a wide range of scales. Unstructured hexahedral or rather mixed (polyhedral) grids are required for this, which, in turn, involves addressing additional technical and research issues (e.g., Minot, Ham, and Pitsch 2005).

Here, we use LES to investigate the flow in the mixer–fuel injector combustor system used in Grinstein et al. (2002), (see Figure 16.6), using both conventional LES and ILES. We make comparisons with experimental velocity data and previous hybrid RANS–LES velocity results.

16.3.1 The triple annular research swirler

Based on a design by GEAE and Goodrich Aerospace typically used in industrial dry low-emission combustors (Pritchard et al. 2002), the model gas-turbine fuel injector presented in Figure 16.6(a) features multiple independent fuel supply lines for efficient fuel distribution and multiple air inlets to obtain suitable combinations of corotating and counterrotating swirling air streams. The combustion air is supplied through the mixer–fuel injector, which is located at the combustor dome, denoted the triple annular research swirler (TARS). The mixer includes three passages equipped with swirlers, outer, α, intermediate, β, and inner, γ, leading into the combustion chamber. The two central coaxial passages feature axial swirlers while the external air passage has radial swirling vanes. Air blast fuel atomizers are distributed between the second and third annuli. Fuel is injected into the inner and outer annuli for efficient mixing, whereas an ordinary pressure atomized pilot is located in the central passage. Different TARS configurations can be obtained by changing the swirlers to different swirling vane angles or different rotating directions. For example, S3045C45 means that the mixer's outer, intermediate, and inner swirler's vane angles are 30°, 45°, and 45°, respectively, and the outer and inner swirler corotate whereas the intermediate swirler counterrotates. The overall length of the TARS is 66.0 mm, and the diameter of the outer swirler exit is $D = 50.8$ mm. The TARS is attached to a round combustor with a length of 0.457 m and a diameter of $2D$. The mass flow through the TARS is $\dot{m} = 0.032$ kg/s, which corresponds to an average velocity of 13.7 m/s through the swirler exit, and a 4% pressure drop through the TARS. Based on the circumferentially and time-averaged velocity data at the TARS exit plane, the swirl (S) and radial (R) numbers are $S \approx 0.40$ and $R \approx 0.04$, respectively, whereas Re $\approx 51,000$. At the University of Cincinnati in the United States, an extensive research program was performed with the TARS to examine lean direct fuel injection (LDI) multiswirl gas turbine combustors (Li and Gutmark 2002a, 2002b, 2005; Pritchard et al. 2002). In this program the velocity was measured by using LDV and stereoscopic PIV, and temperature by thermocouples, and in addition OH* chemiluminescence images are collected.

The TARS geometry and a hexahedral grid was provided by Li and Gutmark at the University of Cincinnati. The grid was originally developed for RANS calculations and covered a quarter section of the TARS and the attached round combustor, and it contained about 500,000 cells. For the purpose of the LES–ILES calculations, we transformed this grid to cover the full geometry, and we refined the grid in critical regions, such as boundary layers. It finally contained about 2.2 million cells; see Figure 16.6(b). Guided by previous RANS (Li and Gutmark 2002a) modeling, we selected a fixed mass flow distribution between the three TARS inlets: the distribution ratio is 0.80:0.15:0.05 for the outer, intermediate, and inner swirlers, respectively. If we assume incompressible flow, with the density of air being 1.125 kg/m^3, this corresponds to the inflow velocities $v_\alpha = 47.9$ m/s, $v_\beta = 11.9$ m/s, and $v_\gamma = 6.0$ m/s, respectively, which are supplied by zero-gradient Neumann conditions for the pressure. At the combustor outlet $\bar{p} = p_0$ and $(\nabla \bar{v})\mathbf{n} = \mathbf{0}$, where p_0 is the pressure of the ambient air, equal to 101.3 kPa. Tests performed to assess the importance of the convective outflow boundary condition used suggest that the flow in the combustor and in the TARS are virtually unaffected by their specifics. On the walls of the TARS and of the combustor, no-slip conditions are used together with the wall model (Fureby et al. 2004). All LES are initiated with quiescent conditions and the unsteady flow evolves by itself. For one flow-through time, 20,000 time steps are typically required.

16.3.2 Computational model

We assume the nonreacting flow through the TARS and in the accompanying combustor is here assumed incompressible. This is a reasonable assumption since the peak Mach number observed in the simulations occurs at the location of the air-blast fuel atomizers, located between the second and third annuli, never exceeds 0.4. Weak compressibility effects are thus present in this region but are not likely to affect the overall flow in the TARS or in the combustor. In the simulations we discuss here, the model and code utilized were the same as used for the submarine hydrodynamics calculations

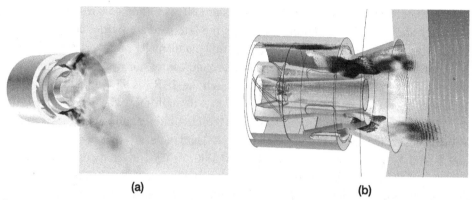

(a)   (b)

Figure 16.7. Flow in a multiswirl gas turbine combustor: (a) Perspective view from inside the combustor toward the TARS outlet plane, showing instantaneous contours of the axial velocity component, and (b) instantaneous velocity vectors colored by the axial velocity component.

(Section 16.2), but limited to two SGS models: the MM and the ILES. Likewise, the ILES model is based on using the second-order FCT scheme discussed in Chapter 4a. We combine the discretized momentum and continuity equations to form a Poisson equation for the pressure. We solve the scalar equations sequentially, with iteration over the coupling terms, and we handle the pressure–velocity coupling with a PISO-type procedure. Moreover, to handle the near-wall effects, we use the wall model discussed by Fureby et al. (2004); also see Chapter 10.

16.3.3 Results for the flow through the TARS

Figure 16.7(a) is a perspective view from inside the combustor toward the TARS outlet plane showing instantaneous contours of the axial velocity component, $\tilde{v}_x$, along the centerplane of the combustor. We observe the highest axial velocity (with a peak value of about $\tilde{v}_x \approx 140$ m/s) in the outermost annulus, in which also the primary fuel is injected. In this outermost annulus the flow is rapidly diverted by the outer swirl vanes, resulting in a strong tangential velocity component. In the intermediate annulus, in which the pilot fuel is injected, we also find high axial velocities ($\tilde{v}_x \approx 80$ m/s). The jets from the intermediate and outermost annuli converge in the conical section of the TARS whereby also a significant radial velocity component is created. In the inner annulus the central recirculation region starts to develop some distance upstream of the central swirler, and it continues into the combustion chamber. It is bounded by swirling motion in a thin layer along the walls of the inner annulus. The shape of the central recirculation region initially follows the shape of the inner annulus and broadens downstream of the inner annulus to reach a width of $0.30D$ around the TARS exit plane and further increases to about $0.60D$ some $0.75D$ downstream of the TARS exit plane, and thereafter it rapidly contracts. Outside of the central recirculation region is the region of an annular jet that essentially is composed of two annular jets, one from each of the intermediate and outer annuli. Specific characteristics of this annular jet are that it initially follows

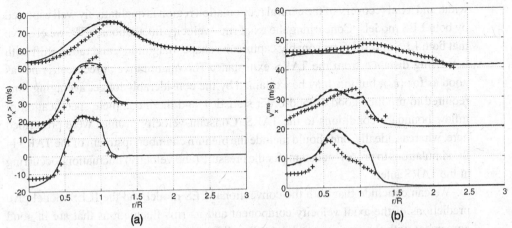

Figure 16.8. Flow in a multiswirl gas turbine combustor: (a) Time-averaged axial velocity component at three cross sections, $x/D = 0.050$, 0.125, and 0.625, downstream of the TARS exit plane; and (b) the corresponding rms axial velocity fluctuations. Note that the curves are shifted 20 and 30 units in the vertical direction to facilitate comparison. Legend: $+$, experimental data; ———, LES using the MM; and $---$, ILES using FCT.

the shape of the conical section just upstream of the TARS exit plane; and it interacts with the boundary layer. Occasionally, it separates from the wall and deflects somewhat toward the centerline of the combustor. The entire flowfield at the TARS exit plane is swirling and strongly three dimensional, with most of the swirling resulting from the outermost annulus. Moreover, the flow in the TARS exit plane contains very strong imprints of the TARS geometry, with eight high-speed regions resulting from the eight swirl vanes in the outer swirler. Further downstream the annular jet breaks up, creating a complex three-dimensional flowfield. Figure 16.7(b) shows the instantaneous velocity vector field at the centerplane of the TARS. The formation of the annular jet is clearly identified, as is the complex three-dimensional nature of the flow, including the unsteady nature of the central recirculation region, in which the velocity fluctuations appear to be rather intense. We observe the highest-velocity fluctuations in the shear layer between the swirling annular jet and the central recirculation region. In the inner annulus the velocity field shows distinct imprints of the four swirlers situated at the location of the smallest cross section.

Figure 16.8 shows comparisons of the first- and second-order statistical moments of the axial velocity component, that is, $\langle v_x \rangle$ and $v_x^{\mathrm{rms}} = \sqrt{\langle (v_x - \langle v_x \rangle)^2 \rangle}$, along three lines across the combustor at $x/D = 0.050, 0.125$, and 0.625, respectively. In Figures 16.8(a) and 16.8(b) the curves are shifted 30 and 20 units in the vertical direction, respectively, to facilitate the comparison. For the time-averaged axial velocity, $\langle v_x \rangle$, in Figure 16.8(a), the agreement among LES, ILES, and the experimental data is satisfactory. The LES and ILES predictions are very close to each other, with ILES producing somewhat broader profiles with peak values that are a few percent lower than those of the LES. Both methods correctly capture the radial shift of the peak values with increasing distance from the exit plane of the TARS and the associated decrease in $\langle v_x \rangle$. Both LES models indicate a stronger recirculation region than is found in the experiments, whereas the width and axial recovery of the central recirculation region is qualitatively well captured by both LES models. Concerning the axial rms–velocity fluctuations, v_x^{rms}, we observe that both LES and ILES reasonably capture the shape, magnitude, and radial shift with increasing distance from the TARS exit plane. The quantitative agreement is not as good as for $\langle v_x \rangle$, but this may be explained by the considerably longer averaging time required to produce good second-order statistics, and the incomplete specification of inflow boundary conditions to the TARS. Constant-velocity profiles were prescribed here, whereas, ideally, one should include the plenum chamber upstream of the TARS in the simulations to properly account for the pressure (and velocity) fluctuations occurring at the TARS inlet.

We can conclude that both the conventional LES model and the ILES model give predictions of the axial velocity component and its rms fluctuations that are in good agreement with the experimental data. The differences between these models are also found to be small. Moreover, the LES–ILES calculations are in better agreement with the experimental data than are the RANS model predictions. Concerning the differences between LES–ILES calculations with and without (Grinstein et al. 2002) considering the TARS within the actual geometry of computational domain, we find that the decrease of the axial velocity in the annulus jet as well as the recovery of the central recirculation region are predicted more accurately when it is included. Coherent structures created inside of the TARS, strongly affect the dynamics of the flow downstream of the TARS exit plane and in the combustor. Methods of replacing the LES in the air–fuel mixing–injectors by simpler models, thus reducing the overall cost of the simulations, are currently being looked into by many research groups. The most promising methods include (a) hybrid RANS–LES models, with RANS being performed in the air–fuel mixing-injectors, and (b) LES carried out within the combustor, and methods based on using linear stochastic estimation or proper orthogonal decomposition of LES or experimental data (Bonnet, Delville, and Glauser 2003; Verfaillie et al. 2006) from separate studies of the air–fuel mixing–injectors to prescribe turbulent inflow boundary conditions for LES within the combustor. These approaches, however, neglect unavoidable couplings across the air-fuel mixing–injectors exit plane, and will thus likely not be able to fully replace whole-system LES studies – particularly when it comes to predicting acoustical or combustion instabilities.

16.4 Flow in an emulated solid rocket motor

Injection-driven flows are of great interest when considering solid-propellant rocket motors (SRMs), as the mass injection resembles the gas injection from a burning solid-propellant surface. Rocket motors emulated in this way show essential flow characteristics found in real motors and thus constitute a useful instrument for both experimental and computational studies of SRM internal ballistics. Of particular interest in such studies is the time-averaged flow and turbulence intensity instabilities, as they are known to be related to the erosive burning process (King 1982; Traineau and Kuentzmann 1984) – a phenomenon whose effects can be dramatic and even disastrous.

Figure 16.9. Emulated SRM flow: Schematic of the emulated rocket-motor geometry and (coarse) grid used in this study.

Many experimental and computational investigations of the injection-driven flow in nozzleless SRMs have been conducted over the years. For a survey of the most distinguished work in this area until 1994, see Liou and Lien (1995) and references therein. Later studies have almost exclusively been on the numerical side and mainly focused on the experimental setup in Traineau, Hervat, and Kuentzmann (1986). The study by Liou, Lien, and Hwang (1998) considered two-dimensional LES without explicit SGS models, letting the numerical dissipation of a modified Godunov scheme act as an implicit (or built-in) SGS model, to study mean quantities and the time-averaged flow and turbulence intensity instabilities. The configuration used was that of Traineau et al. (1986). That work was extended by Liou et al. (1998) to also include the effects of viscosity, compressibility, and inflow turbulence on the flow instabilities. It also included the effect of adding an explicit SGS model – a Smagorinsky (SMG) model extended to compressible flows with a van Driest type of wall damping.

More recently, great effort has been devoted to this problem in the studies of Apte and Yang (2001, 2003, 2004) in order to gain detailed insights in the unsteady flow in the laboratory rig of Traineau et al. (1986). An early study by Apte and Yang (2001) used a two-dimensional LES framework and focused on transition to turbulence and accuracy of the simulation, whereas a later one (Apte and Yang 2003) addressed the effects on energy-exchange mechanisms by the imposition of periodic excitations at different amplitudes and frequencies. In Apte and Yang (2004), a three-dimensional LES, employing a dynamic version of the Smagorinsky model (DSMG) for compressible flows (Moin et al. 1991), yielded very good agreement with the experimental results.

16.4.1 The cold flow case of Traineau et al.

The emulated rocket-motor configuration used in this study is the same as that in the laboratory study of Traineau et al. (1986), and presented schematically in Figure 16.9. The rectangular SRM chamber has an overall length of 0.48 m, a half-height of $h = 0.01$ m, and a width of 0.04 m. The expansion section is 0.032 m long and has a divergence angle of 15°. Air with a temperature of 260 K at a pressure of 314.2 kPa is injected through the upper and lower (porous) walls with a mean mass flux of 13 kg/m^2 s, rendering a mean injection velocity of 3.1 m/s. To facilitate the description of the experimental and LES results, we introduce a coordinate system, $Ox_1x_2x_3$, with x_1, x_2, and x_3 denoting the coordinates in the streamwise, vertical, and spanwise directions, respectively. The origin is placed in the centroid of the head-end wall.

A complication of this laboratory rig, when we are trying to construct numerical simulation models, is that the inflow perturbations caused by the injection through the porous walls are not negligible. We can quantify this in terms of the statistical quantity, $\sigma_w = \sqrt{\langle v_2'^2 \rangle / \langle v_2 \rangle^2}$, where the angle brackets denote a temporal average, the prime denotes a fluctuation, and v_2 is the vertical (or wall-normal) component of the velocity, denoted as the pseudoturbulence. From the experimental study it was found that the inflow perturbations, in terms of pseudoturbulence $\sigma_w \approx 1$, were involved, a value about one order of magnitude larger than that proposed by Beddini (1986).

In the experiments, three successive regimes of flow development were found. In the first regime, $x_1/h < 20$, the mean streamwise velocity profiles are almost equal to the incompressible profiles; and the mean turbulent intensity decreases with increasing x_1/h. This corresponds to the first two regimes in Beddini (1986), and the pseudoturbulence effects appear dominant in this region. In the second regime, $20 < x_1/h < 30$, the mean streamwise velocity profiles start to deviate more and more from the incompressible profiles, whereas the mean turbulent intensity increases substantially; this is similar to the third regime suggested in Beddini (1986). In the last regime, $x_1/h > 30$, the compressibility effects are predominant and the mean streamwise velocity profiles are further flattened and the mean turbulent intensity decreases, while the vertical location of the peak in mean turbulent intensity remains almost the same until the throat is reached.

16.4.2 The computational approach

The fluid dynamic model we use is the NSE for compressible flows, describing conservation of mass, momentum, and energy of a Newtonian fluid obeying Fourier's law of thermal conduction and the ideal gas law. In Chapters 3, 4a, and 5, compressible LES and ILES models were introduced, and here we just summarize the specifics used for the SRM simulations. A finite-volume discretization of the filtered compressible LES equations and the raw compressible NSE results in the following modified equations (Chapter 4a):

$$\partial_t(\bar{\rho}) + \nabla \cdot (\bar{\rho}\,\tilde{\mathbf{v}}) = \tau^\rho + \mathbf{m}^\rho,$$

$$\partial_t(\bar{\rho}\,\tilde{\mathbf{v}}) + \nabla \cdot (\bar{\rho}\,\tilde{\mathbf{v}} \otimes \tilde{\mathbf{v}}) = -\nabla \bar{p} + \nabla \cdot (\bar{\mathbf{S}} - \mathbf{B}) + t^\upsilon + \mathbf{m}^\upsilon,$$

$$\partial_t(\bar{\rho}\,\tilde{e}) + \nabla \cdot (\bar{\rho}\,\tilde{\mathbf{v}}\tilde{e}) = \nabla \cdot (\bar{\mathbf{h}} - \mathbf{b}) + \bar{\mathbf{S}} \cdot \tilde{\mathbf{D}} + \bar{\rho}\varepsilon + \tau^e + \mathbf{m}^e. \qquad (16.2)$$

Here $\mathbf{m}$ represents the commutation error terms, $\boldsymbol{\tau}$ the leading-order truncation errors, $\mathbf{B}$ and $\mathbf{b}$ the SGS stress and flux terms, respectively, and ε the SGS dissipation. These equations are closed here by one of the following explicit SGS models.

1. The first is the DSMG (Moin et al. 1991), in which the SGS stress tensor and flux vector are modeled by $\mathbf{B} = \frac{2}{3}\bar{\rho}k\mathbf{I} - 2\mu_k\tilde{\mathbf{D}}_D$ and $\mathbf{b} = -\mu_k\nabla\tilde{e}$, where $\tilde{\mathbf{D}}_D$ is the

deviatoric part of the rate-of-strain tensor, $k = c_I \Delta^2 \|\mathbf{D}\|^2$ is the SGS kinetic energy, and $\mu_k = \bar{\rho} c_D \Delta^2 \|\mathbf{D}\|$ the SGS viscosity. We evaluate the model coefficients by using a dynamic approach (Zang, Street, and Koseff 1993), with "clipping" of the SGS viscosity both from below and from above.

2. The second is the MM, which here is a combination of the scale-similarity model of Bardina et al. (1980), $\mathbf{B} = \bar{\rho}(\widetilde{\bar{\mathbf{v}} \otimes \bar{\mathbf{v}}} - \bar{\bar{\mathbf{v}}} \otimes \bar{\bar{\mathbf{v}}})$, and $\mathbf{b} = \bar{\rho}(\widetilde{\bar{v}\bar{e}} - \bar{\bar{v}}\bar{\bar{e}})$, observed not to be sufficiently dissipative, and a second-order-accurate monotone convection algorithm, formulated in terms of a flux limiter, Ψ, thus hybridizing a high-order and a low-order (dispersion-free) convection algorithm. Here we have used the monotone (non-total-variation-diminishing) gamma limiter (Jasak, Weller, and Gosman 1999).

3. The third is the ILES, $\mathbf{B} = \rho[\mathbf{C}(\nabla \mathbf{v})^T + (\nabla \mathbf{v})\mathbf{C}^T + \chi^2(\nabla \mathbf{v})\mathbf{d} \otimes (\nabla \mathbf{v})\mathbf{d}]$ and $\mathbf{b} = \rho[\mathbf{C}\nabla e + \chi^2(\nabla e \cdot \mathbf{d})(\nabla \mathbf{v})\mathbf{d}]$ where $\mathbf{C} = \chi(\mathbf{v} \otimes \mathbf{d})$ and $\chi = \frac{1}{2}(1 - \Psi)$ $(\beta^- - \beta^+)$ with Ψ being the flux limiter and $\beta = \beta(\Psi)$. Different flux limiters are discussed in Chapter 4a, including the van Leer limiter (van Leer 1974) and the gamma limiter (Jasak et al. 1999).

We do the spatial discretization by using a second-order-accurate finite-volume method (see Section 16.4 and Hirsch 1984), using hexahedral control volumes (CVs) for the decomposition of the computational domain and a cell-centered, co-located variable arrangement. The time integration uses the four-stage Runge–Kutta scheme by Jameson (e.g., Hirsch 1984), although with the difference that the dissipative terms are calculated at each stage.

16.4.3 Computational details

We discretize the SRM by using two grids with $255 \times 128 \times 32$ and $321 \times 160 \times 40$ CVs in the streamwise, vertical, and spanwise directions, respectively. We cluster the grids toward the injection boundaries, toward the $x_2 x_3$ plane located at the head-end end of the rectangular SRM chamber, and toward the upper and lower walls in the expansion section as well as toward the vertical walls. Moreover, we stretch the grids in other regions with the aim of keeping moderate aspect ratios. The resolution at the injection boundaries given in dimensionless wall units ranges from about 0.05 to 1 in the coarse grid and from about 0.01 to 1 in the fine grid – increasing from the head-end area toward the expansion section. As a result of the surface transpiration (and the consequently smaller wall shear stress), the first cell need not be located as close to the wall as $x_2^+ \approx 1$, which is the recommended value for channel-flow LES. It should thus be possible to use fewer cells in the vertical direction, thereby reducing the computational cost.

We divide the computational boundary into three primary parts: we use no-slip, adiabatic conditions for the head end, the vertical walls, and the upper and lower walls in the expansion section; the conditions for the upper and lower inlets are as already specified. More specifically, we use constant-temperature and zero-gradient pressure conditions, giving the density by means of the ideal gas law. The time-varying inlet

Table 16.1. *Models and meshes for the SRM computations*

Case	Model	Grid
1	LES, MM	coarse
2	LES, MM	fine
3	LES, DSMG	coarse
4	ILES, gamma	coarse
5	ILES, van Leer	coarse

velocity is then given by the prescribed mean mass flux. We use extrapolated values for all variables at the (supersonic) outlet.

We performed simulations both with and without perturbation of the inflow velocity flux, $\sigma_w = 0$, corresponding to the idealized burning of a smooth, nonporous solid propellant, and with perturbation of the inflow flux, $\sigma_w = 0.9$, thus emulating the laboratory configuration, and a more realistic solid propellant. In these simulations, we consider the vertical velocity fluctuations, $v_2'^2$, to be a normally distributed stochastic variable, which in the present study we compute by means of the Box–Müller method (Box and Müller 1958). We compute the uniformly distributed random numbers used in the Box–Müller method with the "Mersenne Twister" (Matsumoto and Nishimura 1998).

We initiate the simulations with the pressure, velocity, and temperature in the rectangular SRM chamber chosen in accordance with the simplified one-dimensional analysis of Traineau et al. (1986), assuming that $\rho = 2.2$ kg/m^3 at the throat together with zero vertical and spanwise velocity components. The initial conditions in the expansion section are, for simplicity, supposed to vary linearly from the throat section to the end of the expansion.

The models and grids used for the computational cases are given in Table 16.1 and the time steps for the coarse and fine grids were 50 ns and 40 ns, respectively; this yielded maximum Courant numbers lower than 0.3 for the whole duration of the simulations. After transition to turbulence, a fairly long time of flow evolution and development is required for the flow to reach a state with stable oscillations in the flow. We observed this by monitoring the velocities in a few "virtual measuring probes." By averaging the mean streamwise velocity in the Ox_1x_2 plane, restricted to the section downstream of the mean velocity transition, one can estimate flow-through times, $\tau_{0.0} = 1.1$ ms and $\tau_{0.9} = 1.7$ ms, for the two levels of pseudoturbulence, respectively. As the sampling times of most of the simulations are rather short, only statistical properties based on mean velocities are shown here. To obtain reliable results for higher-order statistical moments for all simulations, longer sampling times are required.

16.4.4 Results and discussion

We first consider the main flow features, focusing on the vorticity, both for the $\sigma_w = 0.0$ and $\sigma_w = 0.9$ cases. Figure 16.10 shows contours of the instantaneous spanwise

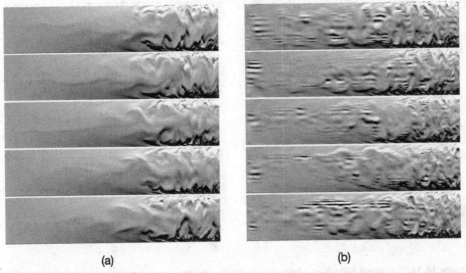

(a)   (b)

Figure 16.10. Emulated SRM flow: Spanwise vorticity, ω_3, on the coarse mesh in $x_1 x_2$ planes located at (from top to bottom) $x_3/h = -1.0$, -0.5, 0.0, 0.5, and 1.0 for (a) $\sigma_w = 0.0$ and (b) $\sigma_w = 0.9$. The panels are magnified five times in the vertical direction and the interval is from -2.5×10^4 s^{-1} (black) to 2.5×10^4 s^{-1} (white).

vorticity, $\omega_3 = \frac{1}{2}(\nabla \times \mathbf{v})_3$, predicted by the MM LES on the coarse grid in various $x_1 x_2$ planes for $\sigma_w = 0.0$ and 0.9, respectively. The figure parts have, for clarity, been magnified five times in the vertical direction. The vortex shedding from boundary separation in the head-end region and the subsequent stretching and breakup of these vortices is clearly noticed for the $\sigma_w = 0.0$ case. They are also seen to strongly interact both with the coherent vortex structures emanating from the injection surface in the midsection of the chamber, with the roll-up vortex structures that are present further downstream, and with the turbulent flow that has developed in the throat region. However, for $\sigma_w = 0.9$, the vortex shedding in the head-end region is not as localized in the vertical direction. Also visible is that the transition region is further upstream in this case, which is fully in line with the previous studies of Liou et al. (1998) and Apte and Yang (2001, 2003). In Figure 16.11 the vertical vorticity $\omega_2 = \frac{1}{2}(\nabla \times \mathbf{v})_2$ as predicted by the MM LES on the coarse grid in various $x_1 x_3$ planes is shown for both $\sigma_w = 0.0$ and 0.9, respectively. As in Figure 16.10, we see that the case with $\sigma_w = 0.9$ has more turbulent structures farther upstream than the case with $\sigma_w = 0.0$. The most notable difference, however, is the distinct corner vortices clearly seen in the upper and lower panels of Figure 16.10(b) and their interaction with the boundary layer on the (noninjecting) back and front walls.

Figure 16.12(a) shows the time-averaged and spanwise-averaged streamwise velocity component, $\langle v_1 \rangle$, for $\sigma_w = 0.9$ at $x_2/h = 0$ and $x_2/h = 0.9$, respectively. We find that the mean velocities, despite the somewhat short sampling time, are reasonably well predicted by the MM LES model as compared with the experimental data. Figure 16.12(a) and 16.12(c) show normalized time-averaged and spanwise-averaged streamwise velocity components at some different streamwise locations for

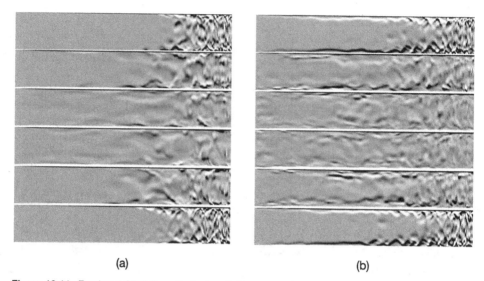

(a)   (b)

Figure 16.11. Emulated SRM flow: Vertical vorticity, ω_2, on the coarse mesh in x_1 x_3 planes located at (from top to bottom) $x_2/h = -0.9, -0.6, -0.3, 0.3, 0.6,$ and 0.9 for (a) $\sigma_w = 0.0$ and (b) $\sigma_w = 0.9$; the interval is from $-2.5 \times 10^4 \text{ s}^{-1}$ (black) to $2.5 \times 10^4 \text{ s}^{-1}$ (white).

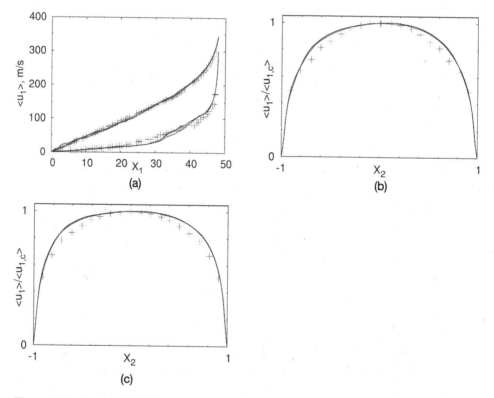

Figure 16.12. Emulated SRM flow: (a) Mean streamwise velocity, $\langle v_1 \rangle$, averaged over $x_3/h \in [-1, 1]$ for $\sigma_w = 0.9$ at $x_2/h = 0$ (upper) and $x_2/h = 0.9$ (lower). Mean streamwise velocity, $\langle v_1 \rangle$, normalized by the streamwise velocity at the centerline, denoted by $\langle v_{1,c} \rangle$, for $\sigma_w = 0.9$ at different streamwise locations: (b) $x_1/h = 38.0$, (c) $x_1/h = 47.0$ on the coarse mesh. Legend: +, experiment; $-----$, MM LES model (coarse mesh); and $\longrightarrow$, MM LES model (fine mesh).

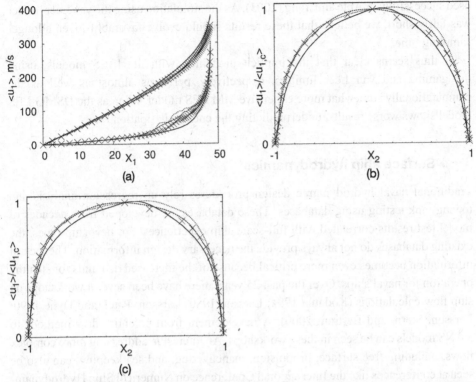

Figure 16.13. Emulated SRM flow: (a) Mean streamwise velocity, $\langle v_1 \rangle$, averaged over $x_3/h \in [-1, 1]$ for $\sigma_w = 0.0$ at $x_2/h = 0$ (upper) and $x_2/h = 0.9$ (lower). Mean streamwise velocity, $\langle v_1 \rangle$, normalized by the streamwise velocity at the centerline, denoted by $\langle v_{1,c} \rangle$, for $\sigma_w = 0.0$ at different streamwise locations: (b) $x_2/h = 38.0$ and (c) $x_2/h = 47.0$ on the coarse mesh. Legend: ×, MM LES: ·····, ILES with van Leer, – – – –, ILES with gamma; and ——, DSMG LES.

$\sigma_w = 0.9$. The measured profiles are well predicted on both grids. The asymmetry around $x_2/h = 0$ indicates that a longer sampling time would be beneficial. Moreover, by comparing these mean velocities with those at times approximately $0.5\tau_{0.9}$ earlier, we find that the simulation results continue to get closer to the experimental results with increasing averaging time.

As the differences between the predictions on the coarse and fine meshes are so small, it also seems evident that grid independence is almost reached. Being fairly confident that the MM LES model performs reasonably well, at least as long as the first-order statistical moments of the velocity are considered, we bridge to the $\sigma_w = 0.9$ cases, using the results from the coarse-grid MM LES (Case 1a) as reference for the calculations with the other LES and ILES models. As we see in Figure 16.13, both ILES models, using the gamma and van Leer limiters, produce results close to the reference model prediction, whereas the results from the DSMG LES model deviate substantially, most notably in the interval $x_1/h \approx 36–46$. For $x_2/h = 0.9$, however, none of the models seem to capture the behavior of the reference solution for $x_1/h \approx 36–44$. The reason for these deviations is not clear, but for the DSMG model the clipping procedure could possibly be the origin, as this model, without clipping, was used successfully in Apte and Yang (2003). As the simulation using the van Leer limiter was fairly short, we believe that these results should evolve favorably, given a longer sampling time.

It thus seems clear that the implicit modeling, with the ILES models, using the gamma and van Leer limiters, respectively, performs almost as well as the computationally somewhat more expensive MM LES model, whereas the DSMG LES model shows worse results, underpredicting the energy dissipation rate.

16.5 Surface ship hydrodynamics

Traditional naval hydrodynamic design procedures rely on regression analysis and towing-tank testing using databases. These databases are developed from decades of model test results correlated with full-scale ship experiences. For new ship types, the existing databases do not always provide the necessary design information. The lack of information becomes even more critical because of the increased demands for stealthy operation for naval ships. Over the past 25 years there have been several workshops on ship flow calculations (Kodama 1994; Larsson 1980: Larsson, Patel, and Dyne 1990; Larsson, Stern, and Bertram 2000). A development from potential flow methods to RANS models can be seen in these workshops. An interest in addressing more complex flows, including free surface, propulsion, maneuvering, and sea keeping, can also be seen at conferences like the International Conference on Numerical Ship Hydrodynamics (2003) and the Symposium on Naval Hydrodynamics (2004). Still, there is much to be done before most physical phenomena are accurately calculated or modeled.

The ability to accurately predict the wave pattern of ships and submarines is of importance for drag assessments, signatures, and secondary effects such as coastal erosion. Furthermore, the behavior of the free surface affects the flow around ships and submerged vessels near the surface. Their motion generates a disturbance on the free surface that is important for the understanding of the flow around these vessels. The free surface affects the location and magnitude of vortices produced by the motion of the vessel. Vortices originating from the bow, appendages, propulsors, and rudders may affect the performance of the propulsion system.

The physical model for surface ship hydrodynamics consists of the incompressible NSE, modified to accommodate a variable density and a separate model for handling the (usually thin) interface between the water and the air. We solve the governing equations here by using LES in combination with an interface-capturing algorithm in a finite-volume framework. This enables simultaneous computation of the air and water flow around the ship, and it includes the evolution of the water surface resulting from the pressure disturbance caused by the ship's motion – the Kelvin waves. The ability to compute such flowfields by use of LES gives the possibility of providing assessments of design parameters, and enables us to study the turbulent wake behind the hull. A ship that moves through the water not only creates the detectable surface wave pattern but also a persistent trace of turbulence, vortices, and waves underneath the sea surface.

This viscous wake, and its interaction with the Kelvin wave pattern, creates a distinct trace for up to 20 km astern of the ship, which is of significant military and civil interest.

In this study we use the 5415 model hull (http://www50.dt.navy.mil/5415/5415n.html) fabricated at the DTMB, Naval Surface Warfare Center, Carderock Division, as an early design variant of the DDG51 Arleigh Burke-class destroyer currently deployed in the U.S. Naval fleet. This hull thus represents a modern surface combatant with transom stern and a sonar dome configuration below the bow. For this particular hull configuration, several experimental studies have been made (e.g., Larsson et al. 2000), which makes it highly suitable for validation of CFD models and verification of numerical results. However, there are significant differences in the laboratory measurement data produced at different towing-tank facilities. The experiments cover several Froude numbers, $\text{Fr} = |\mathbf{v}|^2/(|\mathbf{g}|L)$, but we consider only $\text{Fr} = 0.28$ in this study to avoid breaking waves and their inherited difficulties. This problem will, however, be addressed in future studies.

16.5.1 The computational model

For surface ship flow, the computational model must be able to handle both the flow, described by the NSE, and the motion of the free surface that separates air and water. To model such interfaces several approaches exist that can be divided into either interface-capturing or interface-tracking methods. Here, we use an interface-capturing volume of fluids (VOF) method (Ubbink 1997), in which a passive scalar, α, representing the volume fraction of one phase with respect to the other, is convected with the flow by a separate equation. A finite-volume disctretization of the filtered equations or the raw equations results in the following modified equations:

$$\partial_t(\rho\bar{\mathbf{v}}) + \nabla \cdot (\rho\bar{\mathbf{v}} \otimes \bar{\mathbf{v}}) = -\nabla\bar{p} + \nabla \cdot (\bar{\mathbf{S}} - \mathbf{B}) + \rho\mathbf{f} + \tau^v + \mathbf{m}^v, \quad \nabla \cdot \bar{\mathbf{v}} = 0,$$

$$\partial_t(\alpha) + \nabla \cdot (\alpha\bar{\mathbf{v}}) = \tau^\alpha + \mathbf{m}^\alpha, \tag{16.3}$$

where $\mathbf{v}$ is the velocity, p is the pressure, $\mathbf{m}$ is the commutation error, τ is the truncation error, and $\mathbf{B}$ is the (explicit or implicit) SGS stress tensor that is modeled by use of either

- the MM, $\mathbf{B} = \overline{\bar{\mathbf{v}} \otimes \bar{\mathbf{v}}} - \bar{\bar{\mathbf{v}}} \otimes \bar{\bar{\mathbf{v}}} - 2\nu_k\overline{\mathbf{D}}$ (Bardina et al. 1980), using the OEEVM (Schumann 1975) to compute the SGS viscosity, ν_k, or
- the ILES, $\mathbf{B} = \mathbf{C}(\nabla\mathbf{v})^T + (\nabla\mathbf{v})\mathbf{C}^T + \chi^2(\nabla\mathbf{v})\mathbf{d} \otimes (\nabla\mathbf{v})\mathbf{d}$ (Boris et al. 1992) in which $\mathbf{C} = \chi(\mathbf{v} \otimes \mathbf{d})$ and $\chi = \frac{1}{2}(1 - \Psi)(\beta^- - \beta^+)$, where Ψ is the flux limiter related to the flux reconstruction of the convective terms (see also Chapters 4a).

Moreover, $\rho = \alpha\rho_1 + (1 - \alpha)\rho_2$ is the density: $\mathbf{S} = 2\mu\mathbf{D}$ is the viscous stress tensor, where $\mu = \alpha\mu_1 + (1 - \alpha)\mu_2$ is the viscosity and $\mathbf{f}$ the surface tension, which usually

Figure 16.14. Surface ship flow: Side view of the DTMB 5415 model hull geometry.

can be neglected in ship applications. Here, indices 1 and 2 refer to the water and the air phase, respectively.

We combine the discretized continuity and momentum equations to form a Poisson equation for the pressure, whereas we use a specific reconstruction algorithm for the convective term in the α equation. The CICSAM scheme (Ubbink 1997) is based on the donor–acceptor flux formulation that forms the basis for compressive differencing, whereby downwind data are included through the use of nonlinear flux functions (Lafaurie et al. 1994). CICSAM determines the total flux by merging fluxes from two high-resolution schemes to comply with the convective boundedness criterion (CBC), the hyper-C (HC) scheme (Gaskel and Lau 1988), and the ultimate-quickest (UQ) scheme (Leonard 1991). This hybridization is accomplished by a second nonlinear flux limiter, Γ, that depends on the local orientation of the interface and the direction of flow, given the flux functions α_f^{HC} and α_f^{UQ}, from the HC and UQ algorithms, respectively. To enforce monotonicity and positivity in multiple dimensions, we use directional operator splitting and obtain the total flux by summation of sweeps. Finally, we solve the scalar equations sequentially, with iteration over the coupling terms, and we handle the pressure–velocity coupling with a PISO-type procedure.

16.5.2 Computational configuration

In the present study we place the DTMB 5415 hull (shown in Figure 16.14), which has an overall length of $L = 5.72$ m, in a 26.0 m long, 12.0 m wide, and 3.0 m deep computational domain with 1.0 m of air above the water surfaces, emulating the facility in which the laboratory experiments were performed. We use two computational models: a half-hull model, with a symmetry plane along the centerline, and a full-hull model. For both models we place the hull with the forward perpendicular (FP) at $x = 3.0$ m from the inflow plane, whereas we apply trim and sinkage according to the DTMB 5415 measurements, for a model velocity of 4.01 knots. For these conditions the numerically computed wetted surface, S_{DWL}, deviates only 2% from the experimental value. We use physical values for densities and viscosities for both water and air, which yields Re $=$ 12.6×10^6, based on the length of the hull, the freestream velocity, and the viscosity.

We specify the velocity and the undisturbed free surface location at the inflow plane, and we use Neumann conditions for all other quantities except the pressure, which we hold constant on the outflow plane. On the hull, we use no-slip conditions, whereas we use symmetry conditions on the centerplane ($y = 0$) for the half-hull computations. We use freestream boundary conditions at the lateral outer boundaries as well as on the top and bottom boundaries. A uniform velocity, a constant pressure, and the domain

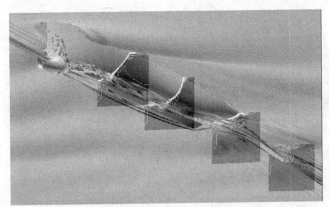

Figure 16.15. Surface ship flow: Subsurface view of the LES flow pattern past the DTMB 5415 hull at Fr = 0.28, showing the surface elevation, streamlines colored with vorticity magnitude, and planes showing normalized axial velocity.

divided into air and water by means of the α field serve as initial conditions from which the flow evolves freely. We discretize the computational domain by using ~3.0 and ~6.0 million hexahedral cells for the full-hull domain. The distance from the hull to the first cell center is such that y^+ ranges between 1 and 30, depending on the local flow velocity; hence, near-wall SGS models (Fureby et al. 2004) will be required.

16.5.3 Results for the towed model LES calculations

Figure 16.15 shows the instantaneous flow around the DTMB 5415 hull. The most prominent feature of the sub-surface flow consists of the developing boundary layer and the coherent vortical structures rolling up just downstream of the widest section of the sonar dome, forming the sonar-dome vortices. The boundary layer thickens with downstream distance from the bow, similarly to that over a flat plate. The curvature of the hull and the associated pressure distribution, however, modify the growth rate. The flat-plate boundary layer is a reasonable model along the parallel midbody section of the hull, but it fails toward the transom stern. The sonar-dome vortices, one on each side of the keel, dominate the flow over the hull and persist throughout the computational domain. Along the hull, the sonar-dome vortices interact with the boundary layer, resulting in redistribution of shear stresses and turbulence intensity, as well as an alteration of the topology and shape of the vortices themselves. The sonar-dome vortices persist far downstream in the wake behind the ship, and they dominate also the near wake, in which these vortical structures slowly lose their intensity and coherence as a result of molecular viscosity, vortex stretching, and turbulence interactions, emulated by the SGS model.

For a clearer understanding of the predominant flow features in the near wake as well as the progress of the wake, a perspective view of the stern is shown in Figure 16.16. In this figure the evolution of the flowfield is shown up to three hull lengths downstream of the stern, visualized by streamlines released randomly on a virtual surface at $y^+ \approx 10$

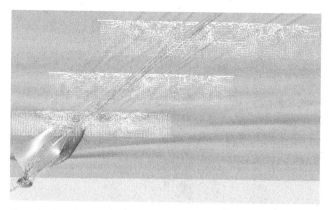

Figure 16.16. Surface ship flow: Perspective view of the stern of the DTMB 5415 hull at Fr = 0.28, showing randomly released streamlines and three planes with secondary velocity vectors showing the vortical structures in the wake.

along the hull. Furthermore, three crossflow planes with secondary velocity vectors are shown to visualize the evolution and broadening of the wake. Close to the hull, emanating from the sonar dome, are again the sonar-dome vortices, but in addition to these, the flow in Figure 16.16 indicates the presence of other vortical structures closer to the hull. These are best represented on the starboard side, where they are seen being formed underneath the hull and intensified at the rising of the stern. These vortical structures are also visible close to the hull in Figure 16.16. In addition, along the ship sides close to the free surface, strong vortices are created and swept toward the centerline at the transom stern. The recirculation zone behind the transom stern generates turbulence and vortical structures, which interact with the free surface in the wake. All these entangled structures sum up to create the complexity of the wake and result in long-lasting structures far behind the hull.

Of significant importance to the flow around a surface ship is the dynamics and shape of the free surface. It is a sharp interface between air and water, where we have a sudden change in density and viscosity. The location of this is not known beforehand and must be computed along with the dynamics of the flow. Hence, it is a challenge to track the water surface accurately. The predictions herein for the wave patterns on the hull and for wave cuts close to the hull are in reasonable agreement with measurement data. The grid coarsens rapidly outside of the hull boundary layer and the grid spacing becomes close to the natural wavelength of the Kelvin wave pattern, which smear out the solution. Typically, a resolution of 20 to 40 cells over the wavelength of the Kelvin waves is needed to accurately capture the shape and behavior of the surface wave pattern. In addition, for the surface waves, time plays an important role since the time-accurate solution produced by LES requires time to transport information through the domain. Thus, the Kelvin wave pattern gradually builds up, starting with the bow and stern waves.

In Figure 16.17, we compare the predicted wave pattern with experimental data by using an isosurface of the volume fraction at $\alpha = 0.5$, colored with height, h/L.

Figure 16.17. Surface ship flow: Comparison of computed (top) and experimental (bottom) wave contours around the DTMB 5415 hull at Fr = 0.28 and Re = 12.6 × 10⁶ for the ILES + WM model.

We observe the smearing of the divergent and stern waves as a result of numerical diffusion, grid stretching, and comparatively large grid spacing, but still the amplitudes and positions of the wave crests close to the hull are in good agreement with data. The difference in wave height also depends on the value of α chosen for the isosurface representation. Because of the constraint on the volume fraction field, the value of α varies from 0 to 1 over three grid cells or equivalently over a distance up to 0.03 m. This is significant considering that the amplitude of the bow wave is only 0.10 m. Note, however, the very good agreement in relative location, phase speed, and shape of the primary waves.

Figure 16.18 shows a comparison of wave profiles at different distances from the hull for the ILES + WM and MM + WM models together with experimental data. On the hull, Figure 16.18(a), the differences between the two models are small, and the most significant difference from the experimental data is observed in the bow region, where all models underpredict the height of the bow wave. Figure 16.18(b) presents the wave

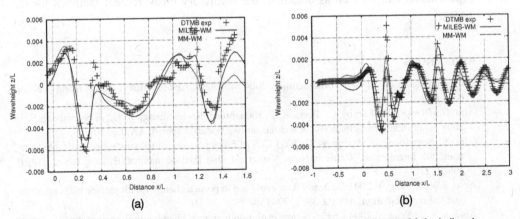

Figure 16.18. Surface ship flow: Experimental and predicted wave cuts along lines at (a) the hull and (b) $y/L = 0.172$, using ILES + WM and MM + WM, respectively, at Fr = 0.28 and Re = 12.6 × 10⁶.

profile at $y/L = 0.172$, the location chosen for comparison in Larsson et al. (2000). The first crest is too high and the second too low. However, the phase and the first wave trough are well predicted. At this distance from the hull the waves are still predicted reasonably well, and the results compare favorably with the best results from the Gothenburg 2000 workshop (Larsson et al. 2000). We designed the grid in the present computations to resolve the boundary layer and the wake, not to resolve the entire wave and flowfield, and thus the wave profile is not expected to reproduce the experimental wave profile far from the hull.

16.6 Concluding remarks

Our aim in this chapter has been to illustrate the successful use of ILES and LES for complex engineering problems. We selected four different flow problems: flow around a submarine, flow in a multiswirl gas turbine combustor, flow in a SRM, and flow around a modern surface combatant. We selected these cases not only to provide insight into the usefulness of these computational methods in terms of increased understanding of the flow physics, but also to provide some qualitative and quantitative comparison with experimental data. The cases selected can typically be broken down in simpler flows, many of which are discussed in Chapters 10 and 11. The overall conclusion is that both LES and ILES are capable of handling such flows – even for rather coarse grids – if appropriate explicit or implicit SGS models are used, and if separate SGS wall models are employed to handle the effects of the unresolved near-wall flow on the outer flow.

Acknowledgments

We gratefully acknowledge financial support for the collection and compilation of data from the Swedish Defense Research Institute (FOI). We thank F. F. Grinstein, E. J. Gutmark, and particularly G. Li for their detailed discussions and sharing of valuable experimental data as well as details of the laboratory measurement facilities for the TARS.

REFERENCES

25th Symposium on Naval Hydrodynamics. 2004 St. John's, Newfoundland, Canada, and previous conferences in this series.

Alin, N., Bensow, R., Fureby, C., Persson, T., Ramamurti, R., Sandberg, W. C., & Svennberg, S. U. 2005 3D unsteady computations for submarine-like bodies. Paper AIAA 05-1104.

Alin, N., Svennberg, U., & Fureby, C. 2003 LES of flows past simplified submarine hulls, presented at the 8th International Conference on Numerical Ship Hydrodynamics, Bussan, Korea, vol. II, 208–222.

Apte, S. V., & Yang, V. 2001 Unsteady flow evolution in porous chamber with surface mass injection, part 1: Free oscillation. *AIAA J.* **39**, 1577–1586.

Apte, S. V., & Yang, V. 2004 Unsteady flow evolution in porous chamber with surface mass injection, part 2: Acoustic exitation. *AIAA J.* **40**, 244–253.

Apte, S. V., & Yang, V. 2003 A large-eddy simulation study of transition and flow instability in a porous-walled chamber with mass injection. *J. Fluid Mech.* **477**, 215–225.

Bardina, J., Ferziger, J. H., & Reynolds, W. C. 1980 Improved subgrid scale models for large eddy simulations. Paper AIAA 80-1357.

Beddini, R. A. 1986 Injection-induced flows in porous-walled ducts. *AIAA J.* **24**, 1766–1773.

Bonnet, J. P., Delville, J., & Glauser, M. N. 2003 The generation of realistic 3D, unsteady inlet conditions for LES. Paper AIAA 2003-0065.

Boris, J. P., Grinstein, F. F., Oran, E. S., & Kolbe, R. L. 1992 New insights into large eddy simulation *Fluid Dyn. Res.* **10**, 199–228.

Box, G. E. P., & Müller, M. E. 1958 A note on the generation of random normal deviates. *Ann. Math. Stat.* **29**, 610–611.

Eriksson, L. E. E., Andersson, L., Lindblad, K., & Andersson, N. 2003 Development of a cooled radial flameholder for the F404/RM12 afterburner, in *Proceedings of the AIAA-ISAB Conference.*

Ferziger, J. H. 1983 Higher level simulations of turbulent flows, in *Computational Methods for Turbulent, Transonic and Viscous Flows*, ed. J.A. Essers. Oxford: Oxford University Press.

Fureby, C. 2000 Large eddy simulation of combustion instabilities in a jet-engine afterburner model. *Combust. Sci. Tech.* **161**, 213–243.

Fureby, C., & Grinstein, F. F. 2002 Large eddy simulation of high Reynolds-number free and wall bounded flows. *J. Comput. Phys.* **181**, 68–97.

Fureby, C., Alin, N., Wikström, N., Menon, S., Persson, L., & Svanstedt, N. 2004 On large eddy simulations of high Re-number wall bounded flows. *AIAA. J.* **42**, 457–468.

Fureby, C., Grinstein, F. F., Li, G., & Gutmark, E. J. 2007 An experimental and computational study of a multi-swirl gas turbine combustion, in Proceedings of the Combustion Institute, **31**, 3107–3114.

Gaskel, P. H., & Lau, A. K. C. 1988 Curvature compensated convective transport: SMART, a new boundedness preserving transport algorithm. *Int. J. Numer. Meth. Fluids* **8**, 617–641.

Grinstein, F. F., & Fureby, C. 2004 LES studies of the flow in a swirl gas combustor, in *Proceedings of the 30th International Symposium on Combustion* 1791–1798.

Grinstein, F. F., Young, T. R., Gutmark, E. J., L. G., Hsiao, G., & Mongia, H. C. 2002 Flow dynamics in a swirl combustor. *J. Turb.* **3**, 30, 1–19.

Groves, N. C., Huang, T. T., & Chang, M. S. 1989 *Geometric Characteristics of DARPA SUBOFF Models*. Report DTRC/SHD-1298-01.

Hirsch, C. 1984 *Numerical Computation of Internal and External Flows*. New York: Wiley, vol. 2, 334, 440. http://www50.dt.navy.mil/5415/5415n.html

Huang, T. T., Liu, H-L., Groves, N. C., Forlini, T. J., Blanton, J., & Gowing, S. 1992 Measurements of flows over an axisymmetric body with various appendages (DARPA SUBOFF experiments), in *Proceedings of the 19th Symposium on Naval Hydrodynamics.*

Hura, H. S., Joshi, N. D., Mongia, H. C., & Tonouchi, J. 1998 *Dry Low Emission Premixed CCD Modeling and Validation*. American Society of Mechanical Engineers, Fairfield, NJ: Report ASME-98-GT-444.

Jasak, H., Weller, H. G., & Gosman, A. D. 1999 High resolution NVD differencing scheme for arbitrarily unstructured meshes. *Int. J. Numer. Meth. Fluids* **31**, 431–449.

Kim, W.-W., Menon, S., & Mongia, H. C. 1999 Large-eddy simulation of a gas turbine combustor flow. *Combust. Sci. Tech.* **143**, 25–36.

King, M. K. 1982 Prediction of solid propellant burning rates in nozzleless motors. Paper AIAA 82-1200.

Kodama, Y., ed. 1994 *CFD Workshop Tokyo 1994 Proceedings*. Tokyo Japan: Ship Research Institute.

Lafaurie, B., Nardone, C., Scardovelli, R., Zaleski, S., & Zanetti, G. 1994 Modeling merging and fragmentation in multiphase flows with SURFER. *J. Comput. Phys.* **113**, 134–147.

Landahl, M. T., & Mollo-Christensen E. 1986 *Turbulence and Random Processes*. New York: Cambridge University Press.

Larsson, L., ed. 1980 *SSPA-ITTC Workshop on Ship Boundary Layers, Proceedings*. Göteborg, Sweden: SSPA, Report No. 90.

Larsson, L., Patel, V. C., & Dyne, G., eds. 1990 *Proceedings, 1990 SSPA-CTH-IIHR Workshop on Ship Viscous Flow.* Göteborg, Sweden: Flowtech International AB, Research Report No. 2.

Larsson, L., Stern, F., & Bertram, V., eds. 2000 *Gothenburg 2000. A Workshop on Numerical Ship Hydrodynamics, Proceedings.* Göteborg, Sweden: Chalmers University of Technology; see http://www.iihr.uiowa.edu/gothenburg2000/index.html.

Launder, D. B., & Spalding, D. B. 1972 *Mathematical Models of Turbulence.* London: Academic Press.

Leonard, B. P. 1991 The ULTIMATE conservative differencing scheme applied to unsteady one-dimensional advection. *Comput. Meth. Appl. Mech. Eng.* **88**, 17.

Li, D.-Q. 1994 *Investigation on Propeller-Ruder Interaction by Numerical Methods*, PhD dissertation, Chalmers University of Technology, Göteborg, Sweden.

Li, G., & Gutmark, E. J. 2002a Experimental and numerical studies of the velocity field of a triple annular swirler, International Gas Turbine Institute: Report GT-2002-30069.

Li, G., & Gutmark, E. J. 2002b Flow field measurement of a triple swirler spray combustor, presented at the Joint Propulsion Conference and Exhibit, June, Indianapolis, in, Paper AIAA 2002-4010.

Li, G., & Gutmark, E. J. 2006 Experimental study of boundary conditions effects on non-reacting and reacting flow in a multi-swirl gas turbine combustor. *AIAA. J.*, **44**, 444–56.

Liou, T.-M., & Lien, W.-Y., 1995 Numerical simulations of injection-driven flows in a two-dimensional nozzleless solid-rocket motor. *J. Prop. Power* **11**, 600–606.

Liou, T.-M., Lien, W.-Y., & Hwang, P.-W. 1998 Transition characteristics of flowfield in a simulated solid-rocket motor. *J. Prop. Power* **14**, 282–289.

Matsumoto, M., & Nishimura, T. 1998 Mersenne twister: A 623-dimensionally equidistributed uniform pseudo-random number generator. *ACM Trans. Mod. Comput. Sim.* **8**, 3–30.

Menon, S. 2001 Subgrid combustion modeling for large-eddy simulations of single and two-phase flows, in *Advances in LES of Complex Flows*, eds. R. Friedrich & W. Rodi. Dordrecht: Kluwer.

Minot, P., Ham, F., & Pitsch, H. 2005 Large Eddy simulations of a complex combustor: Cold flow results. Paper AIAA 2005-0151.

Moin, P., Squires, K., Cabot, W., & Lee, S. 1991 A dynamic subgrid-scale model for compressible turbulence and scalar transport. *Phys. Fluids A* **3**, 2746–2757.

Pattenden, R. J., Turnock, S. R., & Bressloff, N. W. 2004 The use of detached-eddy simulation in ship hydrodynamics, presented at the 25th Symposium on Naval Hydrodynamics, St. John's, Newfoundland, Canada., vol. II, 91–103.

Persson, T., Bensow, R., Fureby, C., Alin, N., & Svennberg, U. 2004 Large eddy simulation of the viscous flow around submarine hulls, presented at the *25th Symposium on Naval Hydrodynamics*, St. John's, Newfoundland, Canada.

Pitsch, H., & Duchamp De Lageneste, L. 2002 Large-eddy simulation of premixed turbulent combustion using a level-set approach, in *Proceedings of the 29th International Symposium on Combustion*, 2001–2008.

Pritchard, B. A., Danis, A. M., Foust, M. J., Durbin, M. D., & Mongia, H. C. 2002 *Multiple Annular Combustion Chamber Swirler Having Atomizing Pilot.* European Patent No. EP1193448A2.

Ramamurti, R., Sandberg, W. C., & Löhner, R. 1994 Evaluation of a three dimensional finite element incompressible flow solver. Paper AIAA 1994-0756.

Sagaut, P. 2001 *Large Eddy Simulation for Incompressible Flows.* Heidelberg: Springer-Verlag.

Schonfeld, T., & Poinsot, T. 1999 Initial and boundary conditions for large eddy simulations of combustion instabilities, in *CTR Annual Research Briefs*. Stanford, CA: NASA Ames/Stanford University, 73.

Schumann, U. 1975 Subgrid scale model for finite difference simulation of turbulent flows in plane channels and annuli. *J. Comput. Phys.* **18**, 376–404.

Spalart, P. R., Jou, W.-H., Strelets, M., & Allmaras, S. R. 1997 Comments on the feasibility of LES for wings, and on a hybrid RANS/LES approach, in *Advances in DNS/LES*, ed. C. Liu & Z. Liu. Columbus, OH: Greyden Press.

Svennberg, U., & Lillberg, E. 2004 Large eddy simulation of the viscous flow around a ship hull including free-surface, presented at the *25th Symposium* on Naval Hydrodynamics, St. John's, Newfoundland, Canada, vol. II, 64–95.

Tachmindji, A. J., & Milam, A. B. 1957 The calculation of the circulation distribution for propellers with finite hub having three, four, five and six blades. DTMB: Report No. 1141.

The 8th International Conference on Numerical Ship Hydrodynamics 2003 Bussan, Korea, and previous conferences in this series.

Traineau, J. C., & Kuentzmann, P. 1984 Some measurements of solid propellant burning rates in nozzleless motors. Paper AIAA 84-1469.

Traineau, J. C., Hervat, P., & Kuentzmann, P. 1986 Cold-flow simulation of a two-dimensional nozzleless solid rocket motor. Paper AIAA 86-1447.

Ubbink, O. 1997 *Numerical Prediction of Two Fluid Systems with Sharp Interfaces*. PhD dissertation, London University.

Van, S.-H, Kim, J., Park, I.-R., & Kim, W.-J. 2003 Calculation of turbulent flows around a submarine for the prediction of hydrodynamic performance, presented at the 8th International Conference on Numerical Ship Hydrodynamics, Bussan, Korea.

van Leer, B. 1974 Towards the ultimate conservative differencing scheme V. A second order sequel to Gudonov's method. *J. Comput. Phys.* **32**, 101, vol. II, 12–21.

Verfaillie, S., Gutmark, E. J., Bonnet, J. P., & Grinstein, F. F. 2006 Linear stochastic estimation of a swirling jet, *AIAA J.*, **44**, 457–680.

Yang, C., & Löhner, R. 2003 Prediction of flow over an axisymetric body with appendages, presented at the 8th International Conference on Numerical Ship Hydrodynamics, Bussan, Korea, vol. II, 233–247.

Zang, Y., Street, R. L., & Koseff, J. R. 1993 A dynamic mixed subgrid-scale model and its application to turbulent recirculating flows. *Phys. Fluids* A **5**, 3186–3196.

17 Large-Scale Urban Simulations

Gopal Patnaik, Fernando F. Grinstein, Jay P. Boris,
Ted R. Young, and Oskar Parmhed

17.1 Background

Urban airflow that is accompanied by contaminant transport presents new, extremely
challenging modeling requirements (e.g., Britter and Hanna 2003). Reducing health
risks from the accidental or deliberate release of chemical, biological, or radiologi-
cal (CBR) agents and pollutants from industrial leaks, spills, and fires motivates this
work. Configurations with very complex geometries and unsteady buoyant flow physics
are involved. The widely varying temporal and spatial scales exhaust current modeling capacities. Crucial technical issues include turbulent fluid transport and boundary
condition modeling, and postprocessing of the simulation results for practical use by
responders to actual emergencies.

Relevant physical processes to be simulated include complex building vortex shedding, flows in recirculation zones, and approximating the dynamic subgrid-scale (SGS)
turbulent and stochastic backscatter. The model must also incorporate a consistent
stratified urban boundary layer with realistic wind fluctuations; solar heating, including
shadows from buildings and trees; aerodynamic drag and heat losses that are due to the
presence of trees; surface heat variations; and turbulent heat transport. Because of the
short time spans and large air volumes involved, modeling a pollutant as well mixed
globally is typically not appropriate. It is important to capture the effects of unsteady,
buoyant flow on the evolving pollutant-concentration distributions. In typical urban scenarios, both particulate and gaseous contaminants behave similarly insofar as transport
and dispersion are concerned, so that the contaminant spread can usually be simulated
effectively on the basis of appropriate pollutant tracers with suitable sources and sinks.
In some cases, the full details of multigroup particle distributions are required. Additional physics includes the deposition, resuspension, and evaporation of contaminants.

17.1.1 The established approach: Gaussian plume models

The contaminant plume prediction technology that is currently in use is based
on Gaussian similarity solutions ("puffs"). This class of extended Lagrangian

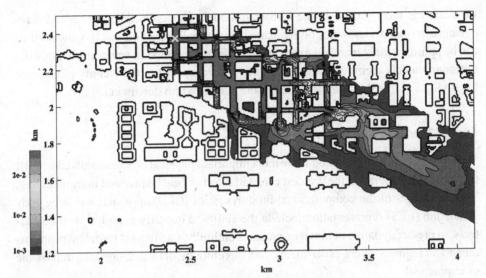

Figure 17.1. FAST3D-CT Washington DC 1-m dosage (kg s m^{-3}) after 12 min. The release sites is marked with a yellow cross. Only a 3-km two portion of the full grid is shown (Pullen et al. 2005).

approximations is only appropriate for large scales and flat terrain. Separated flow and vortex shedding from buildings, cliffs, or mountains is absent. Diffusion is used in puff–plume models to mimic the effects of turbulent dispersion caused by the complex building geometry and by wind gusts of comparable and larger size (e.g., Bauer and Wolski 2001; DTRA 2001; Leone et al. 2001; NOAA 1999).

Detailed comparisons using actual "common use" puff–plume models (e.g., Figures 17.1 and 17.2, from Pullen et al. 2005) show a range of results, depending on how much of the three-dimensional urban boundary layer information from the detailed

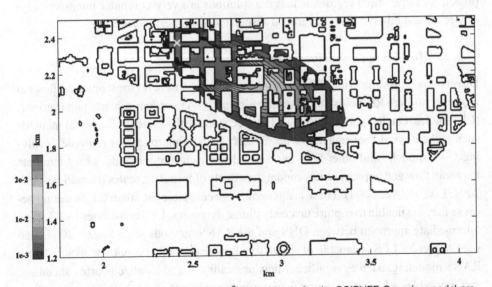

Figure 17.2. Washington DC dosage (kg s m^{-3}) after 12 min for the SCIPUFF Gaussian model prediction (Pullen et al. 2005).

simulation is incorporated in the Gaussian model. Though building-generated aero-dynamic asymmetries cannot be replicated, crosswind spreading and downwind drift can be approximately matched given enough free parameters. In urban areas, however, Gaussian models seem to predict too rapid a lateral spread in the vicinity of a source to provide a plume that is approximately the correct width downwind.

17.1.2 The CFD approach

Since fluid dynamic convection is the most important physical process involved in CBR transport and dispersion, the greatest care and effort should be invested in its modeling. The advantages of the computational fluid dynamics (CFD) approach and large eddy simulation (LES) representation include the ability to quantify complex geometry effects, to predict dynamic nonlinear processes faithfully, and to treat turbulent problems reliably in regimes where experiments, and therefore model validations are impossible or impractical.

17.1.2.1 Standard CFD simulations

Some "time-accurate" flow simulations that attempt to capture the urban geometry and fluid dynamic details are a direct application of standard (aerodynamic) CFD methodology to the urban-scale problem. An example is the finite-element CFD simulations of the dispersion of a contaminant in the Atlanta, Georgia, metropolitan area (Aliabadi and Watts 2002). The model includes topology and terrain data, and a typical mesh contains approximately 200 million nodes and 55 million tetrahedral elements. These are grand-challenge-size calculations and are run on 1024 processors of a CRAY T3E, taking up to a whole day to run. Similar approaches are being used by other research groups (e.g., Camelli and Löhner 2003; Chan 1994). The chief difficulty with this approach for large urban regions is that the solutions are very computer intensive (days or weeks) and involve severe overhead associated with mesh generation.

17.1.2.2 The LES approach for contaminant transport

Direct numerical simulation (DNS) is prohibitively expensive for most practical flows at moderate-to-high Reynolds numbers, and especially so for urban contaminant transport (CT) studies. On the other end of the CFD spectrum are the standard industrial methods such as the Reynolds-averaged Navier–Stokes (RANS) approach, for example, involving $k - \varepsilon$ models, and other first- and second-order closure methods, which simulate the mean flow and approximately model the effects of turbulent scales (Hendricks et al. 2004; Lien and Yee 2004). These are generally unacceptable for urban CT modeling because they are unable to capture unsteady plume dynamics. LES constitutes an effective intermediate approach between DNS and the RANS methods (e.g., Sagaut 2004; also see Chapter 3). LES is capable of simulating flow features that cannot be handled with RANS modeling, such as significant flow unsteadiness and localized vortex shedding, and it provides a higher accuracy than the industrial methods do, at a lower cost than DNS. The main assumptions of LES are these: (i) that transport is largely governed by

Figure 17.3. Contaminant dispersion from an instantaneous release in Times Square, New York City, as predicted by the FAST3D-CT MILES model. The frames show concentrations at 3, 5, 7, and 15 min after release. (See color inset, Plate 20.)

large-scale unsteady convective features that can be resolved; and (ii) that one can undertake the less-demanding accounting of the small-scale flow features by using suitable SGS models. Because the larger-scale unsteady features of the flow govern the unsteady plume dynamics in urban geometries, the LES approximation can capture some key features that the RANS methods and the various Gaussian plume methodologies cannot. Given its potential for computational efficiency, monotone integrated LES (MILES) is ideally suited to CFD-based plume simulation for urban-scale scenarios, an application in which RANS methods are inadequate and classical LES methods are too expensive.

17.2 MILES for urban-scale simulations

The three-dimensional FAST3D-CT MILES model (Boris 2002; Cybyk et al. 1999, 2001) is based on a scalable, low dissipation, fourth-order phase-accurate flux-corrected transport (FCT) convection algorithm (Boris and Book 1973, 1976). Other than changes subsequently described in Section 17.2.7, the specific version of FCT implemented in FAST3D-CT is documented in Boris et al. (1993); also see Chapters 1 and 4a.

A practical example of urban-scale MILES is shown in Figure 17.3, showing contaminant dispersion in Times Square, New York City. The figure demonstrates the typical complex unsteady vertical mixing patterns caused by building vortex and recirculation patterns, and the endangered region associated with this particular release scenario. In particular, the figure depicts the so-called *fountain effect* occurring behind three tall buildings. The fountain effect is the systematic migration of contaminant from ground level up the downwind side of tall buildings followed by continuous ejection into the air flowing over the tops of these buildings. Earlier visualizations of FAST3D-CT simulations of the flow in the Chicago downtown area (2002) also showed this effect. This phenomenon has also been seen in experiments in Los Angeles (Rappolt 2002; Rappolt, private communication, 2004) and has been reported in wind-tunnel studies. It is important because the contaminant can be transported downwind much faster than might otherwise be expected during this process. This effect appears to be driven by arch vortices lying behind the buildings, as in the well-studied problem of flow past a surface-mounted cube (e.g., Hussein and Martinuzzi 1996; also see Chapter 10). To fully characterize this behavior, researchers must conduct further studies addressing issues of building geometry (i.e., aspect ratios) and angle of attack.

As we noted in Section 17.1, FAST3D-CT has models for a number of additional physical processes. The incorporation of specific models for these processes in simulation codes is always a challenge, but it has been accomplished with reasonable fidelity. The primary difficulty is the effective calibration and validation of these physical models, since much of the input needed from field measurements of these processes is typically insufficient or even nonexistent. Further, even though the individual models can all be validated separately, the larger problem of validating the overall simulation code has to be tackled as well.

A typical run with the FAST3D-CT model for a complex urban area of 30 km^2 resolved with 6-m cells takes 24 h on a 20-processor SGI system. This is significantly faster per square kilometer than classical CFD models, because of the savings achieved by MILES as well as other algorithmic improvements. The critical dilemma in the CT application is that unsteady urban-scenario flow simulations are currently feasible – but they are still expensive and require a degree of expertise to perform. First responders and emergency managers on site to cope with contaminant-release threats cannot afford to wait while actual simulations and data postprocessing are carried out.

An operational solution of this problem carries out unsteady CFD simulations in advance and precomputes compressed databases for specific urban areas incorporating suitable assumed weather and a full set of wind conditions and distributed test sources. The relevant information is summarized as Dispersion Nomograf datasets (Boris 2002) so that it can be readily used through portable devices, with sensors providing current observational information regarding local contaminant concentrations and winds. With this new approach, implemented in a system called CT-Analyst (Boris et al. 2002), the accuracy of CFD simulations can be recovered instantly with little loss of fidelity.

17.2.1 Atmospheric boundary layer specification

The planetary boundary layer characterization upstream of the finite computational domain directly affects the boundary-condition prescription required in the simulations.

The weather, time of day, cloud cover, and humidity all determine if the boundary layer is thermally stable or unstable and thus determine the level and structure of velocity fluctuations. Sensitivity studies show that the fluctuating winds affect urban dispersion. The strength of the wind fluctuations along with solar heating are shown to be major determinants of how quickly the contaminant density decreases in time. This in turn is extremely important in emergency applications, because it determines overall dosage.

In FAST3D-CT, the time average of the urban boundary layer is specified analytically with parameters chosen to represent the overall thickness and inflection points that are characteristic of the topography and buildings upstream of the computational domain. These parameters can be determined self-consistently by computations over a wider domain, since the gross features of the urban boundary layer seem to establish themselves in a kilometer or so. However, this increases the cost of simulations considerably.

The important length scales (tens of meters to kilometers) and time scales (seconds to minutes) in wind gusts can be resolved easily by CFD models that accurately resolve the buildings. The gusts therefore, should not be averaged or approximated as steady state. To address this, we include a deterministic model for an evolving realization of these fluctuations as part of the boundary conditions with input parameters to approximate particular atmospheric conditions being simulated. The model we have used is a complicated analytic function defined throughout the computational domain that provides the initial and boundary conditions of a run.

We superimposed three types of motion at several different wavelengths to construct this function. We impressed a coherent shearing motion transverse to the average wind direction, typical of meanders, with a sinusoidal structure. On this we superimposed horizontal pancake vortices at several scales to represent a type of flow that is possible in stratified fluids. The third motion is due to longitudinal vortices with finite vertical and horizontal extent to represent wind-induced hairpin vortices found in typical boundary layers. We made the multiple scales for each of these motions incommensurate, permitting the nonlinear interactions to guarantee an overall chaotic boundary-condition representation. In addition, we included a nonlinear term in all sinusoidal dependences to force a broad spectrum of impressed fluctuations.

The vertical dependence of these resolved-scale fluctuations is a superposition of two functions, one for the unobstructed flow and one to provide additional fluctuations that are due to buildings upstream of the domain. The unobstructed component is largest in the center of the domain away from the ground and below the atmospheric boundary layer. The building component is largest near the tops of the assumed buildings upstream, whose general disposition and height we incorporate in the shape of the average inflow urban boundary layer. When we allow this (turbulent) flowfield to evolve by flowing over 0.5 to 1.0 km of actual city geometry, initial inconsistencies are replaced by a more self-consistent flow. Several research issues remain unresolved in this area, both observationally and computationally. Deterministic (Mayor, Spalart, and Tripoli 2002) and other (Druault et al. 2004) approaches to formulating turbulent inflow boundary conditions are currently being investigated in this context.

17.2.2 Solar heating effects

A ray-tracing algorithm that properly respects the building and tree geometry computes solar heating in FAST3D-CT. The trees and buildings cast shadows depending on the instantaneous angle of the Sun. The geometry database has a land-use variable defining the ground composition. The simulated interaction of these various effects in actual urban scenarios has been illustrated in Boris (2002). The rate that a contaminant is swept out of a city by the winds can vary by a factor of 4 or more as a result of solar heating variations from day to night and as a result of variations in the relative strength of the wind gusts (Boris 2002; Boris et al. 2002).

17.2.3 Tree effects

Although we can resolve individual trees if they are large enough, their effects (i.e., aerodynamic drag, introduction of velocity fluctuations, and heat losses) are represented through modified forest canopy models (Dwyer, Patton, and Shaw 1997), including effects that are due to the presence of foliage. For example, we write an effective drag-force source term for the momentum equations as $\mathbf{F} = -C_d a(z)|\mathbf{v}|\mathbf{v}$, where $C_d = 0.15$ is an isotropic drag coefficient, $a(z)$ is a seasonally adjusted leaf area density, z is the vertical coordinate, and $\mathbf{v}$ is the local velocity.

17.2.4 Turbulent stochastic backscatter

The distribution of SGS viscosity is a distinct feature characterizing the ability of different LES models to capture the underlying unresolved physics, ranging from purely dissipative scalar to tensorial scale-similarity models. An overall positive SGS viscosity implies that energy is transferred from resolvable flow structures toward small, unresolved scales by means of a cascade process (outscatter). Conversely, a negative SGS viscosity implies that energy is overall transferred in the opposite direction by a reverse cascade process, that is, backscatter – which can become important in complex geometries such as these when appreciable turbulent kinetic energy may be present in the *unresolved* scales. Backscatter, both systematic and stochastic, can occur at select wavelengths and for certain nonlinear triads of modes even when the overall cascade corresponds to outscatter. Modeling how the unresolved features of the flow contribute to the large scales through this stochastic backscatter process presents a difficult challenge: How are these effects to be predicted on the basis of the resolvable scale information?

Because of the anisotropic features of the implicit SGS modeling incorporated (Fureby and Grinstein 2002), MILES offers an effective approach for the simulation of the inherently inhomogeneous turbulent flows in complex CT geometries. This SGS modeling is not purely dissipative, and some degree of desirable systematic backscatter is actually incorporated implicitly in MILES (Fureby and Grinstein 2002). Historically, researchers have modeled additional (explicit) backscatter effects by incorporating suitable source terms in the actual flow equations being solved. The approach we use here has some points of contact with the work by Leith (1990), in which stochastic backscatter was modeled through source terms in the momentum equations, and unresolved SGS information was parameterized in terms of a conventional Smagorinsky SGS model. The approach we use here also prescribes such source terms, but it takes advantage of the flux-limiter information computed by the FCT convection algorithms to determine where SGS backscatter information effects are strongest and must be supplied.

When the FCT algorithm detects structure in the flow that it "knows" cannot be resolved on the grid, only a fraction of the antidiffusion flux can be applied. The fraction that cannot be used is an explicit estimate of this unresolved flow and is coupled on the grid scale to the specifics of the fluid dynamic convection. FAST3D-CT uses these "stochastic backscatter fluxes" by pseudo-randomly perturbing the resolved flow velocity in each cell by an amount proportional to the unused flux velocities. The unused high-order momentum flux is accumulated at each grid point during the direction-split convection stages of the integration. We measure this backscatter source in terms of its absolute value summed over all three directions for each time step and suitably normalize it by the density, that is,

$$\eta = \left[\sum (1 - \Gamma) \left| \rho v_f^H - \rho v_f^L \right| \right] \Big/ \langle \rho \rangle.$$

We describe the FCT flux limiter, Γ, in some more detail in Section 17.4 (see also Chapter 4a). Examples of typical distributions of quantity η in the context of a representative FAST3D-CT urban flow simulation are reported in Patnaik et al. (2005). In some geometries, these additional grid-scale fluctuations break symmetries and initiate three-dimensional instabilities by means of stochastic backscatter that otherwise would have to grow up from computer round-off. They also transport small particles and droplets to material surfaces because of unresolved turbulence, even though the resolved flowfield has a zero velocity normal to the walls. This means that particles and droplets can deposit on a ceiling as well as any surface regardless of orientation. Finally, the numerical limiting of the imposed stochastic fluctuations, caused by the nonlinear flux limiter, provides a small additional macroscopic (resolved-scale) transport right where the FCT algorithm has detected subgrid structure. Each of these expected realistic effects requires further theoretical analysis and careful calibration by experiment.

17.2.5 Geometry specification

An efficient and readily accessible data stream is available to specify the building geometry database for FAST3D-CT. High-resolution (1 m or smaller) vector geometry data or gridded LIDAR data are available for many major cities. From these data, we determine building heights on a regular mesh of horizontal locations with relatively high resolution (e.g., 1 m). We can extract similar tables for terrain, vegetation, and other land-use variables. We interrogate these tables during the mesh generation to determine which cells in the computational domain are filled with building, vegetation, or terrain. This masking process is a very efficient way to convert a simple geometric representation of an urban area into a computational grid.

We use this grid-masking approach to indicate which computational cells are excluded from the calculation as well as to determine where suitable wall boundary conditions are to be applied. However, the grid-masking approach is too coarse to represent rolling terrain, for which we substitute a shaved-cell approach. We represent the terrain surface by varying the location of the lower interface of the bottom cell. Even though this results in a terrain surface that is discontinuous, the jump between adjacent cells is small, and operational results show that this approach works well and is far better than grid masking (stair steps) for representing terrain. A more accurate representation of the geometry is possible with the virtual-cell embedding (VCE) approach (Landsberg, Young, and Boris 1994), in which the cell volume and interface areas are allowed to vary. This level of detail begins to approach that of conventional aerodynamics CFD, but it has not been seen to be necessary.

17.2.6 Wall boundary conditions

Appropriate wall boundary conditions must be provided so that the airflow goes around the buildings. It is not possible with the typically available resolution for us to correctly model the boundary layer details on the buildings. Therefore, we use rough-wall boundary layer models (Arya 1998) for the surface stress, that is, $\tau = \rho\, C_D (U_{//})^2$, and for the heat transfer from the wall, $H_o = \rho\, C_p C_H U_{//} (\Theta - \Theta_o)$, where ρ is the mass density, C_H and C_D are coefficients characterizing the roughness and thermal properties of the walls or ground surface, $U_{//}$ is the tangential velocity at the near wall (first grid point adjacent to the wall), C_p is the specific heat at constant pressure, and Θ and Θ_o are the potential temperature at the wall, and near wall, respectively.

17.2.7 Tuning the implicit SGS model for urban street crossings

Historically, flux-limiting (flux-correcting) methods have been of particular interest in the MILES context. As discussed extensively in Chapters 4a and 5, properties of the implicit SGS model in MILES are related to the choice of flux limiter, high- and low-order schemes, and specifics of the implementation of the algorithm (Drikakis 2003; Fureby and Grinstein 2002; Fureby, Grinstein, and DeVore 2005; Rider and Margolin 2003). This corresponds to choosing–adjusting an (explicit) SGS model in conventional LES. We now discuss an approach using the freedom here to control unwanted numerical diffusion through appropriate choice of low-order transport algorithms.

In our simulations of urban areas, a typical grid resolution is 5 to 10 m. While this resolution is adequate to represent the larger features of the city, many of the smaller features are resolved with only one to two cells. This is true of smaller streets found in cities, which are about 20 m wide. Alleyways are even smaller. These smaller streets, represented by only one or two cells in our computations, put a tremendous demand on the numerical convection algorithm not to diffuse and retard the flow. By using the rough-wall boundary conditions discussed herein instead of no-slip boundary conditions, we find that the flow can proceed unhampered down a street only one cell wide. However, if there is another street intersecting the first, we find that the flow tends to stagnate at this intersection. The problem only occurs when we are dealing with streets that are one to two cells wide, and not with wider streets. After careful inspection, we determined that this problem arose as a result of the form of the diffusion term in the low-order solution in the standard FCT algorithm, LCPFCT (Boris et al. 1993), used in the FAST3D-CT code.

The traditional low-order component of FCT introduces numerical diffusion even when the fluid velocity goes to zero (as in the cross street; see Boris et al. 1993). In normal situations, the flux limiter is able to locate an adjacent cell that has not been disturbed by the diffusion in the low-order method and is able to restore the solution to its original undiffused value. However, when the streets are one to two cells wide, the region of high velocity is diffused by low-order transport and there are no cells remaining at the higher velocity. Thus the flux limiter cannot restore the solution in these cells to the original high value.

A solution to this problem is to change the form of the diffusion in the low-order method. In LCPFCT, The algorithmic diffusion coefficient for the low-order scheme is given by $\nu = 1/6 + (1/3)\varepsilon^2$, where the Courant number is $\varepsilon = |U|\Delta t/\Delta x$. Note that ν does not go to zero even when U goes to zero (as in the cross street). The simplest less-diffusive low-order algorithm that ensures monotonicity is the upwind method previously used in the formal MILES analysis (e.g., Fureby et al. 2005), for which the diffusion coefficient is given by $\nu_{upwind} = \frac{1}{2}|\varepsilon|$, which has the desired form for ν (Zalesak 2003, private communication). When the diffusion coefficient in the low-order component of FCT is replaced by ν_{upwind}, the flow no longer stagnates at the intersection of streets. With this modification of the low-order method, we altered the global properties of the transport algorithm sufficiently to address this problem peculiar to underresolved flows in urban areas. This approach also has the added advantage of retaining the fourth-order phase properties of the high-order convection algorithm. In practice, the particular choice of low-order scheme is used only for the momentum equations; the usual low-order scheme is used for the mass density convection, since the density is almost constant everywhere. Lowered numerical diffusion in the cross-stream direction and consequent lowering of numerical diffusion overall allows for significantly larger predicted lateral contaminant spreading and for faster downstream plume propagation. Comparative results of urban simulations using the original and modified low-order component of the FCT algorithm are reported in Patnaik et al. (2005).

17.3 Urban simulation model validation

The goal of validating a numerical model is to build a basis of confidence in its use and to establish likely bounds on the error that a user may expect in situations where the

Figure 17.4. USEPA meteorological wind tunnel: typical configuration of three-dimensional model buildings, as one would see when looking upstream (courtesy of Michael Brown, LANL).

correct answers are not known. Establishing the credibility of the solutions is one of the stumbling blocks of urban CFD simulations. Validation with experiments requires well-characterized datasets with information content that is suitable for unsteady simulation models as well as for the cruder steady-state models. Unfortunately, current full-scale field studies do not provide all this information: the data acquired are typically too sparse to fully characterize the flow conditions; the number of trials is limited; and trials cannot be repeated under the same conditions.

Obtaining full-scale (field) datasets for the inherently complex flows in question is costly and difficult, so the alternate validation approaches at present are to (1) compare urban flow simulations with carefully controlled laboratory-scale wind-tunnel experiments, and (2) carry out joint urban transport and dispersion comparisons for simulations performed by different researchers with their own models (to establish confidence or address inconsistencies). Finally, (3) we also carry out detailed comparisons with actual urban field experimental databases as they become available. In what follows, we demonstrate progress and lessons learned in all three such validation approaches.

17.3.1 Benchmarking with wind-tunnel urban model data

Comparisons with laboratory measurements of flow and contaminant over a simple urban model were made to evaluate and validate the ability of FAST3D-CT to model contaminant transport. Brown et al. (2001) measured velocity distributions and tracer concentrations associated with the flow over an array of cubes in the USEPA wind-tunnel facility (Figure 17.4) under controlled conditions. The experiments were conducted in an open-return wind tunnel, with a working test section of length 18.3 m, width 3.7 m, and height approximately 2.1 m. The wind-tunnel experiment simulated a neutrally stratified atmospheric boundary-layer flow over an array of buildings. The

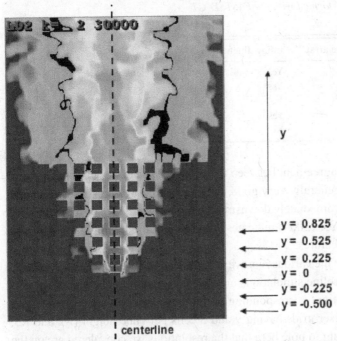

Figure 17.5. Instantaneous distributions of the simulated tracer concentration visualized at the 2-cm-height plane. Release occurred at a location behind the first cube in the vicinity of the centerline plane; flow direction is from bottom to top.

array consisted of 7×11 cubes $(0.15 \times 0.15 \times 0.15$ m) with one cube-height spacing between cubes. The reference velocity at one cube height was 3 m/s. The velocity measurements, made with a pulsed-wire anemometer, consisted of vertical profiles of the mean velocity and turbulence velocity variance in the three coordinate directions. These datasets provided high-quality, spatially dense (but not time-resolved) data. In addition to velocity data, the measured volume fraction data of a C_2H_6 tracer released continuously at the centerline just behind the first cube was also reported. The laboratory profiles used as basic reference for the FAST3D-CT model benchmarking purposes were measured in the vertical symmetry plane of the building array (Figure 17.5).

Previous reported studies used the USEPA wind-tunnel data to test flow simulation modeling aspects – but they did not address their effects on CT. One such study of these data performed by Smith, Brown, and DeCroix (2002) used the HIGRAD code, which is used to predict the evolution of atmospheric phenomena. HIGRAD is second order in time and space and uses a Smagorinsky-type or one-equation turbulent kinetic-energy-based subgrid closure. Advection is done with the MPDATA (nonoscillatory finite-volume) scheme (Chapter 4c). The simulations of Smith et al. (2002) nicely reproduce the mean longitudinal velocity, including the recirculation patterns in the canyons behind the blocks. The turbulent kinetic energy is modeled well, except for some underprediction in the canyons. Another study by Lien and Yee (2004) modeled the USEPA wind-tunnel experiment on the cube array by using a RANS STREAM code. The Kato–Launder model was also used as an alternative to the standard Jones

Table 17.1. *Various runs performed with the FAST3D-CT MILES model*

Run	Grid size (cm)	Cube array?	Inflow fluctuations?
R1	1	Yes	Yes
R2	1.67	Yes	Yes
R3	3	Yes	Yes
R4	1	No	Yes
R5	1	Yes	No

and Launder model. The agreement between the predicted mean velocity profiles and the experimental data is generally very good, with the greatest discrepancy occurring in the recirculation zone immediately downstream of the leeward face of the array. The comparison by Lien and Yee (2004) indicates that a RANS model might be sufficient to predict the *mean flow* features.

Table 17.1 summarizes the various runs we performed with the FAST3D-CT MILES model. We carried out baseline simulations on a $371 \times 350 \times 60$ mostly uniform grid (R1), with a 1-cm resolution, corresponding to 15 cells per cube height; resolution test studies used two coarser grids having 9 and 5 cells per cube height (R2 and R3, respectively). It is important to note here that the resolutions we considered are on the fairly coarse side, if simulations of flow over a single (surface-mounted) such cube are performed. On the other hand, these are resolutions representative of what we can afford to resolve practically in urban simulations relative to typical building dimensions. Results reported in Table 17.1 are mostly from the finer-grid simulations. We also show comparisons with results generated on the coarser grids in selected representative cases to address these resolution issues.

The inflow velocity consisted of a mean profile and superimposed fluctuations prescribed at $y = -0.5$ m (the front of the first cube is located at $y = 0.0$; see Figure 17.5). Including some finite-level fluctuation component at the inflow turns out to be crucial, as we subsequenty show. We included the deterministic model of wind fluctuations discussed in Section 17.2.1 as part of the inflow boundary conditions (at $y = -0.5$ m). We calibrated the strength of the model velocity fluctuations to match the experimentally observed rms values at $y = -0.225$ m. We achieved agreement visually by adjusting amplitude, spatial wavelength, and temporal frequency of the imposed wind fluctuations in the unsteady-wind model in FAST3D-CT. This approach provided a practical approximation to the turbulent inflow boundary-condition specification problem (consistent with the available laboratory data). However, the actual fluctuation modeling and calibration is inherently difficult, given that there is no unique way to prescribe such fluctuations on the basis of the available mean and standard deviation of the velocity components. In particular, it is well known that a simple inflow velocity model based on white-noise perturbations is not adequate to emulate large-scale unsteadiness in the laboratory flows (see e.g., Druault et al. 2004).

Typical instantaneous distributions of a simulated tracer concentration are visualized in a horizontal plane at the 2-cm height as shown in Figure 17.5. The simulated

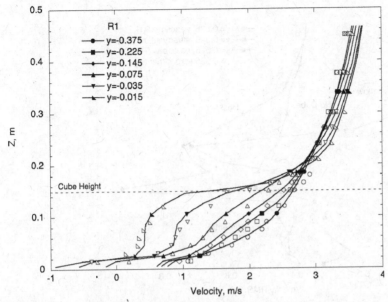

Figure 17.6a. Average streamwise velocity upstream of the first cube. Solid lines are numerical and open symbols are experimental at the corresponding upstream location.

tracer modeled the continuous release in the laboratory experiments at a location behind the first cube on the centerline. The comparisons with experimental results shown in Figures 17.6–17.9 are located in a vertical plane along the centerline. Figure 17.6a shows very good agreement between simulations (solid lines) and experiments for

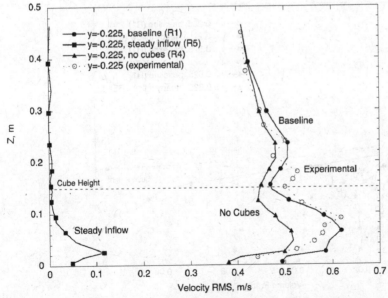

Figure 17.6b. Average rms streamwise velocity fluctuation profiles upstream of the first cube, at the matching location ($y = -0.225$). Note that the cubes enhance fluctuation levels upstream.

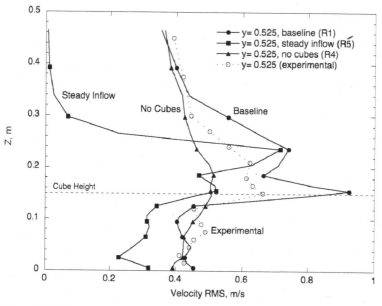

Figure 17.7. Average rms streamwise velocity fluctuations in the canyon between the second and the third cube (second canyon). The absence of inflow fluctuations causes significant underprediction of fluctuation levels.

the time-averaged streamwise velocity at several locations upstream of the cube array. Figure 17.6b focuses on the rms streamwise velocity profiles at the $y = -0.225$ m location (selected as the location for matching with the experiments). By comparison with a case run with steady inflow conditions (no superimposed fluctuations, R5), the

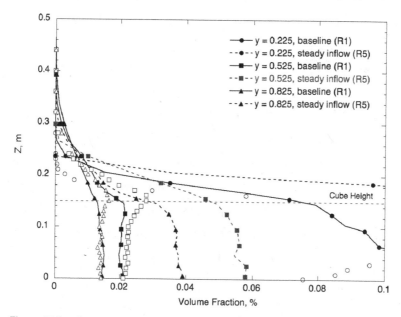

Figure 17.8a. Average concentration profiles are shown at selected stations located in the first three canyons; the sensitivity of the predictions to the lack of inflow velocity fluctuations is addressed.

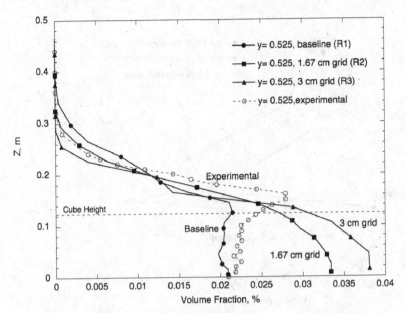

Figure 17.8b. Average concentration profiles are shown at a selected station between the second and third cubes; the sensitivity of the predictions to grid resolution is addressed.

figure clearly shows that the imposed unsteady velocity components account for a large fraction of the velocity fluctuation upstream of the cube array location. Also shown is a curve for rms velocity for flow without the cube arrangement (R4) with exactly the same time-dependent inflow boundary conditions as in the simulations with the

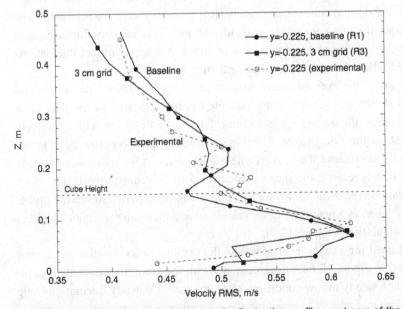

Figure 17.9a. Average rms streamwise velocity fluctuation profiles upstream of the first cube at the matching location for fine (baseline) and coarse resolutions.

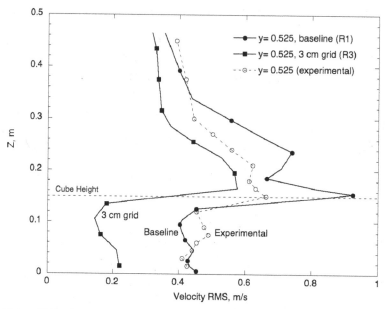

Figure 17.9b. Average rms streamwise velocity fluctuations in the canyon between the second and third cube; sensitivity of the predictions to grid resolution is addressed.

cube array present (R1). Interestingly, there is feedback from the cube array that adds to the prescribed velocity fluctuations, even at a reasonable distance upstream of the first cube.

Figure 17.7 compares rms streamwise velocity fluctuation profiles in the second canyon. Corresponding profiles from the runs carried out with the steady inflow conditions (R5) and without the cube arrangements (R4) are also shown for reference. The comparison shows the influence of the unsteady inflow on the velocity fluctuations in the canyon between the second and third row of cubes. It is apparent that fluctuations are largely due to the cube array up to about 1.5 cube heights.

A comparison of the average tracer concentration profiles from simulations and experiments is shown in Figure 17.8a at selected stations located in the first three canyons. In all cases, the agreement is within a factor of 2 or better, with agreement somewhat worse in the first canyon – perhaps reflecting questions in resolving the precise details of the release there. Agreement gets better as we move downstream. This also may occur because the mean velocity and fluctuations agree better as we move downstream. In the simulations, we find the contaminant to rise somewhat higher in the boundary layer, as a result of higher velocity fluctuations above 1 cube height in the numerical simulations (Figure 17.7).

The crucial need for a finite level of inflow fluctuations in the simulations is very clearly indicated in Figure 17.8a. Disagreements with the laboratory data are significantly larger when steady inflow conditions (R5) are used. Unsteady fluctuations help condition the flow to enable it to resemble the experimental conditions more quickly downstream. Although the unsteady inflow specifications are important, they are not the main controlling factor once the geometry of the buildings has had sufficient chance to influence the flow.

Effects of grid resolution on the average concentration profiles are shown in Figure 17.8b. The higher-resolution simulations slightly overpredict the concentration above 1 cube height; the lowest-resolution case significantly overpredicts the concentration in the canyon. These predictions are consistent with the velocity fluctuation profiles at this same second canyon location (see Figure 17.9b): the high-resolution case captures the fluctuation level in that canyon very well, but overpredicts at and above 1 cube height, suggesting excess mixing in this region – where the higher concentrations levels are observed. On the other hand, the low-resolution simulations have lower fluctuations everywhere, and the tracer tends to stay trapped in the canyon longer (where very low fluctuation levels are also observed). This is exactly the situation in which stochastic backscatter, once calibrated, can be expected to help.

Figure 17.9a shows the rms of the streamwise velocity fluctuations at fine (R1) and coarse (R3) resolutions at the profile-matching location upstream of the cube. The model captures the fluctuation levels at both resolutions quite well; however, we had to increase the amplitude of the fluctuating winds for the coarse resolution. Further downstream, in the second canyon, the fine-resolution case matches the fluctuation level in the canyon very well, but overpredicts it above the cube height (Figure 17.9b). The coarse-resolution case underpredicts the rms fluctuation value throughout. This observation is consistent with the fact that dissipation in MILES decreases as the resolution is increased, and it explains why stronger inflow fluctuations are required for coarser resolutions.

To summarize, we have found very good agreement between the simulations and the laboratory data with respect to the mean velocities, and fair agreement with respect to the rms velocities and the tracer concentrations. While agreement could presumably be improved with better calibration of the unsteady component of the prescribed inflow conditions, this calibration is very difficult since it must be based on laboratory (or field) databases, which typically provide single-point statistics, which is insufficient to characterize the unsteady structure of the flow. However, a particularly valuable insight is that, despite these inherent difficulties in calibrating the inflow boundary conditions, the fluid dynamics within the cube arrangement (i.e., beyond the first canyon) seems to be partially insulated by the flow events in the boundary rows, and thus appears to be less dependent on the exact details of the inflow conditions. The results indicate that reasonable agreement can be achieved with benchmark laboratory data with current MILES CT models using resolutions achievable in actual larger urban contexts.

17.3.1.1 Convergence issues

It is important to note here that *grid independence can only be truly pursued in the context of DNS*. Solutions associated with different resolutions correspond to the selection of correspondingly different values of some characteristic effective Reynolds number. This difficulty is actually inherent to any LES approach (and to ILES in particular), given that the smallest characteristic resolved scale is determined by a resolution cutoff wavelength prescribed by a spatial-filtering process – irrespective of whether the characteristic filter length is or is not itself resolved by grid resolution. Although some sort of agreement on the large-scale dynamics is expected in the high-Reynolds-number limit, faster coherence breakdown with increasing resolution (effective Reynolds number) unavoidably occurs as finer dynamical features are allowed to contribute and affect the larger scales. As a consequence, observations are always affected by the observational LES process itself!

Moreover, this discussion assumes that SGS modeling issues can be meaningfully addressed independently of other crucial aspects controlling LES performance, that is, boundary-condition (supergrid) modeling. The supergrid modeling here involves specific models used at solid (wind tunnel, cube) and open (e.g., inflow) boundaries. In the present case study, it is clear that issues of SGS and supergrid modeling *cannot* be separated. In particular, in addition to the noted limitations (when changing grid resolution), we must cope with the fact that the available reference data are not sufficient to fully characterize the unsteady (inflow) boundary conditions needed to ensure that the (reference) flow realizations are reasonably selected among possible simulated solutions.

17.3.2 Sensitivity to the urban simulation model: Old Town, Stockholm, CT studies

We performed simulations of CT within Stockholm City by using two separate MILES CT models, one using monotonicity-preserving FCT (Boris et al. 1993) and the other based on gamma (Jasak 1996) flux-limiting schemes for the convection terms (also see Chapter 4a). As already introduced, the FCT-based model is the Naval Research Laboratory's FAST3D-CT, and the gamma-based model is the FOI (Swedish Defence Research Agency) FOAM model (Weller et al. 1997). The simulations were carried out independently by the Naval Research Laboratory and the FOI, and they focused on a few islands near the center of the city where the Old Town of Stockholm of medieval origin is located (Figure 17.10). The very complex geometry of the city with its old houses and nonorthogonal distribution of streets is a challenge for dispersion modeling. By studying the transport and dispersion in this area, we can also adopt reasonably simple conditions for the lateral boundary conditions, since these are now all located over water.

The domain was $1750 \times 830 \times 1000 \text{ m}^3$ in size. We solved the flow equations in a finite-volume discretization, with variably sized control volumes (as applicable). The MILES character of the convective discretization arises directly from the gamma and the FCT algorithms. The gamma method is a nonlinear monotonicity-preserving algorithm that mixes upwind and central differencing dependent on the strength of the gradients. Thus, it is similar to FCT except for the specifics of the limiter and details of the high-order scheme (second order in gamma and fourth-order phase accurate in FCT). Both the gamma and FCT algorithms that we used in the model comparisons here involve an upwind method as the low-order scheme – which aids in modeling street crossings. We do time integration with an implicit Euler method treatment

Figure 17.10. FOI FOAM mesh in the center of the Old Town, Stockholm, domain; the geometry of the problem is also shown.

with the gamma method and an explicit two-step predictor–corrector method in FAST3D-CT. We made simulations on a nonuniform, structured, finite-volume mesh consisting of ~1.6M nonuniform cells (FOI FOAM), or ~6.4M finite-volume cells (FAST3D-CT). An illustration of the FOAM mesh in the center of the domain, also showing the geometry of the problem, is presented in Figure 17.10. The smallest cell used in the FOI simulations was less than 1 m; the FAST3D-CT calculations mostly used evenly spaced cells with a smallest cell size of 3 m.

The inflow boundary conditions used a log-linear velocity profile with a "free-atmosphere" wind of 3 m/s (equivalent to a wind of about 2 m/s at a 10-m height) in the West–East direction – fairly typical of a day with gentle winds. We used no superimposed wind fluctuations in either simulation. We used advective extrapolation boundary conditions to allow both mass and pressure perturbations to pass out of the domain. The lower boundary condition in the FOI FOAM model depended on the surfaces: ground and water surfaces as well as roofs are considered no-slip whereas walls are considered free-slip surfaces. In the FAST3D-CT model the material boundary conditions were all similarly prescribed for simplicity here as free-slip with added rough-wall drag models.

No atmospheric stratification was assumed, solar heating was not modeled, and dispersed gases were considered neutrally buoyant. The stochastic backscatter modeling feature in FAST3D-CT was disabled. There was no assumed chemistry included in

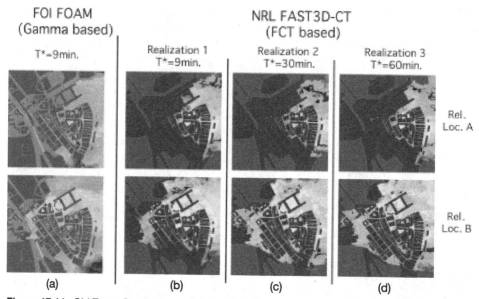

Figure 17.11. Old Town, Stockholm, contaminant-dispersion patterns from the two selected release locations (see Figure 17.10); various different urban realizations are shown in terms of logarithmic distributions of contaminant concentration at a 3-m height, 15 min after release at time T^*.

any of the models used. Deposition modeling was either disabled (FAST3D-CT), or modeled using a deposition velocity formulation with deposition velocities varying between wet and dry surfaces (FOI FOAM). We did not assess the possible effects of these differences quantitatively in this study. The simulations started with the prescribed boundary conditions initializing the flow throughout the computational domain. After some simulated spin-up time, T^*, allowing the flow to develop natural fluctuations, we made releases of a fixed quantity of contaminant from the two locations indicated in Figure 17.10. Releases occurred from relatively small regions and over short time intervals, so that they could be regarded as instantaneous and local. However, we made no effort to make these sources identical in the two runs.

Figure 17.11 demonstrates typical observed contaminant-dispersion patterns, showing logarithmic distributions of the contaminant concentrations at a 3-m height from the two selected release locations grouped on the top (Release A) and bottom (Release B), respectively, 15 min after release. Figure 17.11(a) corresponds to the FOI FOAM run, and Figures 17.11(b)–17.11(d) were generated with the FAST3D-CT model. At 9, 30, and 60 min after initialization in the FAST3D-CT run, any remaining contaminant was zeroed and new releases with the flow as developed up to the current time were injected into the simulations. This is denoted as Realizations 1, 2, and 3; likewise, the FOI FOAM run involved releases with the flow as developed at time $T^* = 9$ min.

By design, we chose the two release locations in distinct areas, between buildings within a very narrow street (Release A), and within a large open area (Release B) as shown in Figure 17.11. Although the release locations were only a few blocks away from each other, the associated dispersion patterns near ground level (at the 3-m-height plane considered) were quite different (Figure 17.11). The contaminant released from the first (congested) location (Release A) tends to have a narrower dispersion plume;

FAST3D-CT   FOI FOAM

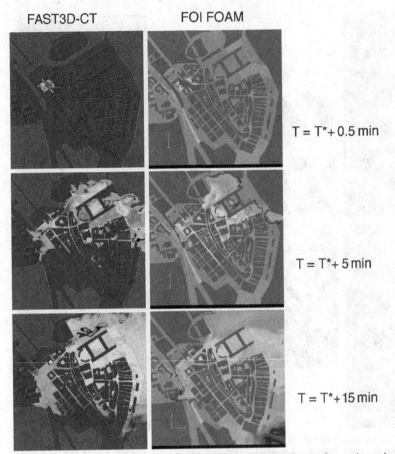

T = T*+ 0.5 min

T = T*+ 5 min

T = T*+ 15 min

Figure 17.12. Old Town, Stockholm, observed dispersion patterns from release location B, indicated in Figure 17.10. Contaminant (log) distributions at a 3-m height, after release at time $T^* = 9$ min, from location B.

for example, it missed the Royal Palace (cf. Figure 17.10) nearly completely – despite being the closest to it. These differences are quite independent of release times, and they are noted similarly with both simulation models, in spite of their significantly different local resolution capabilities. Good agreement on the dispersion patterns between the two different simulations is further demonstrated in Figure 17.12, where snapshots at selected representative earlier times of the Naval Research Laboratory and FOI runs are compared. The main noticeable difference between the two runs is the lower contaminant densities observed in the FOI FOAM run, which are partially attributed to the effect of the deposition terms that were active during the FOI simulation. Slightly larger contaminant spreading predicted by the FOI simulation may be due to its better resolution of the narrow and complicated system of streets in Old Town, Stockholm. This larger spread may also be part of the cause of the lower contaminant levels in the FOI FOAM run.

A preliminary conclusion from these studies is that the predicted large-scale CT patterns can be very sensitive to release locations, as also noted previously in our simulations of CT in downtown Chicago (e.g., Boris et al. 2002; Patnaik et al, 2005).

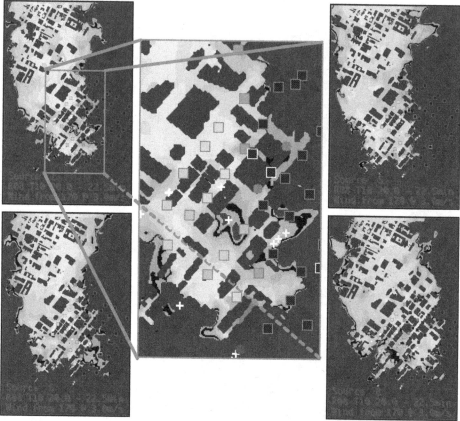

Figure 17.13. Four of the eight FAST3D-CT realizations of a single event are compared with Los Angeles experimental sampler data. Different actual release times with moderate (chaotic) wind fluctuations result in the differences shown.

Moreover, CT appears to be largely driven by the urban geometry and fairly independent of the specifics of the MILES simulation model. The good observed agreement in the first two (straight out-of-the-box) MILES urban simulations in the Old Town, Stockholm, scenario using two different methods not only speaks to the probable veracity of the individual models, but is also a clear indicator of the value of the MILES treatment for this very challenging class of problems.

17.3.3 Los Angeles simulations: Validation with actual urban field data

This section discusses validation of FAST3D-CT, using full-scale field trial data for acute (short-duration) releases in urban settings. Rappolt (2002) conducted a series of short-duration SF_6 releases in downtown Los Angeles, California. The SF_6 was released continuously for 5 min in each trial. Fifty synchronized samplers each took twelve 2.5-min duration samples for a total experimental trial duration of 30 min. The region instrumented was about 1 km^2, as shown by the square sampler locations in the panels of Figure 17.13. We present summary results from one of these field trials in shortened form.

As closely as we could determine, we ran FAST3D-CT for the same conditions as Field Trial 8. We specified moderate wind fluctuations and we set the Sun angle to

Table 17.2. *Summary of results*

Wind 170° at 3 m/s	±20% (%)	0.5–2 × (%)	0.2–5 × (%)	0.1–10 × (%)	Counts
Field trial vs. simulations (avg.)	14.5	39.5	73.8	88.4	1147
One realization vs. simulations (avg.)	19.5	53.6	84.4	90.0	1496

correspond to midmorning. We computed eight independent realizations of the experimental release, all taken from the same fluctuating wind distribution. These realizations correspond to releases 5 min apart in a continuously computed flowfield. We established 5 min as an adequate decorrelation time for the computational experiments.

Cross sections of the contaminant concentration for four of the eight realizations are shown in Figure 17.13 for the tenth sampling interval, 22.5 to 25.0 min. This interval begins 25 min after the SF_6 release commenced. The differences from one realization to another are substantial, as evidenced by a comparison of the four outer panels. Obtaining equivalent multiple realizations of the experimental plumes is not realistically possible at city scale, so there is no direct experimental yardstick for comparing a measured with a computed concentration. This is an acknowledged drawback to field trials. How close two different solutions actually are depends on the natural variability to be expected in the measured contaminant distribution.

By computing multiple realizations, however, we can measure the expected variance computationally, and this provides a quantitative yardstick – as long as background conditions, including the wind fluctuations, are well characterized. When we turned off impressed wind gusts completely in the simulations, we observed only a small reduction in the overall computed concentration, because building vortex shedding still provides turbulence and the boundary layer is unstable from solar heating. In each of the four realizations shown in Figure 17.13, the set of 50 experimental sampler values (squares) is shown for comparison, colored with the same concentration scale as used for the FAST3D-CT simulations. The blowup of one realization in the center of the figure shows this comparison of the experimental and simulated values more clearly.

We applied several of the accepted techniques for comparing a simulation with a single experimental realization (Bendat and Piersol 2000; Hanna, Hansen, and Dharmavaram 2004; Wirsching, Paez, and Ortiz 2005) to these data. We constructed scatter plots on a log-log scale for the experimental measurements plotted versus the ensemble of simulated measurements. The main result is that when one simulated realization is plotted against all the others, as if it were the experiment, the scatter plots are very similar to those showing the experimental data against all the realizations. Congruency counts are also used to compare model results with a single field trial. For the eight computational realizations of a baseline simulation (wind from 170° at 3 m/s), we computed the percentage of simulation data points within 20% of the experimental values, within a factor of 2, within a factor of 5, and within a factor of 10. Table 17.2 summarizes these results for the field trial and for one computational realization compared with the full set of simulated realizations.

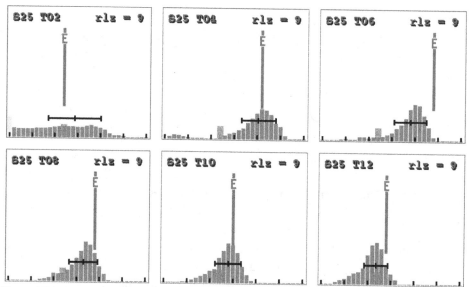

Figure 17.14. Ensemble concentration distribution functions from eight FAST3D-CT simulated realizations at the location of Sampler 25 in run LA 8. This sampler had above-threshold measurements in all 12 sampling intervals (duration 2.5 min); 6 of these are shown here. The vertical lines denoted with "E" indicate the experimental measurement and the short crosshatched bar is the threshold value of 20 parts per trillion used for this comparison. After the first few minutes, the distributions are seen to be approximately normal.

About 160 of the possible 600 field trail values were above the experimental threshold of 20 parts per trillion. The total number of counts is about eight times this number because there were eight realizations. We can see that almost 90% of all observations (center row of Table 17.2) were within a factor of 10, almost 75% were within a factor of 5, but barely 40% were within a factor of 2. Only about 15% of the number pairs were within 20% of each other. Is this good or bad agreement? We performed exactly the same congruency count test by comparing the numerical realizations with the ensemble composed of the set of the remaining seven realizations. In this case, we know each of the realizations is coming from the same distribution, so the lack of perfect agreement must be attributed to the natural variability from one realization to another; it is not the result of any error in the solutions or systematic differences between distributions. A typical result is also given in the bottom row of Table 17.2. Less than 20% of the number pairs are within 20% of each other, a little over 50% within a factor of 2, less than 85% within a factor of 5, and about 90% within a factor of 10.

We conclude that the field-trial-with-simulation congruency comparison is nearly the same as the simulation-with-simulation comparison. There is no reason to expect that any method could give closer congruency between simulation and experiment, given the differences from one simulated realization to another. Therefore, to gain a more quantitative way to look at this comparison, we began to study the distributions of the simulated values in the ensemble of realizations.

Figure 17.14 shows the distributions of simulated values at the location of Sampler 25 for 6 of the 12 sampling intervals (2.5 min in duration), beginning with the SF_6

release and continuing for 30 min. We chose Sampler 25 because it had measured concentrations above threshold for all 12 sampling intervals. The horizontal concentration scale is logarithmic in each small figure, as used in other validation studies because of the wide range of meaningful concentration values. Representative concentration values are collected at the experimental sampling sites in the simulation and at nearby points to build up the relatively continuous distributions shown. The distribution of concentration values collected this way is reasonably Gaussian in each time interval after a few minutes have elapsed. Therefore, the mean and standard deviations of the concentrations are also meaningful. This standard deviation approximates the concentration variance needed to compare simulated and experimental values quantitatively.

The center of the black bar in each small figure is at the mean value, and the bar extends 1 standard deviation on either side of the mean for each local distribution. This particular statistical comparison shows that the experimental data taken by Sampler 25 have a 50% chance of having been drawn from the simulated distributions because the chi-square (χ^2) value is about 13 for 12 degrees of freedom (Abramowitz and Stegun 1964). This particular agreement is not very good because the sampler was very close to the source and thus was above threshold for all 12 sampling intervals.

Most of the other samplers that were above threshold for a few of the sampling intervals showed much better agreement with the χ^2 test. When we took all samplers and time intervals into account, there were 159 degrees of freedom in the experiment and the baseline simulation ensemble showed a 98% agreement probability with Field Trial 8. This means that this level of agreement could be obtained by chance less than 1 time in 50. This analysis approach allows us to derive quantitative (probabilistic) results regarding the validity of the FAST3D-CT simulations.

Furthermore, this quantitative analysis permits parameter variations and system-sensitivity studies. We found that the quality of agreement depends at most weakly on the meteorological wind fluctuations assumed, because the buildings generate most of the turbulence from vortex shedding. We also varied the inflow wind direction and speed from the nominal baseline run at 170° at 3 m/s for Trail 8. Changing the wind 10° in one direction or increasing the speed by 1 m/s gave unacceptably poor results (37% and 2% probability of agreement, respectively). However, changing the wind 10° in the other direction or reducing the speed to 2 m/s gives results in somewhat better agreement with the field data than the baseline simulations (99.7% and 99.9% probability of agreement, respectively). By way of contrast, using only the congruency count test to conduct these sensitivity studies, we could not determine with any confidence that some case other than the baseline simulation conditions might actually be better.

In closing, we summarize the Los Angeles field trial validation studies to date. The multirealization FAST3D-CT simulations with 6-m resolution seem to be virtually indistinguishable from the Los Angeles field trial data. Naturally occurring variations between realizations can be quite large, as a result of building vortex shedding, even when inflow wind gusts are absent, but multirealization CFD simulations provide a way to approximate the missing concentration variability data for scientifically quantitative comparison of datasets. The χ^2 probability approach also gives us a sensitive way to approximate unknown parameters as well as to validate time-dependent CFD models.

17.4 Concluding remarks

Physically realistic, time-dependent, urban CT simulations are now possible but still require some resolution compromises because of time and computer limitations. Detailed time-dependent wind field observations at key locations can be suitably processed to provide initial and boundary conditions, and, at the least, they can be used for global validation. We believe that the building and large-scale fluid dynamics effects that can be captured today govern the turbulent dispersion. However, there is room to improve both the numerical implementation and the understanding of the stochastic backscatter that is being included both implicitly and explicitly. We know that the quality of the spatially and time-varying imposed boundary conditions (i.e., the fluctuating winds) require improvement. Inherent uncertainties in simulation inputs and model parameters beyond the environmental conditions also lead to errors that have to be further quantified by comparison with high-quality reference data. Judicious choice of test problems for calibrating models and numerical algorithms is essential, and sensitivity analysis helps to determine the most important processes requiring improvement.

Despite inherent physical uncertainties and current model trade-offs, it is clearly possible to achieve some degree of reliable predictability. Direct comparisons with field data, such as those from Los Angeles, provide an intuitively more "believable" validation; however, the sparsity of experimental data makes quantitative validation difficult. Wind-tunnel comparisons allow more rigorous validation, but typical available data insufficiently characterized the coherent structures that control dispersion. In the absence of any experimental data, out-of-the-box comparisons of CT results from two different MILES models improve confidence in the results of each.

The FAST3D-CT simulation model can also be used to simulate sensor and system response to postulated threats, to evaluate and optimize new systems, and to conduct sensitivity studies for relevant processes and parameters. Moreover, the simulations constitute a virtual test range for micro scale and nanoscale atmospheric fluid dynamics and aerosol physics, to interpret and support field experiments, and to evaluate, calibrate, and support simpler models.

Acknowledgments

We thank Bob Doyle for helpful technical discussions about and scientific contributions to this effort and Julie Pullen for allowing us to use figures from her research (Pullen et al. 2005). We also thank Michael Brown for providing the USEPA wind-tunnel data, as well as for very helpful discussions. An earlier research version of the FOAM software package (www.openfoam.org) was used. Aspects of the work presented here were supported by the office of Naval Research through the Naval Research Laboratory,

Missile Defense Agency, and the Department of Defense High Performance Computing Modernization Office (HPCMO).

REFERENCES

Abramowitz, M., & Stegun, I. A., eds. 1964 *Handbook of Mathematical Functions*. Washington, DC: National Bureau of Standards.

Aliabadi, S., & Watts, M. 2002 Contaminant propagation in battlespace environments and urban areas. *AHPCRC Bull*. **12**(4). Download from http://www.ahpcrc.org/publications/archives/v12n4/Story3/.

Arya, S. P. 1988 *Introduction to Micrometeorology*. New York: Academic Press.

Bauer, T., & Wolski M. 2001 *Software User's Manual for the Chemical/Biological Agent Vapor, Liquid, and Solid Tracking (VLSTRACK) Computer Model, Version 3.1*. Report NSWCDD/TR-01/83.

Bendat, J. S., & Piersol, A. C. 2000 *Random Data: Analysis and Measurement Procedures*. New York: Wiley, 3rd ed.

Boris, J. P., 2002 The threat of chemical and biological terrorism: Preparing a response. *Comp. Sci. Eng*. **4**, 22–32.

Boris, J. P., & Book, D. L. 1973 Flux-corrected transport I. SHASTA, a fluid transport algorithm that works. *J. Comput. Phys*. **11**, 8–69.

Boris, J. P., & Book, D. L. 1976 Solution of the continuity equation by the method of flux-corrected transport. *Meth. Comput. Phys*. **16**, 85–129.

Boris, J. P., Landsberg, A. M., Oran, E. S., & Gardner, J. H. 1993 *LCPFCT – A Flux-Corrected Transport Algorithm for Solving Generalized Continuity Equations*. U.S. Naval Research Laboratory Washington, DC: Memorandum Report NRL/MR/6410-93-7192.

Boris, J. P., Obenschain, K., Patnaik, G., & Young, T. R. 2002 CT-ANALYST™, fast and accurate CBR emergency assessment, in *Proceedings of the 2nd International Conference on Battle Management*.

Britter, R. E., & Hanna, S. R. 2003 Flow and dispersion in urban areas. *Annu. Rev. Fluid Mech*. **35**, 469–496.

Brown, M. J., Reisner, J., Smith, S., & Langley, D. 2001 *High Fidelity Urban Scale Modeling*. Los Alamos National Laboratory: Report LA-UR-01-1422.

Camelli, F., & Löhner, R. 2003 Assessing maximum possible damage for release events, in *Seventh Annual George Mason University Transport and Dispersion Modeling Workshop*., George Mason University, Fairfax, VA.

Chan, S. 1994 *FEM3C – An Improved Three-Dimensional Heavy-Gas Dispersion Model: User's Manual*. Lawrence Livermore National Laboratory, Livermore, CA: UCRL-MA-116567 Rev. 1.

Cybyk, B. Z., Boris, J. P., Young, T. R., Lind, C. A., & Landsberg, A. M. 1999 A detailed contaminant transport model for facility hazard assessment in urban areas. Paper AIAA 99-3441. 37th AIAA Aerospace Sciences Meeting, Reston, VA.

Cybyk, B. Z., Boris, J. P., Young, T. R., Emery, M. H., & Cheatham, S. A. 2001. Simulation of fluid dynamics around complex urban geometries. Paper AIAA 2001-0803. 39th AIAA Aerospace Sciences Meeting Reston.

Drikakis, D. 2003 Advances in turbulent flow computations using high-resolution methods. *Prog. Aero. Sci*. **39**, 405–424.

DTRA. 2001 *The HPAC User's Guide. Hazard Prediction and Assessment Capability, Version 4.0*. Washington, DC: Defense Threat Reduction Agency.

Druault, Ph., Lardeau, S., Bonnet, J. P., Coiffet, C., Delville, J., Lamballais, E., Largeau, J. F., & Perret, L. 2004 Generation three-dimensional turbulent inlet conditions for LES. *AIAA J*. **42**, 447–456.

Dwyer, M. J., Patton, E. G., & Shaw, R. H. 1997 Turbulent kinetic energy budgets from a large-eddy simulation of airflow above and within a forest canopy. *Boundary-Layer Meteorol*. **84**, 23–43.

Fureby, C., & Grinstein, F. F. 1999 Monotonically integrated large eddy simulation of free shear flows. *AIAA J*. **37**, 544–556.

Fureby, C., & Grinstein, F. F. 2002 Large eddy simulation of high Reynolds-number free & wall-bounded flows. *J. Comput. Phys*. **181**, 68–97.

Grinstein, F. F., Fureby C., & DeVore, C. R. 2005 On MILES based on flux-limiting algorithms. *Int. J. Numer. Meth. Fluids* **47**, 1043–1051.

Hanna S. R., Hansen, O. R., & Dharmavaram S. 2004 FLACS air quality CFD model performance evaluation with Kit Fox, MUST, Prairie Grass, and EMU observations. *J. Atmos. Environ.* **38**, 4675–4687.

Hendricks, E., Burrows, D. A., Diehl, S., and Keith, R. 2004 Dispersion in the downtown Oklahoma City domain: Comparisons between the joint urban 2003 data and the RUSTIC/MESO models, in *5th AMS Symposium on the Urban Environment*. American Meteorological Society Boston, MA.

Hussein, H. J., & Martinuzzi, R. J. 1996 Energy balance for turbulent flow around a surface mounted cube place in a channel. *Phys. Fluids* **8**, 764–780.

Jasak H. 1996 *Error Analysis and Estimation for the Finite Volume Method with Applications to Fluid Flows*. PhD dissertation, Imperial College.

Landsberg, A. M., Young, T. R., & Boris, J. P. 1994 An efficient parallel method for solving flows in complex three-dimensional geometries. Paper AIAA 94-0413. 32nd AIAA Aerospace Sciences Meeting, Reston, VA.

Lien, F. S., & Yee, E. 2004 Numerical modelling of the turbulent flow developing within and over a 3-D building array, Part 1: A high-resolution Reynolds-averaged Navier–Stokes approach. *Boundary-Layer Meteorol.* **112**, 427–466.

Leith, C. E. 1990 Stochastic backscatter in a subgrid-scale model: Plane shear mixing layer. *Phys. Fluids A2* **3**, 297–299.

Leone, J. M., Nasstrom, J. S., Maddix, D. M., Larsen, D. J., & Sugiyama, G. 2001 *LODI User's Guide, Version 1.0*. Livermore, CA: Lawrence Livermore National Laboratory.

Mayor, S. D., Spalart, P. R., & Tripoli, G. J. 2002 Application of a perturbation recycling method in the large-eddy simulation of a mesoscale convective internal boundary layer. *J. Atmos. Sci.* **59**, 2385–2395.

NOAA. 1999 *ALOHA Users Manual*. Washington, DC: Chemical Emergency Preparedness and Prevention Office. Available for download at http://www.epa.gov/ceppo/cameo/pubs/aloha.pdf.

Patnaik, G., Boris, J. P., Grinstein, F. F., & Iselin, J. 2005 Large scale urban simulations with FCT, in *High-Resolution Schemes for Convection-Dominated Flows: 30 Years of FCT*, ed. D. Kuzmin, R. Lohner, & S. Turek. New York: Springer, 105–130.

Pullen, J., Boris, J. P., Young, T. R., Patnaik, G., & Iselin, J. P. 2005 A comparison of contaminant plume statistics from a Gaussian puff and urban CFD model for two large cities. *Atmos. Environ.* **39**, 1049–1068.

Rappolt, T. J. 2002 *Measurements of Atmospheric Dispersion in the Los Angeles Urban Environment Summer 2001*. Tracer ES&T Project Report 1322. Tracer ESOT, San Marcos, CA.

Rider, W. J., & Margolin, L. G. 2003 From numerical analysis to implicit subgrid turbulence modeling. Paper AIAA 2003-4101.

Sagaut, P. 2005 *Large Eddy Simulation for Incompressible Flows*. New York: Springer, 3rd ed.

Smith, W. S., Brown, M., & DeCroix, D. 2002 Evaluation of CFD simulations using laboratory data and urban field experiments, in *4th AMS Symp. Urban Env.* Symposium in the Urban Environment, AMS, Boston MA.

Weller, H. G., Tabor, G., Jasak, H., & Fureby, C. 1997 A tensorial approach to CFD using object oriented techniques. *Comp. Phys.* **12**, 620–631.

Wirsching, P., Paez, T., & Ortiz, K. 2005 *Random Vibrations Theory and Practice*. New York: Wiley.

18 Outlook and Open Research Issues

Fernando F. Grinstein, Len G. Margolin, and William J. Rider

In this final section, we summarize the contributions to this book and briefly discuss new and open issues.

18.1 Précis

Our goal in this book has been to introduce a relatively new approach to modeling turbulent flows, which we term implicit large-eddy simulation (ILES). Simply stated, the technique consists of employing a fluid solver based on nonoscillatory finite volume (NFV) approximations and allowing the numerical truncation terms to replace an explicit turbulence model. NFV techniques have been a mainstream direction in the broader computational fluid dynamics community for more than 25 years, where they are known for their accuracy, efficiency, and general applicability. The application of NFV methods to turbulent flows has been more recent, but it already has produced quality results in a variety of fields.

Despite these computational advantages and simulation successes, the turbulence modeling community has been slow to accept the ILES approach. We hope to promote this acceptance by the present gathering of individual contributions of ILES pioneers and lead researchers, providing a consistent framework for and justification of this new approach.

There are several paths that contribute to the justification of the ILES approach to simulating turbulent flows. Practical demonstrations of capability are a necessary component, and they constitute the main content of this volume. Simulations in various chapters of this book, ranging from canonical flows with theoretical outcomes to more complex flows that have been investigated experimentally, serve to verify and validate the ILES approach. Further, these results show that ILES is competitive with classical large-eddy simulation (LES) approaches in terms of accuracy while offering advantages in computational efficiency and ease of implementation.

A second path to justifying ILES is to establish the connections to conventional LES models; we touched on this in Chapter 2, where we discussed the historical relationship of the development of nonoscillatory methods for advection to Smagorinsky models for turbulence, which both evolved from the concept of artificial viscosity. In Chapters 4a and 5, we made this relationship more quantitative; by employing modified equation analysis (MEA), we were able to demonstrate the similarity of the truncation terms of NFV schemes to explicit LES models. In particular, we showed that the MEA of many NFV schemes have the form of "mixed models" of conventional LES.

In Chapter 2, we described a third and more theoretical path to justification. There we argued that the equations that govern the evolution of the average velocity of a finite-sized fluid parcel are the Navier–Stokes equations augmented by new terms that arise from averaging the nonlinear advective terms. We then derived an infinite series expansion for these extra terms and evaluated the lowest-order terms to derive the augmented, finite-scale Navier–Stokes equations. Next, we compared these finite-scale equations to the modified equation of a particular NFV scheme. The essential congruence of finite-scale equations and the modified equations, in terms of their dynamical properties and in terms of their energy dissipation, provides a rationale for ILES – that ILES works because it is solving the appropriate model equations.

There are many open issues to address on the path to building a more complete framework for ILES. We list and briefly discuss a few of these below. We have grouped these issues into four categories – numerics, connections and extensions, analysis, and verification and validation – recognizing that there is some overlap among these. We also note that many of these issues transcend ILES and represent research areas for conventional LES as well.

18.2 Numerical issues

ILES has been demonstrated to be effective in a wide range of applications. This is despite the use of a variety of underlying methods incorporating essentially different nonlinear principles to produce nonoscillatory results. Despite advances in our understanding of nonoscillatory methods stemming from modified equation analysis (MEA), an a priori design of an ILES method to produce particular model properties remains elusive. One of the key challenges in developing new ILES methods is to understand the relationship of the discrete nonoscillatory principles to the physical properties of the overall ILES method. In Chapter 5, we described some progress in this direction, identifying some commonalities among the schemes that are found to be effective in ILES modeling and the crucial differences that make some NFV methods ineffective for ILES. In the following subsections, we list some remaining issues.

18.2.1 Choosing an NFV algorithm – is there an optimal ILES scheme?

The NFV methodologies described in Chapter 4 yield substantially equivalent, but not identical, solutions to a given problem; see, for example, Chapter 4a for comparisons of the evolution of the Taylor–Green vortex with various schemes. It is unlikely that one method will prove superior for all problems, or even for all possible questions that might be posed in a particular problem. This raises the issue of how to choose an optimal method for a particular problem.

Major distinctions among the implicit subgrid scale (SGS) models implied by distinct NFV methods are related to their behavior in nonsmooth regions of a flow, and near flow discontinuities. Nonoscillatory behavior can be enforced by a variety of design principles of varying natures, whose individual properties must be ascertained by suitable testing through verification and validation. Further, separate principles and methods can be blended to produce even more complex behavior.

In Chapters 4a and 5, we used MEA to compare a variety of different NFV methods. We showed that apparently all of the methods successfully used for ILES share the common presence of the self-similar term in their modified equation. We also showed that each method had dissipative terms, which varied in their dependence on cell size. In computational experiments, we found that strictly monotonicity-preserving schemes better reproduced the low-order moments of the flow while less restrictive sign-preserving schemes were better able to reproduce the variability of the flow. As noted in Chapter 2, and also discussed in Chapter 8, each approach has a priori advantages, and the quality of the associated implied solutions depends on the specific questions that a simulation is meant to address, and the available means to validate its merits.

It remains to relate the relative strengths of different algorithms to the details of their modified equations. Indeed, the dissipative properties of many NFV schemes are most manifest in regions of steep gradients in the flow, regions where the Taylor series expansions upon which MEA is based may not be convergent. This suggests a need for additional tools for analysis that may more quantitatively describe the dissipative features of individual methods.

18.2.2 Reverse engineering: Can one design new schemes that are optimal for a particular problem, or seek to emulate a particular SGS model?

A more proactive approach to numerics might lie in constructing new NFV schemes that embody desired features. First attempts at this approach have focused on constructing algorithms that correspond (in the sense of the modified equation) to known explicit SGS LES schemes. An example of this approach is specifically elaborated upon in Chapter 6, where the authors demonstrate how deconvolution and numerical flux function can be tuned to achieve particular implicit SGS modeling. Such an approach is similar to the reverse-engineering ideas proposed in Margolin and Rider (2005), where the modified equation of a NFV scheme is matched to an evolution equation with a specified SGS model term. Here also, such a strategy will be simplified when better analysis tools for characterizing the properties of discrete equations become available. Additional numerics design goals, such as capturing a particular physical property or performance feature targeting a specific class of problems, will likely require the combined use of MEA with actual tests in relevant case studies (e.g., as in the Taylor–Green vortex studies in Chapter 4a).

18.2.3 Mimetic schemes and the connection to second law

In Chapter 2, we demonstrated the similarity of the finite-volume equations of the two-dimensional Navier–Stokes equations to the corresponding MEA of MPDATA, and raised a more fundamental question: Is the numerical constraint of no oscillations somehow equivalent to the physical laws governing the increase of entropy and irreversibility? NFV methods ensure that entropy increases *on a cell-by-cell basis* in a simulation (see Merriam 1987); that is, NFV methods are sufficient, but not necessary, to ensure compliance with the second law of thermodynamics. Are there advantages to be gained by reducing the constraint of nonoscillatory solutions? There is some evidence that loosening the nonoscillatory constraints has a positive impact on turbulence modeling, as evidenced in the results given in Chapter 5.

In general, the practice of building physical laws into the numerical methods is known as mimetic (or compatible) differencing. One example that has already been introduced is that of flux-form (finite-volume) approximation; by requiring that the flux of any quantity into a computational cell is exactly the negative of the flux leaving the adjacent cell, one ensures exact local conservation of the fluxed quantity – mass, momentum, energy, and so on. As noted in Chapter 2, the importance of flux-form methods was noted more than forty years ago, and they form the basis of most numerical programs today. Mimetic differencing extends well beyond conservation laws (see Margolin, Shashkov, and Smolarkiewicz 2000 for additional examples). What other physical properties might be built into our ILES solvers?

18.2.4 Algorithmic details – synchronization and direction splitting

Among the important issues that we need to understand more deeply are several details of a numerical method that may be overlooked by a casual analysis. Systems of equations require some degree of coordination in how nonoscillatory principles are applied to the separate advective fluxes, such as mass, momentum, energy, scalars, and the like. In some methods, this is achieved through a combination of characteristic analysis of the system and Riemann solvers. In other cases, this is produced through synchronization strategies across the system of equations. This is related to the variables being limited in a coordinated fashion (say between density and velocity).

A closely related issue is how multidimensionality is produced numerically. Many methods are intrinsically one dimensional and are applied in more than one dimension in a dimensionally split fashion. Unsplit methods that do not apply dimensional splitting have been shown to have advantages in reducing the mesh imprinting of a solution. This is especially true in the case of systems of equations such as incompressible flow, where the flow is multidimensional as a result of a strong system constraint.

It is not clear at this time how these issues change the results, but the evidence is substantial that such details matter greatly in the quality of results. Synchronization choices directly affect the implicit SGS modeling. For example, not enough synchronization can lead to fluctuating derived quantities (i.e., backscatter), whereas overly coordinated treatments can lead to excessive dissipation. More complicated questions that surround the use of nonuniform meshes, as in adaptive mesh refinement (AMR) schemes and operator splitting for coupled physical processes are, at this time, largely unexplored.

18.2.5 Initial and boundary conditions

Beyond the issues of SGS modeling, there are critical issues to be addressed in practical simulations that relate to the treatment of the unresolved flow features at the so-called supergrid-scale (SPGS) level, that is, beyond the computational boundaries; such information must be prescribed for closure of the flow equations in numerical simulations. SPGS models provide the set of initial and boundary conditions that ensure unique, well-posed solutions; also see Chapter 3, this volume, and Grinstein (2004, 2006). From this perspective, it is clear that the observational (simulation) process is inherently affected (determined) by this (unresolved) SPGS information prescription. Because actual boundary condition choices select flow solutions, emulating particular flow realizations demands precise characterization of their initial (e.g., inflow) and other relevant conditions (e.g., of asymptotic flow or at solid and facility boundaries). This flow characterization issue is especially challenging when laboratory realizations are involved, because the reported information is typically insufficient.

SPGS modeling is required because in studying space–time developing flows, we must initialize the simulated solution and we can investigate only a finite spatial portion of the flow. Numerical boundary condition models must be consistent numerically and physically to ensure well-posed solutions. Moreover, to ensure that specifically desired flow realizations are simulated, we must make sure that the boundary condition models are capable of (1) emulating feedback effects from (presumed) virtual flow events outside of the finite-sized computational domains at open (subsonic) boundaries, (2) enforcing appropriate near-wall flow and energy transfer dynamics at solid boundaries, (3) prescribing effective unsteady turbulent inflow (or initial) conditions, and (4) minimizing spurious numerical reflections at all computational boundaries.

The mathematical process that led to the finite-volume equations in Chapter 2 is not directly applicable to boundary conditions. However, as noted, the physical issues involved in the analysis are similar, namely, issues of numerical consistency and difficulties in mathematically prescribing the actual physical models. For example, in compressible regimes, inherently different reflectivities are implied at open boundaries by continuous and finite-scale boundary condition formulations. Furthermore, wall boundary conditions must represent the effects of thin boundary layers unresolved on the length scales of the averaging, which are responsible for dissipating energy, for promoting energy transfer between scales, and for generating flow anisotropy.

Idealized fundamental problems, such as the decay of turbulence simulated in a box domain with mathematically well-defined periodic boundary conditions, may appear to avoid confronting questions of boundary conditions. Actually, such periodic box simulations involve all the difficult issues already discussed. It is now well established that for sufficiently long simulation times, the integral scale of turbulence will eventually saturate since the larger simulated scales cannot have unaffected growth beyond the box size, and will eventually distort the characteristic power law of the turbulence decay (e.g., Wang and George 2002). Moreover, it is also now recognized that turbulent flows *remember* their initial conditions (e.g., George and Davidson 2004, and references therein). As a very particular consequence, starting with the typical availability of single-point statistical data, there is no unique way to reconstruct a three-dimensional unsteady velocity field with turbulent eddies to define realistic turbulent inflow boundary conditions; such data have been shown to be typically insufficient to parameterize turbulent inflow boundary conditions for LES of inhomogeneous flows (e.g., Druault et al. 2004). Examples of such boundary condition implementation difficulties are discussed in Chapters 16 and 17.

The possible sensitivity of ILES (more generally, of any LES) results to initial conditions is an important unsettled issue. If the additional information contained in the initial conditions of the smaller and SGS spatial scales can significantly alter the evolution of the larger scales of motion and practical integral measures, then the utility of any LES for their prediction as currently posed is dubious and not rationally or scientifically justifiable. For example, the sensitivity of scalar mixing to initial conditions has been reported (e.g., Slessor, Bond, and Dimotakis 1998) and is believed to have important effects in the case of active mixing subject to Rayleigh–Taylor instabilities (e.g., Ramaprabhu, Dimonte, and Andrews 2005 and references therein).

Finally, most practical problems involve inhomogeneous flows and often invoke complex geometries with curved walls and corners. We touch upon typical boundary condition issues involved in the applications throughout the volume, specifically in Chapters 3, 10–12, and 16–17, where practical issues and strategies are discussed. SPGS modeling for ILES (or LES) remains inherently difficult with many largely unresolved issues.

18.3 Connections and extensions

In several places, especially Chapters 2, 4a, and 5, we have pointed out the connections between conventional LES and ILES in terms of the equations that they solve. This is hardly surprising, given the success of both approaches in simulating turbulent flow, and indeed may be expected at the simple level of dimensional analysis. Nevertheless, there are identifiable differences in both the implementation and the interpretation of results. These should explored for their potential benefit to both approaches. Furthermore, many flows of practical interest feature coupled processes that introduce additional physical scales of length and time. The design of ILES models to incorporate such coupled processes may require explicit SGS models beyond those implicit in the numerical algorithms.

18.3.1 Tensor invariance and mesh imprinting

One important difference between LES and ILES concerns the role of the computational mesh. In LES, one typically emphasizes the independence of the equations, particularly that of the explicit SGS models from the mesh. However, in ILES, the mesh is analogous to an experimental apparatus; there one expects the simulation results to depend on the choice of mesh. Although at first glance this may seem a philosophical point, it has important practical implications. In particular, the implicit SGS models of ILES depend both on the length scales *and the geometry* of the computational mesh.

In many applications of meteorology and aerodynamics, one needs to employ very complicated meshes, including irregular domains, large-aspect-ratio cells, and very nonuniform zoning; see, for example, Chapters 14 and 16. As noted in Chapter 4a, anisotropies introduced by eventual nonuniform (e.g., AMR) gridding directly reflect as contributions to the implicit SGS tensor, and this is an additional way in which *good or bad* SGS physics can be implicitly designed in an otherwise suitable ILES algorithm.

18.3.2 Coupled physics

Flows that feature coupled processes may occur strictly in the context of hydrodynamics, for example, compressibility, rotation, stratification, etc. Alternately, these may include additional regimes of physics in which the flow features are not dominated by convection, such as combustion, radiation transport, plasma physics, and magnetohydrodynamics. Using appropriate NFV methods for additional advective equations may suffice in many cases, although proper synchronization of the equations will be required (see the previous section). The presence of new physical scales will generally imply the need to add specialized explicit SGS models to ILES. So, in the case of compressible transport, no additional terms appear to be necessary (see, e.g., Chapters 4a, 5, and 11). However, in the case of strong rotation, the situation is less clear (e.g., Yang and Domaradzki 2004). When other physical processes are considered, the situation is similarly uncertain. For example, explicit models for advection–condensation have been found to be necessary (Grabowski and Smolarkiewicz 1999) in meteorology.

Assuming such explicit SGS modeling is needed (to be determined for each class of problems), we envision building quite efficiently on ILES by incorporating models of the physics of SGS-driven phenomena as suitable source terms in the appropriately modified conservation equations. The ILES approach provides an ideal framework in this context by decoupling resolved scales and SGS' in a simple implicit fashion through the NFV numerics, while allowing for large-scale dynamics to be captured well and efficiently.

ILES transport, combined with explicit (resolved-scale) models for multispecies, physics-based diffusive transport, thermal conduction, and global chemistry, has been found to work well in low-subsonic convectively dominated reactive flows (e.g., Chapter 8 and references therein). At a fundamental level, we need to establish how ILES treats turbulent mass-density and species-concentration fluctuations, the role of which is crucial for Kelvin–Helmholtz, Rayleigh–Taylor, and Richtmyer–Meshkov driven instabilities, and which is increasingly important for higher Mach numbers. A basic question is how ILES (or any LES) works in predicting the integral consequences of small-scale mixing. Major unresolved issues of ILES of scalar mixing involve both the effectiveness of implicit SGS scalar models provided by representative NFV methods and their ability to resolve effects of initial and boundary conditions on mixing.

18.4 Analysis issues

One of the conclusions of Chapters 2 and 5 is that the Navier–Stokes model may not be the most appropriate model (i.e., set of equations) upon which to base a numerical simulation of turbulent flow. The consideration of the dynamics of finite volumes of fluid, necessitated by the discrete nature of simulation, transcends numerics and has recently been discussed in a more analytic framework (see Brenner 2005). The results in Chapter 2 are suggestive, but are far from constituting a complete theory.

18.4.1 Modified equation analysis

The use of MEA has allowed some understanding of how ILES modeling is related to classical LES subgrid models. Nonetheless, the limitations of MEA have begun to become apparent. A key aspect of NFV methods is their treatment of unresolved regions of the flow (see, e.g., the discussion in Chapter 4a). Where the flow is underresolved, that is, in regions of steep gradients, the limiters are invoked and they radically influence the nature of the solution. In these regions, the algorithm may not be amenable to expansion as a convergent Taylor series. This is a direct consequence of a small radius of convergence. Alternate procedures for analysis in a series expansion, such as those using Padé approximants (Baker 1975), typically have much larger convergence domains and could provide greater applicability of MEA in underresolved regions of a flow field to analyze the particular numerical mechanisms operating in that regime. Although such techniques have not yet been evaluated in the MEA context, their use might allow for improved understanding on how the NFV methods work when the limiters are active.

18.4.2 Finite-scale analysis

The presence of certain terms in both the finite-scale Navier–Stokes equations derived in Chapter 2 and the modified equations of many NFV algorithms provides a rationale for the success of the ILES approach. However, both the finite-scale equations and the modified equations have an infinite number of terms, with differences at all orders of the expansion parameter. A particular example is the presence of *explicitly* dissipative terms in the modified equations that are absent in the finite-scale equations. Further understanding of the properties of the finite-scale equations, and especially the truncated versions of these equations, would be interesting from a theoretical standpoint. Are the solutions of the truncated finite-scale equations unique? Do they have an inertial manifold? What is its dimension?

The extension of the finite-scale analysis to the compressible flow equations is mathematically straightforward. However, there are issues of interpretation of the pressure and equation of state for a volume that has correlated, as opposed to random, molecular velocities (i.e., unresolved eddies) and that may not be in local thermodynamic equilibrium.

18.5 Verification and validation

Understanding the capabilities and limitations of a numerical simulation tool would seem to be a prerequisite to successful application. Verification is the process of assuring that one is solving the equations correctly, and it is usually addressed through convergence studies and analytic test problems. Validation is the process of demonstrating that one is solving the appropriate equations, and is usually addressed by comparison with experiment. A further issue that is currently attracting attention is the quantification of the uncertainty of a numerical simulation.

The issues raised below are not solely those of ILES, but are inherent in any attempt to simulate turbulent flows numerically. Nevertheless, we believe it is useful to call these out.

18.5.1 Convergence

In convergence studies, we compare a sequence of numerical simulations of a single problem on meshes of increasingly finer resolution. Errors are estimated in each simulation by comparison with *truth*. Here *truth* is defined as the analytic solution of the problem (if such exists) or as the limit of very fine resolution (e.g., direct numerical simulation). There are several issues to expose here.

First, turbulent flows are stochastic in nature. We can only expect convergence of statistically averaged quantities, not of the detailed flow results. This is widely recognized in the turbulence community, which typically focuses on quantities such as energy spectra and higher moments of the velocity field, velocity gradient field, etc., on mean flow and transport.

Second, we remark that the finite-scale equations, which we have asserted represent the most appropriate model for the finite volumes of fluid within a computational cell, depend on the resolution. Consequently, the solutions also depend on the mesh; that is, there is no unique truth in convergence studies. One bypass to this issue is to always "average" the results of finer resolutions to the coarsest mesh before constructing the statistical measures to be analyzed. In this way, one is comparing solutions that are all elements of the same mathematical solution space. A more elegant method to postprocess the energy spectra simulated at different resolutions is described in Margolin, Smolarkiewicz, and Wyszogrodzki (2002).

From a more philosophical point of view, we note that the specification of a particular problem for a convergence study requires not only geometry and model equations but also initial and boundary conditions. Higher-resolution simulations will require more initial conditions. In other words, there are many different problems on the fine mesh that correspond to the specified problem on the coarsest mesh. This is a different but equivalent way to view the lack of uniqueness of *truth*. The common choice that there is no energy at the smallest scales may be convenient, but is unphysical; see Drikakis, Margolin and Smolarkiewicz (2002) for a discussion.

18.5.2 Validation

The question of the influence of small-scale, unresolved initial conditions must be addressed by validation. The chapters in the book section on Verification and Validation illustrate a wide range of applications where it has been possible to establish confidence in the ILES approach through comparisons with relevant available data and theoretical studies. In contrast, the chapters in the last section on Frontier Flows correspond to the inherently difficult simulation of complex systems for which experimental data may be too difficult or impossible to obtain and direct numerical simulation too expensive to consider, thus implying that achieving some amount of practical predictability through modeling and simulation is of utmost importance. Quantifying the uncertainty of such simulations remains an open issue at this time. However, the fact that the ILES is automatic in the use of an appropriate NFV scheme and does not require, or allow, user-specified parameters would appear to be a step in the right direction.

Ultimately, the further development and acceptance of the ILES methodology depends on the continued success of researchers in applying this approach to complex problems of academic and practical interest. We hope this book will be an aid and an incentive for those interested in exploring new and potentially useful tools to simulate turbulent flows.

REFERENCES

Baker, G. A. 1975 *Essentials of Padé Approximants*. New York: Academic Press.

Brenner, H. 2005 Kinematics of volume transport. *Physica D* **349**, 11–59.

Drikakis, D. 2003 Advances in turbulent flow computations using high-resolution methods. *Prog. Aerosp. Sci.* **39**, 405–424.

Drikakis, D., Margolin, & L. G., Smolarkiewicz, P. K. 2002 On 'spurious' eddies. *Int. J. Numer. Meth. Fluids* **40**, 313–322.

Druault, P., Lardeau, S., Bonnet, J. P., Coiffet, F., Delville, J., Lamballais, E., Largeau, J. F., & Perret, L., 2004 Generation of three-dimensional turbulent inlet conditions for large-eddy simulation. *AIAA J.* **42**, 447–456.

Foias, C., Sell, G., & Temam, R., 1988 Inertial manifolds for nonlinear evolutionary equations. *J. Diff. Eqs.* **73**, 309–353.

Foias, C., Manley, O., & Temam, R. 1988 Modelization of the interaction of small and large eddies in two-dimensional turbulent flows. *Math. Mod. Numer. Anal.* **22**, 93–114.

Garcia-Archilla, B., Novo, J., & Titi, E. S. 1998 Postprocessing the Galerkin method: A novel approach to approximate inertial manifolds. *SIAM J. Numer. Anal.* **35**, 941–972.

George, W. K., & Davidson, L. 2004 Role of initial conditions in establishing asymptotic flow behavior. *AIAA J.* **42**, 438–446.

Grinstein, F. F., & Fureby, C. 2002 Recent progress on MILES for high Reynolds-number flows. *J. Fluids Eng.* **124**, 848–886.

Grabowski, W. W., & Smolarkiewicz, P. K. 1999 CRCP: A cloud resolving convection parameterization for modeling the tropical convecting atmosphere. *Physica D* **133**, 171–178.

Grinstein, F. F., & Fureby, C. 2004 From canonical to complex flows: Recent progress on monotonically integrated LES. *Comput. Sci. Eng.* **6**, 37–49.

Grinstein, F. F, ed. 2004 Boundary conditions for large eddy simulation. *AIAA J.* **42**, 437–492.

Grinstein, F.F. 2006 On integrating numerical and laboratory turbulent flow experiments, presented at the 36th AIAA Fluid Dynamics Conference, June 5–8, San Francisco, CA. AIAA Paper 2006–3048.

Jones, D. A., Margolin, L. G., & Titi, E. S. 1995 On the effectiveness of the approximate inertial manifold-a computational study *Theor. Comput. Fluid Dyn.* **7**, 243–260.

Jones, D. A., Margolin, L. G., & Poje, A. C. 2002 Accuracy and nonoscillatory properties of enslaved difference schemes. *J. Comput. Phys.* **181**, 705–728.

Margolin, L. G., Shashkov, M., & Smolarkiewicz, P. K. 2000 A discrete operator calculus for finite difference approximations. *Comput. Meth. Appl. Sci. Eng.* **187**, 365–383.

Margolin, L. G., Smolarkiewicz, P. K., & Wyszogrodzki, A. A. 2002 Implicit turbulence modeling for high Reynolds number flows. *J. Fluids Eng.* **24**, 862–867.

Margolin, L. G., & Rider W. J. 2002 A rationale for implicit turbulence modeling. *Int. J. Num. Meth. Fluids* **39**, 821–841.

Margolin, L. G., & Rider, W. J. 2005 The design and construction of implicit LES models. *Int. J. Numer. Meth. Fluids* **47**, 1173–1179.

Margolin, L. G., Titi, E. S., & Wynne, S. 2003 The postprocessing Galerkin and nonlinear Galerkin methods – A truncation analysis point of view. *SIAM J. Numer. Anal.* **41**, 695–714.

Merriam, M. L. 1987 Smoothing and the second law. *Comput. Meth. Appl. Mech. Eng.* **64**, 177–193.

Pope, S. 2000 *Turbulent Flows*. New York: Cambridge University Press.

Pope, S. B. 2004 Ten questions concerning the large eddy simulation of turbulent flows. *New J. Phys.* **6**, 35.

Ramaprabhu, P., Dimonte, G., & Andrews, M. J. 2005 A numerical study of the influence of initial perturbations on the turbulent Rayleigh–Taylor instability. *J. Fluid Mech.* **536**, 285–319.

Sawford, B. L., & Hunt, J. C. R. 1986 Effects of turbulence structure, molecular diffusion and source size on scalar fluctuations in homogeneous turbulence. *J. Fluid Mech.* **165**, 373–400.

Slessor, M. D., Bond, C. L., & Dimotakis, P. E., 1998 Turbulent shear-layer mixing at high Reynolds numbers: Effects of inflow conditions. *J. Fluid Mech.* **376**, 115–138.

Wang, H. L., & George, W. K. 2002 The integral scale in homogeneous isotropic turbulence. *J. Fluid Mech.* **459**, 429–443.

Yang, X., & Domaradzki, J. A. 2004 Large eddy simulations of decaying rotating turbulence. *Phys. Fluids* **16**, 4088–4104.

Index